Basic Botany

Basic Botany

Second Edition

Arthur Cronquist
New York Botanical Garden

1817

HARPER & ROW, PUBLISHERS, New York
Cambridge, Philadelphia, San Francisco,
London, Mexico City, São Paulo, Sydney

Sponsoring Editor: Steven P. Heckel
Project Editor: Holly Detgen
Designer: Emily Harste
Senior Production Manager: Kewal K. Sharma
Compositor: Black Dot Computer Typesetting Corp.
Printer and Binder: Halliday Lithograph
Cover Photograph: Emily Harste

Basic Botany, Second Edition

Library of Congress Cataloging in Publication Data
Cronquist, Arthur.
　Basic botany.
　Includes index.
　1. Botany. I. Title.
QK47.C8948 1981　　　　581　　　　　　　81-4138
ISBN 0-06-041429-4　　　　　　　　　　　AACR2

Contents

Part IV   POPULATIONS AND COMMUNITIES   523

Preface

Subatomic particles, atoms, molecules, cells, organisms, populations, and communities represent progressively more complex levels of organization and activity in nature. In theory it should be true that if we had a perfect understanding of subatomic particles, and a perfect intelligence, we could predict everything about the nature and behavior of atoms; that if we understood everything about the nature and behavior of atoms, we could predict everything about the nature and behavior of molecules; and so on up to the level of communities. In fact, neither our information nor our intelligence is up to such a task. A proper understanding of any of these levels of organization requires some knowledge about other levels, but each level has its own body of factual data and its own principles, which can only be discovered and understood by study of that level.

The more remote the level, the less direct its bearing on the level being studied. An understanding of molecular behavior is very important to students of cellular metabolism, but much less so to students of communities, who must be more concerned with the behavior of organisms. Furthermore, an understanding of levels beneath the one being principally studied is more important than an understanding of those above. It is more important for the student of photosynthesis to understand subatomic physics than for the subatomic physicist to understand photosynthesis.

This book is concerned with plants, from the level of cells and cellular metabolism to the level of their role in biotic communities. We pay only enough attention to physics and abiotic chemistry to help us understand the plants. Because an understanding of each level is founded on the level beneath, we begin at the bottom. After two short chapters to set the scene, we proceed from physics and chemistry to cells, thence to organisms, and thence to populations and communities.

The great advances in understanding of molecular genetics during the 1950s and 1960s misled even some scientists to suppose that this kind of molecular biology is essentially the whole of biology, but now molecular genetics has been incorporated into the larger whole, and progressively more attention is being devoted to ecology—the relationship of organisms to their environment. It must be so. Our civilization can survive only if we begin to think of ourselves as an integral part of the global ecosystem, an intricate network with complex and often unforeseen responses to disturbance of any one thread.

As in the first edition, this edition of *Basic Botany* tries to reach a plateau of understanding about each phase of the subject, and frequently to peer into the future, calling attention to what remains to be learned. No initial knowledge beyond normal high school study on the part of the student is assumed, but I have not hesitated to go as deeply into any subject as I think useful. Both facts and interpretation are stressed, in the belief that facts without interpretation are meaningless, and interpretation without facts is poorly understood and soon forgotten. Due attention is paid to the historical background for present-day concepts, and it may be hoped that the student will gain some knowledge about the nature of science and scientific progress, as well as about botany.

In this second edition I have tried to take account of pertinent new information and ideas in all phases of botany, and I have paid particular attention to bringing the exposition of photosynthesis and molecular genetics up to date. Even with all its uncertainties of detail, photosynthesis now begins to look like a coherent story. The pieces fit. I have tried also to update the suggested readings. Finally, I have added a key for the identification of common street trees. Regardless of what other botanists may expect, the folks at home expect a botany student to know the trees.

Many people have helped me in the preparation of this book. Some readers may recognize the source of a particular idea or bit of information, even when I myself have forgotten. I recall that particular paragraphs for the first edition were written or revised after discussions or correspondence with or lectures by Edward S. Ayensu, Harlan Banks, T. M. Barkley, Tyge Christensen, James Doyle, Peter Edwards, Richard Eyde, Leo Hickey, Max Hommersand, Francis Huber, Daniel Janzen, Dale Jenkins, Paul J. Kramer, Alberto Mancinelli, Lynn Margulis, A. E. Mirsky, Peter Raven, Gary Smith, G. L. Stebbins, F. C. Steward, William Stimson, and Armen Takhtajan. Many others have helped me with the second edition, and some have gone over considerable segments of the text with me. I owe especial thanks to Steven Carpenter, Neal Farber, Martin Gibbs, Richard M. Klein, Yoke Wak Kow, and Karl Niklas for advice on particular parts of the second edition. To all of these, and to others whose contribution escapes my memory, I give my thanks, and I apologize in advance to those who contributed to my thinking but whose names have been inadvertently omitted. As always, the advice received has gone through my own filter, and the people who helped me are not responsible for what I say.

My continuing thanks go to Herman Becker, José Cuatrecasas, Peter Edwards, T. H. Everett, Alan Graham, Peter Hepler, Walter Hodge, Arthur H. Holmgren, Lloyd Hulbert, Thomas Jensen, E. J. Kohl, Gordon McBride, Mary

Parke, B. L. Turner, the National Audubon Society, and the U.S. Forest Service for providing numerous photographs that are carried over from the first edition to the second. Mrs. M. A. Brown, copyright owner of W. H. Brown's book, *The Plant Kingdom*, has kindly allowed me to use a number of Brown's illustrations in both editions. My photographs from the Soviet Union were taken on a trip sponsored by Akademia Nauk and the National Academy of Sciences and kindly arranged by Academician Armen Takhtajan of the Komarov Botanical Institute. Many other photographs and electron micrographs, some new, are individually acknowledged where they appear. Several of the new electron micrographs were provided by Myron C. Ledbetter. Irving Zimmerman has consistently made the best possible prints from the many negatives I have entrusted to him. Many of the line drawings originally made by Charles C. Clare have been carried over into the second edition, and Robin Jess has made a number of new ones, notably those in the key to street trees.

I would like to thank D. C. Jackson, Gene Pratt, Ronald A. Pursell, and George H. Ward for reviewing the revision plans and providing valuable information at the time I began to revise *Basic Botany*. David E. Fairbrothers, Richard Hebda, Mark Schedlbauer, Aaron Sharp, Warren Wagner, and Robert Wilce reviewed the final stages of the complete manuscript of the second edition and provided helpful evaluations to guide me in preparing the finished work. Again, what is written has come through my own filter, and these gracious reviewers are not responsible for what I say.

Most of all I am indebted to my wife, Mabel, who has been a most useful professional consultant on writing style, who has stimulated my thinking, who has provided the domestic milieu conducive to my work, and who has done much of the typing and preliminary editing. Her hand in the book is not always readily visible, but can clearly be seen on p. 575.

Arthur Cronquist
Senior Scientist
New York Botanical Garden

PART I
INTRODUCTION

Pinus longaeva, Great Basin bristlecone pine, on Wheeler Peak in eastern Nevada. This species attains the greatest known age of any tree, about 5000 years. [Photo by Willy Albrecht.]

Chapter 1

Botany as a Science

Agave (left foreground) and *Beaucarnea* in Mexico, with
Volcan Orizaba in the background.

THE NATURE OF SCIENCE

The Kinds of Science

Botany is the science that treats of plants. It is a branch of **biology,** the science that deals with living things, both plant and animal. Biology, in turn, may be considered as one of the **natural sciences,** which deal with nature—in its broadest sense. Some other natural sciences are chemistry, physics, geology, and astronomy. The natural sciences, exclusive of biology, may be called the physical sciences. In addition to the natural sciences there are the social sciences, such as economics and sociology, and the exact sciences, notably mathematics.

Such an organization of the kinds of science is convenient, but not wholly satisfactory for all purposes. The boundaries between the sciences are not sharp, and many special fields may include parts of two or more other sciences. Thus genetics (the study of heredity) and ecology (the study of the relationship between organisms and their environment) are parts of biology that transcend the distinction between botany and zoology. Molecular biology, which has undergone spectacular advances in recent years, is nearly as much chemistry as biology. It is but a short step from molecular biology, which focuses on the role of chemical compounds and chemical changes in life processes, to biochemistry, which focuses on the chemical structure of these same compounds.

The different sciences are, furthermore, so interdependent that a proper understanding of any one requires a knowledge of several others. A good botanist must know some zoology, chemistry, physics, mathematics, and geology. An advance in any field of science may prove useful or necessary for further progress in some other field. For example, the developments of the past several decades in the knowledge of radioactive elements have made possible the great upsurge of molecular biology. Any of a number of elements used by a plant or animal can be supplied in radioactive form, and the subsequent history of these tagged or labeled atoms can be followed because of their radioactivity.

In contrast to the organization given above, science may be divided into **pure science,** in which knowledge is sought for its own sake, without regard to its immediate economic importance, and **applied science,** in which an attempt is made to put knowledge to practical use. Efforts now being made to find out just how a plant manufactures food from raw materials are an example of pure science. A solution to this problem in pure science may permit the eventual application of this knowledge to the practical problem of producing food for human use. Today's minor bit of scientific knowledge may tomorrow be an essential part of some new development in pure or applied science.

The Concepts of Science

It is commonly said that science is organized knowledge, but that is not a complete definition. Inherent in any concept of science is the search for objective truth or understanding, without regard for what would be pleasant or convenient to believe. Also inherent is the necessity for the facts or data to be sufficiently numerous and reliable to permit a solid core of agreement among virtually all serious students. Individual taste and value judgments, which are so important in the arts, are minimized in science. Ideally, all scientists working on a given problem should arrive at the same conclusion, and when experiment is possible, the results should be the same no matter who performs the experiment.

Scientists in all fields are continuously trying to add to the body of knowledge and understanding. The expansion of scientific knowledge involves both the discovery of previously unknown facts and the formulation of generalizations based on those facts. A scientific guess aimed at explaining certain observed facts is called a **hypothesis.** When additional information indicates that the guess is probably correct, the hypothesis may attain the status of a **theory.** If the theory proves to be correct and consistently applicable to a certain type of situation, it becomes a **principle** or **law.** The term law is applied especially to generalizations that can be used to predict accurately the results of the interaction of different factors or forces. Mendel's laws of heredity, taken up later in this book, are examples of scientific laws, although their applicability is now known to be more limited than was once believed.

Frequently, particularly in the observational as opposed to the experimental phases of science, it is convenient to continue to refer to a universally accepted principle as a theory. The **cell theory,** which states that living matter is characteristically organized into small, complex, definite bodies called cells, is an example of such a principle. The fact that some organisms (such as slime molds) are not composed of cells does not impair the usefulness of the cell theory in enabling us to understand the structure of living matter; indeed the exceptions to the cell theory are best understood in the light of knowledge concerning cellular organisms.

Another scientific principle is the theory of **organic evolution.** It states that organisms of different kinds are related through a common ancestry, that the existing kinds of organisms have arisen from one kind of simple organism or a relatively few kinds of simple organisms, and that the whole process is very gradual and consists of small steps. The mechanisms of organic evolution are partly but not wholly understood; they are discussed in Chapter 30. The principle of organic evolution is no longer a matter of contention among scientists.

All scientific concepts, be they hypotheses, theories, principles, or supposed laws, are subject to re-examination, modification, or rejection as additional information is obtained. Often the additional information necessitates adjustment or refinement of existing ideas, so that successive concepts tend to represent ever-closer approximations to the ultimate truth.

The Scientific Method

The methods employed by scientists in an effort to solve a problem have so much in common that the term **scientific method** is customarily used. In essence, the scientific method consists of (1) the acquisition of a body of facts pertaining to the problem; (2) the formulation of one or more hypotheses based on these facts; and (3) the testing of each hypothesis, either by the acquisition of new data of critical importance, or, when possible, by one or more experiments designed to determine whether a prediction based on a hypothesis is correct. The experiment, or the additional data, should be of such nature that if they confirm the applicability of one hypothesis, they exclude the alternatives. If no hypothesis is borne out by the third step, then the whole process is repeated, or the last two steps are repeated until a satisfactory conclusion is reached.

Most scientific generalizations, particularly in the biological sciences, are subject to exception. In this book we shall give primary attention to the usual or typical conditions and only passing (if any) mention to the unusual. The attentive reader will note the frequent use of qualifying words that admit exceptions to the statements made.

THE NATURE OF LIFE

The Characteristics of Life

We have said that biology is the science that deals with life and botany is the branch of biology that deals with plants. All of us know, in general, what is meant by "life" and "plant," but neither of these terms is precisely definable. There is no sharp boundary between plants and animals, or even between living and nonliving things.

The most characteristic features of life are **growth, reproduction,** and the complex series of chemical processes known collectively as **metabolism.** Of these, metabolism is the most important, since in any living thing growth and reproduction may be postponed or even eliminated, but if metabolism stops, the organism is no longer alive (except for the apparently suspended animation under some conditions in certain bacteria and spores). Living matter, called **protoplasm,** is (at least ordinarily) a complex organized mixture of proteins, fats, and many other substances, suspended in water, the whole in a state of continuous chemical activity. Any one of these substances, when isolated from the others, is not alive; it is the chemically active system that is alive.

The viruses, of which more will be said later, do not have a complete metabolic system of their own. They depend on the metabolic system of the host cell, which they convert to their own use. Thus the general statements here about life and living systems are only partly applicable to viruses.

One of the processes of metabolism is the breakdown of complex substances, called foods, into simpler substances, with energy being released as a result. The breakdown of foods may be divided into (1) **digestion,** a preliminary step that makes them more soluble or more readily diffusible without significantly changing the amount of stored energy, and (2) **respiration,** a further step in which some or all of the energy that was stored in the food is released. Respiration usually involves the use of oxygen and the formation of carbon dioxide as one of the waste products. As used by biologists, the term respiration refers to the actual chemical processes within a living organism that result in the release of energy from food, not merely to the breathing or exchange of gases that accompanies it. Some of the energy released in respiration is used to keep the protoplasm of a cell in constant motion; some of it is used in the formation, from raw materials and digested foods, of new compounds that become part of the protoplasm; and about half of it is transformed immediately into heat. In warm-blooded animals the heat released in respiration is used to maintain a body temperature higher than the usual temperature of the outside environment; in most other kinds of organisms the heat is largely wasted. The formation of new protoplasm from raw materials and digested foods is called **assimilation.** Digestion, respiration, and assimilation are among the essential processes of metabolism. The most continuous of these processes is respiration. With minor exceptions, continuous respiration is characteristic of all living things; if respiration completely stops, the organism dies. (It should be noted, however, that a state of suspended animation can be obtained in some kinds of cells by quick freezing and drying, under conditions far more extreme than exist in nature.)

Changes in the environment influence metabolic activities, and particular changes tend to call forth particular responses. This capacity to respond to stimuli, called **irritability,** is one of the most important characteristics of protoplasm. In both plants and animals responses to stimuli often result in movement, of either the whole organism or some part of it, and in plants they often also affect the direction of growth. Some types of plant response to stimuli are discussed in Chapter 26.

Protoplasm is constantly wearing out, breaking down, and being replaced by new protoplasm. When protoplasm is formed faster than it breaks down, the organism grows. Some or all of the cells increase in size, and these growing cells divide periodically, each into two identical daughter cells. As a result, the organism appears to grow from within, rather than by merely adding material externally. Growth by **intussusception,** as this type of growth is called, is characteristic of living things. Nonliving things may swell by the absorption of water or other liquids (e.g., a raisin) or may grow by accretion, the mere addition of material of the same sort to what is already there (e.g., a hailstone, or a chemical crystal), but they do not grow by intussusception. *If any one thing can be said to be the ultimate criterion of life, it is the ability to transform and organize food by chemical processes into body substance which in turn uses food in similar fashion.*

The Origin of Life

Many years ago it was generally believed that living things frequently originate from nonliving ones, that is, by **spontaneous generation.** Cheese and old rags left undisturbed in a dark closet for some days were supposed to generate mice; meat exposed to the open air was supposed to generate maggots; etc. The concept of spontaneous generation still survives in popular superstition, as, for instance, that a horsehair in a watering trough will turn into a wireworm.

Experiments by various scientists dispelled most of these ideas, but until well into the nineteenth century microorganisms were widely believed to arise spontaneously. The Italian biologist Lazzaro Spallanzani (1729–1799) demonstrated in 1768 that organisms did not reappear in cultures that were sterilized by boiling and then sealed by fusion of the mouth of the glass container. Schulz and Schwann in 1838 improved on Spallanzani's demonstration by permitting air to enter the sterilized flasks through red-hot tubes which killed any micro-organism borne in the air. These experiments attracted little attention, however. It remained for the French biologist and chemist Louis Pasteur (1822–1895, Fig. 1.1) to deliver the death blow to the theory of continuing spontaneous generation. Apparently unaware of the work of his predecessors, he improved on their experiments (1862) by permitting air to enter the sterilized flasks through slender tubes twisted into such shapes that particles floating in the air could not fall into the flasks (Fig. 1.2). His work was well publicized, and the principle that all life arises from pre-existing life soon became standard scientific doctrine.

Spallanzani's experiments bore practi-

Fig. 1.1   Louis Pasteur (1822–1895), versatile French biologist and chemist, famous for his work on the cause and prevention of human and animal diseases and for his studies on fermentation, spontaneous generation, etc. [Courtesy of the New York Botanical Garden.]

cal fruit long before the idea of spontaneous generation was fully discredited among biologists. François Appert, a French chef, founded the canning industry in 1804 and made a fortune from it. The French government awarded him a

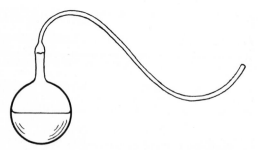

Fig. 1.2   Flask of a type used by Pasteur in his experiments on spontaneous generation.

substantial prize in 1810 for making public his process, which consisted of sterilizing the food by boiling it and then sealing it in sterilized glass containers.

The ultimate origin of life is still not fully understood, but most biologists now tentatively accept the idea of an original spontaneous generation at the molecular level, under environmental conditions quite different from those that exist today. For the last 3 billion years, at least, the principle that all life arises from pre-existing life seems to be correct.

THE NATURE OF PLANTS

Once we have arrived at the level of complexity of the most primitive existing organisms, which are microscopic and consist of single cells, three principal modes of nutrition are possible. The organism can make its food; it can absorb food from its surroundings; or it can eat (ingest) its food.

Organisms that make their own food have developed a series of features related to that way of life and are called **plants.** Organisms that eat their food have developed other features related to that way of life and are called **animals.** Both plants and animals have given rise to organisms that neither make nor eat their food, but instead absorb it from their surroundings. Food-absorbing organisms are classified as plants or animals according to their probable evolutionary relationship as indicated by their structure and chemical processes. Some food-absorbing organisms, such as the fungi and some of the bacteria, probably originated prior to the divergence of plants and animals into wholly separate evolutionary lines.

from raw materials with the aid of the green pigment chlorophyll. It is not motile, and it has neither a nervous system nor an excretory system. It is usually branched, with many similar parts. The size of the plant, the number of branches, and the number of organs of each kind are variable and subject to considerable modification by the environment. Increase in size is due largely to cell divisions in certain limited regions called meristems, and growth tends to continue indefinitely. The cells are bounded by a nonliving wall composed largely of cellulose, a complex carbohydrate. During cell division, the protoplasm is divided into two parts by the formation of a cell plate, as described in Chapter 5.

A typical animal has no chlorophyll. It eats its food rather than making it. It is motile and has both a nervous and an excretory system. It has a fairly definite form and size that are relatively little affected by the environment, and the number of organs of each kind is definite and usually not large. Increase in its size is due to cell divisions in all parts of its body, and growth stops when a certain size is reached. The cells, or most of them, are bounded by a flexible, living, proteinaceous membrane. Those cells which give a degree of rigidity to the body are surrounded by calcium salts, as in the vertebrates, or have a chitinous wall, as in the insects and other arthropods. (Although chitin is chemically more like cellulose than like protein, it does resemble protein and differs from cellulose in having nitrogen as an essential constituent.) During cell division the protoplasm is constricted or furrowed into two parts without the formation of a cell plate. (In both plants and animals there is ordinarily a precise duplication and division of one of the parts of the protoplasm, the nucleus, as explained in Chapter 5.)

All these features by which plants differ from animals are subject to exception, some more so than others; but among the more complex organisms there is rarely any difficulty in distinguishing a plant from an animal. The Venus's-flytrap (Fig. 1.3), a green plant that gets some of its food by snapping its leaves shut on unsuspecting insects, is still obviously a plant; dodder (Fig. 1.4), a twining non-green plant that parasitizes other plants

Fig. 1.3 Venus's-flytrap, *Dionaea muscipula*, an insectivorous plant native to North and South Carolina. (About half natural size.) [Photo by Allan D. Cruickshank, from National Audubon Society Collection/Photo Researchers, Inc.]

Fig. 1.4   Dodder, *Cuscuta*, a twining, yellowish, parasitic plant lacking chlorophyll. It obtains its food through special absorbing organs (haustoria) which enter the tissues of the plant it grows on. (Slightly less than natural size.) [Photo by Hugh Spencer, from National Audubon Society Collection/Photo Researchers, Inc.]

and absorbs its food from them, would never be confused with an animal; and tapeworms, although they absorb rather than eat their food, would scarcely be confused with plants.

As one considers progressively more simple organisms, the distinction between plants and animals becomes progressively more difficult. The fungi are generally considered to be plants, although their cell walls are usually chitinous rather than being composed of cellulose. Many of the single-celled algae swim freely in water, and some of these are

commonly studied in elementary courses in zoology as well as in botany. Among the most notable of these intermediate forms is *Euglena* (Fig. 1.5), which usually has chlorophyll and makes its own food, but which also has a gullet. Apparently *Euglena* itself does not ordinarily ingest food through the gullet, but some very similar and closely allied organisms without chlorophyll do. The cells of *Euglena* are bounded by a flexible, living, proteinaceous membrane; they have a simple excretory system; and the beginning of a nervous system is indicated by an eyespot that is sensitive to light. It is hard to deny that any organism that contains chlorophyll is a plant, yet in other respects these organisms certainly resemble animals; and some of the close relatives of *Euglena* that lack chlorophyll are by all criteria animals.

The classification of organisms into the plant kingdom and the animal kingdom, based on the structural and chemical characters attendant on their mode of nutrition, is natural, useful, and for the most part not difficult, but the cleavage between the two kingdoms is not absolute. Many of the chlorophyll-bearing unicellular organisms, which are treated by botanists as plants, are so obviously related to unicellular, nonchlorophyllous, food-eating organisms that zoologists regularly treat them as animals in spite of their chlorophyll. Once past the single-celled stage, however, the evolutionary development of plants and animals appears to have followed wholly separate channels, with no cross-connections.

Biologists dissatisfied with the vagueness of the traditional distinction between the plant and animal kingdoms have repeatedly tried to resolve the difficulty by recognizing three or more kingdoms of

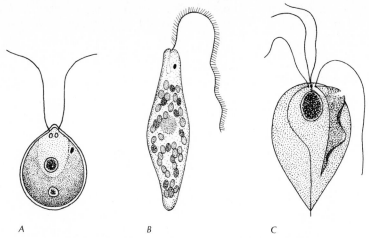

A   B   C

Fig. 1.5   Some single-celled motile organisms. (A) *Chlamydomonas,*
which has an eyespot and a definite cell wall and makes
its own food. (× 1200.) (B) *Euglena,* which has an eyespot
and makes its own food but lacks a cell wall and has a mouthlike
gullet at one end. (× 500.) (C) *Trichomonas,* which lacks a cell wall
and absorbs its food from its surroundings. (× 3500.) All three
organisms are commonly treated as protozoa (animal kingdom) by
zoologists; *Chlamydomonas* and *Euglena* are also treated as algae
(plant kingdom) by botanists.

organisms instead of only two. None of
these schemes has been widely accepted,
perhaps because all of them are subject to
the same problems as the scheme they
propose to replace.

There is some justification for classifying all organisms into two great groups,
the **eukaryotes** and **prokaryotes,** on the
basis of the complexity of organization of
their protoplasm (see Chapter 11). The
bacteria and blue-green algae, with relatively simple protoplasmic organization,
constitute the prokaryotes, and all other
organisms collectively constitute the eukaryotes. Such a classification, however,
leaves the vast majority of kinds of organisms in one kingdom, which must then be
arranged into two or more major groups
with vague boundaries between them.

SUGGESTED READING

Cohen, J. E., Irreproducible results and the
breeding of pigs (or nondegenerate limit
random variables in biology). *BioScience*
26:391–394, 1976. Whimsical but mathematically sound iconoclasm.

Farley, J., *The spontaneous generation controversy from Descartes to Oparin,* Johns Hop-
kins Press, Baltimore, Md., 1977. A scholar-
ly treatment.

Heslop-Harrison, Y., Carnivorous plants, *Sci.
Amer.* **238**(2):104–115, February 1978.

Kuhn, T. S., *The structure of scientific revolutions,* 2nd ed., University of Chicago Press,
1970.

Pasteur, L., Memoir on the organized corpuscles which exist in the atmosphere. Examination of the doctrine of spontaneous generation, in M. L. Gabriel and S. Fogel (eds.),
Great experiments in biology, Prentice-Hall,

Englewood Cliffs, N.J., 1955. An abridged translation of the original paper in *Ann. de Chim. et de Phys.* **64**:1–110, 1862.

Ponnamperuma, C., *The origins of life*, Dutton, New York, 1972. Designed for the layman or the beginning student.

Popper, K. R., *Conjecture & refutations: The growth of scientific knowledge*, Harper &

Row, New York, 1968.

Russell, B., *Scientific outlook*, Norton, New York, 1962. A general consideration by a well-known philosopher.

Smith, C. U. M., *The problem of life, An essay in the origins of biological thought*, Halsted Press (Wiley), New York, 1976.

Chapter 2

The Classification of Plants

Drimys winteri, in the family Winteraceae, one of the most archaic families of flowering plants. [Photo courtesy of R. M. Schuster.]

PRINCIPLES OF CLASSIFICATION

The diversity among living things is almost infinite. No two individuals are exactly alike in all details. Sometimes the differences between two individuals of very unlike appearance can be bridged by a series of other individuals, each differing so slightly from the next that there is no appreciable gap in the series. On the other hand, the variability among living things as a whole is not continuous but exhibits many distinct gaps of various sizes. The job of the taxonomist, or systematic biologist, is to organize the knowledge of the diversity and variability among organisms into a system of classification that best reflects the totality of their similarities and differences. Most taxonomists believe that such a system will necessarily reflect evolutionary relationships. If a scheme of classification based on the presently available and necessarily limited information is in accord with evolutionary relationships, then new information, as it is acquired, will tend to fall into place and be compatible with the scheme; otherwise not. Probable relationships are therefore taken into account in the development of a scheme of classification.

The Species

Any taxonomic unit of classification is called a **taxon** (pl., taxa). The basic taxon is the **species** (pl., species). The word species is taken over from Latin, in which it means a particular kind or sort. In modern biological usage the species are the smallest groups of individuals which can be recognized by ordinary methods as groups and which are consistently and persistently different from other groups. Among sexually reproducing organisms, the members of a single species are usually capable of interbreeding freely, whereas interbreeding between members of different species is prevented or restricted by natural causes. In other words, a species is a particular kind of plant or animal that retains its distinctness from other kinds in nature over a period of many successive generations.

Some species are more variable than others. Most biologists find it helpful to divide many of the more variable species into subspecies and/or varieties, which are populations within the species that are persistently different enough to merit some notice but are connected to each other by numerous intermediate individuals. Differences that are the direct result of response of the individual to the environment are generally considered beneath taxonomic notice, as are also those hereditary differences that are customarily reassorted in successive generations because of interbreeding. Errors have been made, and will continue to be made, as to whether certain differences really mark separate, self-perpetuating natural populations or whether they are merely differences between individuals in the same interbreeding population; but these errors are corrected when enough information becomes available.

The origin of one species from another, or the divergence of a single species into two or more, is usually a slow process which takes many generations. When two populations of a common origin have achieved a degree of divergence that permits recognition of essentially all the individuals as belonging to one or to the other population, with few or no exceptions or intermediates, they are considered to have become distinct species. Since evolution is a continuing process, there are some taxa which are now sufficiently distinct so that some taxonomists

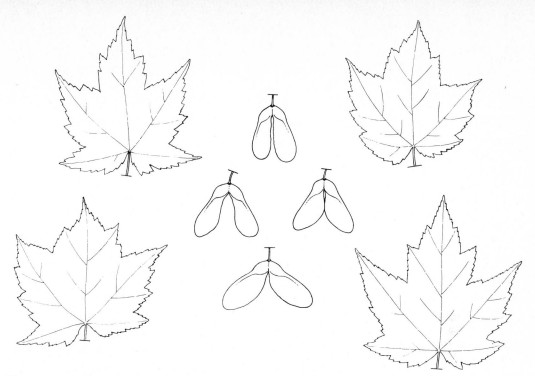

Fig. 2.1 Leaves and winged fruits of two species of maple, showing variation within a species and differences between species. (*Top*) Red maple, *Acer rubrum*. (*Bottom*) Norway maple, *Acer platanoides*.

call them species, but which are so closely related and so evidently connected by intermediate individuals that other taxonomists prefer to call them parts of one species. There is thus room for difference of opinion as well as error in determining the limits of species, but taxonomists are agreed that the species is a fundamental natural unit.

Supraspecific Groups

The levels of classification above the species show the degree of similarity and evolutionary relationship among species and groups of species. A group of similar species constitutes a **genus** (pl., genera); a group of similar genera constitutes a **family;** a group of similar families constitutes an **order;** a group of similar orders constitutes a **class;** and a group of similar classes constitutes a **division.** The division is the highest (most inclusive) category regularly and necessarily used in the classification of the plant kingdom, although many botanists prefer to group the divisions into optional categories called **subkingdoms.** It is customary and useful to classify all organisms into only two kingdoms, the plant kingdom and the animal kingdom, even though (as noted in Chapter 1) the cleavage between the two is not sharp. Except that the division of plant taxonomy is equivalent to the **phylum** of animal taxonomy, the terms used for the essential taxonomic groups are the same for both plants and animals.

The ending -*phyta* on a botanical scientific name indicates a division; the ending -*ales* indicates an order; and the ending -*aceae* indicates a family. Because of long-established custom that antedates the present rules of botanical nomenclature, the names of a few families and higher taxa take endings different from any of the foregoing ones. The ending for classes differs in different groups: usually -*phyceae* in the algae, -*mycetes* in the fungi, and -*opsida* in the higher plants. The classification of the American elm is as follows:

Kingdom: Plantae
Division: Magnoliophyta
Class: Magnoliopsida
Order: Urticales
Family: Ulmaceae
Genus: *Ulmus*
Species: *Ulmus americana*

The number of natural groups between the division and the species may or may not coincide with the number of categories given in the foregoing system of classification, but none of the categories is omitted. For example, the maidenhair tree, *Ginkgo biloba* (Fig. 2.2), is so distinct from all other living species that it is put into a class by itself. Among living plants the class Ginkgoöpsida contains only the order Ginkgoäles; the order Ginkgoäles contains only the family Ginkgoäceae; the family Ginkgoäceae contains only the genus *Ginkgo;* and the genus *Ginkgo* contains only the species *Ginkgo biloba.* On the other hand, if more taxonomic categories are needed than the usual list provides, others can be inserted, usually with the prefix *sub-.* A subkingdom is a taxon between a kingdom and a division; a subdivision is a taxon between a division and a class; a subclass is a taxon between a class and an order; and so on.

SCIENTIFIC NAMES

Use of Latin

In Roman and medieval times Latin was the language of learning in Europe. Those who wrote at all wrote in Latin, regardless of the language they generally spoke.

Fig. 2.2 *Ginkgo biloba*, the maidenhair tree, the only modern representative of its botanical class, regarded as a "living fossil". (*Left*) Habit. [Photo by Eric J. Hosking, from National Audubon Society Collection/Photo Researchers, Inc.] (*Right*) Leaves and fleshy seeds. [New York Botanical Garden photo.]

Those who wrote about plants naturally used Latin names.

The use of Latin for scientific names has been continued, in spite of the virtual abandonment of Latin in other writing, because of several advantages. Plants are not limited by national or linguistic boundaries, and if botanists in different countries are to understand each other's work, a plant should have the same name wherever it grows. Even within a single country, common names are often unsatisfactory or confusing. The same common name may be applied to several different plants in different parts of the country or even in the same region, and the same plant may have several different common names in different regions or even in one region. A great many plants have no common names at all.

It is true that some common names coincide, at least in part, with the genera and species recognized by botanists. Such common names as spruce, fir, pine, hemlock, ash, maple, elm, hickory, goldenrod, ragweed, and violet correspond in large part to botanical genera, and these names are readily converted to the binomial system (see below) by the addition of another word in each case to indicate the particular species. Longleaf pine, jack pine, slash pine, sugar pine, and Norway pine are common names for some individual species of the genus *Pinus*; and American elm, slippery elm, cork elm, and English elm are common names for some individual species of the genus *Ulmus*.

Even the common names that seem to conform to taxonomic groups cannot be depended on to do so, however. Most of the species that are called violets belong to the botanical genus *Viola*, which is a member of the family Violaceae, but the African violet belongs to the genus *Saintpaulia* of the family Gesneriaceae. Most of the species that are called pines belong to the genus *Pinus*, but the name pine is often loosely used for any of several different related genera, and the Australian pine is *Casuarina*, which belongs to a different botanical division from the true pines. Cabbage, brussels sprouts, broccoli, kale, kohlrabi, cauliflower, turnips, and black mustard all belong to the genus *Brassica*, a relationship that might be expected from the flavor but never from their English common names.

The Binomial System

The scientific name of any species of plant or animal consists of two words, usually Latin or of Latin form. The first word indicates the name of the genus to which the species belongs, and the second indicates the particular species of that genus. *Acer* is the scientific name of the genus that includes all the maples. The sugar maple is *Acer saccharum;* the Norway maple is *Acer platanoides;* and the red maple is *Acer rubrum.*

The second of the two words making up the name of a species is called the *specific epithet.* A particular specific epithet may be used only once within a given genus, but it may be used repeatedly in different genera. The specific epithet *rubrum* has been used not only in the genus *Acer,* but also in *Allium* (the onion genus), *Chenopodium* (the goosefoot genus), and many other genera. By itself *rubrum* is merely a Latin adjective meaning red; it is not the name of a plant. As a scientific name, *rubrum,* or any other

specific epithet, has meaning only if it is linked with the name of a particular genus. Because the name of a species consists of two words, scientific biological nomenclature is said to follow a *binomial system.*

Nomenclature

It is a general principle of botanical nomenclature that each kind of plant can have only one correct scientific name, and each name can be used for only one kind of plant. In order to minimize the possibility of confusion resulting from the inadvertent use of the same name for two different species, the author's name, or more often an abbreviation of it, is frequently added after the specific epithet. Thus the American elm, mentioned earlier as *Ulmus americana,* is more formally called *Ulmus americana* L. The L. stands for Linnaeus, the author of the name. In general, when two or more names have been used for the same species, or when the same name has been applied to two different species, it is the first name or usage that must stand, and subsequent ones are rejected.

Improvements in our understanding of the limits of species and genera sometimes necessitate changes in nomenclature, and scientific names are not always as stable and free from confusion as might be desired. The difficulty is minor, however, compared to that which would attend any attempt to organize common names into a comprehensive and scientifically accurate system.

HISTORICAL SUMMARY

The binomial system of nomenclature, which is now used for both plants and animals, was established by the Swedish botanist and naturalist Carl Linnaeus

Fig. 2.3 Carl Linnaeus (1707–1778), Swedish botanist and naturalist, who founded the binomial system of nomenclature.

(1707–1778, Fig. 2.3). His *Species Plantarum*, published in 1753, in which he attempted to name and describe all the species of plants in the world, is now the formal starting point for the scientific nomenclature of higher plants. Prior to Linnaeus' work, the botanical names of genera were customarily single Latin words, just as now, but the names of species were really short Latin descriptions, usually of several words, which attempted to give the characters by which the species could be recognized.

The convenience of the binomial system as opposed to the polynomial system, and the thoroughness with which Linnaeus covered the known kinds of plants,

soon led to the general adoption of both his classification and his system of nomenclature. Linnaeus' classification of groups above the rank of genus has been abandoned, but the binomial system of nomenclature remains, together with most of his genera and species.

Linnaeus grouped all plants into some 24 convenient classes. He freely admitted that this organization did not always put the most similar things together, but he felt that knowledge was not then adequate to permit the establishment of a comprehensive natural system. The development of a **natural system,** in which plants are classified according to the totality of their similarities and differences, has occupied the attention of post-Linnaean (and some pre-Linnaean) taxonomists to this day, and it is still not satisfactorily completed.

For more than a century after the publication of the *Species Plantarum*, biologists in general believed that each species had been separately created. The concept of organic evolution, which rapidly gained scientific acceptance after the publication of Charles Darwin's monumental *On the Origin of Species* (see Chapter 30) in 1859, gave new meaning to taxonomy. The progress that had been made toward a natural classification was so fully compatible with the principle of evolution that taxonomists in general soon accepted the concept and made it their own. Taxonomists differ as to how best to take account of evolutionary relationships in trying to establish a natural system of classification, but they are agreed that it is these relationships which make a natural system theoretically possible.

PRESENT CLASSIFICATION

The subkingdoms, divisions, and classes of plants recognized in this book are listed in the accompanying table. English names for some of the groups are given in parentheses.

Subkingdon I. Thallobionta (thallophytes)
    Division 1. Schizophyta (prokaryotes)
        Class 1a. Schizomycetes (bacteria)
        Class 1b. Cyanophyceae (blue-green algae)
    Division 2. Rhodophyta (red algae)
        Class 2a. Bangïophyceae
        Class 2b. Florideophyceae
    Division 3. Chlorophyta (green algae)
        Class 3a. Chlorophyceae
        Class 3b. Charophyceae (charoids)
    Division 4. Euglenophyta (euglenoids)
        Class 4a. Euglenophyceae
    Division 5. Cryptophyta (cryptomonads)
        Class 5a. Cryptophyceae
    Division 6. Pyrrophyta
        Class 6a. Desmophyceae (desmomonads)
        Class 6b. Dinophyceae (dinoflagellates)
    Division 7. Chrysophyta
        Class 7a. Chloromonadophyceae (chloromonads)
        Class 7b. Xanthophyceae (yellow-green algae)
        Class 7c. Chrysophyceae (golden algae)
        Class 7d. Bacillariophyceae (diatoms)
    Division 8. Phaeophyta (brown algae)
        Class 8a. Isogeneratae
        Class 8b. Heterogeneratae
        Class 8c. Cyclosporae
    Division 9. Fungi (fungi)
        Class 9a. Myxomycetes (slime molds)
        Class 9b. Chytridiomycetes (chytrids)
        Class 9c. Oomycetes
        Class 9d. Zygomycetes
        Class 9e. Ascomycetes (sac fungi)
        Class 9f. Basidiomycetes (club fungi)

Subkingdom II. Embryobionta (embryophytes)
    Division 10. Bryophyta (bryophytes)
        Class 10a. Anthocerotopsida (horned liverworts)
        Class 10b. Marchantiopsida (liverworts)
        Class 10c. Bryopsida (mosses)

Division 11. Rhyniophyta (rhyniophytes)
    Class 11a. Rhyniopsida
Division 12. Psilotophyta (psilotophytes)
    Class 12a. Psilotopsida
Division 13. Lycopodiophyta
    Class 13a. Lycopodiopsida (club mosses)
    Class 13b. Isoëtopsida
Division 14. Equisetophyta
    Class 14a. Hyeniopsida
    Class 14b. Sphenophyllopsida
    Class 14c. Equisetopsida (horsetails and scouring rushes)
Division 15. Polypodiophyta (ferns)
    Class 15a. Polypodiopsida
Division 16. Pinophyta (gymnosperms)
    Subdivision 16a. Cycadicae
        Class 16a1.   Lyginopteridopsida (seed ferns)
        Class 16a2.   Bennettitopsida (cycadeoids)
        Class 16a3.   Cycadopsida (cycads)
    Subdivision 16b. Pinicae
        Class 16b1.   Ginkgoöpsida (ginkgo)
        Class 16b2.   Pinopsida (conifers)
    Subdivision 16c. Gneticae
        Class 16c1.   Gnetopsida
Division 17. Magnoliophyta (angiosperms; flowering plants)
    Class 17a. Magnoliopsida (dicotyledons)
    Class 17b. Liliopsida (monocotyledons)

In addition to the formal taxonomic groups as given above, it is sometimes useful to refer to recognizable groups that may or may not be considered to have formal taxonomic standing. Thus it may be convenient to refer to all organisms (plant and animal) other than the Schizophyta as the **eukaryotes,** referring to their detailed protoplasmic structure; and indeed some systems of classification recognize the **prokaryotes** (Schizophyta) and eukaryotes as proper kingdoms of organisms. All of the Thallobionta except the fungi, bacteria, and viruses, may collectively be called the **algae,** and this group, too, has been treated as a formal taxon in some systems of classification. All of the divisions of Embryobionta, except the Bryophyta, may collectively be called the **vascular plants,** in reference to their well-developed conducting system, and some systems of classification recognize this huge group as a single division **Tracheophyta.** The divisions Pinophyta and Magnoliophyta collectively constitute the **seed plants,** or **phanerogams,** as opposed to all other divisions of plants, which collectively constitute the **cryptogams.** Those cryptogams that have a well-developed water-conducting system are often called **vascular cryptogams,** or **pteridophytes.** The meaning of these terms is further discussed in Chapter 16.

The **viruses** have been omitted from the foregoing table and discussion. They are usually considered in connection with the bacteria, to which they may (or may not) be related, but their proper classification is still highly controversial.

The probable relationships among the divisions and classes of plants are shown in the accompanying diagram (Fig. 2.4). Diagrams of this sort are commonly known as **phylogenetic trees.** Such diagrams are useful, but they should all be received with some reservation. No man has witnessed the events of geologic time, and the fossil record, especially of plants, has many gaps and ambiguities. Furthermore, the diagrams necessarily involve some oversimplification. Our chart shows all other plants as being descended from the bacteria. This is very probably correct in the sense that if we had the common ancestor to all other plants, we would consider it to be a bacterium, but on the other hand no presently living kind of bacterium even approximates that hypothetical ancestor. The diagram also masks a certain doubt concerning the ancestry of

the fungi, discussed in Chapter 14. Concepts of evolutionary relationships among some groups of plants have changed considerably over the years, a fact which has prompted the comment that phylogenetic trees, having no roots, are easily blown over. Still, they are useful if one does not crawl too far out on a limb.

Most of the divisions and classes of plants are individually characterized in subsequent chapters, but it may be useful to indicate here in a general way the contents of some of the groups.

The **thallophytes** consist of the algae, fungi, bacteria, and viruses. (The special status of the viruses is discussed in Chapter 11.) In general, the thallophytes have a plant body that is not clearly divided into roots, stems, and leaves; they lack complex sex organs; and they do not have well-developed conducting tissues, or at least they do not have water-conducting tissues. Although some of the brown algae are very large and complex, most thallophytes are relatively small and simple plants. Indeed many of them are microscopic. The bacteria and viruses, for example, are microscopic and relatively simple in structure.

Aside from the bacteria and viruses, those thallophytes which have the green pigment chlorophyll and make their own food from raw materials are called **algae;** those which lack chlorophyll and do not make their own food are called **fungi.** (The minor exception of some colorless algae is not important at this level of study.) Algae release oxygen as a byproduct of food manufacture. Some few of the bacteria make their own food, but they do not release oxygen as a result.

Most algae (Fig. 2.5) are water plants, although some occur on land, where they may, for example, form a thin, green coating on the trunks of trees or flower

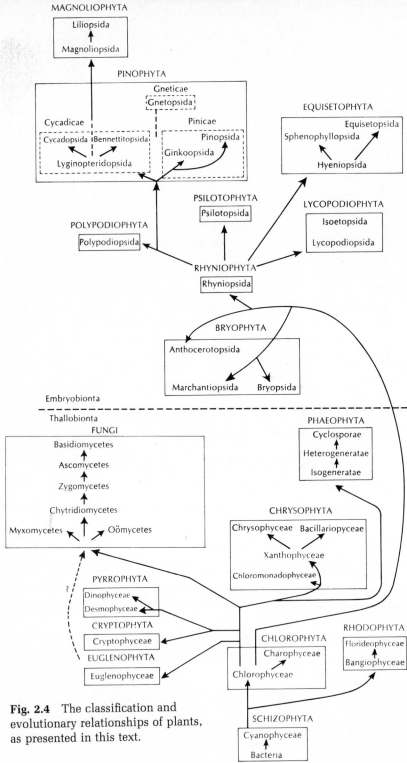

Fig. 2.4 The classification and evolutionary relationships of plants, as presented in this text.

 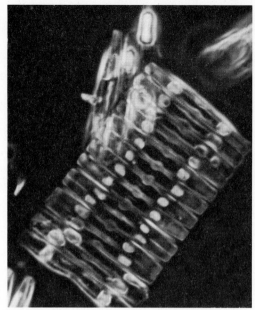

Fig. 2.5 Algae. (*Left*) *Laminaria hyperborea*, a kelp, in a marine inlet near Fort William, Scotland. [Photo by Claude Weber.] (*Right*) *Fragilaria*, a colonial diatom. (× about 800.) [Photomicrograph courtesy of Elsa O'Donnell-Alvelda.]

pots. All freshwater algae are fairly simple in structure. They consist of long filaments, or small plates, or individual blobs which may collectively form a green scum on ponds or make the water in an aquarium green. Most plants of the open ocean and of the intertidal zone are algae. Some oceanic algae are small and inconspicuous like freshwater algae, but others, such as the kelps (belonging to the brown algae) are large, conspicuous, and complex.

Although nearly all algae have the green pigment chlorophyll, not all of them are visibly green. Many of them have other pigments that mask the green color of the chlorophyll. The names for some of the groups reflect the usual presence of these other pigments. Most of the larger marine algae, for example, belong either to the Phaeophyta (brown algae) or the Rhodophyta (red algae).

The **fungi** (Figs. 2.6 and 2.7) consist of the molds, mildews, mushrooms, toadstools, and similar organisms. Many of them are parasitic, causing diseases especially of plants. Many others live on dead organic material that they absorb from the soil or other surroundings. Some of them form mutually beneficial partnerships with certain algae or with the roots of trees and other ordinary land plants.

The **embryophytes** make up most of the obvious vegetation of the land and are often called the **land plants.** Some of them do grow in water, however, as for example cattails and water lilies. Features of the life cycle are much used to differentiate the various groups of embryophytes, as indicated in later chapters. Aside from the matter of life cycles, the smaller and simpler embryophytes, which lack well-developed conducting tissues, are the

Fig. 2.6 A minnow infested with the water mold *Saprolegnia*. [New York Botanical Garden photo.]

bryophytes (Fig. 2.8). Mosses and liverworts are bryophytes. Although simpler in structure than other embryophytes, the bryophytes are more advanced than the thallophytes in having specialized, complex sex organs.

The remainder of the embryophytes

Fig. 2.7 *Amanita muscaria*, the fly amanite, a highly poisonous mushroom. Note the "death cup" at the base, found on many poisonous (and some edible) mushrooms. [Courtesy of the Department of Plant Pathology, Cornell University.]

Fig. 2.8 *Hypnum crista-castrensis*, a common moss in the boreal forest, taken in northern Minnesota. [U.S. Forest Service photo.]

Fig. 2.9 Deer fern, *Blechnum spicant*, in Oregon. [Photo by Willy Albrecht.]

typically have well-developed conducting tissues and are collectively called **vascular plants.** They usually have well-differentiated roots, stems, and leaves. The ferns (Polypodiophyta) (Fig. 2.9) and fernlike plants (Rhyniophyta, Psilotophyta, Lycopodiophyta, and Equisetophyta) do not have seeds, but produce tiny, scarcely visible, usually wind-distributed reproductive bodies called spores. The **gymnosperms** (Pinophyta) have seeds but not flowers. Pines (Fig. 2.10) and other conifers are some typical gymnosperms.

Fig. 2.11 Flowering twig of coffee tree, *Coffea arabica.* [Photo by W. H. Hodge.]

The **angiosperms,** or **flowering plants** (Magnoliophyta) (Fig. 2.11), have both seeds and flowers and are also more advanced and complex than the gymnosperms in certain structural features. The angiosperms not only have more species than all the other groups of plants combined, they also dominate the vegetation of most of the land surface. In some places, however, gymnosperms form considerable forests and make up a prominent or even dominant part of the vegetation. In north-temperate regions the needle-leaved trees are gymnosperms, but the ones with broader, more ordinary leaves are angiosperms. Grasses, most farm crops, and garden and wild flowers are also angiosperms.

Fig. 2.10 Longleaf pine, *Pinus palustris,* in the Choctawhatchee National Forest, Florida. [U.S. Forest Service photo by E. S. Shipp.]

The scheme of classification and probable evolutionary relationships here presented is not fixed and immutable for all time. There is room for some difference of opinion on how to organize our knowledge into the most generally useful scheme, although most of the same major groups and concepts of relationships show up in nearly all arrangements. Accumulation of new information about the algae, in particular, is forcing some changes in concepts that have not yet been fully integrated into a comprehensive scheme. It may become necessary to recognize one or more additional classes of algae to accommodate the new information. Within the brown algae (Phaeophyta), it may be predicted that the classes Isogeneratae and Heterogeneratae will be combined, and the nomenclature of the two remaining classes regularized. It is not at all likely, however, that the taxonomic deck will be reshuffled. Any generally acceptable new scheme will certainly build upon and incorporate large parts of the old.

NUMBER OF SPECIES OF PLANTS

The number of species of plants in the world has not yet been accurately determined, both because many species remain to be discovered and because some species have inadvertently been named and described more than once. Probably not many bona fide new species of vascular plants remain to be discovered in temperate and boreal regions, but almost every collecting expedition into the tropics brings back its quota of plants not previously known. Doubtless many species of fungi remain to be discovered even in temperate regions.

Rough estimates of the number of known species are given for most of the divisions and classes of plants in the following chapters. They may be summarized here as about 1,000 bacteria, 25,000 algae, 40,000 fungi, 20,000 bryophytes, 10,000 pteridophytes, 500 gymnosperms, and 230,000 angiosperms. Adding these figures together we obtain a total of 326,500 species. Such a figure gives an impression of greater accuracy than it actually has, however, because the estimates in the various groups are not on the same level of approximation. The 500 species of gymnosperms are insignificant in relation to the probable error in the estimate of 230,000 angiosperms. For practical purposes the figure of 326,500 known species may therefore be rounded off at about 325,000. It is possible that when all the species of plants in the world are known, the figure may approach half a million.

SUGGESTED READING

Jeffrey, C., *An introduction to plant taxonomy,* Churchill, London, 1968. A very good short paperback explaining general taxonomic principles.

Stafleu, F. A., *Linnaeus and the Linnaeans,* International Association for Plant Taxonomy, Utrecht, 1971.

Chapter 3

Some Physical and Chemical Background

Cymbalaria muralis, the Kenilworth ivy, growing on a wall at Ostia Antiqua, the port of ancient Rome. The specific epithet *muralis* is the adjectival form of Latin *murus,* "wall". There are plants for every habitat.

UNITS OF MEASUREMENT

Measurements of length of ordinary visible things are conveniently expressed by the **metric system** or, in English-speaking countries, by the antiquated system of inches, feet, yards, etc. One **meter** is about 39.37 inches, a little more than 1 yard; 1 millimeter, or 1/1000 meter, is about 1/25 inch. The word *meter* is often abbreviated as m and the word *millimeter* as mm. One thousand meters make up a kilometer, abbreviated km.

Another scale of measurement is necessary for objects of microscopic or ultramicroscopic size. The units of this scale represent a further subdivision of ordinary metric units. One one-thousandth of a millimeter is called a **micron,** or **micrometer,** often abbreviated as μ. One one-thousandth of a micron is called a **millimicron,** or **nanometer,** abbreviated nm. One tenth of a nanometer is 1 **Angstrom** unit, designated Å. Thus 1 Å is 0.0000001 mm.

SUSPENSIONS, SOLUTIONS, AND COLLOIDS

The chemical and physical properties of any mixture of water and another substance vary partly according to the size of the particles of the substance in the water. In general, if solid particles in water are more than about 0.1 micron (100 millimicrons) in diameter, the mixture is cloudy or turbid, and the particles will eventually settle out. Such a mixture is called a **suspension.**

If the particles dispersed in water are smaller than about 1 millimicron in diameter, corresponding to the size of ordinary molecules or ions, the mixture is transparent and the particles do not settle out. Such a mixture is called a **solution.** The term solution may also be applied to

similar liquid mixtures that do not contain water, and to mixtures of gases. In a liquid solution the substance whose particles are dispersed separately from each other is called the **solute,** and the substance in which the solute is dispersed is called the **solvent.** The solute is dissolved in the solvent. The most common solvent is water. The molecules of many substances in aqueous solution become separated into electrically charged, chemically active particles called **ions.**

If the particles dispersed in water are between 1 and 100 millimicrons in diameter, corresponding in size to very large individual molecules or to groups of smaller molecules, the mixture may be either turbid or transparent, but the particles do not settle out. Such a mixture is called a colloidal suspension, a colloidal solution, or simply a **colloid.** Colloidal particles are often electrically charged, and some of them tend to accumulate molecules of water or other substances on their surfaces. A colloid that tends to adsorb molecules of water on the surfaces of its particles is said to be **hydrophilic** (water loving). The swelling of a substance as a result of the adsorption of water by hydrophilic colloids is referred to as **imbibition.**

A colloid containing a relatively small amount of water is semisolid and is called a **gel.** A colloid containing a larger amount of water is liquid and is called a **sol.** A gel can often be changed to a sol by raising its temperature, even without adding water, and the reverse change may occur when the temperature is lowered. Ordinary gelatin dessert is a colloidal gel which also contains dissolved sugar and flavoring.

Liquids that do not readily mix, such as oil and water, can form a physically stable mixture if the particles of one of the constituents are of colloidal size. Any colloidal suspension of a liquid is called an **emulsion,** and the same term is commonly applied to any colloidal suspension of fat in water or water in fat, regardless of whether the pure fat is liquid or solid. Cow's cream is an emulsion, as is butter.

SOME PROPERTIES OF WATER

Water is the most abundant constituent of living matter (protoplasm), generally constituting more than 90 percent by weight. It is just as important a part of protoplasm as the more complex organic molecules to which so much attention is devoted, and life (as we know it) cannot exist without it. It is the medium in which the more complex substances are dissolved or suspended, providing the proper milieu for the chemical reactions of life. It also takes direct part in many of these reactions, being formed in some and used up in others.

Much of the biological importance of water results from its unique physical and chemical properties. Its specific heat, heat of vaporization, and heat of melting are all unusually high. It is a good conductor of heat compared to other liquids and most solids (except metals). It is fairly transparent to visible radiation, but more opaque to the longer, heat-bearing wavelengths. The hydrogen and oxygen atoms of water are so strongly bound together that only one molecule in about 10 million is at any one time dissociated into hydrogen and hydroxyl ions. Water has a much higher surface tension and viscosity than most other liquids, because of the high internal cohesive forces among its molecules. It is a good solvent both for **electrolytes** (substances which separate into ions in solution) and for nonelectrolytes, and it tends to bind strongly to the surface of many other substances.

Many of the special properties of water relate to the fact that simple **covalent bonds,** about which we learn in elementary chemistry, are not the only forces affecting the hydrogen and oxygen atoms that make up the molecule. There is a lesser, or secondary, attraction between the hydrogen atoms of one molecule and the oxygen atoms of nearby molecules. The action of these weak bonds, or **hydrogen bonds** as they are called, tends to produce a regular, three-dimensional latticework structure of water molecules. In ice this latticework is fixed and stable; in liquid water it is more shifting and labile. Hydrogen bonds can also be formed by other hydrogen-containing substances, but it is water that we are concerned with here.

The mutual attraction of hydrogen and oxygen atoms from different molecules of water illustrates what is called a **dipole** effect. Water is **polar,** or electrically asymmetrical, because it has, in effect, positively and negatively charged areas. Its two hydrogen atoms are attached to adjacent points on the oxygen atom, rather than on opposite sides. In sharing an electron with oxygen, each hydrogen atom acquires in effect a weak positive charge, so that it is slightly attracted to negatively charged molecules (or parts of molecules) elsewhere. Likewise, in accepting two electrons from hydrogen (in the conventional sharing arrangement), oxygen acquires a weak negative charge, so that it is weakly attracted to positively charged molecules (or parts of molecules) elsewhere. The water molecule thus has a positively charged (hydrogen) end and a negatively charged (oxygen) end; it is dipolar (Fig. 3.1).

The high solubility of electrolytes in water reflects the fact that ions are attracted to the positive and negative areas on

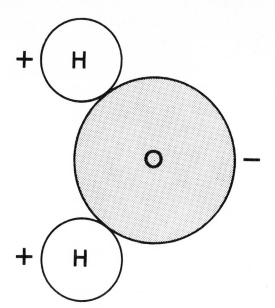

Fig. 3.1 Diagram of water molecule, showing-dipolarity.

water molecules, forming dipole bonds. A positively charged ion (**cation**) will attract a coating of water molecules, with the negatively charged (oxygen) end inward; a negatively charged ion (**anion**) will like-wise attract a coating of water molecules, with the positively charged (hydrogen) end inward (Fig. 3.2). The ions thus tend to be kept afloat and out of close contact with each other.

The high solubility of many nonelectrolytes in water reflects the formation of hydrogen bonds between hydrogen atoms of the solute and oxygen atoms of water. Thus not only ions but also protein molecules and cell surfaces in general tend to be coated with a layer of water molecules, because either the hydrogen or the oxygen of water is attracted to specific atoms in the other substance. The same molecule of water may have its oxygen atom attracted to a hydrogen (or other) atom of some other kind of molecule, and its two hydrogen atoms individually attracted to oxygen (or other) atoms of other kinds of molecules. These weak bonds do not cause the separation of hydrogen and oxygen atoms in one molecule of water from each other, but they do influence the patterning and movement of various kinds of molecules and ions in solution.

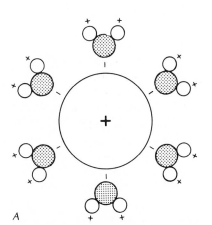

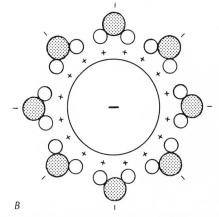

Fig. 3.2 Diagram of cation (A) and anion (B), each with a shell of hydrogen-bonded water molecules.

DIFFUSION

In any substance, be it solid, liquid, or gas, the molecules are in constant motion. In a solid the individual molecules move within a very limited range, but in a liquid or gas the individual molecules (or ions) are free to move throughout the solution. The molecules or ions of each substance eventually become uniformly distributed throughout the solution, even if they are introduced at only one point. The spreading of molecules or ions in solution is called **diffusion.** Any substance in solution tends to diffuse from a region of its own high concentration toward a region of its own lower concentration, until a uniform distribution is achieved. After the solute has attained a uniform distribution throughout the solution, its molecules continue to move, but their movement does not change the equilibrium of the solution; there is no net movement in any direction nor any change in concentration.

EFFECT OF TEMPERATURE ON CHEMICAL CHANGES

The speed at which molecules and ions move increases with the temperature; temperature is in fact merely a measure of the speed of motion of these particles. Since any chemical reaction requires that different kinds of molecules or ions come into contact with each other, the rate at which these reactions take place also varies with the temperature. In the absence of other controlling or modifying factors, *the rate of a chemical reaction approximately doubles with each increase of 10°C (18°F) in temperature.* This relationship was discovered by the Dutch physical chemist Jacobus Henricus van't Hoff (1852–1910). Like many other natural laws, **van't Hoff's law** operates within a limited set of conditions. Often there is a minimum temperature below which the reaction does not take place, and sometimes there is a maximum as well. In a living system the principle of increase in rate of reaction according to the temperature is often modified or obscured by the action of other complex factors, but the tendency still exists.

CARBON CHEMISTRY

At the temperatures prevailing on earth, one element, carbon, differs from all others in the ability of its atoms to link with each other in chains, rings, and other structures to which atoms of other elements are also attached, thus forming a large number of different compounds. It was once thought that these complex carbon compounds could be produced only by living organisms; carbon chemistry is still called **organic chemistry,** even though a great many organic compounds can now be produced in the laboratory and many of these do not even exist in nature.

When a chemical compound composed of two or more elements is produced, some electrons that previously belonged to one atom seem to be shared with the other atom, and this sharing of electrons in what is called a **covalent bond** holds the atoms together into a molecule. The number of electrons that may be given or received by an atom in such an arrangement is referred to as the **valence** of the element. If electrons are given by the element, its valence is positive; if they are received by the element, its valence is negative.

In relatively simple compounds carbon has a valence of plus (+) 4 or minus (−) 4. In more complex compounds it is difficult to distinguish the plus from the minus valences because the carbon atoms are linked to each other. In organic chemistry, then, it is customary to think of each carbon atom as having four bonds without worrying about whether these are positive or negative. On such an assumption it is possible to produce **structural formulas** for many organic compounds, with each atom represented by the symbol for its element and each bond represented by a line. The formula of carbon dioxide, a molecule that consists of one atom of carbon and two of oxygen, may be written as CO_2 (C for carbon, O for oxygen) or as

$$\begin{array}{c} O \\ \| \\ C \\ \| \\ O \end{array}$$

Methane, or marsh gas, may be written as CH_4 (H for hydrogen) or as

$$\begin{array}{c} H \\ | \\ H-C-H \\ | \\ H \end{array}$$

Formaldehyde is CH_2O, or

$$\begin{array}{c} H \\ | \\ C=O \\ | \\ H \end{array}$$

The reader will note that in each of these formulas hydrogen has a valence of one and oxygen two. These are the characteristic valences of these two elements.

Often organic compounds will have one atom of hydrogen and one of oxygen collectively linked to a carbon atom by a single bond. Since hydrogen, like carbon, can have either a plus or a minus valence, whereas oxygen characteristically has a minus valence, it would be possible to show the oxygen atom connected to the carbon atom by one bond and to the

hydrogen atom by the other. However, it is customary to write the OH as a single group, omitting the bond between the hydrogen atom and the oxygen atom in the diagram. Thus the structural formula for formic acid is written

$$\begin{array}{c} O \\ \| \\ H-C \\ \backslash \\ OH \end{array} \quad \text{instead of} \quad \begin{array}{c} O \\ \| \\ H-C \\ \backslash \\ O-H \end{array}$$

Organic acids regularly have a carbon atom that shares two of its bonds with an oxygen atom, one with an OH group, and one with some other atom (most often another carbon atom). The acid effect is produced by the partial dissociation (in water) of the hydrogen atom, without the shared electron, from the OH group. Such **hydrogen ions,** as they are called, are characteristic of all acids, organic or inorganic. If one wishes to write the formula of formic acid without diagramming its structure, one writes HCOOH, the terminal COOH portion of the formula giving clear indication that the substance is an organic acid. The formula of acetic acid, a slightly more complex compound, may be written as CH_3COOH, or it may be diagrammed as

$$\begin{array}{cc} H & O \\ | & \| \\ H-C-C & \\ | & \backslash \\ H & OH \end{array}$$

The formula of propionic acid may be written as C_2H_5COOH or diagrammed as

$$\begin{array}{ccc} H & H & O \\ | & | & \| \\ H-C-C-C & \\ | & | & \backslash \\ H & H & OH \end{array}$$

Although it is customary and convenient to show many structural formulas in

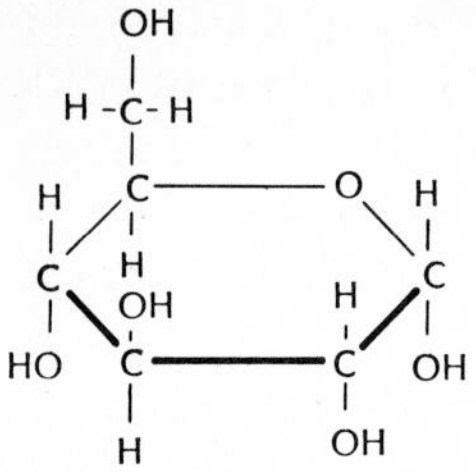

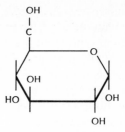

Fig. 3.3 Structural formula of α-glucose. The heavy line indicates the near side of the ring.

Fig. 3.4 Structural formula of α-glucose, shown in simplified form. This form is intended to convey exactly the same information as the form used in Fig. 3.3.

only two dimensions, careful studies by X-ray diffraction and other methods show that the molecules are actually three-dimensional. A methane molecule, for example, is tetrahedral, with a carbon atom in the center and a hydrogen atom at each corner of the tetrahedron. In many of the more complex compounds it is necessary to show the three-dimensional shape in order to clarify the structure. Glucose (Fig. 3.3) is an example.

In showing the structure of more complex compounds it is often useful to show the bonds connecting the carbon atoms, but not the carbon atoms themselves. Any corner (where bonds meet) in the diagram that is not otherwise marked is then assumed to hold a carbon atom, and each of these carbon atoms is assumed to have enough hydrogen atoms (or sometimes other kinds of atoms) attached to it to bring its total bonds to four (Fig. 3.4). Comprehension of this type of diagram comes easily after one has become accustomed to seeing diagrams with all of the atoms indicated.

OXIDATION-REDUCTION AND FREE ENERGY

Many kinds of chemical reactions involve changes in energy relations. Energy may be required for a reaction, or it may be liberated in the reaction. The most common and important of such energy-changing reactions involve either the uptake or the release of oxygen. In general, any process that uses free oxygen and forms oxides, such as carbon dioxide or water, as end products releases chemical energy; whereas any process in which chemically bound oxygen is transformed into free oxygen absorbs energy. The use of oxygen with resultant release of energy is called **oxidation.** The opposite process, in which oxygen is released and energy is taken up, is called **reduction.** This simple concept is useful and adequate for some purposes, and we here use it to lead into the expanded and more technical concept of oxidation and reduction which chemists have come to prefer. This procedure is also in harmony with the historical development of these concepts.

Chemists now find it useful to think of oxidation and reduction as different aspects of the same chemical reaction. Whenever something is oxidized, something else is reduced, and vice versa. Oxidation-reduction reactions involve a change of effective valence, either by sharing or transfer of electrons. The electron acceptor acts as an oxidizing agent and is itself reduced, whereas the electron donor acts as a reducing agent and is itself oxidized. For example, free molecules of elemental oxygen and elemental hydrogen have no effective valence; their potential valence is not being used. When hydrogen burns in oxygen to form water, the potential valence of each element is fully realized. Hydrogen shares one of its electrons with oxygen, and thus goes from an effective valence of zero to plus one; its positive valence is increased, and it is therefore by definition oxidized. Oxygen accepts one electron from each of two hydrogens, thus going from an effective valence of zero to minus two; its negative valence is increased (or one can think of the change as a *decrease* in positive valence), and it is therefore by definition reduced. Hydrogen, in donating an electron, acts as a reducing agent and is itself oxidized. Oxygen, in accepting two electrons, acts as an oxidizing agent and is itself reduced.

The most common biological oxidation-reduction reactions are carried out by the transfer of pairs of hydrogen atoms from the substance being oxidized (the **hydrogen donor**) to the substance being reduced (the **hydrogen acceptor**). Each hydrogen atom in these reactions is carrying its own donatable or sharable electron, and thus the transfer of hydrogen atoms is functionally equivalent to the transfer of electrons. The ultimate hydrogen acceptor is typically oxygen, although there may be several intermediate acceptors along the way.

The capacity (in terms of energy) to do work is often referred to as **free energy**. The most important sorts of free energy from a biological standpoint are the energy of molecular motion (measured as heat), and the energy that can be released in (or used to promote) oxidation-reduction reactions. Typically (and most importantly from the standpoint of plant physiology) the liberation of free energy involves the use of free oxygen and the breakdown of complex substances to simpler substances such as carbon dioxide and water; but sometimes it involves other sorts of changes, or the mere rearrangement of atoms within existing molecules. When a complex substance is oxidized, its free energy may be released as chemical energy that can be used to promote oxidation-reduction reactions or other reactions that require energy, or it may be released as heat. This simple description of the concept of free energy is adequate for our purposes at this level of study, but it must be considerably refined in the study of chemical thermodynamics.

FOODS

Any chemical compound that can be broken down by a living organism with release of energy is called a **food**. Such a breakdown of food is always an oxidation-reduction reaction in which the food is partly or completely oxidized. Complete oxidation of the food requires free oxygen, but a partial oxidation which releases some of the free (stored) energy can often be accomplished by rearrangements that do not require any input of oxygen.

Most foods synthesized by plants may be classified into three general groups: **carbohydrates, fats,** and **proteins.** Typical fats and carbohydrates are composed wholly of carbon, hydrogen, and oxygen; proteins contain nitrogen also, and usually one or more other elements (especially sulfur and often also phosphorus) as well. The simpler compounds from which foods are made, such as carbon dioxide, water, and various inorganic salts, are regarded as raw materials rather than as foods. The "plant food" that one buys at garden supply houses actually contains mineral nutrients that are necessary for plant growth; these mineral nutrients are not themselves foods, as the term is used by biologists.

Carbohydrates

In carbohydrates there are characteristically twice as many atoms of hydrogen per molecule as there are of oxygen. One of the commonest carbohydrates is **glucose** (also called dextrose or corn sugar). Glucose exists in two slightly differing forms called α-glucose and β-glucose; β-glucose differs from α-glucose only in that the carbon atom at the farthest right, as shown in Fig. 3.3, has the OH above and the H below the plane of the ring, instead of vice versa.

Glucose is called a **hexose** sugar because it contains six carbon atoms. Some other hexose sugars, such as fructose (also called levulose) and mannose, also have the formula $C_6H_{12}O_6$, but have the atoms somewhat differently arranged. **Pentose** sugars, containing five rather than six carbon atoms, are also of frequent occurrence and have the general formula $C_5H_{10}O_5$. Simple sugars having three, four, or seven carbon atoms are less common. Any simple sugar, whether it has three, four, five, six, or seven carbon atoms, may be called a **monosaccharide.** A **disaccharide** is a sugar formed by the combination of two similar or different monosaccharides, the combination occurring in such a way that a molecule of water, H_2O, is released in the process. One OH group from each of the two monosaccharide molecules is affected in such a way that both of the hydrogen atoms combine with one of the oxygen atoms to form a molecule of free water, and the remaining oxygen atom is bonded to each of the two monosaccharides, linking them into a disaccharide. Without resorting to a struc-

tural diagram, a typical example of such a process may be written

$$2C_6H_{12}O_6 \rightarrow C_{12}H_{22}O_{11} + H_2O$$

Under proper conditions this reaction may be reversed, and two molecules of monosaccharide may be produced from one molecule of disaccharide, using up a molecule of water in the process. Many other chemical compounds may be broken down into similar compounds in somewhat similar fashion, with water being used in the process. Such a chemical reaction is called **hydrolysis.**

The most common disaccharide is cane sugar, or sucrose. It is formed from one molecule of glucose and one molecule of fructose and may be hydrolyzed to yield the original glucose and fructose molecules. Maltose is a disaccharide formed by the combination of two molecules of glucose.

Polysaccharides are formed by the union of several or many monosaccharide molecules in the same way that two monosaccharides are linked to form a disaccharide. Each such linkage (called a **glycoside linkage**) results in the production of one molecule of water. A hexose polysaccharide may thus be thought of, in one way, as a series of $C_6H_{10}O_5$ groups linked to an original $C_6H_{12}O_6$. No matter how many monosaccharide molecules (called **glycoside units** or **glucose residues**) are combined to make the polysaccharide, the proportions of carbon, hydrogen, and oxygen never quite reach that of $C_6H_{10}O_5$ because the complete polysaccharide molecule always contains two additional atoms of hydrogen and one of oxygen. A generalized formula for a hexose polysaccharide may therefore be written as $(C_6H_{10}O_5)_n \cdot H_2O$. The n in the formula stands for any number, and in many polysaccharides it is evidently a very high number, sometimes as much as 3,000 or even 10,000. Two very common hexose polysaccharides are starch, a reserve food that actually consists of a mixture of two different kinds of hexose polysaccharide molecules, and cellulose, the principal constituent of the cell walls of most plants. The formation of complex molecules by combination of a number of simpler molecules of the same or closely related kinds is called polymerization, and the resulting compounds are called **polymers.**

Some compounds that do not have exactly twice as many hydrogen atoms as oxygen atoms per molecule are still considered to be carbohydrates because of their close similarity to other carbohydrates; and indeed there is no inherent and precise division between carbohydrates and other organic compounds.

Fats

Fats differ from carbohydrates in the arrangement and proportion of atoms in the molecules, always having fewer oxygen atoms in relation to the number of hydrogen atoms than do carbohydrates. They have a higher free-energy content per unit of weight than do carbohydrates, and more oxygen is required to oxidize or respire them.

Fats are formed by the chemical combination of **glycerol** (glycerin), a kind of alcohol, with certain kinds of organic compounds known as **fatty acids.** Glycerol has the following structure:

$$
\begin{array}{c}
H \\
| \\
HO-C-H \\
| \\
HO-C-H \\
| \\
HO-C-H \\
| \\
H
\end{array}
$$

In the formation of a fat, three molecules of fatty acid (the same or different acids) are combined with one molecule of glycerin, and three molecules of water are released. If the convenient symbol R is used in each case to represent the majority of the structure of the fatty acid, the reaction may be shown as in Fig. 3.5.

It will be seen that the chemical linkage of fatty acids to glycerin is accomplished in similar fashion to the linkage of monosaccharide molecules to form a disaccharide. In each case the formerly separate molecules are linked by an atom of oxygen at a point where each of the separate molecules had carried an OH, and a molecule of water is formed in the process. As might therefore be expected, fats can be hydrolyzed to produce glycerol and fatty acids, just as polysaccharides can be hydrolyzed to form monosaccharides.

Fats that are liquid at ordinary temperatures are often referred to as **oils,** but many of the substances called oils in general speech are of a quite different chemical nature.

There are many compounds which, like fats, are insoluble in water, but are soluble in certain organic solvents such as acetone, ether, chloroform, and carbon tetrachloride. These, together with fats, are known as **lipids.** Many lipids are formed in part from fatty acids and thus have some chemical similarity to fats, but some others are quite different in structure. An important group of lipids contains phosphorus as well as fatty acids; these are known as **phospholipids.** Some other lipids have fatty acids in combination with carbohydrates or with protein, thus breaking down the usual distinctions among these three types of foods.

Proteins

Proteins differ from carbohydrates and fats in that they always contain nitrogen.

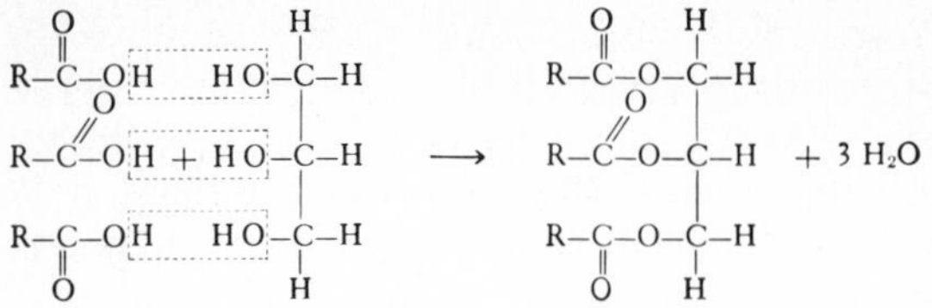

Fig. 3.5 Formation of fat from glycerin and fatty acids.

They usually also contain sulfur, frequently phosphorus, and sometimes other elements as well. Proteins can be hydrolyzed to yield **amino acids,** which are organic acids having one or more **amine,** or NH_2, groups in place of hydrogen atoms. Most proteins are so complex that satisfactory structural formulas have not been worked out for them, but the structure of the individual amino acids is known. The structure of the simplest amino acid, called **glycine,** or aminoacetic acid, is shown in Fig. 8.7. Some other amino acids are much more complex.

Protein molecules are usually very large, and many individual amino acid molecules go into the formation of each protein molecule. Twenty principal kinds of amino acids are involved, but not every protein molecule contains all of these kinds. The same 20 amino acids occur regularly in the proteins of at least the vast majority of kinds of organisms, both plant and animal. They are alanine, arginine, aspartic acid, cystine, glutamic acid, glycine, histidine, hydroxylysine, hydroxyproline, isoleucine, leucine, lysine, methionine, phenylalanine, proline, serine, threonine, tryptophan, tyrosine, and valine. Some kinds of organisms also incorporate one or more other kinds of amino acids into some of their proteins, but we need go no further with these exceptions than to note that they exist.

Plants also produce a considerable number of other amino acids that are not incorporated into proteins but remain free in the cytoplasm. In some plants these **free amino acids,** as they are called, are poisonous or unpalatable, and help to discourage predators. Otherwise their function is still largely speculative.

There is an almost infinite number of different proteins, varying in the number and kind of amino acids used, their ratios to each other, and their positions within the molecules. The amino acids that make up a given protein may be linked together in a single long chain, or in two or more chains with occasional covalent bonds (often involving sulfur) connecting the links of different chains. These **polypeptide** chains, as they are called, are not straight, but are variously contorted into three-dimensional structures which are characteristic for each protein. The configuration of the molecule, which is apparently governed by the amino acid sequence, is at least as important as its amino acid composition in determining its chemical activity. The determination of the complete amino acid sequence of a protein and the determination of the three-dimensional configuration of the polypeptide chain(s) are difficult and time-consuming processes which have as yet been completed for only a relatively few proteins. The first complete determination of the sequence of amino acids in any protein was presented for insulin in 1953 by the English biochemist Frederick Sanger (1918–   ) and his associates. Insulin is a rather small protein consisting of only 51 amino acid units arranged in two chains.

Proteins that are formed solely by the linkage of amino acids, and which thus yield only amino acids on hydrolysis, are called **simple proteins.** Proteins that yield one or more other compounds in addition to amino acids when hydrolyzed are called **conjugated proteins.** Nearly all enzymes that exist in plants and animals are proteins or occur in conjugation with proteins. Most proteins, in turn, are enzymatic, although there is commonly a small proportion of nonenzymatic, structural protein in cell walls.

ENZYMES, COENZYMES, AND ATP

Many chemical reactions between two substances occur whenever the two substances are brought together in solution. Other chemical changes require the intercession of a third substance, which takes part in the reaction but is not used up by it. Such substances, which facilitate chemical reactions between other substances, are called **catalysts.** Most of the chemical reactions characteristic of life depend on catalysts. Theoretically, a catalyst merely changes the speed of a chemical reaction, without causing anything that would not happen anyway at a slower speed. Practically, many biological catalysts are so essential to the reactions they govern that these reactions do not occur at any detectable rate without them.

Most of the catalysts produced by living organisms are simple or conjugated proteins. The part of the conjugated protein that is not amino acid may be closely bound to the remainder of the molecule, as in the phycobilin pigments of certain algae, or more loosely bound and detachable, as is chlorophyll. In either case it is known as a **prosthetic group.** Both of the examples here given as prosthetic groups belong to the class of compounds known as porphyrins, but there are also many other kinds of prosthetic groups. A catalyst that is a simple or conjugated protein is called an **enzyme.**

Finally, there is a group of biological catalysts that must cooperate with enzyme proteins to produce their effect, but are not themselves persistently attached to any particular protein. These catalysts, always consisting of molecules much smaller than proteins, are called **coenzymes.** The compound known as NAD (for nicotinamide adenine dinucleotide), for example, is a coenzyme.

In practice there is often no clear line between coenzymes and enzymes with loosely attached prosthetic groups, or between coenzymes and substances such as various metallic ions which seem merely to provide the proper milieu for the action of enzymes and coenzymes. The categories are mentally convenient even though only arbitrarily limited.

There may be on the order of 10,000 different kinds of enzymes in a single cell, totaling perhaps 500 million molecules. Inasmuch as the structure of only a very few enzymes is known, it is impossible to classify them effectively according to their chemical constitution. Instead it is customary to classify and name them according to their effect. The suffix -*ase* is attached to a word which indicates either the substrate (the substance acted upon) or the kind of chemical change that is governed. Thus all enzymes that govern hydrolysis (or the reverse change) may be called **hydrolases.** Hydrolases that act on carbohydrates are called **carbohydrases;** carbohydrases that act on very large carbohydrate molecules (consisting of several or many monosaccharide units) are called **polysaccharases;** polysaccharases that act on starch and glycogen (but not cellulose) are called **amylases;** and polysaccharases that act on cellulose (but not starch and glycogen) are called **cellulases.** These categories may be further subdivided.

Most enzymes appear to be more or less highly specific, governing a particular sort or set of related sorts of changes in one or a few particular sorts of molecules. Some kinds of maltase, for example, such as the one found in the fungus *Aspergillus,* govern the hydrolysis of maltose (a disaccharide composed of two units of α-glucose) alone, being ineffective on other kinds of disaccharides and even on polysaccharides composed wholly of α-glucose units (i.e., α-glucosides). Some other so-called maltases found in other organisms work on maltose and any other α-glucoside, and might more properly be called α-glucosidases. Lipases (which govern the hydrolysis of fats) are in general much less specific. Any of many lipases can apparently govern the hydrolysis of any of a wide range of fats. Some few enzymes are known to catalyze two or more totally unrelated reactions, but these are clearly the exception rather than the rule. Enzyme specificity is, in effect, a measure of the structural specialization that an enzyme requires in its substrate. All enzymes apparently require a considerable degree of such specialization, but some are more demanding than others.

Many enzymes catalyze a particular reaction in either direction, according to the conditions, tending toward an equilibrium in which a certain percentage of the reactants remains on each side of the equation. In theory all enzymatic reactions should be reversible, but here again there are practical factors that prevent the theory from being fully realized. Reactions that involve any significant change in free energy balances can generally be catalyzed only in the direction in which free energy is liberated, unless energy in the proper form is available.

A great many of the necessary chemical reactions of protoplasm require the input of chemical energy. The standard carrier of this energy, in both plants and animals, is a coenzyme known as **ATP,** or adenosine triphosphate (Fig. 3.6). The triphosphate part of the name refers to the fact that the molecule carries three phosphate groups as part of its structure. It readily gives up the third phosphate group, forming **ADP** (adenosine diphosphate) and phosphate ion. The release of the third phosphate group also releases a relatively large amount of energy, some of which can be transferred into energy-requiring reactions. The covalent bond by which this third phosphate group is attached to the ATP molecule is referred to as a **high-energy bond.** It is often represented by the symbol ~, instead of the simple — used for ordinary bonds.

In giving up its third phosphate group, ATP can provide the necessary chemical energy for a wide array of energy-requiring reactions in the protoplasm. The energy is in effect supplied in prepackaged units of a given size. If the reaction can be energized by one such unit (the high-energy bond), it can proceed; otherwise not. Any excess energy in this prepackaged unit, not required by the reaction being fostered, is lost as heat. The system therefore tends to run down, trans-

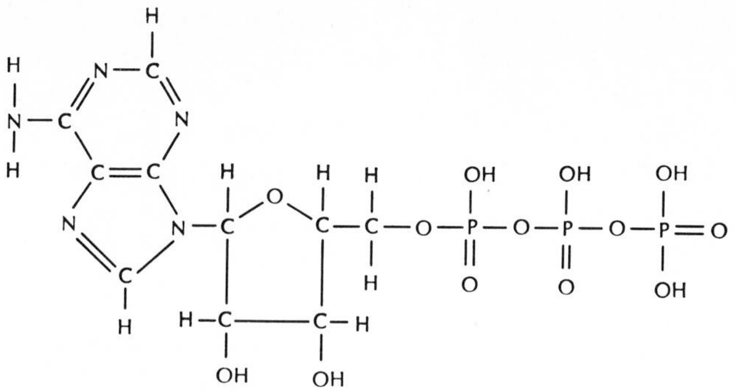

Fig. 3.6   Structure of ATP.

forming a certain amount of energy into heat each time the high-energy phosphate bond is broken.

ATP can be regenerated from ADP and phosphate ion by supplying the requisite amount of energy in the proper way. As we shall see in subsequent chapters, ATP is generated in the process called photosynthesis, but most of this ATP is used immediately to provide the energy necessary for the formation of carbohydrate. When the carbohydrate is subsequently broken down in the process called respiration, much of the stored energy is used to generate ATP from ADP and phosphate ion. *Photosynthesis and respiration may collectively be considered as a way of generating the ATP necessary for metabolic work.*

Energy from ATP is transmitted more or less directly into many metabolic processes, but it may also be transmitted through accessory carriers that also have high-energy phosphate bonds. Two such accessory carriers are the compounds known as uridine triphosphate (UTP) and guanosine triphosphate (GTP). Under some circumstances ADP can break another high-energy phosphate bond, forming AMP (adenosine monophosphate) and

liberating additional energy along with another phosphate ion. We shall see that certain protoplasmic reactions can also be energized by coenzymes known as NAD and NADP.

RADIANT ENERGY, THE SPECTRUM, AND PIGMENTS

Radiant energy is energy which can be transmitted through space in a form that can be compared to waves. Wavelengths of radiant energy range from a low of less than 0.0001 nm (nanometer) for cosmic rays, to a high of 1 km or even more, for some electric waves. Various sorts of radiant energy are commonly referred to as rays, such as X-rays, ultraviolet rays, light rays. The nature of radiant energy is not fully understood, and some of its aspects are better rationalized by treating it as particles instead of waves.

The energy reaching the earth's surface as solar radiation varies in wavelength from about 300 nm to about 2600 nm. Wavelengths from about 390 to about 750 nm make up the visible spectrum. The color of light varies according to its wavelength. The shortest visible rays are violet, and the longest are red, the series running

VISIBLE SPECTRUM

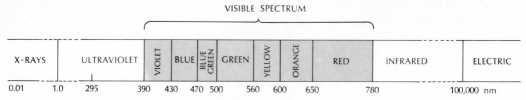

| X-RAYS | ULTRAVIOLET | VIOLET | BLUE | BLUE GREEN | GREEN | YELLOW | ORANGE | RED | INFRARED | ELECTRIC |

0.01   1.0   295   390   430 470 500   560   600   650   780   100,000 nm

Fig. 3.7 The spectrum of radiant energy; figures represent nanometers (nm).

from violet to blue, blue-green, green, yellow, orange, and red. White light, such as sunlight, has a mixture of all of these wavelengths. Under ordinary conditions about 40 percent of the radiant energy that reaches the earth's surface is in the visible spectrum. Solar wavelengths too short to be seen are called *ultraviolet*, and those too long to be seen are called *infrared*. The major part of the spectrum of radiant energy is shown in Fig. 3.7.

Radiant energy, of whatever wavelength, seems to be delivered in particles of a finite (though very small) size. These ultimate, indivisible units of radiant energy are called **quanta** (sing., quantum). The amount of energy per quantum varies inversely with the wavelength: the longer the wavelength, the less energy per quantum. A quantum of light is called a **photon.**

All substances absorb and reflect (or transmit) light differentially, absorbing more quanta of some wavelengths than of others. Substances with markedly differential light-absorbing properties are called **pigments,** especially if they occur in relatively small amounts in another substance or body whose color they affect. Chlorophyll, carotene, and anthocyanin are some familiar plant pigments. Some other plant pigments, such as cytochrome, generally occur in such small amounts as to escape the eye. The absorption of light by a pigment may be directly related to its function in the protoplast, as is true of chlorophyll and carotene, or it may have no functional importance, as appears to be true of anthocyanin and cytochrome. (Anthocyanin in flowers attracts insect pollinators, but that is another matter.)

It is often useful to consider the light-absorbing capacity of a pigment in terms of an **absorption spectrum.** This is commonly shown as a simple graph, with the wavelengths of light on the horizontal

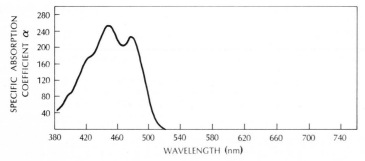

Fig. 3.8 Absorption spectrum of β-carotene. [Data of Zscheile et al.]

axis and the amount of absorption on the vertical axis. There are diverse ways of measuring the amount of absorption, but the methods and the units of measurement are immaterial for our purposes. Given a method and a unit of measurement, the important thing is the differences in absorption at different wavelengths. The absorption spectrum of β-carotene is shown in Fig. 3.8.

SUGGESTED READING

Astin, A. V., The standards of measurement, *Sci. Amer.* **218**(6):50–62, June 1968.

Battista, O. A., Colloidal macromolecular phenomena, *Amer. Sci.* **54**:151–173, 1965.

Changeux, J. P., The control of biochemical reactions, *Sci. Amer.* **212**(4):36–45, April 1965.

Fruton, J. S., *Molecules and life: Historical essays on the interplay of chemistry and biology,* Wiley-Interscience, New York, 1972.

Lambert, J. B., The shapes of organic molecules, *Sci. Amer.* **222**(1):58–70, January 1970.

White, E. H., *Chemical background for the biological sciences,* Foundations of Modern Biology Series, Prentice-Hall, Englewood Cliffs, N.J., 1964.

PART II
CELLS

Euglena, a one-celled "plant", much enlarged.
[Photomicrograph courtesy of Elsa O'Donnell-Alvelda.]

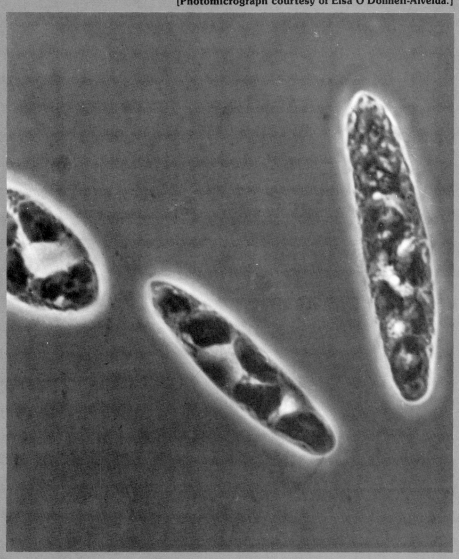

Chapter 4

The Basic Structure and Function of Cells

Young cell of root tip of corn. (× about 5000.) [Electron micrograph courtesy of W. Gordon Whaley.]

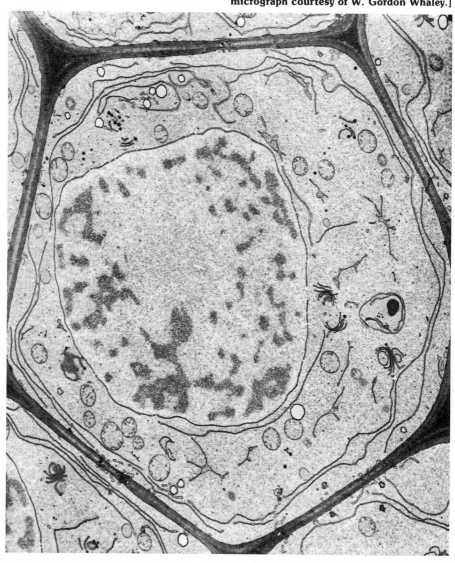

SOME HISTORICAL BACKGROUND

We have noted that a development in one field of science may be necessary to further advance in some other field of science. The invention of the compound microscope in 1590 by Zacharias Jansen, a spectacle maker of Middleburg, Netherlands, was one of the most important events in the history of biology. It had long been known that a double-convex lens forms an enlarged visual image, but only a few diameters magnification could be obtained with most such lenses. Jansen found that by using two lenses, properly spaced, the magnified image produced by the first lens was again magnified by the second. His own microscopes were relatively crude, but they demonstrated the principle that enabled subsequent workers to make much more useful instruments.

The nature of light limits the magnification that can be obtained with a microscope of the type invented by Jansen. With ordinary present-day equipment this is a little more than 1000 diameters, and the most specialized accessory equipment does not extend the profitable range of magnification to more than about 3000 diameters. The **resolving power** of the microscope, that is, its ability to produce distinct images of objects that are close together, is inherently limited by the wavelength of light. For practical purposes, structures less than about 200–300nm in diameter cannot be distinguished with microscopes using visible light. This is, of course, still much better than the unaided human eye, which generally cannot distinguish objects less than about 80 (exceptionally down to about 25) microns in diameter.

If a beam of electrons instead of a beam of light is thrown on the object to be examined, then with proper apparatus an image can be obtained that is profitably enlargeable to 100,000 diameters or more, with a resolving power now down to about 1.4 Å.

The essential principle of the electron microscope (often abbreviated EM) was discovered by a German physicist, Hans Busch, in 1924, and the electron microscope was developed during the 1930s. The design was much improved in the 1940s by the Radio Corporation of America, and with continuing improvements the instrument has come into widespread scientific use (Fig. 4.1). It has obvious advantages over conventional microscopes in the much greater magnifications

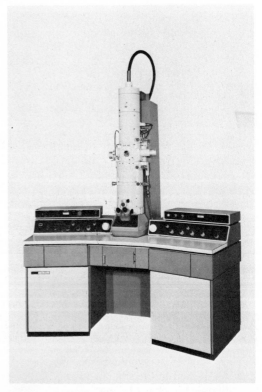

Fig. 4.1   An electron microscope. [Courtesy of JOEL U.S.A., Inc.]

that can be obtained, but manipulation and maintenance require considerable skill. It cannot easily be used on ordinary living material, because it operates in a high vacuum.

Two types of electron microscopes are now produced: the original type, in which the electrons pass through the object examined, and another type, in which the electrons bounce off the object examined. The former type is designated as a transmission electron microscope (TEM) and the latter as a scanning electron microscope (SEM). Objects to be examined by SEM must first be coated with a thin layer of electron-dense material, such as gold. The SEM is especially useful in showing the form and surface of three-dimensional objects, such as pollen grains.

One of the first microscopists was Anton van Leeuwenhoek (1632–1723), a dry goods merchant and, later, custodian of the city hall in Delft, Netherlands. He used a simple rather than a compound microscope (Fig. 4.2), but he ground his lenses far more carefully than did any of his contemporaries. One of his instruments, preserved at Utrecht, magnifies about 280 diameters, which is extraordinarily good for a single lens, even by modern standards. He was probably the first man to observe bacteria, yeasts, and protozoa, among other things. Many of his observations became known through his

Fig. 4.2 Three Leeuwenhoek microscopes, showing front, side, and back views. The lens is mounted in the small opening in the plate. The position of the needle point, which holds the object to be examined, is adjusted by the controls shown in the side and back views. (Slightly less than natural size.) [Copyright by Rijksmuseum voor de Geschiedenis der Natuurwetenschappen, Leiden.]

Fig. 4.3 A microscope used by Robert Hooke. [Courtesy of the Medical Museum of the Armed Forces Institute of Pathology.]

long correspondence with the British Royal Society.

Robert Hooke (1635–1703), a versatile British physicist, mathematician, and architect, devoted some time to improving and using the compound microscope (Fig. 4.3). His *Micrographia* (Fig. 4.4), published in 1665, included a description and illustration of the structure of cork. He introduced the term *cell* for the minute

cavities into which the cork was regularly partitioned.

Subsequent investigators found that other parts of plants, and of animals as well, were also divided into cells. The German botanist Matthias Schleiden (1804–1881) (Fig. 4.5) and his zoologist friend Theodor Schwann (1810–1882) were impressed by the essential similarity of the structure of plant and animal tissues, and together they formulated, in 1838, the concept that all tissues, and indeed all organisms, are composed of cells. We now know that very small organisms often consist of only a single cell and that some kinds of organisms, notably many of the fungi, are not divided into separate cells; but their idea was basically sound, and the **cell theory**, as it is called, is now one of the fundamental concepts of biology.

The cork cells observed by Hooke had no living contents, because cork cells die on reaching maturity and their protoplasm disintegrates. Even when living cells were studied, however, early observers apparently failed to recognize their protoplasmic contents, and such works as the *Anatomy of Plants*, published in 1675 by the Italian professor of medicine Marcello Malpighi (1628–1694), are concerned entirely with cell walls rather than cell contents.

The first clear recognition of the living contents of a cell may have been made by the Italian botanist Bonaventuri Corti (1729–1813), who in 1772 observed protoplasmic streaming in cells of *Chara*, an alga. In 1833 the Scots botanist Robert Brown (1773–1858) described the nucleus as a regular component of cells, but he did not understand its significance. The French zoologist Félix Dujardin (1801–1860) proposed the term *sarcode* in 1835

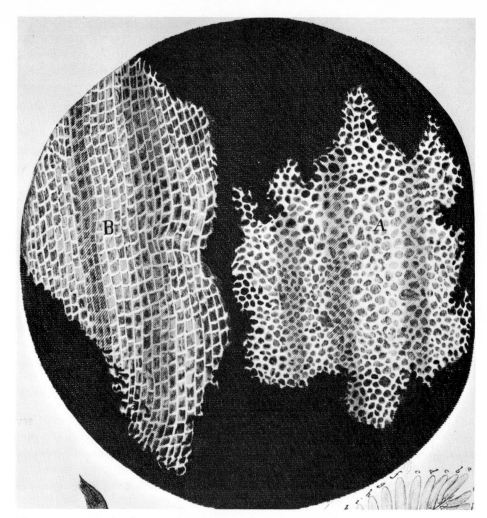

Fig. 4.4 A part of Robert Hooke's *Micrographia*.

for the living substance of the cells of some simple animals he was studying, but his term was not generally adopted. The Czech zoologist Johannes Purkinje (1787–1869) used the term *protoplasm* in 1840 for the formative substance of young animal embryos, and in 1846 the German botanist Hugo von Mohl (1805–1872) applied this same term to the living contents of plant cells. The German botanist Ferdi-

nand Cohn (1828–1898) suggested in 1850 that the protoplasm of plant and animal cells is essentially similar. In 1861 the German zoologist Max Schultze (1825–1884) set forth the doctrine (the protoplasm doctrine) that each unit of organization of living things consists of a mass of **protoplasm**, including a nucleus. Such an organized unit of protoplasm is now called a **protoplast.**

Fig. 4.5   Matthias Schleiden (1804–1881), German botanist, who with Theodor Schwann proposed the cell theory.

The role of the cell nucleus as the "vehicle of heredity" was tentatively suggested by the German naturalist and philosopher Ernst Haeckel (1834–1919) in 1866. The soundness of Haeckel's speculation was independently and almost simultaneously confirmed in 1884–1885 by four other Germans—three zoologists and one botanist—in 1884–1885.

The concept of the cell has thus changed considerably from Robert Hooke's cavities surrounded by walls. Now we think of a cell in living tissues as a protoplast which (in plants) is surrounded by a wall, or, in dead tissues or cells, as the wall itself. The empty space enclosed by the wall of a dead cell is now called the **lumen** and is of little scientific interest.

The reader should understand that any historical summary, such as the one just presented, is oversimplified and to that extent possibly misleading. Scientific concepts spring from diverse sources and depend on the accumulation of facts and ideas produced by many different workers. Often the person who gets the credit for a major advance is the one who puts the last brick into place or who marshalls the evidence for an idea that has been floating around for some time without proof. When the time is ripe for a new idea or interpretation, then if one person does not think of it, another soon will, and it is often difficult to apportion the credit properly. In this book we mention names and dates more to enable the student to understand the nature of scientific progress than to honor the memory of particular scientists. Some names would be the same in any useful historical summary, but others would be changed by a change in emphasis or approach.

STRUCTURE OF CELLS

Cells are extremely varied in size and shape, and they often have specialized functions related to these differences. Some bacteria are only about 200 nm in diameter, which is hardly large enough to be observed with the most powerful conventional microscopes. The cells that make up the juice sacs of citrus fruits are among the larger plant cells. The slender fiber cells of some plants are as much as 50 cm long, and the nerve cells of some of the higher animals may reach a length of more than 1 m. The largest cells of all, in terms of volume, are the yolks of the eggs of some of the larger birds, such as the ostrich. Most cells are so small as to be only barely, if at all, visible to the naked eye, but large enough to be studied in fair detail under microscopes giving 100 to 400 diameters magnification.

Although cells are typically very small

in terms of human eyesight, they are very large in terms of molecular capacity. It has been estimated that an ordinary plant cell of the type called parenchyma contains about 100 million protein molecules, a figure that may also be expressed as 10^8. For every protein molecule in such a cell, there may be about 10^8 or 10^9 molecules of water. Great numbers of molecules of other kinds are also present in any living cell. A large tree may have as many as 10^{14} cells, but a great many of these cells are dead, without living contents.

The size of cells is restricted by the ratio of their surface area to volume. The need of cells for nutrients, for respiratory exchange of gases, etc., tends to be proportional to their volume, that is, to the cube of the radius in a spherical cell. The ability of the cell to take in nutrients and to exchange gases tends to be proportional to its surface area, that is, to the square of the radius in a spherical cell. Increase in cell size above the ordinary measurements soon results in a disparity between the needs and the means of obtaining the supplies, and the size of cells is thus limited. Plant cells can partly avoid this restriction by growth in length without any increase in diameter, or by having a highly vacuolated protoplast (see below), so that increase in cell volume does not cause a commensurate increase in the amount of protoplasm. It should not be surprising that there is an inverse rough correlation between the size of a cell and its rate of metabolic activity: The larger the cell, the more sluggish its life processes.

A plant cell (Fig. 4.6) characteristically consists of (1) a **protoplast** which encloses (2) a **central vacuole** and is enclosed by (3) a **cell wall**. The central vacuole is not empty but contains the **cell sap.** The protoplast is the living part of the cell. It

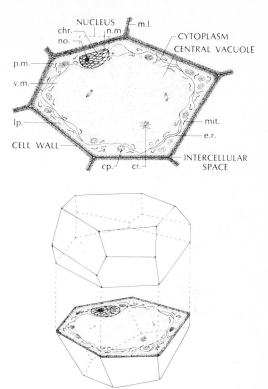

Fig. 4.6 A typical plant cell, much enlarged: chr., chromatin; cp., chloroplast; cr., crystal; e.r., endoplasmic reticulum; lp., leucoplast; mit., mitochondrion; m.l., middle lamella; n.m., nuclear membrane; no., nucleolus; p.m., plasma membrane; v.m., vacuolar membrane. For purposes of illustration the plasma membrane is shown slightly removed from the cell wall, instead of closely appressed to the wall as it would be in a normal cell.

consists of a complex, organized mixture of proteins, lipids, and other substances in colloidal suspension, plus some materials in true solution. It is usually divisible into two parts: (1) a mass of **cytoplasm,** in which is embedded (2) a distinctive body called the **nucleus.** Both the cytoplasm and the nucleus contain distinctive organelles which can be recognized with conventional microscopes, as well as smaller organized particles which can be seen only by careful electron microscopy. Commonly 75–90 percent of the cell, by weight, consists of water. In some kinds of cells, the proportion of water even exceeds 95 percent. In terms of volume, a typical plant cell consists of a bit more than 90 percent vacuole, less than 5 percent protoplasm, and about 5 percent cell wall.

Prokaryotic cells, characteristic of the bacteria and blue-green algae, have a less complex organization than the eukaryotic cells found in other plants and animals. Except as otherwise specifically indicated, our discussion here is based on eukaryotic plant cells.

The Cell Wall

The cell wall is ordinarily composed of one or two layers. The thin outer layer is called the **primary wall.** The thicker inner layer, when present, is formed after the outer layer and is called the **secondary wall.** Both the primary and the secondary wall are usually composed largely of **cellulose,** a complex polysaccharide that can be broken down into β-glucose by hydrolysis. It typically forms the framework around which the other wall substances are deposited.

The chemical formula of cellulose can be written $(C_6H_{10}O_5)_n \cdot H_2O$. The number represented by n is between 1,000 and 3,000, or possibly, sometimes up to 10,000. The β-glucose units (residues) that make up the cellulose molecules are apparently linked together in a long, simple, unbranched chain. The chains are organized into groups to form slender cellulose rods or ribbons called **microfibrils.** Each microfibril is on the order of 8–30 nm wide, well below the resolving power of conventional microscopes. The microfibril itself has a complex internal organization, consisting partly of dense, crystalline subrods called **micelles** and partly of seemingly unordered molecular chains. The individual cellulose chains are longer than the crystalline micelles, and one chain may pass through several consecutive micelles within its microfibril.

The cellulose microfibrils, which make up the principal component of the cell wall, form an irregular latticework, with many small openings and scattered longer ones (Fig. 4.7). Often the microfibrils are further grouped into macrofibrils as much as 0.5 micron in diameter, which can readily be seen with conventional microscopes.

The capillary spaces among the cellulose microfibrils of the wall are commonly filled by other substances, notably **cellulosans** and **pectic substances.** Cellulosans, or pentosans, are polysaccharides that differ from cellulose in being formed of pentose sugars rather than of glucose, which is a hexose sugar. **Pectin** and some other compounds, collectively called **pectic substances,** are polymers of either pentosans or hexosans or both, and they further differ chemically from cellulose in that some or all of the simpler molecules obtained by hydrolysis have the projecting

$$
\begin{array}{c}
\text{H} \\
| \\
-\text{C}-\text{OH} \\
| \\
\text{H}
\end{array}
$$

group of glucose replaced by a —COOH group. This feature makes pectic substances hydrophilic, and it also permits them to combine with metals such as calcium and magnesium. It is the hydrophilic property of pectin that makes it

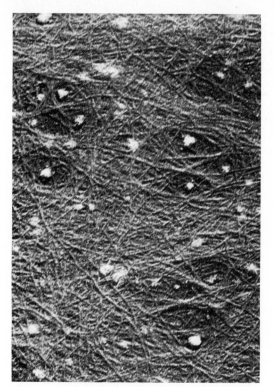

Fig. 4.7 Electron micrograph of cell wall in the root tip of an onion, showing loosely interwoven micelles of cellulose, with scattered openings of various sizes. Remains of plasmodesmata can be seen here and there as white blobs. This cell has a primary wall but no secondary wall. (× 34,000.) [Courtesy of Flora Murray Scott.]

useful in the manufacture of jams and jellies. Pectic substances are amorphous, having no apparent structure even in electron microscopic preparations.

The cellulosans, pectic substances, and some other compounds related to pectic substances have certain properties in common and are often collectively referred to as **hemicelluloses.** It is not always easy to determine to which of the subcategories a particular hemicellulose belongs. Hemicelluloses are in general more readily soluble than cellulose, and indeed one definition of hemicelluloses is that they are polysaccharides which are extractable with 17.5 percent NaOH.

In addition to occurring with cellulose in cell walls, pectic substances commonly form a thin layer, called the **middle lamella,** which holds adjacent cells together and is not regarded as part of the wall of either cell. The middle lamella between young cells is composed largely of pectin, which is soft and jellylike. As the cells mature, much of the pectin is commonly changed to calcium pectate and magnesium pectate, forming a much harder cement that holds the cells firmly together.

The walls of many cells, particularly those of wood, also contain considerable amounts of **lignin.** Lignin is a high polymer of several compounds derived from phenyl propane (Fig. 4.8). It consists wholly of carbon, hydrogen, and oxygen, but unlike cellulose and hemicellulose it is not of carbohydrate nature. Polymerization to form lignin occurs in all directions of space, rather than in linear fashion to form simple chains as in cellulose. The phenyl-propane units are cross-linked by benzyl-ether groups and become ionically bound to the cellulose in the cell wall. When lignin is produced, it ordinarily infiltrates in succession the middle lamella, the primary wall, and especially the

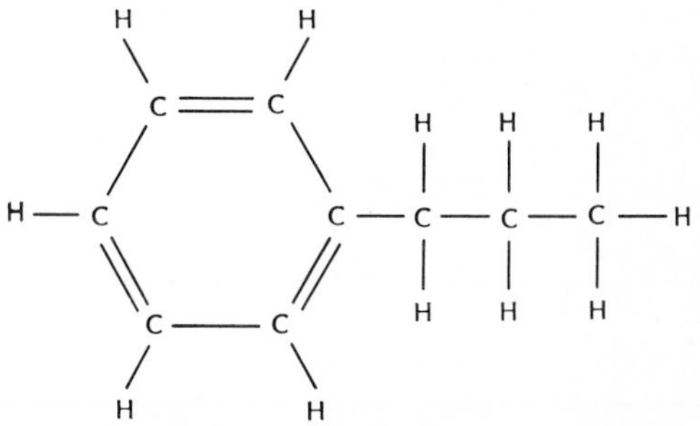

Fig. 4.8 Structure of phenyl propane.

secondary wall. The cell wall is then said to be lignified. Lignification increases the firmness, hardness, strength, and wettability of the wall.

A small amount of structural (i.e., nonenzymatic) protein is generally incorporated into cell walls. The wall is generally also saturated with water. Still other substances that occur in the cell walls of some kinds of plants are discussed under the plant groups in which they are found. Some of these other substances are silica, cutin, suberin, and various sorts of waxes, tannins, gums, and mucilages.

The Central Vacuole

The protoplast commonly occupies essentially the whole space enclosed by the cell wall in young cells, often with scattered small vacuoles. In mature plant cells, however, the protoplast usually forms a thin layer, just within the cell wall, which encloses a large central vacuole. The vacuole commonly constitutes about 90 percent of the volume of the cell, or even more. The material contained in the vacuole is called the **cell sap.** It consists of water, with a variety of other substances in true and colloidal solution, and it often also contains small crystals of various sorts. Because of its colloids, the cell sap is usually slightly viscous. Thin strands of cytoplasm often traverse the vacuole.

The central vacuole is usually regarded as a nonliving inclusion of the cell, just as the wall is customarily regarded as a nonliving boundary to the cell. That view is merely a convenient oversimplification and should not be taken too literally. As we have already noted, no one constituent of a living cell is alive when considered by itself. It is true that in general the contents of the vacuole do not appear to be in a state of chemical activity, as are the constituents of the protoplast proper, and indeed some of the vacuolar constituents are simply waste products. On the other hand, there is some transfer of materials back and forth between the vacuole and the cytoplasm; in particular, some food reserves that are stored in the central vacuole of some cells can be reincorporated into the cytoplasm. Furthermore, the central vacuole is necessary, in those cells which have it, for the normal functioning of the protoplast, and circumstances that cause withdrawal of water from the vacuole lead to reduction of protoplasmic activity and eventually (if prolonged and severe) to death.

Some kinds of pigments that give a characteristic color to certain plant parts occur in vacuoles, although other pigments are borne in the cytoplasm. Notable among the dissolved pigments occurring in the vacuoles of some cells, especially of higher plants, are the **anthocyanins** (Fig. 4.9). These range in color from blue or violet to purple or crimson, and they are responsible for the colors of many flowers and some other plant parts. The anthocyanins, together with the **anthoxanthins** (a chemically related group of pigments that range from pale yellowish to yellow, orange, or orange-red) are called **flavonoid**

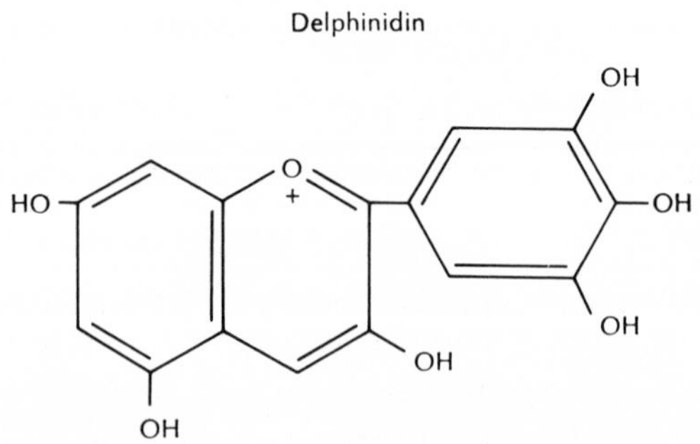

Fig. 4.9 Structure of delphinidin, a typical anthocyanin.

pigments. Aside from their function as flower pigments, the function of flavonoid pigments is obscure. Anthocyanins are further briefly discussed in Chapters 21 and 26. There are also some colorless flavonoids that may serve a defensive function, making the plant unpalatable to potential predators.

The Cytoplasm

The cytoplasm of a typical plant cell consists of a relatively clear, more or less viscous fluid (the **hyaloplasm**), an endoplasmic reticulum, and certain specialized bodies (organelles), notably plastids, mitochondria, golgi bodies, and ribosomes. It may also contain tiny droplets, granules, and vacuoles, which are chemically relatively inactive and are therefore customarily regarded as nonliving inclusions rather than as part of the living cytoplasm. The organelles and nonliving inclusions in the hyaloplasm give the cytoplasm a granular appearance under microscopic examination in life. The hyaloplasm is often referred to as the soluble part of the cytoplasm, in contrast to the organelles and nonliving inclusions.

The compartmentalization of much of the cytoplasm into various kinds of organelles fosters the simultaneous occurrence of different (and potentially antagonistic) sets of chemical reactions in the cell and increases efficiency by permitting the local concentration of sets of enzymes necessary for particular reaction sequences.

The cytoplasm is bounded externally by a **plasma membrane** and internally by a **vacuolar membrane.** Normal, healthy cells are ordinarily in a state of **turgor** (seemingly overfilled), so that the cytoplasm is pressed closely against the cell wall. In multicellular plants the cytoplasm is commonly continuous through numerous tiny holes in the cell wall. These connecting strands of cytoplasm are called **plasmodesmata** (Fig. 4.10).

The cytoplasm of a typical cell is in constant motion. Most often the hyaloplasm seems to stream around and around, carrying the organelles with it. Even the nucleus may be caught up in the stream, though it generally moves at a relatively slow rate. This streaming motion of the cytoplasm is called **cyclosis.** Some parts of the cytoplasm may move at a rate as high as 20 microns per second, while adjacent parts may be still or even move in a contrary direction. The mechanism and significance of cyclosis are still controversial and scarcely understood. It does seem likely that the resultant mixing facilitates proper metabolism. Cyclosis appears to depend at least partly on ATP as the source of energy.

The **plasma membrane** (Fig. 4.11) is commonly 7–10 nm thick. It consists of two parallel layers of phospholipids, with interspersed molecules of protein. Phospholipids differ from fats in that one of the three fatty acids associated with glycerol is replaced by a radical (a coherent submolecular group of atoms) containing phosphorus as an important constituent. The phosphate end of the phospholipid is soluble in water and is said to be hydrophilic; the fatty acid end is insoluble in water and is said to be hydrophobic. The phospholipid molecules of the plasma membrane are organized to form a sort of sandwich in which the hydrophilic phosphate heads of the molecules are at the outside (like the bread of a sandwich) and the hydrophobic tails point inward (forming the filling). The hydrophilic and hydrophobic layers of the plasma membrane together form a barrier to the passage of

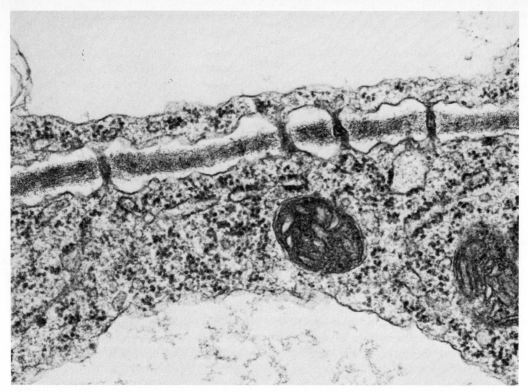

Fig. 4.10 Portion of two adjacent cells of tomato leaf, × 50,000. Four plasmodesmata are shown; mitochondria are visible at right and right center of lower cell; the numerous dark dots are ribosomes. [Electron micrograph courtesy of Myron C. Ledbetter, Biology Department, Brookhaven National Laboratory.]

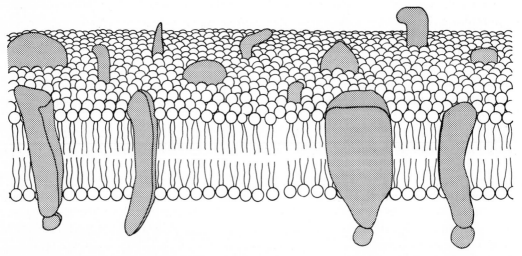

Fig. 4.11 Diagrammatic representation of the structure of the plasma membrane. Phosphate heads of phospholipids shown as empty circles. Protein molecules stippled.

58

both water-soluble and fat-soluble substances.

Passage of molecules and ions through the plasma membrane is governed at least in large part by the molecules of protein that are scattered throughout. The proteins are of diverse sorts. Each sort is responsible for the passage of one or a few related substances (aside from water, discussed in Chapter 6). Each of these protein molecules extends completely through the plasma membrane, sticking out at both ends. Ordinarily the two ends are hydrophilic, and the central part hydrophobic. Some of these *integral* protein molecules are attached to one or another kind of *peripheral* protein molecule at the inner surface of the membrane. The outer end of the integral protein molecule often bears a short polysaccharide chain.

The precise way in which the protein molecules promote the passage of materials through the plasma membrane is uncertain. Probably it often involves a temporary chemical combination with the substance that is in transit. It was at one time thought that the protein molecules might rotate, carrying the attached materials of other sorts from one side of the membrane to the other. This now seems highly unlikely, partly because of the hydrophilic-hydrophobic properties of both the protein and the phospholipid molecules of the membrane, partly because the polysaccharide chains attached to the integral proteins appear to be always on the outside of the membrane, and the peripheral proteins always on the inside.

The **vacuolar membrane** is essentially similar to the plasma membrane, perhaps with a smaller proportion of protein. Some substances that pass freely through the plasma membrane are restrained by the vacuolar membrane, and vice versa.

Each of the many smaller vacuoles scattered through the cytoplasm is also bounded by a membrane of similar structure.

The **endoplasmic reticulum** (Fig. 4.12) is a set of irregularly joined and perforated membranes that permeate the cytoplasm and appear in ultramicroscopic section to form a loose network. It is said to be **lamellar** in structure, that is, it consists mainly of thin plates, or lamellae, rather than filaments or threads. In contrast to the plasma membrane and vacuolar membrane, each of which is a "unit membrane," the endoplasmic reticulum is clearly a double membrane, composed of two discrete unit membranes, each about 8 nm thick. The two unit membranes are commonly separated by a liquid-filled region often about twice as thick as one of the unit membranes. The double membrane of the endoplasmic reticulum completely encloses the liquid within, apparently without open-ended connections to the soluble part of the cytoplasm. Any material that passes from inside to outside, or vice versa, apparently must pass through one of the two unit membranes.

The nature of the endoplasmic reticulum (often abbreviated ER) is still incompletely understood, and indeed its very existence was unsuspected until about 1950, when cytologists began to make extensive use of the electron microscope. It is evidently continuous at intervals with the outer layer of the nuclear membrane, but the functional significance of the connection is still doubtful. It is difficult to reconcile the free movement of organelles in cyclosis with the existence of a physically stable ER that is continous with the nuclear membrane, and this difficulty has given rise to the thought that the ER is physically unstable, constantly changing in form and position but always

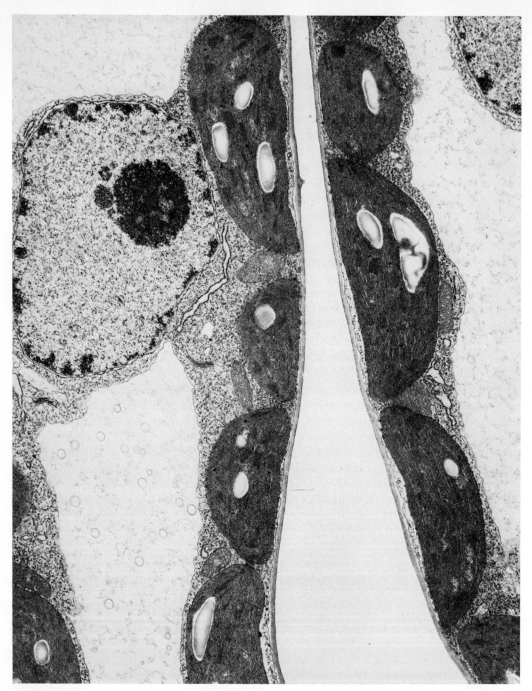

Fig. 4.12 Portions of two cells in palisade tissue of tomato leaf, × 10,000. Both cells show chloroplasts (large, dark bodies, with included pale grains of starch), mitochondria (much smaller, not so dark, ellipsoid bodies), and part of the central vacuole. Cell at left shows nucleus, bounded by a perforated nuclear membrane, a strip of endoplasmic reticulum to the right and below the nucleus, and nucleolus (large, dark body in the nucleus). [Electron micrograph courtesy of Myron C. Ledbetter, Biology Department, Brookhaven National Laboratory.]

retaining the structure of an irregularly perforated double membrane. The changes in form and structure in a mass of soap bubbles might provide a useful comparison.

The cytoplasm of eukaryotic plants contains a number of specialized, differentiated bodies called **plastids.** (Plastids are in general absent from animal cells.) Colorless plastids are called **leucoplasts;** green plastids are called **chloroplasts;** and colored plastids other than chloroplasts are called **chromoplasts.** Since a chromoplast is, literally, a colored plastid, a chloroplast could be considered as one type of chromoplast; but because chloroplasts are so different from other chromoplasts, it is customary and convenient to discuss them separately. Plastids are highly varied in form, but are most commonly shaped like a double-convex lens, a disk, or a sphere. They are bounded by a double membrane that appears to be structurally rather similar (though certainly not identical in detail) to the endoplasmic reticulum and the nuclear membrane.

Plastids of seed plants generally develop from smaller precursors called **proplastids.** Proplastids are colorless, self-reproducing bodies which have a continuous identity distinct from the rest of the cytoplasm. The details of their reproduction are still obscure, but it appears that they merely pinch in two. It is doubtless significant that they contain a small amount of DNA (deoxyribonucleic acid), a critical nuclear component that under proper circumstances undergoes self-replication. Proplastids are commonly of undistinguished or amoeboid shape; they are usually somewhat less than 1 micron (often 0.4–0.9 micron) thick, but their apparent precursors may be as small as 20 nm. Proplastids are bounded by a double membrane, but they are less complex internally than mature chloroplasts. Among the seed plants the several kinds of mature plastids typically arise by growth and differentiation of some of the proplastids.

In most of the algae, proplastids are apparently absent, and the mature plastids reproduce by division (self-replication). Many of the bryophytes and vascular cryptogams are more or less intermediate between the algae and the seed plants in plastid ontogeny, having one or several relatively large proplastids. In any case, plastids of all kinds of plants apparently arise only by division of previously existing plastids or proplastids. They do not arise *de novo* in the cytoplasm.

Chloroplasts are characterized by the presence of a green pigment, **chlorophyll.** In the living cell, chlorophyll and some associated pigments occur in conjugation with proteins (as the prosthetic groups of enzymes), but they can be extracted from dead cells without the protein. Extracted chlorophyll of higher plants (Embryobionta) commonly exists in two slightly differing forms, called chlorophyll *a* (Fig. 4.13) and chlorophyll *b*. The formula for extracted chlorophyll *a* · is $C_{55}H_{72}N_4O_5Mg$; chlorophyll *b* differs in having one more atom of oxygen and two fewer of hydrogen. Still other variants of chlorophyll occur among some of the algae and bacteria.

The core of the chlorophyll molecule consists of one atom of magnesium (Mg), surrounded by four nitrogen atoms, to which are attached rings and chains of carbon, hydrogen, and oxygen. A more complex but chemically more useful and accurate way to consider the structure of chlorophyll is to start with a **pyrrole,** which is a ring of four carbon atoms and one nitrogen atom. Four pyrroles chemically linked together form a **tetrapyrrole;**

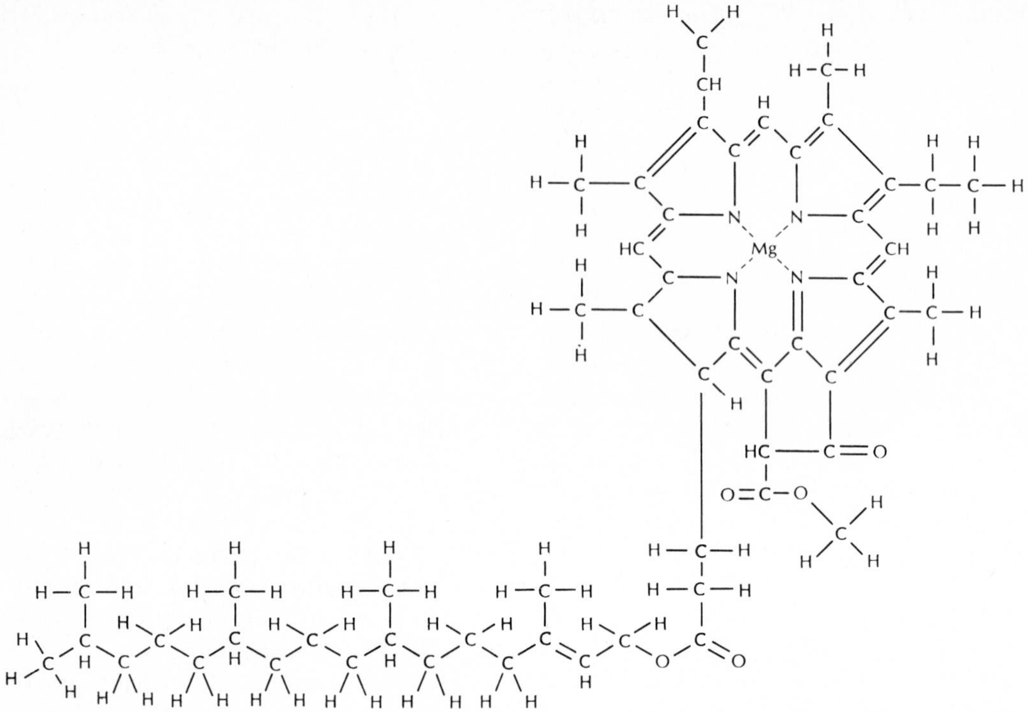

Fig. 4.13   Structural formula of chlorophyll *a*, the porphyrin head at the right, the phytol tail at the left.

when a tetrapyrrole itself forms a closed ring, the substance is a **porphyrin.** A **metalloporphyrin** has a metallic atom in the center. Chlorophyll is a metalloporphyrin that contains magnesium.

Porphyrins are the most ubiquitous reactive parts of enzymes in general. The heme part of the well-known enzyme hemoglobin, which occurs in red blood cells of animals and is essential for respiration, is an iron porphyrin. We shall see that iron porphyrins are basic to respiration in plants as well as animals. Iron is also used, in some way as yet not fully understood, in the formation of chlorophyll, but it is not present in chlorophyll itself. According to recent phylogenetic speculation, the first significant metalloporphyrins were iron porphyrins active in

anaerobic respiration, and the first magnesium porphyrin was the result of an ancient metabolic accident—an accident that prepared the way for the evolution of higher forms of life.

Chlorophyll is always accompanied by one or more associated pigments that are not green. In the higher plants (Embryobionta), two yellow pigments, **carotene** and **xanthophyll,** are customarily associated with chlorophyll, but their presence is masked by the more abundant chlorophyll. Carotene is the name for several closely allied compounds, all having the formula $C_{40}H_{56}$, but differing slightly in the arrangement of the atoms in the molecule (Fig. 4.14). The xanthophylls differ from carotene in having one or several atoms of oxygen in the molecule; a common xan-

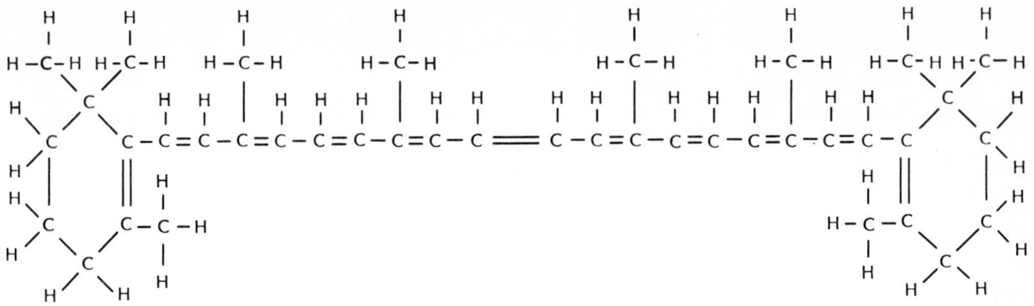

Fig. 4.14 Structure of β-carotene.

thophyll has the formula $C_{40}H_{56}O_2$. Carotene, xanthophyll, and some other mostly yellow to red or brownish pigments have certain structural and chemical features in common, and are called **carotenoid pigments.** Carotenoid pigments sometimes occur without chlorophyll, and they furnish the color of many chromoplasts.

In some algae chlorophyll is accompanied and more or less masked by brownish, red, or blue pigments. Some of these pigments are carotenoids, whereas others are of a different type called **phycobilins.** It is often convenient to use the term chloroplast for any chlorophyll-bearing plastid, regardless of its actual color, and to restrict the term chromoplast to colored plastids that do not contain chlorophyll. That practice is adopted in this text.

Chloroplasts are the most complex plastids. They are further considered in Chapter 7, which deals with photosynthesis.

Some other specialized parts of the cytoplasm, the **mitochondria** (Fig. 4.15), are colorless but so different from colorless plastids that they must be considered separately. Mitochondria, like plastids, are variable in shape; most often they are crooked rods. In spite of their small size (up to about 5 microns long, seldom longer), they have a complex, organized

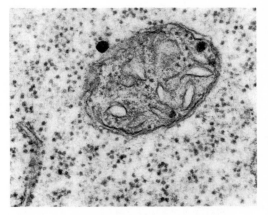

Fig. 4.15 Mitochondrion in the common bean, *Phaseolus vulgaris*, with ribosomes (dark dots) and a bit of endoplasmic reticulum (lower left). (× about 63,000.) [Electron micrograph by Peter K. Hepler.]

structure. They are bounded by a double membrane and contain a profusion of internal plates or tubular processes (cristae) of varied shape, formed by intrusion of the inner layer of the membrane. Mitochondria therefore have a very large surface area in relation to volume. Unlike plastids, mitochondria are characteristic cell organelles of nearly all eukaryotes, both plant and animal.

Mitochondria contain some DNA and are apparently strictly self-reproducing. In the higher plants, at least, they originate chiefly by division of tiny precursors in the cytoplasm, followed by growth and differentiation. These mitochondrial precursors are in the size range of proplastids (20–50 nm in diameter), but in good EM preparations the two kinds of organelles can now be distinguished even at this early stage. Although both plastids and mitochondria are self-reproducing and maintain a degree of hereditary continuity, their heredity is also influenced by nuclear DNA. Control of the hereditary features of mitochondria is shared by the mitochondrial and the nuclear DNA. The same is true of plastids.

The number of mitochondria per cell varies with the organism and kind of cell. In ordinary plant cells it is commonly in the range of several hundred to a few thousand, but in some very small cells there may be only a few mitochondria or even only one.

In addition to plastids and mitochondria, the cytoplasm also contains scattered minute bodies called **dictyosomes** or **golgi bodies** (Fig. 4.16), after the Italian zoologist Camillo Golgi (1844–1926), who in 1898 reported the existence of what are probably these same structures in the nerve cells of barn owls. Golgi bodies are in the size range of small mitochondria (commonly 1–3 microns wide), and they are even more difficult to observe with ordinary techniques and conventional microscopes. Observations with electron microscopes have revealed that golgi bodies of plants characteristically consist of about four to seven parallel platelets, without a collective bounding membrane. Each platelet is actually a flattened vesi-

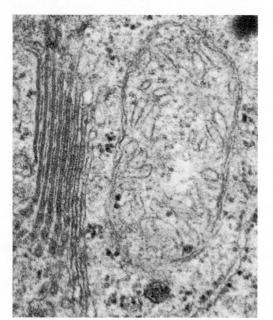
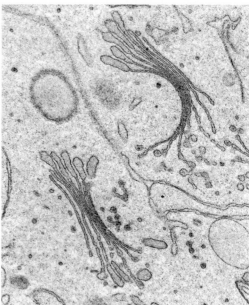

Fig. 4.16 (*Left*) Golgi body (*left*) and mitochondrion (*right*) in root tip cell of timothy, *Phleum pratense.* (× about 70,000.) [Electron micrograph by Peter K. Hepler.] (*Right*) Golgi body, pinching off vesicles, in root tip cell of corn, *Zea mays.* (× about 32,000.) [Electron micrograph courtesy of M. Dauwalder, W. Gordon Whaley, and Joyce Kephart.]

cle, roughly comparable to a segment of the endoplasmic reticulum in that it has two layers (unit membranes) separated by a fluid interior. Like mitochondria, golgi bodies appear to be standard constituents of the cytoplasm of eukaryotic organisms in general, but their detailed structure is somewhat different in higher animals than in plants. Probably several hundred golgi bodies occur in an ordinary plant cell.

It may be significant that the endoplasmic reticulum, the plastids, the mitochondria, the platelets of the golgi bodies, and the nuclear membrane either consist of or are enclosed by double membranes that have some features in common. The evolution of eukaryotic organisms from the ancestral prokaryotes is still incompletely understood, but the prokaryotes notably lack the double membrane that marks these eukaryotic organelles.

Much smaller than the mitochondria and golgi bodies and correspondingly more difficult to observe are numerous cytoplasmic bodies called **ribosomes** (Fig. 9.8). These can be seen only in the most delicately fixed EM preparations. In the best preparations they appear as compound structures, each composed of two joined, unequal spheroids, somewhat flattened in the contact zone. Whereas golgi bodies and mitochondria can be measured in microns, ribosomes are measured in nanometers. Ribosomes are commonly about 20 nm in diameter. Unlike mitochondria, they probably do not have a highly complex internal structure; it is the outer surface that is chemically active. Many ribosomes occur free in the cytoplasm, but others are attached to the endoplasmic reticulum. Ribosomes also occur within the nuclei, chloroplasts, and mitochondria. It has been estimated that there may be on the order of 500,000

ribosomes in an ordinary plant cell. Even the bacterium *Escherichia coli*, with cells much smaller than those of most eukaryotes, has on the order of 6000 ribosomes per cell. Ribosomes take their name from the fact that they contain ribonucleic acid (RNA). The nature and function of ribonucleic acid are discussed in Chapter 9.

Cytoplasm of eukaryotes, in general, also contains numerous very small structures called **microtubules,** with the form of straight, hollow threads or pipes. These are on the order of 25 nm in diameter and of undetermined length, some of them being at least several microns long. The tubules themselves are compound structures which generally consist of about 13 (in plants) microfibrils in a ring sheath surrounding a hollow core. We shall see in Chapter 5 that the spindle fibers seen in cell division consist of clusters of microtubules.

Several morphologically similar but metabolically differentiated kinds of cytoplasmic organelles are collectively known as **microbodies.** These are ellipsoid bodies 0.2–0.6 micron long, bounded by a unit (not double) membrane. Those that contain digestive enzymes (hydrolases) are called **lysosomes.** Those that contain enzymes relating to the metabolism of 2-carbon compounds, notably glycolate (a photosynthetic product) and acetate, are coming to be called **peroxisomes,** because their set of enzymes include some which destroy hydrogen peroxide. Some of the peroxisomes have also been called **glyoxysomes,** because the glyoxylate cycle for producing more complex compounds from acetate occurs within them.

All normal plant cells (except those of bacteria and blue-green algae) apparently contain leucoplasts, mitochondria, ribosomes, microtubules, and probably microbodies, golgi bodies, and an endoplasmic reticulum. Indeed all of these structures except leucoplasts appear to be characteristic of eukaryotes in general, both plant and animal. Not all plant cells contain chloroplasts and chromoplasts, however.

The Nucleus

The nucleus of the cell is embedded in and completely surrounded by the cytoplasm. Its volume is typically only about 2 percent of that of the whole protoplast, but it contains about 10 percent of the protein. A typical, spherical nucleus is commonly on the order of 10 microns thick.

The nucleus is ordinarily bounded by a **nuclear membrane,** which is a double membrane morphologically much like the double membrane of the endoplasmic reticulum. Each layer of the nuclear membrane is on the order of 8 nm thick, and the two layers are separated by 10–30 nm of fluid. The outer layer is often continuous at intervals with the endoplasmic reticulum, but the inner layer remains largely intact. The nuclear membrane apparently has scattered pores about 40 nm wide.

Within the nucleus is the **nuclear sap,** or **karyolymph,** in which are embedded the **chromonemata** and a more or less spherical body called the **nucleolus** (or often two or more nucleoli). The nuclear sap, unlike the cell sap of the central vacuole, is chemically active and is regarded as living substance.

The **chromonemata** are slender, thread-like bodies, commonly 20–30 nm thick, which are so twisted and tangled that they appear to form a network. The network appearance is illusory, inasmuch as the individual chromonemata retain their identity throughout. During cell division (see Chapter 5) the chromonemata coil and contract, forming much shorter and thicker bodies, called **chromosomes**. Another approach is to say that the chromonemata are interphase chromosomes, the interphase being the time between one cell division and the next. From a functional standpoint this is surely correct, but from a descriptive standpoint it is useful to retain the distinction between chromonemata and chromosomes.

The number of chromosomes in a cell varies according to the kind of organism, but except for the regular and alternating changes during the reproductive cycle, described in Chapter 10, it is usually constant for each individual species. The number may be as low as 1 or more than 1000, but numbers of 6 to 50 are more common.

The word chromosome is derived from two Greek words: *chroma*, "color," and *soma*, "body." The name was originally proposed because some of the dyes used in preparing parts of plants and animals for microscopic examination have a particular affinity for these structures, which therefore appeared as colored bodies within the cell. The chromosomes consist in large part of a complex series of substances called **chromatin.** The chromatin consists of **nucleoproteins,** which are conjugated proteins with some complex organic acids called **nucleic acids.** The individual units of chromatin, called **genes,** seem to be arranged in linear sequence on the chromosomes, although this is almost certainly an oversimplification. The structure of genes is discussed in Chapter 9, which deals with molecular genetics.

All of the many different kinds of nucleic acids contain nitrogen and phosphorus, as well as carbon, hydrogen, and oxygen. Most nucleic acids can be divided into two general groups: **ribonucleic acid,** commonly called **RNA,** and **deoxyribonucleic acid,** commonly called **DNA.** One of the blocks that goes into the formation of RNA is ribose, a pentose sugar. DNA differs from RNA, among other respects, in that a riboselike molecule containing less oxygen than true ribose is used in place of ribose in the formation of the DNA molecule. DNA occurs principally in the chromosomes, of which it is the essential constituent, but as we have seen it is also present in the self-reproducing cytoplasmic organelles called plastids and mitochondria. RNA, on the other hand, occurs chiefly in the cytoplasm and in the nucleolus, being present also in small amounts in the nuclear sap.

The **nucleolus** is a more or less spherical body within the nucleus, apparently without a bounding membrane. It is associated with a particular region on a particular chromosome, called a nucleolar organizing chromosome, or simply a **nucleolar organizer.** The nucleolar organizer appears to pass through the nucleolus. The nuclei of some kinds of cells have two or more nucleolar organizers and a corresponding number of nucleoli.

The nucleus of eukaryotes, as here described, is conventionally said to be **vesicular,** that is, it forms a definite vesicle bounded by a membrane. We shall see that prokaryotes (bacteria and blue-green algae) have DNA-bearing bodies that have been compared to nuclei but are less complex in structure and not bounded by a membrane. Whether these prokaryotic structures should be called nuclei is in part a matter of definition, discussed in Chapter 11. In any case, the vesicular nucleus is a feature of eukaryotes only, as are also such cytoplasmic organelles as plastids, mitochondria, and golgi bodies.

FUNCTIONS OF THE INDIVIDUAL PARTS OF CELLS

The Cell Wall

The cell wall gives shape and some degree of rigidity to the cell. Unless it is impregnated with waterproofing materials, as is true, for example, of cork cells, it presents no significant barrier to the diffusion of water and other materials in solution. Some of the hemicelluloses of the cell wall often serve as stored food, being resorbed into the cytoplasm in time of need. The middle lamella holds adjacent cells together. One of the effects of calcium shortage in plants is that the pectin of the middle lamella fails to be converted into calcium pectate, and the cells thus fail to stick together.

The Nucleus

The nucleus, and specifically the genes of the nucleus, govern most of the hereditary characteristics of the cell. In general, it may be said that the DNA of the genes governs the formation of RNA molecules which migrate out into the cytoplasm and there, in turn, directly or indirectly govern the formation of molecules of various cytoplasmic enzymes and structural proteins. Thus the kind and amount of enzymes in the cytoplasm are controlled mainly by the genes. Since the chemical activities and, indirectly, the physical properties of the cell depend largely on the enzymes present, this explains the important role the nucleus has long been known to play in controlling the nature of the whole cell. A more detailed consideration of the action of DNA and RNA is given in Chapter 9.

The function of the nucleolus has long been controversial, and many diverse suggestions have been made and discarded in the past. By the late 1950s it began to be evident that the nucleolus is somehow involved in RNA synthesis, and by the late 1960s it had become clear that the nucleolus is concerned specifically with the packaging of ribosomal RNA into units that are apparently assembled in the cytoplasm into functional ribosomes.

The Cytoplasm

The manufacture, storage, and digestion of foods are carried out in the cytoplasm, although some foods may also be stored in the central vacuole or in the cell wall. Respiration is largely a cytoplasmic function, involving especially the mitochondria. The formation of living substance is carried out in both the nucleus and the cytoplasm, each having its own role.

Photosynthesis occurs in the chloroplasts. Some of the principal steps of respiration ordinarily occur in the mitochondria. Many cytoplasmic enzymes are formed in association with ribosomes attached to the endoplasmic reticulum, whereas others are formed in association with ribosomes free in the hyaloplasm. Grains of starch, oil, or protein are often formed and stored in leucoplasts, which may thus be distinguished and named as **amyloplasts, elaioplasts,** or **protein-oplasts,** respectively. The only function of chromoplasts, so far as known, is to produce colored tissues, which may be significant in attracting animals, as in the skin or flesh of many fruits. The final packaging of many diverse kinds of cytoplasmic products (especially carbohydrates concerned in cell-wall formation) occurs in the golgi bodies. Entry to or departure from the cell of water and dissolved substances is governed in part by the plasma membrane and vacuolar membrane and in part also by the kind and amount of dissolved and colloidal particles in both the cell sap and the cytoplasm. In addition to harboring ribosomes, the endoplasmic reticulum may also serve as a passageway for distribution of various materials within the cytoplasm and between the cytoplasm and the nucleus. The microtubules play an important role in cell division, in the orientation of cellulose deposited in cell walls, in the transport of materials within the cell, and in several other processes. Digestion of certain foods, especially carbohydrates, may occur in the lysosomes, and important processes relating to acetate and glycolate occur in the peroxisomes. Most other metabolic processes have not been clearly demonstrated to be associated with particular parts of the cytoplasm.

METABOLISM AND RELATED PROCESSES

The formation of living substance from raw materials is called **assimilation.** Assimilation is the phase of metabolism toward which all other metabolic processes are normally directed. Even in mature organisms, the formation of some new protoplasm is necessary for continued life, inasmuch as existing protoplasm is constantly wearing out and breaking down. Living matter is chemically active and unstable and is therefore subject to attrition by the alteration or disintegration of its molecules.

Those phases of metabolism, such as photosynthesis and assimilation, that result in making more complex substances from simpler ones are called **anabolism.** Anabolism generally requires an input of energy. Those phases of metabolism, such as digestion and respiration, that result in making simpler substances from more complex ones are called **catabolism.** Catabolism generally results in the liberation of stored energy. For purposes of discussion and understanding, the division of metabolism into anabolism and catabolism is often useful, but the distinction within the living cell is not precise. Both anabolic and catabolic processes occur constantly and simultaneously, and some of the intermediate products of anabolic processes are the same as some of the intermediate products of simultaneously occurring catabolic processes.

Photosynthesis and Respiration

Osmosis, photosynthesis, respiration, and the synthesis of complex metabolites are parts of metabolism that are discussed in separate chapters and are therefore excluded from detailed consideration here. We note here only some very general features about some of these processes.

Photosynthesis is the process by which plants make food from raw materials, using light as the source of energy. The raw materials of photosynthesis are usually carbon dioxide and water, and glucose is ordinarily the principal end product. A generalized, oversimplified equation for photosynthesis is

$$6CO_2 + 6H_2O + \text{light energy} \rightarrow C_6H_{12}O_6 + 6O_2$$

Glucose and other immediate products of photosynthesis are the building blocks from which other, often much more complex substances are made, either directly or through a chain of intermediate substances.

Respiration is the process by which foods are broken down within a cell, with some of the energy being released into the metabolic system of the cell. Many different substances can be respired, but glucose is the most common and typical one.

A generalized, oversimplified equation for respiration of glucose is

$$C_6H_{12}O_6 + 6O_2 \rightarrow 6CO_2 + 6H_2O + \text{heat and metabolic (chemical) energy}$$

It may thus be seen that photosynthesis and respiration, taken collectively, form a mechanism for transforming light energy into metabolic energy. The process is not highly efficient in absolute terms. Much of the light energy received by the plant escapes being used, and much of the energy liberated in respiration is lost in the form of heat. Nevertheless, with some insignificant exceptions involving chemosynthesis among certain bacteria, photosynthesis is the sole eventual source of metabolic energy for all organisms, both plant and animal. Only plants can make food from raw materials. Anything that does not make its own food must eat it or absorb it from the substrate. This brings us back eventually to plants. *The food chain begins with photosynthesis.* Furthermore, green plants on one hand plus all *heterotrophic* organisms (organisms that do not make their own food) on the other make up a balanced system by which foods and raw materials are maintained in continuous cycle, and relatively stable proportions of oxygen and carbon dioxide are maintained in the atmosphere. In a subsequent chapter we shall study the nitrogen cycle in some detail, but each of the elements necessary for plant growth could as well be considered in a similar way.

Food Storage

If all the glucose formed by a cell during photosynthesis were maintained in that form within the cell, it would interfere with metabolism in many ways. One important way would be to upset the osmotic relations discussed in Chapter 6.

In fact, glucose is seldom accumulated in large quantities; instead it is usually transformed into insoluble, chemically and osmotically inactive reserves. The most common of these reserve foods into which glucose is transformed for storage is starch.

Starch is a mixture of two different polysaccharides, called **amylose** and **amylopectin.** A molecule of amylose is formed by the polymerization of about 300 to 1000 molecules of α-glucose into a simple chain, and a molecule of amylopectin (Fig. 4.17) is formed by the polymerization of about 1000 to 3000 molecules of α-glucose in a different way into a branched chain. Amylose turns deep blue when a solution of iodine is applied to it; amylopectin turns red to purple; the resulting color of iodine-treated starch is a deep violet-purple. No other common natural substance gives such a color when treated with iodine, and iodine is commonly used as a test for the presence of starch.

Starch commonly occurs in the form of definite grains filling certain leucoplasts (called amyloplasts) in the cytoplasm (Fig.

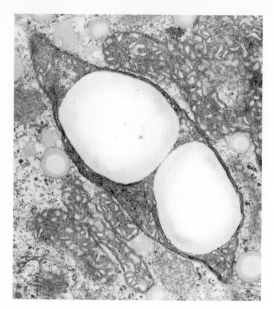

Fig. 4.18   Amyloplast with two included starch grains, several mitochondria nearby, in endosperm of *Haemanthus katheriniae*, a South African blood-lily. (× about 18,000.) [Electron micrograph by Peter K. Hepler.]

4.18). The shape and size of starch grains differ in different kinds of plants, partly according to differences in the amylopectin fraction. Often the grains have evident concentric layers. A small proportion—commonly about 1 percent by weight—of each starch grain consists of substances other than amylose and amylopectin. These other substances, which commonly include water, fatty acids, and phosphate compounds, may be chemically combined with the starch or merely associated with it, according to the circumstances and the kind of plant.

Another form in which food is stored is fat. The formation of fats from carbohydrates requires a series of chemical changes, of which the final step is the combination of one molecule of glycerin with three

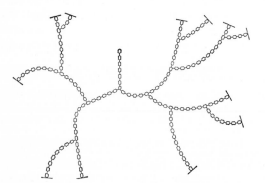

Fig. 4.17   Structural diagram of part of an amylopectin molecule. Each small block indicates a glucose residue.

molecules of fatty acid (see Chapter 3). Energy is used in the formation of glycerin and fatty acids from carbohydrates and is retained in the molecule when these are combined to form fats. The respiration of fats, therefore, releases more energy than the respiration of the same amount (by weight) of carbohydrates. Fats are commonly stored as globules in certain leucoplasts (called elaioplasts) or free in the cytoplasm. Fats and other lipids also occur in colloidal form as part of the living protoplasm.

Food may also be stored in the form of protein. Stored protein may exist as small crystals or granules in certain leucoplasts (called proteinoplasts) or free in the cytoplasm or cell sap. It is also frequently found in semiliquid globules, alone or with fats. Stored protein is commonly used as a building material for new protoplasm rather than for respiration. As might therefore be expected, it is especially likely to be found in specialized structures, such as seeds, which must draw on reserve foods for growth.

Still other substances are sometimes stored as food. In some seeds, in particular, hemicelluloses are withdrawn from the cell wall and respired during germination. Sweet corn is sweet because much of the carbohydrate in the grain is stored as sugar rather than as starch.

Digestion

Digestion is a chemical process by which insoluble or not readily diffusible foods are rendered more soluble or more readily diffusible. It is ordinarily a hydrolysis reaction, in which water is used and large molecules are broken down into smaller molecules. A small amount of energy is released in digestive hydrolysis, but it is released in packages too small to be captured metabolically (i.e., it cannot generate the formation of ATP from ADP and phosphate ion), and the quantity is so negligible that it is customary to say that digestion does not involve the use or release of energy.

The chemical reactions in digestion are essentially the opposite of the reactions by which the substances digested were originally formed from less complex substances. Polysaccharides eventually yield monosaccharides when digested, although the process may be interrupted at any intermediate state, as, for example, at the disaccharide level. The digestion of fats yields glycerin and fatty acids, and the digestion of simple proteins yields amino acids. The chemical equation for the digestion of sucrose may be written as follows:

$$\underset{\text{sucrose}}{C_{12}H_{22}O_{11}} + H_2O \rightarrow$$

$$\underset{\text{glucose}}{C_6H_{12}O_6} + \underset{\text{fructose}}{C_6H_{12}O_6}$$

As we noted earlier, glucose and fructose have the same empirical formula, but there are slight differences in the way the atoms are arranged. The equation for the digestion of maltose, another disaccharide, is exactly the same as that for sucrose, except that two molecules of glucose (and no fructose) are produced.

Digestion, like other chemical processes in the cell, requires particular enzymes. Two different enzymes are required to digest the amylopectin fraction of starch, one for the detachment of successive ordinary links in the chain and another for the branch points. The enzymes involved in the digestion of starch are collectively called **amylase** and have also been called **diastase.**

ELEMENTS NECESSARY FOR PLANT GROWTH

From the preceding discussion of cell structure and metabolism, it can be seen that several different chemical elements are necessary to the proper function or to the very existence of plants. Ten of these elements are ordinarily needed in sufficient quantity for their necessity to be easily demonstrated by trying to grow plants in an otherwise satisfactory artificial environment from which the test element is excluded. These are carbon (C), hydrogen (H), oxygen (O), nitrogen (N), sulfur (S), phosphorus (P), magnesium (Mg), potassium (K), iron (Fe), and calcium (Ca). All of these, with the probable exception of calcium, are believed to be necessary for all plants. The necessity of calcium for some of the fungi, algae, and bacteria is doubtful, and it may not be necessary for every individual cell in higher plants. A simple mnemonic device to keep all these ten elements in mind is the phrase "C HOPK'NS CaFe Mighty good," in which each element is represented by its symbol, except magnesium, which is represented by "Mighty good."

Four other elements, aluminum (Al), chlorine (Cl) silicon (Si), and sodium (Na), are usually present in considerable quantity in plant cells. The necessity of these for plants in general remains to be demonstrated, but sodium is necessary to the growth of certain marine algae; silicon is necessary for the normal development of diatoms (a group of algae with silicified cell walls), although they can be grown in culture without it; and there is some evidence indicating that traces of both chlorine and aluminum are necessary for at least some of the higher plants.

In addition to the ten **macrometabolic elements** (i.e., those necessary in appreci-

able quantities for metabolism) and the four other elements usually present in much greater quantity than necessary, several other elements are necessary in minute amounts for the growth of most (or all) plants. These are known as **trace elements.** All of the currently known trace elements are toxic if present in the cell in more than minute amounts. Very careful experimental work is required to demonstrate the necessity of these trace elements, since they are usually present as impurities in the other substances used to set up the nutrient medium; and even if they are rigorously excluded from the nutrient medium, the plant may already have enough to last for the duration of the experiment. It is sometimes necessary to grow plants in the artificial nutrient medium for several consecutive generations before the effects of lack of the test element appear.

Six trace elements are now generally admitted to be necessary for the growth of most or perhaps all plants. These are boron (B), cobalt (Co), copper (Cu), manganese (Mn), molybdenum (Mo), and zinc (Z). Gallium (Ga) also seems to be necessary as a trace element for some kinds of plants, and further work may well enlarge the list of plants for which it is necessary. In some of the bacteria, at least, vanadium (V) can substitute for molybdenum. It is probable that there will be additions in the future to the list of necessary trace elements.

One other element, selenium (Se), is necessary in considerable quantity for the growth of a few kinds of plants (Fig. 23.5), but is toxic to many others. It is quite probable that each of several other elements may be required (perhaps only in traces) by some plants but not by others. Students aware of the fluoridation of public water supplies in many localities may already be acquainted with the concept of necessary trace elements that are poisonous in larger quantities.

Functions of the Necessary Elements

Most of the essential elements have several or many functions in plant metabolism. Some of these functions are fairly well understood, especially for the macrometabolic elements, but others are more obscure. Some functions are merely inferred from the symptoms of lack. Some of the functions of the necessary elements are noted in the following paragraphs.

Carbon is the essential element for all organic compounds, and a very large proportion of these also contain hydrogen and oxygen. All proteins contain nitrogen, and many of them (including some essential ones) contain sulfur.

Phosphorus is an essential constituent of nucleic acids. It also plays an essential role in energy transfer within the cell (including both photosynthesis and respiration) because it is a necessary constituent of ATP. Phospholipids, which make up a major part of the plasma membrane, contain some phosphorus.

Magnesium is present in chlorophyll and is also necessary for plants that do not contain chlorophyll. Magnesium is involved in some way in phosphorus metabolism, and magnesium ions are necessary for the proper function of some of the essential enzymes.

Iron is an essential constituent of some of the enzymes necessary for ordinary respiration. It is also necessary for the formation of chlorophyll, although it is not present in the chlorophyll molecule.

Calcium is involved in nitrogen metabolism. It also enters into a number of compounds in the cell, probably includ-

ing nucleoproteins. Multicellular plants ordinarily contain calcium in the middle lamella; it is the reaction with calcium (and often also magnesium) that changes the pectic substances of the maturing middle lamella from a jelly to a cement. For reasons not yet fully understood, calcium deficiency leads to the breakup of the nuclear membrane, plasma membrane, and vacuolar membrane.

Potassium is certainly necessary for growth, but, unlike the other macrometabolic elements, it is not definitely known to be a component of essential organic compounds. Potassium ions appear to be necessary to the action of certain enzymes. Sodium can partly, but not wholly, replace potassium in metabolism.

Probably all of the trace elements (plus those that are necessary only in traces but occur in larger amounts) are either contained in or necessary to the formation of certain enzymes. Boron is involved in calcium metabolism and has other functions as yet not fully understood. When the supply of boron is insufficient, phenolic acids accumulate, causing necrosis (local death of tissues) and ultimate death. Cobalt is a constituent of vitamin B_{12}, an enzyme that appears to be necessary for nearly all organisms. Copper is present in some of the respiratory enzymes and in one of the enzymes (plastocyanin) necessary for green plant photosynthesis. Manganese is involved in iron metabolism and nitrogen metabolism, and manganese ions are necessary for the action of certain enzymes in photosynthesis and others in respiration. Molybdenum is involved in nitrogen metabolism. The fixation of atmospheric nitrogen by some bacteria requires the presence of molybdenum or vanadium (either one will do). Zinc is necessary to the synthesis of indole acetic acid (an important growth-regulating substance) and is probably present in some of the essential respiratory enzymes.

Selenium belongs to the same chemical group as sulfur and can substitute for it in certain compounds. It probably substitutes for sulfur in some of the proteins of those plants for which it is an essential element, and its toxic properties for other organisms probably depend on this same sort of substitution.

SUGGESTED READING

Bradbury, S., *The evolution of the microscope*, Pergamon Press, New York, 1967.

Bryan, J., Microtubules, *BioScience* **24**:701–711, 1974.

Dobell, C., *Antony van Leeuwenhoek and his "little animals,"* Staples Press, London, 1932. A biography and appraisal, with many excerpts from his letters to the Royal Society.

Fox, C. F., The structure of cell membranes, *Sci. Amer.* **226**(2):30–38, February 1972.

Gabriel, M. L., and S. Fogel (eds.), *Great experiments in biology*, Prentice-Hall, Englewood Cliffs, N.J., 1955. A collection of some of the classic papers that established important new ideas. Well worth the student's attention in connection with this and later chapters.

Ledbetter, M. F., and K. R. Porter, *Introduction to the fine structure of plant cells*, Springer-Verlag, Berlin, New York, 1970. Beautiful electron micrographs.

Lodish, H. F., and J. E. Rothman, The assembly of cell membranes, *Sci. Amer.* **140**(1):48–63, January 1979.

Mirsky, A. E., The discovery of DNA, *Sci. Amer.* **218**(6):78–88, June 1968.

Neutra, M., and C. P. Leblond, The golgi apparatus, *Sci. Amer.* **220**(2):100–107, February 1969.

Rustad, D. C., Pinocytosis, *Sci. Amer.* **204**(4):120–130, April 1961.

Thomas, L., *Lives of a cell. Notes of a biology watcher*, Viking Press, New York, 1974. Good, thought-provoking reading.

Wiebe, H. H., The significance of plant vacuoles, *BioScience* **28**:327–331, 1978.

Chapter 5

Cell Division

Series of photomicrographs showing mitosis in living, unstained cells of endosperm of *Haemanthus katheriniae,* a South African blood-lily. [Courtesy of Andrew S. Bajer; some of these photomicrographs were published by Andrew S. Bajer and Jadwiga Molè-Bajer in *Chromosoma* **27:**452–453, 1969, copyright by Springer-Verlag, Heidelberg.]

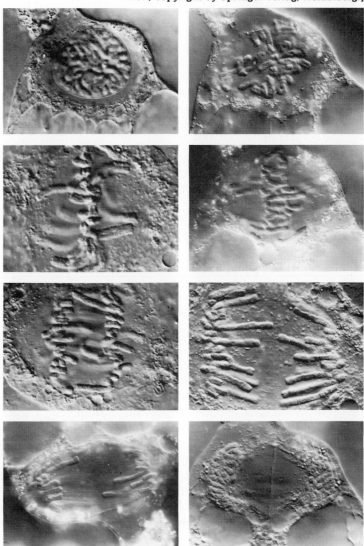

MITOSIS

Introductory Summary

Increase in number of cells is brought about by division of existing cells. In typical cell division the chromatin material of the nucleus is precisely divided into two equal parts, and the cytoplasm is more roughly divided into two essentially similar parts, giving two daughter cells that are initially alike.

The process that results in the precise division of the chromatin is called **mitosis,** and the process that results in the division of the cytoplasm is called **cytokinesis.** With the notable exception of many fungi and some algae, mitosis and cytokinesis are usually closely associated, and the term *mitosis* is often loosely used to cover the whole process of mitotic cell division, including cytokinesis. Among animals, and among some plants, cytokinesis is ordinarily accomplished by the simple constriction, or furrowing, of the cytoplasm into two parts; but among most plants it (cytokinesis) ordinarily occurs by the formation of a distinct transverse membrane, the **cell plate,** which eventually becomes the middle lamella between the two daughter cells.

Cell division was noted by a number of observers between 1830 and 1840, and in 1847–1848 the German botanist Wilhelm Hofmeister (1824–1877) described much of the mitotic process, but without understanding its significance. A more complete understanding of the essential features of mitosis developed from the separate contributions of several men during the years 1873 to 1884. Subsequent improvement in understanding has consist-ed of refinement and the filling in of details.

The time occupied by the mitotic process varies according to the kind of organ-ism, the kind of cell, the temperature, and other environmental conditions, but it is usually from 1/2 to 2 hours. Mitosis may occur at any time of the day or night, but in some kinds of cells of some kinds of plants there is a tendency toward a daily rhythm, with more mitoses occurring at one time of day than at the others.

It is customary and convenient to divide mitosis into four stages, called, in order, the **prophase,** the **metaphase,** the **anaphase,** and the **telophase.** This is merely a convenient mental organization of a process that is continuous from beginning to end (Fig. 5.1).

It should be emphasized that the division of the chromatin material that occurs in mitosis is qualitatively as well as quantitatively equal. Mitotic cell division is a means of increasing the number of cells without changing their hereditary potentialities.

The Prophase

The **prophase** is the part of mitosis in which the chromonemata are modified into much shorter and thicker chromosomes that can be recognized under the microscope as separate bodies. The change in shape is mainly or wholly a result of compound coiling of each chromonema. The coiling process has been roughly compared to what happens when a piece of string is twisted until it coils upon itself. Each chromosome, as seen at late prophase, has its own characteristic shape and consists of two equal, parallel bodies, called **chromatids.**

The nucleolus generally disappears during late prophase. In some kinds of plants the nucleolus retains its identity throughout, merely pinching in two during early anaphase.

During late prophase a characteristic bipolar structure called (because of its

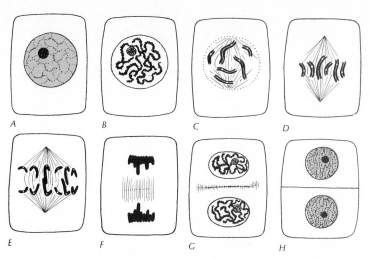

Fig. 5.1   Mitosis. Series of diagrams showing mitosis in a cell with six chromosomes. *A*, interphase; *B*, early prophase; *C*, late prophase; *D*, metaphase; *E*, anaphase; *F*, late anaphase verging toward telophase; *H*, interphase.

shape) the **spindle** is formed in or around the nucleus. The spindle is composed of numerous longitudinally oriented fibrils, each consisting of (one to) several or many (sometimes more than 100) microtubules in a cluster. The larger clusters are visible with the light microscope as the spindle fibers. Recent studies by J. R. McIntosh (1938–   ) and Peter K. Hepler (1936–   ) indicate that just before prophase the microtubules are largely concentrated near the nucleus, but randomly oriented and not clustered (Fig. 5.2). During the prophase, the microtubules become oriented parallel to each other (Fig. 5.3) forming the familiar spindle, and many of the microtubules become clustered in groups to form the visible **spindle fibers.** Probably some new microtubules grow out from the poles (or from a tiny organelle called a centriole, in those cells which have centrioles) and contribute to the formation of the spindle.

Toward the end of the prophase, more or less coincident with the formation of the spindle, the so-called **tractile fibers** begin to form. Each tractile fiber consists of a cluster of microtubules, much as do the spindle fibers. Two tractile fibers develop from each chromosome; one leads from one chromatid to one pole of the spindle, and the other leads from the other chromatid to the other pole of the spindle. The fiber grows in length, probably by the addition of new material at the base (the point of origin) and progresses eventually to the spindle pole.

The formation of each tractile fiber generally begins on the chromosome at a specialized point called the **centromere** (Fig. 5.4). The centromere may lie at any point along the length of the chromosome, having the same position in corresponding chromosomes of different cells in the same or different individuals of any given kind of plant. In some kinds of plants, however, including many members of the sedge family (Cyperaceae) and rush fami-

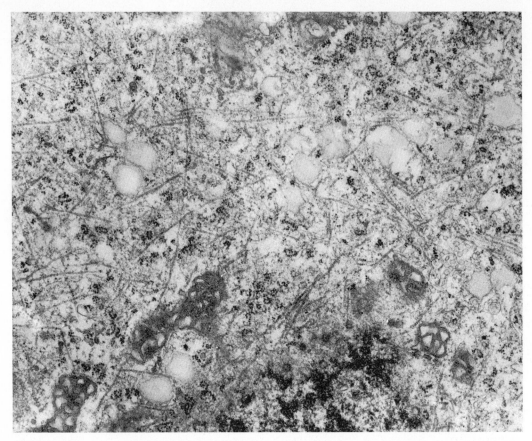

Fig. 5.2 View just outside the nuclear membrane, showing unoriented microtubules in early prophase in endosperm of *Haemanthus katheriniae*, a South African blood-lily. (× about 23,000.) [Electron micrograph courtesy of Peter K. Hepler.]

ly (Juncaceae), as well as the green algal genus *Spirogyra* and some other algae, there is no definite centromere, and the microtubules attach to the chromosome over much or all of its length. The chromosomes of such plants are said to have *diffuse* centromeres.

The Metaphase

The **metaphase** (Fig. 5.5) is the stage of mitosis during which the chromosomes that have taken shape in the prophase become (and for a time remain) arranged in an **equatorial plate** (i.e., a plane lying across the middle of the cell). The arrangement of the chromosomes into an equatorial plate occurs relatively rapidly after the tractile fibers begin to form. Presumably the developing tractile fibers have something to do with the positioning of the chromosomes into a plate, but the details are still obscure.

The Anaphase

The **anaphase** (Fig. 5.6) is the stage of mitosis during which the two chromatids

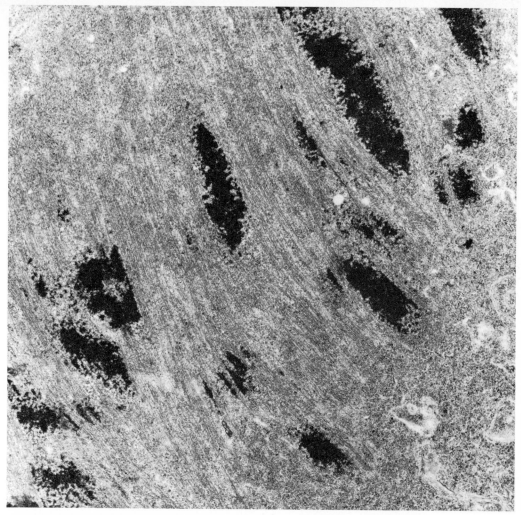

Fig. 5.3 View just outside the nucleus, showing longitudinally oriented microtubules (beginning of spindle formation) at end of prophase in endosperm of *Haemanthus katheriniae*. (× about 3500.) [Electron micrograph courtesy of Peter K. Hepler.]

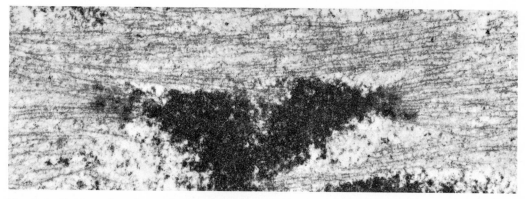

Fig. 5.4 Portion of nucleus at end of prophase in endosperm of *Haemanthus katheriniae*, showing microtubules of the spindle and microtubules developing from the centromere of the chromosome. (× about 12,000.) [Electron micrograph courtesy of Peter K. Hepler.]

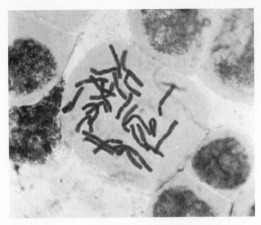

Fig. 5.5 Face view of metaphase plate in root tip squash of *Orontium aquaticum*, an eastern American member of the family Araceae, showing 26 chromosomes (13 pairs). [Photo courtesy of John Grear.]

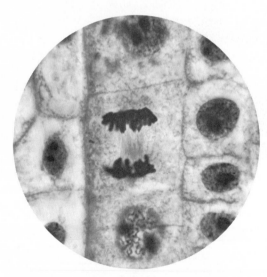

Fig. 5.6 Section of onion root tip, showing a cell in late anaphase. [Courtesy CCM: General Biological, Inc., Chicago.]

of each chromosome separate and move toward opposite poles of the spindle. The mechanism that leads to the separation has received a great deal of study but is only beginning to be understood. Usually the separated parts present the appear-ance of being dragged by the tractile fibers, with the ends lagging behind the centromere.

As soon as the two chromatids of a chromosome become separated during the anaphase, each chromatid is considered (as a matter of terminology) to have ad-vanced to the rank of a chromosome.

The spindle often elongates during the anaphase, so that the poles to which the chromosomes migrate become more wide-ly separated. As the chromosomes ap-proach the poles, the tractile fibers pro-gressively shorten. It has been suggested that the tractile fibers are progressively depolymerized and dissipated at the pole-ward end in a process essentially the reverse of their initial formation, but it is still debatable whether this shortening and dissipation of the fibers directly caus-es the migration of the chromosomes or whether other forces are primarily respon-sible.

The Telophase

The **telophase** is the stage of mitosis during which each of the two groups of chromosomes (one group at each pole of the spindle) is formed into a distinct nucleus similar to the nucleus that existed at the beginning of the mitotic process. The processes in telophase are in large part the reverse of those that occur in prophase. A nuclear membrane appears around the group of chromosomes. The chromosomes uncoil and are transformed into the slender chromonemata, which form a quasi-network. The nucleolus re-appears. As the daughter nuclei develop, the spindle becomes shorter and broader and eventually disappears.

The Interphase

The time between the end of the telophase and the beginning of the next prophase is often called the **interphase.** Some of the preliminaries essential to the next mitosis occur during the interphase, but the interphase is not itself a part of mitosis.

Mitosis is commonly completed in about two hours, but the interphase takes much longer. Even in rapidly growing tissues, in which cell division is frequent, the interphase seldom lasts less than 12 hours. The upper length of time for the interphase is indefinite. Mature cells that have not divided for many days (or even years) can sometimes be induced to undergo mitosis, under the proper conditions.

Each chromonema (interphase chromosome) consists of a tremendously elongate strand of DNA (a double helix, as described in Chapter 9), closely bound to some proteins of a type called **histone.** The histone proteins form an integral part of the structure during interphase as well as during mitosis.

Cytologists find it useful to think of the interphase as consisting of three stages. During the second stage each chromonema is duplicated, the DNA being replicated in the way described in Chapter 9. The members of such a pair of chromonemata remain closely associated throughout the third stage of the interphase, and they become the chromatids of a condensed chromosome that is formed during the next mitosis. The first and third stages of interphase are defined simply as the stage that precedes and the stage that follows DNA replication.

CYTOKINESIS

In both plants and animals cytokinesis commonly begins about the middle of anaphase and is completed by or shortly after the end of the telophase. In most plants cytokinesis occurs by the formation of a **cell plate,** sometimes called a **phragmoplast.** The spindle becomes shorter and broader, and it evidently undergoes a chemical change, as indicated by a change in the stains or dyes that affect it in preparations for microscopic examination. A thin, fluid membrane, the cell plate, is then formed across the equator of the spindle. The spindle continues to shorten and widen, and the cell plate widens until the cell is completely divided into two parts. The remnants of the spindle then disappear.

Electron microscope studies indicate that the formation of the cell plate begins with the concentration and coagulation of numerous tiny vesicles around the spindle fibers in the equatorial region of the cell (Fig. 5.7). Each of these vesicles is about 100 nm in diameter. The vesicles are believed to be formed by or from the golgi bodies. Within the hollow platelets (called *cisternae*) of the golgi bodies, glucose is converted into various sorts of pectic substances. The cisternae then give rise (by fragmentation or otherwise) to tiny vesicles which are believed to migrate through the cytoplasm and be the same vesicles that appear about the spindle fibers in the equatorial region. Initially the vesicles congregate about the spindle fibers, so that the spindle fibers seem to thicken. This thickening can readily be seen in good light-microscope preparations. Continued arrival of more vesicles results in the coalescence of all the vesicles to form the cell plate, which consists mainly of pectic substances. There are some lingering uncertainties about this running interpretation, but it is plausible and provides adequately for many individual bits of information that do not fit into any other recognized interpretative pattern.

As soon as the cell plate has widened so as to extend from wall to wall, cytokinesis is completed, and the cell plate is

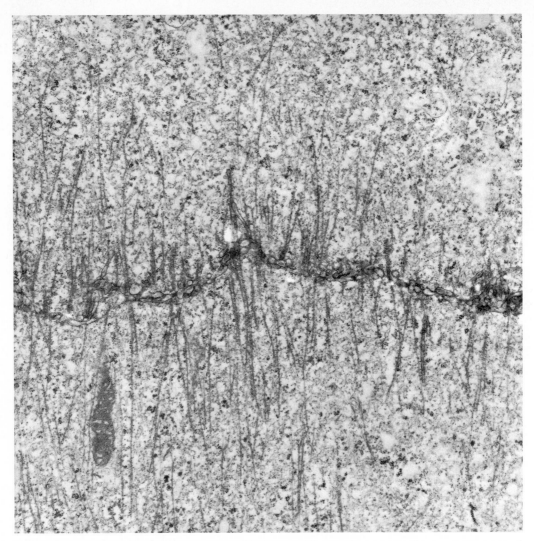

Fig. 5.7 Formation of cell plate from spindle fibers and cytoplasmic vesicles in endosperm of *Haemanthus katheriniae.* (× about 11,000.) [Electron micrograph courtesy of Peter K. Hepler and W. T. Jackson, from *J. Cell Biology* **38**:444, 1968.]

thereafter referred to as the *middle lamella.* The principal constituent of the middle lamella at this early stage is pectin. As we noted in Chapter 4, much of the pectin later combines with calcium and magnesium, and it may also become impregnated with other substances.

Among animals, and among those plants that do not have definite cell walls (chiefly flagellated, unicellular algae), cytokinesis usually does not involve the formation of a cell plate but results instead from a seemingly simple constriction or furrowing of the cytoplasm, with the cell membrane becoming progressively indented or impressed until division is complete. In a few kinds of algae cytokinesis proceeds by a combination of furrowing and the formation of a small cell plate.

WALL FORMATION

Soon after the completion of cytokinesis, a primary cell wall, usually composed largely of cellulose, is generally deposited on each side of the middle lamella, with the result that the plasma membrane of each cell lies against the cell wall rather than against the middle lamella. In many cells a secondary cell wall, usually also composed in large part of cellulose, is laid down against the primary wall. The partition between two adjacent cells (Figs. 5.8, 5.9) then consists, in order, of:

1. the secondary wall of one cell
2. the primary wall of that cell
3. the middle lamella
4. the primary wall of the second cell
5. the secondary wall of the second cell

As noted in Chapter 4, the partition, regardless of whether a secondary wall is present or not, usually has numerous small perforations through which the cytoplasm of adjacent cells is continuous.

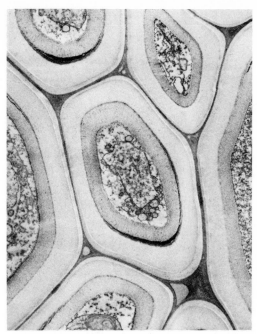

Fig. 5.9 Maturing cells of phloem fibers of black locust, *Robinia pseudoacacia*, showing intercellular spaces, middle lamella, primary wall, and secondary wall, as diagrammed in Fig. 5.8. (× 7,000.) [Electron micrograph courtesy of Myron T. Ledbetter, Biology Department, Brookhaven National Laboratory.]

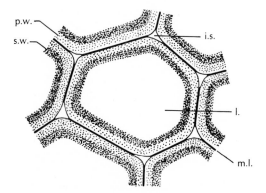

Fig. 5.8 Diagram showing wall layers of adjacent cells: i.s., intercellular space; l., lumen; m.l., middle lamella; p.w., primary wall; s.w., secondary wall.

Some of these openings exist in the newly formed cell plate (Fig. 5.7) and are probably maintained throughout the formation of the cell wall; that is, the cellulose microfibrils are deposited in a pattern that does not obstruct the existing openings. In some kinds of plants, however, notably some of the algae, plasmodesmata are also initiated *de novo* by dissolution of bits of the mature cell wall.

The cellulose that makes up the major part of the cell wall is believed to be synthesized (from glucose) and packaged into the characteristic microfibrils by the golgi bodies. The hemicelluloses that are incorporated into the cell wall along with the cellulose are apparently delivered to the developing wall in vesicles derived from the golgi bodies in the same way as the components of the cell plate. The lignin that later impregnates some cell walls is probably also manufactured in the golgi bodies.

AMITOSIS

Nuclear division that does not follow the pattern of mitosis is called **amitosis.** The nucleus may seem simply to constrict and separate into two or more parts, without the formation of typical chromosomes or a spindle. Among the eukaryotes amitosis is rare and occurs only under exceptional circumstances. Sometimes it is associated with degenerating cells or tissues.

Cell division among prokaryotes, which do not have a vesicular nucleus, is also by definition amitotic. Cells of prokaryotes transmit a full set of DNA to each of the two daughter cells formed in division, but they have no spindle and nothing comparable to mitotic chromosomes. Their DNA is organized into elongate strands comparable to the chromonemata of eukaryotes, but without the associated protein. Cell division among prokaryotes is further discussed in Chapter 11.

SUGGESTED READING

Bajer, A., Fine structure studies on phragmoplast and cell-plate formation, *Chromosoma* (Berlin) **24**:383–417, 1968.

Bajer, A., and J. Molè-Bajer, Formation of spindle fibers, kinetochore orientation, and behavior of the nuclear envelope during mitosis in endosperm, *Chromosoma* (Berlin) **27**:448–484, 1969.

Hepler, P. K., and W. T. Jackson, Microtubules and early stages of cell-plate formation in the endosperm of *Haemanthus katherinae* Baker, *J. Cell Biol.* **38**:437–446, 1968.

Mazia, D., The cell cycle, *Sci. Amer.* **230**(1):55–64, January 1974.

Sloboda, R. D., The role of microtubules in cell structure and cell division, *Amer. Sci.* **68**:290–298, 1980.

Voeller, B. (ed.), *The chromosome theory of inheritance*, Appleton-Century-Crofts, New York, 1968. Reprints of important papers marking historical steps along the way to an understanding of heredity, with commentary by Voeller.

Chapter 6
The Absorption of Water and Solutes

Germinating radish seedling (× about 15) showing numerous root hairs. Root hairs, discussed in Chapter 20, are the principal water-absorbing structures for most higher plants. [Photo by Hugh Spencer from National Audubon Society Collection/Photo Researchers, Inc.]

OSMOSIS

Normal cells are typically in a state of **turgor;** that is, the protoplast pushes firmly against the enclosing cell wall, as if the cell were overfilled. The material in "excess" supply, causing the turgor pressure, is water. If, on the other hand, water is in short supply, metabolic activities in general are likely to be seriously disrupted. The conditions and mechanisms governing the absorption and loss of water by the cell are therefore of vital importance. We shall see that the differential permeability of certain membranes is fundamental to the maintenance of an adequate water supply in the cell.

One of the most important features of the plasma membrane (and the vacuolar membrane) is that it constitutes a partial barrier to the passage by diffusion of materials in solution. Water passes through it relatively freely. Some solutes pass freely, others more slowly, and others not at all. Such a membrane, which permits the free passage by diffusion of some materials and restricts the passage of others, is commonly called a **semipermeable membrane,** although the term **differentially permeable membrane** would be more precisely descriptive.

Another important characteristic of differentially permeable membranes of plant cells is that the permeability changes, so that a substance which is allowed to pass through at one time may be restrained at another. It was at one time thought that these changes of permeability might reflect changes in the physical structure or condition of the membrane. It now seems much more likely that they reflect changes in chemical activity of the proteins in a physically relatively stable membrane.

Because water and some solutes frequently move from outside a cell into the

central vacuole, or vice versa, it is sometimes convenient to think of the cytoplasm itself as a membrane separating the central vacuole from the region outside the cell. Such a movement involves diffusion through both the plasma membrane and the vacuolar membrane, with passage through the intervening portion of the cytoplasm being accomplished partly by diffusion and more especially by means of the constant streaming motion (cyclosis) of the cytoplasm itself. The vacuolar membrane is in some respects a more effective and discriminating barrier than the plasma membrane.

If pure water is placed in one half of a U-shaped container and an equal amount of a solution of cane sugar is placed in the other half, the two halves being separated by a completely permeable membrane as shown in Fig. 6.1, the proportions of sugar and water eventually become uniform throughout the container. Molecules of both substances pass through the membrane in both directions, but there is a net movement of sugar into the side of the container that had held pure water, and a net movement of water into the side of the container that had held the original sugar solution. Each of the two substances, sugar and water, diffuses from a region of

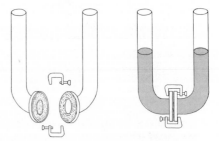

Fig. 6.1 Osmotic apparatus described in the text. (*Left*) Separated parts. (*Right*) Sugar solution in equal concentration on both sides of the membrane after a period of diffusion.

its own high concentration toward a region of its own lower concentration, and the level of the liquids on the two sides of the container remains essentially unchanged throughout the process.

In a similar experiment using a membrane that is permeable to water and impermeable to sugar (Fig. 6.2), the results are different. The sugar is retained in one half of the container, and only the water moves from an area of its own higher to its own lower concentration. The level of the pure water falls, and the level of the sugar solution rises. If the membrane remains unbroken and its differential permeability remains unimpaired, the level of the sugar solution may rise a considerable distance before the increased pressure of the sugar solution against the membrane results in an equilibrium, with as many molecules of water passing in one direction as the other. If the membrane merely retards the passage of sugar, rather than completely preventing it, the level of the sugar solution will at first rise, as more water enters it than leaves, and then it will fall as the equalization of sugar concentrations on the two sides of the membrane proceeds. Eventually·the liquid will stand at the same level

and have the same concentration of sugar on both sides of the membrane.

In the experiment in which the membrane wholly prevents the passage of sugar, the level of the sugar solution can be prevented from rising by inserting a plug in the tube down to the level of solution (Fig. 6.3). Water will then enter the sugar solution until the pressure of the solution against the membrane is built up enough to cause the passage of equal amounts of water in both directions. The same results will be obtained if the two halves of the container merely have different concentrations of sugar molecules, or even if different kinds of solutes are used, so long as these solutes cannot pass through the membrane and so long as the concentration of water in the two solutions is not the same.

An oversimplification which might help one to understand this process is to consider that whenever a molecule of water hits the membrane (and we have seen that all molecules are in constant motion), it passes through, whereas whenever a particle of solute hits the membrane, it is restrained. A given amount of pure water will obviously contain more water molecules than the same amount of a sugar solution. (By actual test, for example, 1 cup of water and 1 cup of sugar

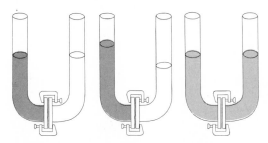

Fig. 6.2 Osmotic apparatus, showing changes when equal amounts of water and sugar solution are separated by a semipermeable membrane, as described in the text.

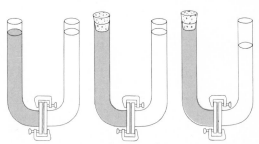

Fig. 6.3 Osmotic apparatus developing pressure, as described in the text.

yield about 1 5/8 cups of sugar solution.) Therefore, under the conditions of the experiment, more molecules of water will hit and pass through the membrane into the sugar solution than in the reverse direction. With the volume of the sugar solution prevented from increasing, a pressure is built up within the solution, because more molecules occupy a given space than had previously occupied that space. When the pressure becomes so great that as many molecules of water hit the membrane from one side as the other, an equilibrium is established.

The diffusion of water through a semipermeable membrane is called **osmosis;** and the pressure exerted on a semipermeable membrane as a result of osmosis is called **osmotic pressure.** Obviously, the actual osmotic pressure exerted by the cell sap of the central vacuole against the enclosing cytoplasm will be greater if the cell is immersed in pure water than if the surrounding medium contains any solutes. Also, other things being equal, the greater the concentration of solutes in the cell sap (and thus the less the concentration of water), the greater the actual osmotic pressure. The presence of hydrophilic colloids in the cell sap will still further increase the osmotic pressure, since the colloidal particles adsorb water molecules on their surfaces and thus decrease the tendency for water to move back out of the cell.

The possible osmotic pressure—that is, what the pressure would be if the cell, otherwise unchanged, were immersed in pure water with the volume of its protoplast held constant—has considerable physiological importance as a measure of the water-absorbing and water-holding capacity of the cell. By long custom this possible osmotic pressure rather than the actual pressure is meant in

speaking of the osmotic pressure of a particular cell. The possible osmotic pressure is usually expressed in terms of atmospheres (atm), that is, the pressure of the atmosphere at sea level. There is a wide variation in possible osmotic pressures among the cells of different kinds of plants; figures of 5 to 20 atm are common.

Students of osmotic phenomena often find it convenient to think in terms of the ability of pure water to move into a solution from which it is separated by a semipermeable membrane, rather than of the ability of the solution to attract and hold water. In these terms, water has a greater **osmotic potential,** or capacity for net diffusion through the membrane, than any aqueous solution with which it might be compared. As compared to water, the solution has a negative osmotic potential. Osmotic potential can be measured in exactly the same way as the possible osmotic pressure; it is the same thing, but with a minus sign. Another term often used in this same context is **diffusion pressure deficit.** The solution has a diffusion pressure deficit as compared with pure water. Being stated as a deficit to begin with, the figure does not carry a minus sign.

Since the volume of a plant cell is held essentially constant by the cell wall, the osmotic pressure exerted by the central vacuole on the cytoplasm ordinarily forces the cytoplasm against the cell wall. The actual pressure of the cytoplasm against the cell wall as a result of osmotic forces is referred to as **turgor pressure.** The turgor pressure of cells is often so great that the cytoplasm would burst if the cell wall were not present.

The actual power of a cell to absorb water is limited by the fact that its volume is held nearly constant by the cell wall, together with the fact that water is not highly compressible. As the cell becomes more and more turgid, its water-absorbing capacity becomes less and less. When the turgor pressure balances the osmotic potential, the cell is fully turgid and can absorb no more water.

The difference between the osmotic potential and the turgor pressure is often called the **water potential.** Like osmotic potential, it carries a minus sign. Thus a cell that has an osmotic pressure of 15 atm has an osmotic potential of −15 atm. If it is partly turgid, so that the turgor pressure is, for example, 10 atm, it has a water potential of −5 atm. If immersed in pure water (and if the osmotic potential does not change), it will continue to absorb water until the turgor pressure reaches 15 atm. It will then be fully turgid, and no longer able to absorb water. Given the same initial conditions of turgor and osmotic potential (10 atm turgor pressure, −15 atm osmotic potential), and placed in a solution that has an osmotic potential of −5 atm, it will neither take on nor lose water, since the difference between its own osmotic potential (−15 atm) and that of the surrounding solution (−5 atm) is exactly balanced by the turgor pressure. The same cell, immersed in a more concentrated solution, with an osmotic potential of say −10 atm, will lose water osmotically until its turgor pressure is enough lower than its osmotic potential to balance the osmotic potential of the surrounding medium.

Precise figures in examples such as these may be misleading, because loss of water from a cell automatically tends to increase its water potential, and the intake of water tends to decrease the water potential (i.e., to bring the figure nearer to zero). The principle, however, is simple enough. The cell takes in or loses water osmotically, according to the circumstances, until its water potential (the difference between osmotic potential and turgor pressure) equals the osmotic potential of the surrounding medium.

If the concentration of solutes in the water surrounding a cell is raised to a level appreciably higher than that of the central vacuole (which means, also, that the concentration of water in the vacuole is greater than that in the surrounding medium), osmosis tends to proceed in reverse direction, and water leaves the cell (unless hindered by hydrophilic colloids in the cell sap). If the process continues, the vacuole diminishes in size and the protoplast shrinks away from the cell wall (Fig. 6.4). The cell is then said to be plasmolyzed, or in a state of **plasmolysis.** A plasmolyzed but still living cell obviously exerts no turgor pressure, but its osmotic potential may be very high. If a more normal environment is restored, water re-enters the cell and turgor is re-established, provided that plasmolysis has not been carried so far or continued so long as to kill the cell. The differential

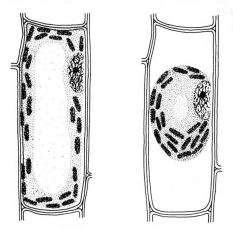

Fig. 6.4 Cell of a leaf of *Elodea*, a water plant; *(left)* normal and *(right)* plasmolyzed. (× 600.)

permeability of the cytoplasmic membrane depends on life, and when the cell dies the cytoplasm becomes wholly permeable.

A somewhat more sophisticated approach to the mechanism of osmosis involves the concept of **free energy.** In this context, free energy may be thought of as the energy of molecular (or ionic) motion. Given a particular volume of liquid or gas, the amount of free energy is determined by the number of molecules (or ions), their molecular weight, and their speed of motion. In the absence of barriers or membranes limiting movement, the net movement of molecules or ions of a particular kind is along a free-energy gradient, from a region of higher free energy into a region of lower free energy, until ultimately an equilibrium is established and the free energy of the substance in question is uniformly distributed. Students familiar with thermodynamic principles may see these changes as resulting in increased entropy, in accord with the second law of thermodynamics.

Since we are here concerned with the movement of water only, we may think in terms of the free energy of the water in a solution. We have noted that osmosis is the diffusion of water through a membrane that is permeable to water but not as permeable to molecules or ions of some or all of its solutes. Free energy of water is lowered by the presence of solutes, decrease in temperature, decrease in pressure, and surface attractions such as those exerted by hydrophilic colloids. Solutes decrease the free energy of water because they take up space that would otherwise be occupied by water molecules, and because water may form dipole bonds with the molecules or ions of the solute. Decrease in temperature lowers free ener-

gy by reducing the speed at which molecules move (indeed, as we have noted, temperature is essentially a measure of the speed of molecular motion). Surface attractions reduce free energy by reducing the freedom of movement of water molecules. Although water is not highly compressible (i.e., its volume decreases only slightly with increasing external pressure), even a small change in the volume occupied by a given number of molecules will correspondingly change the free energy per unit of volume.

In more precise studies it is also necessary to note that the cytoplasmic membrane does impede the passage of water to some extent, though less than it does the passage of most solutes. Not every molecule of water that hits the membrane gets through. Changes in the precise chemical (and possibly physical) state of the cytoplasmic membrane therefore influence the rate of osmosis.

The relatively free flow of water through the plasma membrane has not yet been fully reconciled with current concepts of the structure of the membrane. Possibly the protein molecules in the membrane are permeated by channels large enough to permit the passage of water. This would, if true, have some bearing on the function of the protein molecules in promoting an essentially one-way flow of solutes. It is even doubtful that the classical concept of osmosis as a molecule by molecule process of diffusion is actually correct. There may be a mass flow of ultramicroscopic streams of water through the membrane. At such levels of size it is hard to draw the distinction. In any case, the differentially permeable membranes of plant cells are much more complex in structure and function than the nonliving chemical membranes used to demonstrate osmosis in classroom experiments.

In the foregoing paragraphs the emphasis has been on osmosis as a physical process, the only strictly metabolic contribution being the maintenance of an effective semipermeable membrane. This is not the whole story. A high rate of water uptake is commonly correlated with a high rate of respiration. For some years many botanists believed that this expenditure of energy might permit the absorption of water against the concentration gradient. That view has now been abandoned. The association of use of energy with uptake of water reflects active absorption of solutes, as explained below.

ABSORPTION OF SOLUTES

It is clear that mineral salts, in contrast to water, can be and often are absorbed against the concentration gradient. Metabolic activities of the cell interfere with normal diffusion processes so that certain ions, under certain conditions, enter the cell far more readily than they leave it. Particular minerals may thus be accumulated in concentrations many times higher than their concentration in the surrounding medium. The accumulation reflects in part the removal of the ions from solution or their transformation into forms that do not pass readily through the plasma membrane, but it also reflects the direct expenditure of metabolic energy to retain ions that would otherwise diffuse out of the cell.

Experiments under conditions of moderate or slow respiration have usually indicated that the cytoplasmic membrane is not very readily permeable in either direction to most mineral solutes. Under conditions of rapid respiration, however,

it evidently permits the entry of many solutes while restricting their exit. It is thus clear that a directional difference in the permeability of the cytoplasmic membrane is maintained by the expenditure of respiratory energy, but the means by which the energy exerts its effect remains obscure. It may be said that the respiratory energy makes up for the free-energy deficit of the internal solutes in contrast to the corresponding solutes of the surrounding medium, but this answer still needs a detailed explanation. It seems likely that the energy is provided by the breakdown of ATP to ADP and phosphate ion and that the plasma membrane is intimately involved in the use of ATP energy to absorb ions, but we still need to know just how the energy is used.

The demonstrated use of respiratory energy in the absorption of solutes does help to explain why rapid absorption of water is also correlated with rapid respiration. A concentration of mineral solutes in the cell sap adequate to encourage rapid osmosis can only be achieved and maintained by the expenditure of respiratory energy.

It should be noted that the expenditure of respiratory energy in the absorption of solutes does not encourage the absorption of all solutes equally. On the contrary, some ions are more vigorously absorbed than others, and different kinds of plants sometimes differ greatly in their affinity for various kinds of ions. It thus seems likely that chemical as well as strictly physical actions are involved in the respiratory uptake of mineral solutes. This is in accord with current concepts of the structure of the plasma membrane, which imply chemical combination with membrane proteins as a means of transit. These matters are further discussed in Chapter 23.

SUGGESTED READING

Koslowski, T. T., *Water metabolism in plants,* Harper & Row, New York, 1964.

Kramer, P. J., *Plant and soil water relationships,* McGraw-Hill, New York, 1969.

Meidner, H., and D. W. Sheriff, *Water and plants,* Tertiary Biology Series, Wiley, New York, 1976. A bit more technical than Sutcliffe.

Sutcliffe, J., *Plants and water,* Studies in Biology No. 14, Edward Arnold, London, 1968. Clearly written, scientifically accurate.

Chapter 7

Photosynthesis

Chloroplast of timothy grass, *Phleum pratense* (× about 21,000.) [Electron micrograph by W. P. Wergin, University of Wisconsin, courtesy of E. H. Newcomb.]

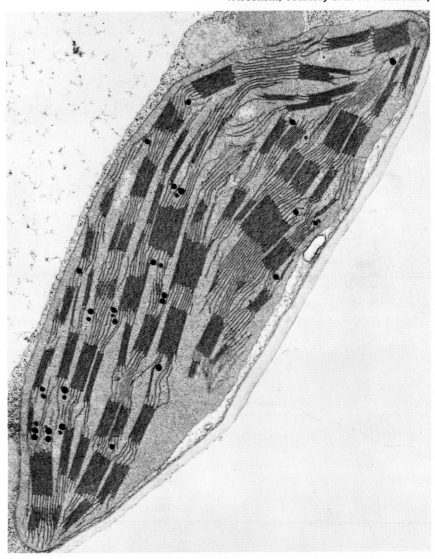

Introductory Summary

Photosynthesis is the process by which light energy is captured and transformed into chemical energy which is used in the formation of photosynthate—complex, energy-storing substances, typically hexose carbohydrates such as glucose. Some of the carbohydrate is used directly as building blocks for more complex substances, but most of it is respired, with chemical energy being released into the metabolic system of the cell (see Chapter 8). Thus *photosynthesis and respiration collectively form a mechanism for transforming light energy into metabolic energy.*

Light energy is absorbed by chlorophyll and accessory pigments (carotenoids and, when present, phycobilins) that are associated with chlorophyll, typically in complex cell organelles called **chloroplasts.** The absorbed light energy is passed into the photosynthetic system in a complex series of reactions following on the transfer of "excited" electrons from chlorophyll to certain other substances which thereby become highly reactive.

The hydrogen in photosynthate is derived from water (in green plant photosynthesis), and the carbon is derived from carbon dioxide. The oxygen in photosynthate is also derived from carbon dioxide, and the oxygen that is released in typical photosynthesis is derived from water. In addition to being assimilated in the process, hydrogen plays a basic role in the transfer of energy which is the essential function of photosynthesis.

The overall process of photosynthesis as carried on by algae and higher plants (i.e., green plant photosynthesis, as the term is here used) is usually visualized in the equation:

$$6CO_2 + 6H_2O \xrightarrow{\text{visible light}} C_6H_{12}O_6 + 6O_2$$

A slightly closer approximation, which takes into account the fact that water is formed as well as being used in photosynthesis, is:

$$6CO_2 + 12H_2O \xrightarrow{\text{visible light}} C_6H_{12}O_6 + 6O_2 + 6H_2O$$

In photosynthetic bacteria, as contrasted to other photosynthetic organisms, the overall process can be represented as:

$$6CO_2 + 12H_2A \xrightarrow{\text{visible and far red light}} C_6H_{12}O_6 + 12A + 6H_2O$$

In this equation H_2A is an arbitrary symbol for a suitable hydrogen donor, such as H_2S, H_2, or various organic compounds that can provide hydrogen for the process. We should note at this point that only a few kinds of bacteria, living in certain specialized habitats, are photosynthetic. The vast majority of bacteria are heterotrophic; that is, they do not make their own food.

The foregoing equations are useful in providing a generalized concept of photosynthesis in terms of raw materials, energy input, and end products, but they are vastly oversimplified and misleading if taken literally. A long and complex series of reactions is required to turn the raw materials on the left side of the equation into the end products on the right side.

Most of the discussion here will center on photosynthesis in algae and higher plants, referred to for convenience as green plant photosynthesis, in which molecular oxygen is released as a by-product. This process is far more common than bacterial photosynthesis and is of much greater significance to living organisms in general. The bacterial process, on the other hand, is important in helping us to understand the evolutionary origin and development of photosynthesis.

The capture of light energy in green plant photosynthesis is now conceived to proceed through two concurrent systems: System I, which has been inherited with relatively minor change from bacterial ancestors, and System II, which has been superimposed on System I. System II, which is necessary for the formation of free oxygen in photosynthesis, does not occur in the bacteria. Systems I and II collectively constitute the **light reactions.**

Energy captured in the light reactions is used in the **dark reactions** to promote the formation of carbohydrate. The dark reactions do not *require* darkness, but they can proceed in the dark after light energy has been captured in the light reactions.

The chemistry of photosynthesis has received a great deal of study during the past several decades, and progress in our understanding has been particularly rapid since about 1950. The availability of radioactive isotopes for experimental studies has been especially instrumental in this progress. Much remains to be learned, however, and some of what we do know has yet to be fitted into a coherent and generally acceptable scheme. In particular, the precise pathways of electron transport and the series of steps preceding the evolution of free oxygen are still controversial and insufficiently understood. Continuing study will undoubtedly lead to changes in some of our ideas about photosynthesis.

It has long been known that photosynthesis is the eventual source of metabolic energy for all organisms, both plant and animal, with the minor exception of a small amount of food produced by certain chemosynthetic bacteria. Green plants make their own food. Animals eat plants, or eat other animals that have eaten plants. *The food chain begins with photosynthesis in plants.*

Relatively recently it has come to be believed that photosynthesis is also the principal source of atmospheric oxygen. Prior to the origin of oxygen-releasing photosynthesis in primitive algae some 2 or 3 billion years ago, the earth's atmosphere apparently had little or no free oxygen. It was only after a considerable supply of oxygen had been released into the atmosphere, possibly through an excess of photosynthesis over respiration, that the way was opened for the evolutionary origin and diversification of complex, multicellular organisms that use free oxygen in respiration. Even more recently, this view has been challenged, and it has been suggested that radiation-induced breakdown of water at the outer limits of the atmosphere is a major or principal source of atmospheric oxygen. The matter is still controversial.

Carbon dioxide, an essential raw material for typical photosynthesis, makes up only a small fraction of the atmosphere, currently (1981) about 335 parts per million (0.0335 percent). The effect of variation in carbon dioxide content of the air on photosynthesis is discussed in Chapter 24, and the possible long-term environmental effects are discussed in Chapter 31.

CHLOROPHYLL

Chlorophyll is a general term for several closely related kinds of green pigments which have in common the capacity to absorb light energy and pass it along to other molecules as excitement energy of electrons. As noted in Chapter 4, chlorophylls are metalloporphyrins with an atom of magnesium at the center of the porphyrin ring.

Chlorophylls are green because they absorb light chiefly toward the two ends of the visible spectrum, transmitting or reflecting most of the light of intermediate wavelengths. They absorb most of the red and blue light and transmit or reflect most of the blue-green, green, and yellow light to which they are exposed. The several kinds of chlorophylls found in plants other than bacteria differ only slightly in their absorption spectra. Most of the bacterial chlorophylls differ markedly from the nonbacterial chlorophylls in that the absorption peak near the red end of the spectrum is shifted into the infrared. The

absorption spectra of chlorophylls *a*, *b*, and bacteriochlorophyll are shown in Fig. 7.1.

The essential chlorophyll in all green plant photosynthesis is chlorophyll *a*. Chlorophyll *a* receives light energy both directly and through transfer from other chlorophylls and accessory pigments. The other chlorophylls and accessory pigments (such as carotenoids) can absorb light energy, but this absorbed energy must be channeled through chlorophyll *a* before it can be passed into the rest of the photosynthetic system. In green plant photosynthesis it is only chlorophyll *a* that can effectively transfer excitement energy of electrons to other molecules that are not themselves photoreceptive.

Among the photosynthetic bacteria, bacteriochlorophyll apparently plays a role comparable to that of chlorophyll *a* in

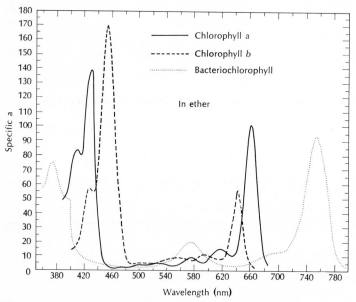

Fig. 7.1 Absorption spectra of chlorophylls *a*, *b*, and bacteriochlorophyll. The vertical axis of the chart represents a complex measurement of light absorption, the mathematics of which do not concern us here.

green plants. Various accessory chlorophylls may or may not be present, according to the species, but only bacteriochlorophyll can effectively pass the absorbed light energy to compounds that are not directly photoreceptive.

The structural formula for extracted chlorophyll a has been given in Chapter 4. It should be recognized, however, that extracted chlorophyll bears about as much relationship to chlorophyll in the chloroplast as a detached carburetor does to a complete automobile motor. Functioning chlorophyll always occurs in loose conjugation with a protein. Efforts to determine the nature of this conjugated protein are still at an early stage. It is quite possible that different molecules of chlorophyll a in a single chloroplast are attached to slightly different sorts of proteins; and it is very probable that there is some difference in chlorophyll-bound proteins in different kinds of plants. It appears that at least some of the chlorophyll-bound protein in spinach has a molecular weight of about 11,000, with six molecules of chlorophyll bound to each molecule of protein. The functional role of chlorophyll-bound protein is still largely unknown.

Most green plants have an accessory chlorophyll in addition to chlorophyll a. In the green algae and higher plants this accessory chlorophyll is always chlorophyll b. In the diatoms, pyrrophytes, cryptomonads, and brown algae the accessory chlorophyll is chlorophyll c, and in the red algae it is chlorophyll d. One or more other kinds of accessory chlorophylls may or may not be present in some other kinds of algae; the evidence is still inconclusive. The blue-green algae, some of the red algae, and most or all of the yellow-green algae and golden algae have only chlorophyll a, without any accessory chlorophylls.

The chemical differences among these several kinds of chlorophyll are relatively small. Chlorophyll b, for example, differs from chlorophyll a only in the substitution of a CHO group for a CH_3 group at one point on the periphery of the porphyrin ring. It may well be that the different kinds of chlorophyll are attached to slightly different kinds of protein, but any such differences remain to be elucidated.

Even chlorophyll a itself occurs in several different forms with different properties and functions. It is still uncertain to what extent these differences reflect different states of aggregation, different sorts of attached proteins, or other factors as yet unspecified. Some of these forms function in both System I and System II, and some only in one or the other. The absorption peak of the chlorophyll a associated with System I is at a somewhat longer wavelength than that associated with System II. The absorption spectrum for chlorophyll a, as given in Fig. 7.1, is really an average reflecting the proportions of the various forms in the mixture.

Chlorophyll b functions in both systems, but mainly in System II. It transfers its absorbed energy directly to chlorophyll a, which is always more abundant. Ratios of chlorophyll a to b from about 2:1 to 5.5:1 are common among the higher plants, with the lower ratios coming mostly from species adapted to shade conditions and the higher ones from alpine species or other species adapted to bright sunlight. Many of the green algae have lower ratios, in the range of 1.3:1 or 1.5:1.

ACCESSORY PIGMENTS

Pigments that absorb light energy and transmit it to chlorophyll in the photosynthetic process are called **accessory pig-**

ments. Chlorophylls *b*, *c*, and *d* might also be considered to be accessory pigments, since they must transfer their absorbed light energy to chlorophyll *a* if it is to be used. Customarily, however, chlorophylls are excluded from the concept of photosynthetic accessory pigments.

Chlorophyll is always accompanied by one or more accessory pigments that absorb light in a part of the spectrum where chlorophyll does not absorb well. The most important of these accessory pigments are the **carotenoids.**

One or more (usually more) carotenoid pigments (i.e., carotenes and xanthophylls) always occur in association with chlorophyll. The general nature of carotenes and xanthophylls has been indicated in Chapter 4, and the structure of β-carotene is given in Fig. 4-13.

β-Carotene is the principal carotene associated with chlorophyll in all photosynthetic organisms except the bacteria and some few of the algae. It is still present even in those algae in which it is not the principal carotene, but it is unknown among the bacteria. Lutein, zeaxanthin (both with the empirical formula $C_{40}H_{56}O_2$), and violaxanthin ($C_{40}H_{56}O_4$) are among the most common xanthophylls; lutein is regularly present in higher plants and in the green algae, as well as in some other algae. Many of the algae contain special xanthophylls that are not known in other groups. The common yellow carotenoids absorb light mainly between about 400 and 500 nm wavelength (Fig. 3.8) in the violet, blue, and blue-green range of the spectrum.

The metabolic roles of the carotenoids are still under active consideration, and investigators using different methods have sometimes come to different conclusions. It is now clear, however, that they consistently have at least two major functions relating to photosynthesis: to transfer absorbed light energy to chlorophyll, and to prevent the photo-oxidation (oxidative destruction by light) of chlorophyll. The uniform presence of β-carotene in nonbacterial photosynthetic systems, as contrasted to the less regular distribution of the other carotenoids, suggests that β-carotene may have some special function of its own. The nature of this other function, if any, is still wholly speculative.

Another class of accessory pigments, called **phycobilins** (Fig. 7.2), occurs in addition to the carotenoids in red algae, blue-green algae, and a few other algae. The phycobilins are open-chain tetrapyrroles, tightly bound to a protein.

Phycobilins may be divided into two general types, according to their absorption spectra and consequent color; each of these types may be further subdivided. Red phycobilins are called **phycoerythrins,** and blue ones are called **phycocyanins.** Usually both types occur together, in varying proportions according to the species and the environmental conditions. Both phycocyanins and phycoerythrins absorb well in the portion of the spectrum where neither chlorophyll nor the common carotenoids are very effective. The phycocyanins have an absorption peak at about 600 or 625 nm (in different extracted forms), in the orange-yellow or orange part of the spectrum (Fig. 7.3). The phycoerythrins as extracted have their principal absorption peak near 500 nm in one form and near 575 in another. Thus they span the green range of the spectrum (500–560 nm). The absorption of green light by phycoerythrin is especially important because under most conditions green light penetrates sea water much better than light of other colors (wavelengths). Red algae that grow in

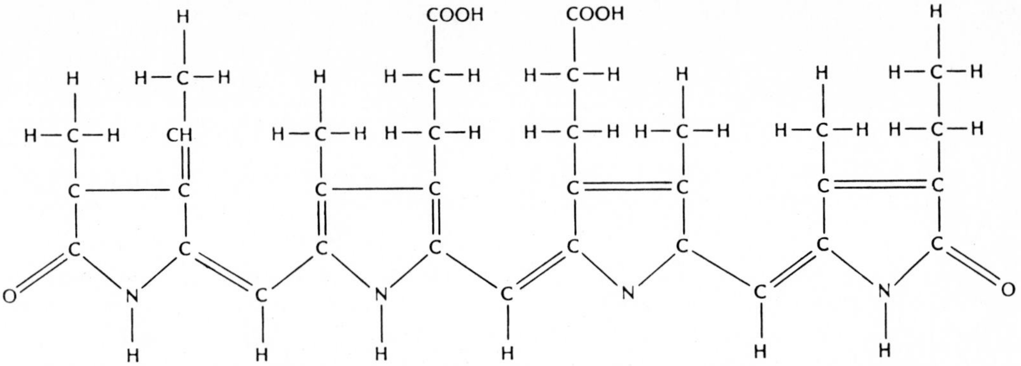

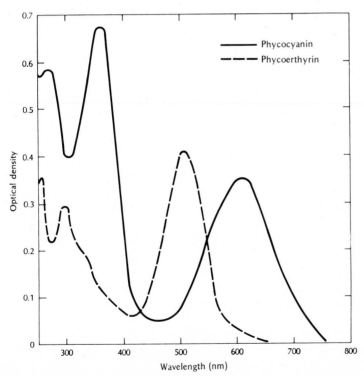

Fig. 7.2 Structural formula for the prosthetic group of phycocyanin, a phycobilin pigment.

relatively deep water commonly have a high proportion of phycoerythrin in relation to other photosynthetic pigments.

In addition to absorbing light energy and passing it on to chlorophyll *a*, phycobilins act as carriers of energy from carotenoids to chlorophyll *a*. The sequence is from carotenoids to phycoerythrin to phy-

Fig. 7.3 Absorption spectra for phycobilins. [After O'hEocha.]

cocyanin to chlorophyll a. In the absence of phycobilins, carotenoids transfer their absorbed energy directly to chlorophyll a.

CYTOCHROME

Cytochromes are metalloporphyrins with an atom of iron at the center. The fact that iron can change from a divalent (ferrous) to a trivalent (ferric) state and back again appears to be of prime importance in the transfer of energy with which cytochrome is characteristically associated. Cytochromes occur in all living organisms, with the possible exception of a few of the simpler bacteria.

Most cytochromes mediate the transfer of energy into the reaction of adenosine diphosphate (ADP) with phosphate ion, converting ADP into ATP (adenosine triphosphate). ATP, in turn, readily reverts to ADP, giving up its third phosphate ion and liberating chemical energy into the economy of the cell. Some kinds of cytochromes mediate this sort of energy transfer as part of the process of respiration, others as part of the process of photosynthesis.

Cytochrome, like chlorophyll, is a group name for a series of closely related compounds, not all of which are found in any one kind of organism. The classification and nomenclature of cytochromes are still in a state of flux. On the basis of their relative activity in oxidation-reduction reactions and slight differences in their absorption spectra, they are usually classified into three general groups, called the a type, b type, and c type. Each of these types can be further subdivided. Cytochromes of type a are always (so far as is known) respiratory in the strict sense, requiring the use of free oxygen. Some b-type cytochromes are photosynthetic, and others are respiratory (at least in the broad sense); the same is true of c-type cytochromes. Any given molecular species of cytochrome has only one of these functions, however, being either respiratory or photosynthetic, but not both. The b-type photosynthetic cytochrome in algae and higher plants is called cytochrome b_6. The c-type photosynthetic cytochrome in algae and higher plants is also called cytochrome f.

Like chlorophyll, cytochrome occurs in loose conjugation with a protein. The complete protein structure of respiratory cytochrome c has recently been worked out for a number of kinds of plants and animals, and small differences in the structure are beginning to be used as an aid in deciphering phylogenetic relationships among major groups.

CHLOROPLASTS

The chloroplasts of vascular plants are highly organized. Generally each chloroplast consists of a number of microscopically barely recognizable units, called **grana,** embedded in a relatively less differentiated mass called the **stroma,** the whole enclosed by a double membrane about 30 nm thick. The chloroplasts are commonly ellipsoidal or lens shaped; their size generally ranges from 4 to 6 microns along the longest axis, but some are as long as 10 microns. The number of grana per chloroplast varies from as few as 10 to as many as several hundred, but is most commonly from 40 to 60.

Each granum (Fig. 7.4) consists of a series of plates, called *lamellae* (or, more specifically, **thylakoids**), stacked on top of each other like a stack of coins. The thylakoids are mostly 0.3–0.6 microns across, corresponding to the diameter of the granum, but some of them extend throughout the stroma and connect the

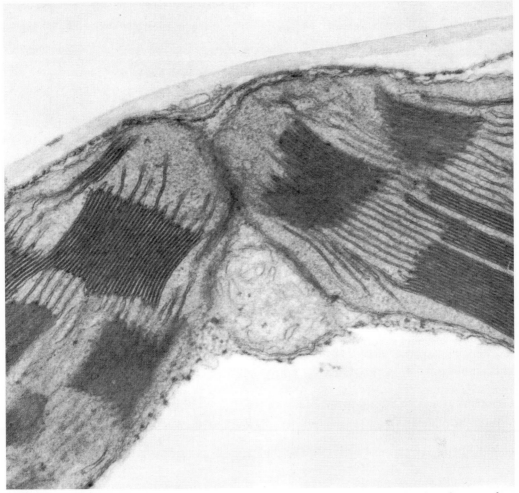

Fig. 7.4 Portions of two chloroplasts from cotyledon of tomato, with a mitochondrion in the angle between them. Several grana and thylakoids connecting adjacent grana can be seen. (× about 53,000.) [Electron micrograph courtesy of Thomas E. Jensen.]

grana to each other. The thylakoids can readily be recognized in electron micrographs, but they are much too thin to be distinguished by light microscopy. The number of thylakoids in a granum ranges from as low as 2 to as high as 100, but is most commonly about a dozen or so.

An individual thylakoid (Fig. 7.5) appears in cross section to consist of two similar, slightly separated layers. In three dimensions, however, the two layers are continuous to form a closed unit, roughly comparable to a nearly empty sack with sealed ends, or to a nearly empty balloon. The space between the two layers is filled mainly with water and solutes.

Each of these two apparent layers of the thylakoid has a complex structure with several sublayers. The organization may be roughly compared to that of a bread

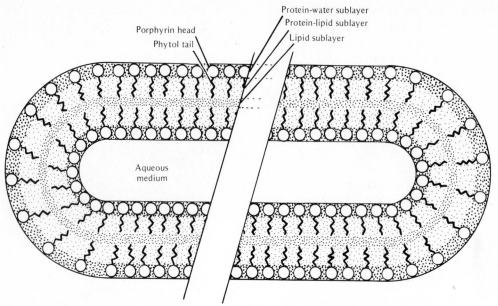

Fig. 7.5   Diagrammatic representation of a cross section of a thylakoid.

and butter and jam sandwich. The outer sublayer on each side (in the position of the bread) is an aqueous (watery) medium containing proteins and some other dissolved substances. A protein-lipid sublayer lies inside the protein-water sublayers (in the position of the butter on each slice of bread). In the center is a sublayer of lipid (in the position of the jam).

It may reasonably be supposed that the hydrophilic porphyrin head of the chlorophyll molecule is in the protein-water part of the thylakoid, and that the hydrophobic phytol tail anchors it in the protein-lipid part. The carotenoid molecules, being soluble in lipid but not in water, are presumably in the lipid layer.

The chloroplast also contains a partial genetic system, with fine strands of DNA, comparable to bacterial "chromosomes", permeating the stroma. These are most easily demonstrated in the proplastid stage, but they persist throughout the life of the chloroplast. Ribosomes are also scattered through the stroma, and it is presumed that these are manufactured within the chloroplast under the control of the chloroplast DNA. It is of some interest that in some structural details the ribosomes of chloroplasts are more like bacterial ribosomes than like the ribosomes of the hyaloplasm and endoplasmic reticulum. The chloroplasts are not wholly independent genetically, however, because their structure and activity are influenced by both nuclear genes and chloroplast genes. Mitochondria, discussed in Chapter 8, have a similarly mixed genetic system.

The demonstration that chloroplasts have their own partial genetic system, combined with the knowledge that they reproduce by division (see Chapter 4), has led some botanists to speculate that they originated as **endosymbionts,** that is, single-celled algae that were captured by protozoa and took up residence inside the protozoan cell. There are, however, formidable difficulties to this hypothesis, as we shall see in Chapter 12.

Algal chloroplasts are somewhat less highly organized than chloroplasts of vascular plants. They have thylakoids that may be organized into small stacks, but they have no evident grana. Some students prefer to say that the whole algal chloroplast consists of a single granum. In some of the more archaic green algae, such as *Chlamydomonas,* the chloroplast occupies more than half the volume of the cell and probably contains a considerable proportion of nonphotosynthetic substance. In the blue-green algae, which are prokaryotic, the thylakoids are scattered through the outer region of the protoplast. In spite of these differences, the individual thylakoids in the various groups of algae appear to be similar in structure and composition to those of higher plants, with the reservation that there are some differences among the secondary chlorophylls and accessory pigments.

Chloroplasts of many of the algae contain a special, proteinaceous, starch-accumulating center, called a **pyrenoid.** With the notable exception of the Anthocerotopsida and Isoëtopsida, these are completely lacking from the chloroplasts of higher plants. Under conditions of rapid photosynthesis, starch does accumulate in chloroplasts of higher plants, but it is not associated with anything resembling a pyrenoid, and it is usually transferred out of the chloroplast as soon as conditions permit. It has been suggested that the pyrenoid is the first evolutionary answer to the problem of removing soluble photosynthate (such as glucose) from the photosynthetic reaction system. With increasing evolutionary specialization in the structure of the chloroplast, associated with increasing efficiency in photosynthesis, the pyrenoid itself becomes an anachronism that must be disposed of—as out of place as bare-footed peasants in a modern winery.

Quantasomes and Reaction Centers

Considerable attention has been devoted to the question of the size of the ultimate photosynthetic unit in the chloroplast. Just how many molecules, and of what kinds, must be associated in what sort of organization to permit the transfer of absorbed light energy from chlorophyll to the next compound in the reaction system?

The answer appears to be that the ultimate photosynthetic unit is a segment of a lamella, containing some 200 to 300 molecules of chlorophyll *a* and a usually much smaller number of molecules of accessory pigments. Apparently only a single one of these 200 to 300 molecules of chlorophyll can actually pass absorbed light energy on to the next agent in the reaction. The remainder of the chlorophyll, called *bulk* chlorophyll, absorbs light energy both directly and from accessory pigments, but it must pass that energy through the single "active" molecule in order to have any photosynthetic effect. The "active" molecule of chlorophyll is always (in green plants) a special form of chlorophyll *a.* The bulk chlorophyll consists of the remainder of the chlorophyll *a,* plus any secondary chlorophylls that are present. The bulk chlorophyll and accessory pigments have been compared to an energy-gathering **antenna** for the active chlorophyll. The active chlorophyll can absorb light energy directly, but it gets most of its supply from the other molecules.

The ultimate photosynthetic unit, containing probably some 200 to 300 molecules of chlorophyll plus all the associated light-gathering apparatus, is called a **quantasome.** An individual thylakoid (lamella) of a chloroplast is made up of many such quantasomes, presumably organized into only one or two layers. There are separate quantasomes, of similar size, for Systems I and II.

The single active molecule of chlorophyll a in a System I quantasome has an *in vivo* absorption peak near 700 nm and is commonly called chlorophyll a 700 or P 700 (the P for pigment). The comparable active molecule of chlorophyll a in a System II quantasome has an *in vivo* absorption peak near 680 nm and may be called chlorophyll a 680 or P 680. In a later paragraph we shall see that the relatively long wavelength of the absorption peak of the single active molecule of chlorophyll a in the quantasome, in comparison with the other chlorophylls, correlates with its function as the receptor of light energy gathered by the antenna (bulk chlorophyll and accessory pigments).

In electron micrographs of disrupted thylakoids the quantasome (or what has by some students been thought to be the quantasome) appears as an oblate spheroid about 10×20 nm. These electron micrograph figures are in the right size range to provide for the proper number of chlorophyll molecules.

There is some uneasiness among students of photosynthesis about the existence of quantasomes as physically stable structures with distinct limits. They may actually be confluent and have only functional rather than morphological reality. Therefore, many investigators prefer to think of each molecule of P 700 or P 680 as a **reaction center** surrounded by an antenna with indefinite boundaries.

The photosynthetic bacteria, being prokaryotic, do not have organized chloroplasts, but they do have observed quantasomes in the same general size range as those of higher plants. According to the kind of bacterium, these quantasomes may be scattered singly through the protoplast, or they may be aggregated in groups of two or more. In some, such as species of *Rhodospirillum*, an aggregate of these quantasomes has the form of a small lamella.

THE PHOTOSYNTHETIC PROCESS

Any attempt at the present time to present a coherent picture of the photosynthetic process at the molecular and submolecular level must consist of a mixture of established fact, probability, informed speculation, and guesswork. There is no comprehensive theory that provides for all the known facts and explains the complete series of chemical reactions and electron transport.

It is customary and convenient to think about photosynthesis in terms of the **light reactions** and **dark reactions.** The light reactions are the early steps in the process, which are so closely tied to the absorption of light that they stop immediately when the light is shut off. Some of the important products of the light reactions, notably ATP and NADPH, are used in the dark reactions to take up carbon dioxide and produce complex new substances, typically glucose. The dark reactions do not require darkness, but because they are several steps removed from the light-gathering process, they can continue for some time after the light has been shut off. (The nature of NADPH is explained on p. 110.)

The Light Reactions

The splitting of water is a fundamental early step in green plant photosynthesis. Two molecules of water yield one molecule of oxygen (O_2), four protons (hydrogen ions), and four electrons. It is not at all clear how this is accomplished. Current thinking implicates an unknown compound called (for the sake of a name) Z, which scavenges the electrons from the hydrogen atoms in water and passes them on to P 680 at the reaction center in System II photosynthesis. Manganese ions are somehow involved in the process. The oxygen molecules are released into the medium, from which they eventually escape as a gas. The hydrogen ions (or others like them) are eventually taken on by NADP, along with electrons that have passed through both System II and System I, to form NADPH. The electrons also govern the formation of ATP during their passage through System II. Eventually the hydrogen is passed into newly formed carbohydrate, formed in the dark reactions.

We shall say no more here about the evolution of oxygen from water that has been split, but it is possible to give a fairly good account of electron transport in the light reactions.

The electrons in chlorophyll and its associated pigments are conceived to be in a state of continuous motion, as indeed are all electrons in all substances. A rough comparison might be made between such electrons and a set of tireless horses racing endlessly around a track, each horse following its leader and being trailed by another. When a photon (a quantum of light) hits one of these electrons in chlorophyll or an accessory pigment, the electron absorbs energy and is said to be *excited*. In this excited state it jumps the track and assumes an orbit with a larger radius. It is convenient to refer to the whole molecule that bears such an excited electron as being excited.

Excited molecules in general may give up their excitation energy in any of several ways. The two that primarily concern us in the study of photosynthesis are (1) the transfer of the excitation energy from an electron of one molecule to an electron of an adjacent molecule and (2) the migration of the excited electron itself from one molecule to another. We shall see that both of these processes take place, at different stages, in the photosynthetic pathway.

The energy per quantum of light varies according to the wavelength; the longer the wavelength, the less energy per quantum. In general a package of excitation energy (called an **exciton**) can be transferred from one molecule to another (without actual flow of electrons) only if the absorption peak of the acceptor molecule is at the same or greater wavelength than that of the donor molecule; that is, the exciton can be transferred only to a molecule that requires the same or a lesser amount of energy to become excited. Similar principles, relating to oxidation-reduction potentials, govern the transfer of the excited electrons themselves. There is a sort of molecular peck order that determines the direction of electron flow.

Some excitation energy is lost each time an excited electron is transferred from one kind of temporary host molecule to another. Under proper conditions some of this energy lost in transfer can be used to facilitate some other chemical reaction that requires an input of energy; otherwise it is generally lost as heat. Sooner or later the excited electron loses all of its excitation energy and returns to the normal or "ground" level of energy.

One way that these facts, and some others, can be fitted into a reasonably coherent scheme is shown in Fig. 7.6, which is explained in the following paragraphs. The reader should understand that some parts of the scheme are subject to change, and that through the use of zigzag arrows we pass over certain series of processes that are not yet well understood or are inconvenient to present here. Nevertheless, the diagram provides a convenient focus for an attempt to present a coherent picture of some of the things now known or believed about photosynthesis.

Typical photosynthesis, leading to the production of carbohydrate, requires that light energy be transformed into two different forms of chemical energy. One is the energy stored in ATP; the other is the energy carried by NADP or (in bacteria) by NAD when hydrogen and excited electrons are attached to it. The nature of NAD and NADP are discussed in subsequent paragraphs. The important thing at the moment is that a supply of hydrogen ions as well as a supply of excited electrons is necessary to convert them into carriers of energy. In bacteria the hydrogen is obtained from sources such as H_2S, from which it is relatively easily extracted. In green plants the hydrogen is extracted from water by the input of energy from System II. The essential function of System II is to provide the necessary supply of hydrogen. We shall see that System II also generates ATP, but that is in a sense an ancillary benefit; the only thing System II contributes that System I cannot do by itself is to split water as a source of hydrogen. For purposes of discussion, it is convenient to begin with System II.

In System II of photosynthesis, excitons are transferred, without actual electron flow, from the antenna to the molecule of P 680 (chlorophyll *a* 680) at the reaction center. Any individual exciton may have been received (as a quantum of light) by an electron on a molecule of bulk chlorophyll or by an electron on one of the accessory pigments, but the important thing from our standpoint just now is that it reaches the molecule of P 680 at the reaction center. According to the standard thermodynamic principles previously noted, the exciton migrates to pigments with absorption peaks at progressively longer wavelengths. P 680, having the longest wavelength of all the light-gathering pigments in the System II quantasome, acts as an energy trap, or sink, for excitons. An electron of the molecule of P 680 at the reaction center may also be excited directly by receiving a photon (or so we believe), but because of the much greater number of molecules of bulk chlorophyll and accessory pigments these molecules receive most of the energy. They form, as we have noted, the light-gathering antenna for the single molecule of P 680 at the reaction center.

Once the excitement energy has reached P 680, the method of its transfer changes. P 680 emits a stream of excited electrons, rather than passing on the excitement energy without electron transfer. These electrons have been obtained, as we have noted, by robbing them from hydrogen during the splitting of water. The excited electrons are passed on from P 680 to some other molecule that cannot absorb light directly, and then from one kind of molecule to another, in a molecular peck order determined in part by what is called redox potential (relative ability to reduce another substance by passing an electron on to it). The sequence appears to be: a substance called Q, plastoquinone, cytochrome *f*, plastocyanin, and P 700 of System I. Quinones are fairly small mole-

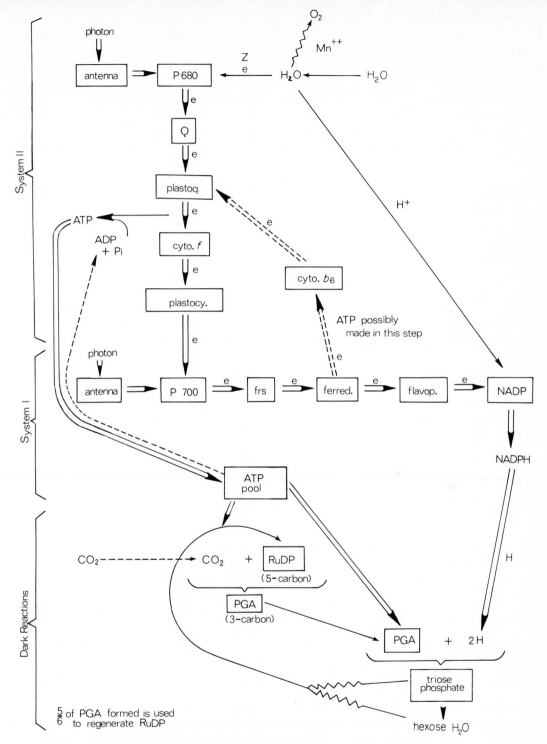

Fig. 7.6 Diagram showing some features of photosynthesis, partly speculative.

cules consisting of one or more benzene rings that collectively have two atoms of attached hydrogen, often also with hydrocarbon side chains. For the chemically inclined student they may be defined as diketones in which the carbon atoms of the carbonyl group are part of the ring structure. Q is a quinone chemically allied to plastoquinone. Plastocyanin is a copper-containing protein.

In passing from plastoquinone to cytochrome *f*, the traveling electron gives up some of its excitation energy in a package that is captured in the formation of ATP from ADP and phosphate ion. There is probably another point along the way where ATP is formed, but it has not been securely identified.

The formation of ATP in association with electron transport in photosynthesis is called **photophosphorylation.** Being several steps removed from the actual trapping of light by chlorophyll and carotenoids, photophosphorylation can continue briefly in the dark after the light supply is shut off. Thus it might technically be considered a dark reaction, but for the purposes of discussion it is more conveniently treated with the light reactions.

The ATP formed by photophosphorylation can at least theoretically be used in any of several ways. The individual molecule of ATP is not branded and herded irrevocably to a particular corral. Because of their simple spatial position, however, most of the ATP molecules formed by photophosphorylation generally channel their energy into subsequent photosynthetic reactions, these necessarily being "dark reactions". The most important of these dark reactions is the set, known as the **Calvin cycle,** which leads to the formation of carbohydrate.

Postponing a consideration of the dark reactions, we turn now to System I.

Knowledge of the existence of two cooperating systems in green plant photosynthesis stems from a series of experiments first reported in 1956 by the American plant physiologist Robert Emerson (1903–1959, Fig. 7.7). He found that exposure of chloroplasts to light of two different wavelengths in the proper parts of the spectrum yielded more photosynthate than the same total amount of light at either wavelength alone. Specifically, he found that providing light in both the red (650–685 nm) and the far red (above 685 nm) portions of the spectrum gives a higher yield than either one alone. The sum of x plus x, provided in two different wavelengths, was greater than the sum of

Fig. 7.7   Robert Emerson (1903–1959), whose work led to the discovery of System II photosynthesis. [Photo courtesy of University Archives, University of Illinois.]

x plus x all in the same wavelength. This **enhancement effect,** as it was called, led to the thought that two different kinds of chlorophyll, with somewhat different absorption peaks, cooperate in photosynthesis. Further experimentation and thought by Emerson and others led to the present concept of Systems I and II.

The linkage of System II to System I here presented is known as the **series formulation.** It envisages the initial excitation of electrons in System II, the stepwise loss of excitation energy as the electrons travel through a series of acceptors, and delivery of the partially "spent" electrons, lacking much of their excitement energy, to P 700 of System I. Taken into System I, the electrons are boosted to a higher level of excitement and emitted into one or another of two series of reactions that we shall discuss. The series formulation emerged in the mid 1960s from the work of a number of different students of photosynthesis. Highly controversial in the beginning, it has now become widely accepted.

The light-gathering steps of System I are essentially similar to those of System II. Carotenoids, chlorophyll b, bulk chlorophyll a, and sometimes phycobilins form an energy-gathering antenna that eventually transmits absorbed light energy to a single molecule of chlorophyll at the reaction center. In System I this molecule at the reaction center is P 700 (chlorophyll a 700).

The molecule of P 700 at the reaction center in System I emits a stream of excited electrons that are passed in a series from one acceptor to another. Probably the first such acceptor is a ferredoxin-reducing substance of unknown nature, now called simply frs. From there the electron goes to ferredoxin (a nonheme iron protein). From ferredoxin it has two

possible pathways: one leads through a series of acceptors back to P 700, producing two molecules of ATP along the way. The other leads into a flavoprotein and thence into NADP.

If the electron takes the route leading back to P 700, it is passed from ferredoxin to cytochrome b_6, thence probably to plastoquinone, thence to cytochrome f, thence probably to plastocyanin, and finally back to P 700. Possibly there are other as yet unknown intermediaries. In one or more of the steps along this route, the traveling electron promotes the formation of a molecule of ATP from ADP and phosphate ion, that is, it passes some of its excitement energy into the formation of ATP. The most probable sites for the formation of this ATP are the steps between ferredoxin and cytochrome b_6 and between plastoquinone and cytochrome f. This route is called **cyclic photophosphorylation,** because it delivers the electron back to the P 700 whence it came. In contrast, the **noncyclic photophosphorylation** of System II is associated with a one-way flow of electrons into System I.

Cyclic photophosphorylation is the standard way to make ATP in ordinary bacterial photosynthesis. It is much less important in green plant photosynthesis, occurring mainly when System II does not deliver a sufficient flow of electrons to absorb all the excitation energy of System I. If System II is somehow depressed or shut off, System I can turn its energy into cyclic photophosphorylation.

If the electron emitted from P 700 does not take the route of photophosphorylation, it is passed on through a flavoprotein to a molecule of NADP. The NADP also takes on hydrogen ions that were formed by the splitting of water in System II, producing NADPH.

At this point we must discuss something of the nature of NADP. NADP is a conventional acronym, not a chemical formula, for a compound known chemically as nicotinamide-adenine dinucleotide phosphate. It differs from NAD, nicotinamide-adenine dinucleotide (the structure of which is shown in Fig. 8.1), in having three instead of two phosphate groups in the molecule. In its oxidized or rest state, before accepting additional electrons and hydrogen, it bears a single positive charge. It is sometimes useful to indicate this charge with a plus sign, $NADP^+$. In its role as a carrier of energy, a molecule of $NADP^+$ takes on two electrons and one proton (hydrogen ion); in this reduced state it is electrically neutral and may be called NADPH. In terms of balances in the photosynthetic process, the two electrons taken on by $NADP^+$ are derived originally from the two hydrogen atoms in a molecule of water. One of the two protons (hydrogen ions) formed in the splitting of a molecule of water is in effect reunited with its external electron when it is taken on by $NADP^+$ in the formation of NADPH. The other proton is eventually reunited with its external electron when NADPH later gives up two electrons and one proton to an acceptor molecule of some other kind. The acceptor molecule accepts the two electrons and one proton from the NADPH, plus one proton from the surrounding medium. In effect, the acceptor molecule accepts two hydrogen atoms, each of which is carrying excitation energy in its external electron. (Here as elsewhere an individual electron or proton may be diverted to some other channel instead of following the course described, but the balances come out right. Two hydrogen atoms are, in effect, extracted from a water molecule in the first stage of System II, and two hydrogen atoms are delivered eventually into the dark reactions, in which they are used in the formation of photosynthate.)

The light reactions of green plant photosynthesis, as here presented, lead to the formation of ATP, the production of reducing power in the form of NADPH, and the generation of free oxygen (which is a waste product). Systems I and II cooperate to produce these results. Light is the source of energy, and water is the source of the hydrogen attached to NADPH. The use of ATP and the reducing power of NADPH to fix carbon dioxide in the form of carbohydrate (or other complex substances that can be respired) is a function of the dark reactions.

The light reactions of photosynthesis do not necessarily lead to the formation of carbohydrate. The system can be blocked at the point of linkage of the light reactions to the dark reactions. Early attempts by botanists to foster photosynthesis in isolated chloroplasts led, in effect, to the production of the light reactions but not the dark reactions.

The Dark Reactions

The reactions involving the use of ATP and NADPH in the assimilation of carbon dioxide can take place just as well in the dark as in the light, because at this stage the light energy absorbed by the chlorophyll and associated pigments has been converted into chemical energy. These reactions are therefore generally known as the **dark reactions.** They are still considered to be a part of photosynthesis in the broad sense.

There are several pathways by which the hydrogen of NADPH can be used, along with ATP energy, to assimilate carbon from carbon dioxide into newly formed organic compounds. One way leads to the production of glycolate, a relatively simple organic acid with the structural formula

$$HO-\overset{\overset{\displaystyle H}{|}}{\underset{\underset{\displaystyle H}{|}}{C}}-\overset{\overset{\displaystyle O}{\|}}{C}-OH$$

We shall have occasion to refer to glycolate again in connection with photorespiration. A much larger proportion of the available photosynthetic energy generally goes into the formation of glucose, and simplified equations for photosynthesis reflect this process alone.

One of the most important facts about the photosynthetic assimilation of carbon dioxide is that it is a cyclic process. The same kinds of compounds are used and produced over and over again. The only net input is that of carbon dioxide, reactive hydrogen, and chemical energy from ATP. The net output is typically glucose and water. One of the key compounds in this cycle is ribulose diphosphate (often abbreviated RuDP), and the cycle is therefore often called the ribulose diphosphate cycle.

Another important fact about such use of carbon dioxide in the formation of glucose, and one that is now firmly established, is that the carbon dioxide is taken on one molecule at a time, reacting with a 5-carbon compound (ribulose diphosphate) rather than being condensed directly into new carbon compounds. The American biochemist Melvin Calvin (1911   , Fig. 7.8) received the Nobel prize in 1961 as a result of his work in the formulation and proof of this concept, which is further explained below. The ribulose diphosphate cycle is therefore sometimes called the **Calvin cycle.**

In the presence of the proper enzymes (called carboxylases), carbon dioxide re-

Fig. 7.8 Melvin Calvin (1911–   ), student of the dark reactions in photosynthesis.

acts with ribulose diphosphate (a 5-carbon compound), forming a 6-carbon compound that immediately splits into two molecules of a 3-carbon compound, phosphoglyceric acid (commonly abbreviated as PGA). No input of energy is required for these two steps in the process; in fact, some energy is released.

In continuation of the dark reactions, PGA reacts (through enzymatic intermediaries) with NADPH that has been formed in the light reactions, forming a triose (3-carbon) sugar with attached phosphate. This reaction requires energy from ATP. The third phosphate group of the ATP is attached to the triose, carrying some of its energy with it, and ADP is released. Inasmuch as the assimilation of one molecule of carbon dioxide results in the formation of two molecules of PGA, 2 ATP are required to deal with the assimilation products of one molecule of carbon dioxide at this stage.

Two molecules of triose phosphate can be converted into a single molecule of glucose or other hexose sugar without further significant expenditure of energy. Glucose is a typical end product of photosynthesis. In the running explanation of the preceding few sentences we have used the general term triose phosphate rather than more specific terms, because two different kinds of triose phosphate are involved. These are glyceraldehyde-3-phosphate and phosphodihydroxyacetone (keto triose phosphate), which have exactly the same atomic constituents in slightly different spatial arrangements. The enzyme triose phosphate isomerase catalyzes a readily reversible reaction between the two, producing an equilibrium with only a small part of the triose in the aldehyde form. It is the glyceraldehyde-3-phosphate that is converted into glucose.

The regeneration of the ribulose diphosphate that serves as the original acceptor of carbon dioxide is a complex process, involving 3-, 4-, 5-, 6-, and 7-carbon sugars. A considerable part of the triose that is formed in the reaction discussed above is diverted into the production of ribulose diphosphate, rather than converted into glucose, and some of the hexose (probably fructose rather than glucose) is also fed into the process. The published diagrams purporting to represent this cyclic use and regeneration of ribulose diphosphate are so complex that they scarcely seem to belong in an elementary textbook; furthermore, there is evidently some variation in detail among different kinds of plants. Simple arithmetic shows that 5/6 of the PGA that is formed must eventually be used in the regeneration of ribulose diphosphate, rather than being available for other metabolic processes. If all the PGA not required for the regeneration of ribulose diphosphate were converted into hexose, the relationship could be expressed thus:

$$C_1 + C_5 \rightarrow C_5 + 1/6 \text{ hexose}$$

Only one step in the regeneration of RuDP requires the input of energy from ATP. This is the formation of RuDP from its immediate precursor, a pentose (5-carbon) sugar. Adding this necessary energy to the 2 ATP seen to be required at an earlier step, we see that 3 ATP are required in the RuDP cycle to provide for the assimilation of a single molecule of carbon dioxide. This is well within the postulated capacity of ATP production in the light reactions, which even at a minimum should produce one ATP for each of the four electrons that are extracted from water (two molecules of water) for each molecule of carbon dioxide taken on.

In reacting with PGA to form triose, NADPH gives up its extra hydrogen and electrons and reverts to NADP. This NADP is then available again as an acceptor of hydrogen and electrons in the light reactions of photosynthesis. The hydrogen given up by NADPH may or may not go *directly* into the triose. In any case, the hydrogen in typical photosynthate is eventually derived from water and the carbon and oxygen from carbon dioxide.

Inasmuch as the assimilation of one molecule of CO_2 results in the formation of 2 PGA and each PGA takes on 2 H from NADPH, it is evident that 4H (derived originally from molecules of water) are used in dealing with one molecule of assimilated CO_2. It is equally clear that adding 4H to CO_2 does not give us the proportions of these three elements that exist in carbohydrate: We would have

CH_4O_2 rather than CH_2O. This discrepancy is explained by the fact that a net of one molecule of water is produced in the process of converting two molecules of PGA into one molecule of hexose. Omitting extraneous complications, the net result is

$$CO_2 + 4H \rightarrow (CH_2O) + H_2O$$

Thus two molecules of water (providing the 4H) must be split to reduce one molecule of CO_2, but one molecule of water is formed in the reduction of PGA. One molecule of water is produced for every two that are split. This fact may be expressed by modifying the simple equation

$$6CO_2 + 6H_2O \rightarrow C_6H_{12}O_6 + 6O_2$$

to read

$$6CO_2 + 12H_2O \rightarrow$$
$$C_6H_{12}O_6 + 6H_2O + 6O_2$$

We shall see in a subsequent paragraph that the necessity to extract four atoms of hydrogen from water to reduce one molecule of carbon dioxide relates to the fact that eight quanta of light are necessary for the fixation of one molecule of carbon dioxide.

We should note at this point that the surplus PGA formed in the RuDP cycle is not necessarily all converted into glucose; PGA is a normal intermediate in the system of glucose utilization. Instead of reacting with NADPH, it can continue in the normal respiratory cycle, with the release of energy into the metabolic system, leading eventually to the formation of water and carbon dioxide. Or it can be used as a building block in the formation of amino acids, fats, and other protoplasmic constituents. Under starvation conditions, the surplus PGA formed in the dark reactions may be largely used in respiration and the formation of protoplasmic constituents, with little or no accumulation of glucose. We must also remember that 5/6 of the PGA that is formed in the dark reactions is used eventually to regenerate RuDP; it is only the other 1/6 (referred to above as surplus) that is available for any other use at all.

The carbon dioxide used in photosynthesis is absorbed from the air (or from the water, by submerged aquatic plants) by diffusion and taken into solution in the cytoplasm. In higher plants it typically enters the leaf through minute openings called stomates, as described in Chapter 23. In most plants the carbon dioxide dissolved in the cytoplasm reacts more or less directly with RuDP, as described. In a scattering of kinds of higher plants, however, the carbon dioxide reacts initially with 3-carbon acids in the cytoplasm, forming 4-carbon acids. These eventually revert to the 3-carbon form by giving up carbon dioxide, which enters the Calvin cycle in the normal manner. This 4-carbon route for carbon dioxide has two variants, conventionally known as crassulacean acid metabolism (CAM) and C_4 photosynthesis. These processes are discussed in Chapter 24.

Glucose, the principal end product of typical photosynthesis, is highly soluble in water. The accumulation of any considerable amount of glucose in the cell would therefore upset osmotic balances and tend to disrupt the metabolic system. We should therefore not be surprised to find, as we do, that in most plants glucose is converted into insoluble storage products as fast as it is formed. The most common of these storage products is starch. Some kinds of plants, however, especially some of the algae, produce oils or other kinds of storage products instead of starch.

We have noted in an earlier paragraph that in many of the algae conversion of glucose into starch takes place in a special part of the chloroplast, the pyrenoid. The chloroplasts of higher plants in general do not have pyrenoids, and any starch storage in the chloroplast is strictly temporary. The synthesis of starch from glucose and its subsequent storage take place mainly in leucoplasts of a special kind, called *amyloplasts*. The chloroplasts retain the mechanism for starch formation, however. Under proper experimental conditions isolated chloroplasts can carry on the complete photosynthetic process, from the reception of light to the formation of glucose, and can then turn that glucose into starch.

We have for convenience and in accordance with custom discussed the RuDP cycle of carbon dioxide assimilation as a part of photosynthesis. Such a viewpoint is satisfactory when one is dealing only with algae and higher plants. It is less useful in dealing with the bacteria. Many bacteria can assimilate carbon dioxide and produce carbohydrate without using light energy. The necessary energy is obtained by manipulating various sorts of compounds already present in the substrate. Some of these bacteria do not have the RuDP cycle and require an organic substrate, such as acetate, as a necessary ingredient for the assimilation of carbon dioxide. Others, such as *Micrococcus denitrificans* and *Thiobacillus denitrificans*, have the complete RuDP cycle for carbon dioxide assimilation. ATP is the source of energy, just as in higher plants, and NADP furnishes the reactive atoms of hydrogen. Yet these bacteria do not use light energy, and there is no indication that their ancestors ever did so. For these and other reasons it has been suggested that the RuDP cycle for carbon dioxide assimilation originated piecemeal in ancient bacteria, before the origin of photosynthesis itself. The attachment of the RuDP cycle to photosynthesis was a great advance, making possible the efficient use of photosynthetic energy without dependence on complex substrates that might be in short supply. The photosynthetic bacteria and some of the more archaic algae can be induced to revert to the use of acetate in assimilating carbon dioxide, but under ordinary conditions they use the more efficient RuDP cycle almost exclusively.

BACTERIAL PHOTOSYNTHESIS

In our discussion of photosynthesis by green plants we have referred repeatedly to features in which bacterial photosynthesis is different. It seems appropriate to summarize the similarities and differences here.

Photosynthesis in bacteria resembles green plant photosynthesis in that chlorophyll, carotenoid pigments, and cytochrome cooperate to transform light energy into chemical energy that is used to assimilate carbon dioxide by the RuDP cycle and produce new organic compounds, typically carbohydrates. It further resembles green plant photosynthesis in that much of the captured light energy is transferred into ATP by photophosphorylation; NAD (differing from NADP in having one less phosphate group) serves as an intermediate hydrogen acceptor in bacterial photosynthesis, just as NADP serves in green plant photosynthesis. A quinone participates in bacterial photosynthesis, possibly in the same way as in green plant photosynthesis. The excited electrons emitted from the reaction center in bacteria have two possible routes, just as in System I of green plants.

One route leads to cyclic photophosphorylation, which is the only method of photophosphorylation in typical bacterial photosynthesis. The other route leads to the formation of reducing power and the acceptance of hydrogen by NAD. All of these similarities reflect the fact that System I photosynthesis in bacteria has been carried over into higher plants, with relatively minor changes.

The principal differences between bacterial photosynthesis and green plant photosynthesis reflect the fact that the bacteria lack System II. Bacteria do not split water in photosynthesis. Bacterial photosynthesis therefore does not result in the evolution of oxygen. It furthermore requires a suitable hydrogen donor, such as gaseous hydrogen, hydrogen sulfide, or various other sulfur or organic compounds. The hydrogen provided by this external hydrogen donor is the eventual source of hydrogen for bacterial photosynthesis.

It may be of some interest that several genera of blue-green, green, red, and brown algae have been shown to retain the ancestral capacity to use substances other than water (such as H_2S or H_2) as hydrogen donors in photosynthesis, under special anaerobic conditions. Under these conditions no oxygen is produced.

We have already noted that bacteria do not have highly organized chloroplasts comparable to those of most algae and higher plants. Only the light reactions of photosynthesis (and indeed only System I) are clearly associated with the chlorophyll-bearing organelles of bacteria. The dark reactions in photosynthetic bacteria have been reported to take place away from these chlorophyll-bearing organelles, but the matter is still controversial. In any case, the photosynthetic bacteria do carry on the dark reactions; the controversy is about *where* they do it.

The pigments in bacterial photosynthesis are comparable but not identical to those in green plant photosynthesis. Bacteriochlorophyll appears to be the functional homolog of chlorophyll *a*. Its absorption spectrum is notably different, however, in that the peak toward the red end of the spectrum is shifted into the infrared, with a maximum between 800 and 900 nm. Other chlorophylls in addition to bacteriochlorophyll are present or not, according to the kind of bacterium. Carotenoid pigments are regularly present in photosynthetic bacteria, but they are chemically slightly different from those of higher plants. There is no reason to suppose that the functions of bacterial carotenoid pigments in photosynthesis are significantly different from those of the carotenoids of higher plants.

Several other enzymes in bacterial photosynthesis are slightly different from their homologs in green plant photosynthesis. Photosynthetic electron transport in bacteria involves a *c*-type cytochrome different from the cytochrome *f* of green plants. The NAD of bacterial photosynthesis has one less phosphate group than the NADP of green plants. The quinone of bacterial photosynthesis is ubiquinone rather than plastoquinone.

A unique kind of photosynthesis carried on by *Halobacterium* is discussed in Chapter 11.

QUANTUM REQUIREMENTS

It is now generally agreed, after long controversy, that eight quanta (ultimate, indivisible units of light energy) are required to reduce one molecule of carbon dioxide and produce one molecule of oxygen in green plant photosynthesis. It appears that this requirement is split four and four: four quanta for System I and four for System II. The series formulation is admirably adapted to explaining a balanced quantum requirement in Systems I and II, inasmuch as it visualizes the excitation of electrons in System II, and their delivery to System I for re-excitation.

Presenting the photosynthetic reduction of carbon dioxide in its simplest terms, four hydrogen atoms are required to reduce one molecule of carbon dioxide:

$$CO_2 + 4H \rightarrow (CH_2O) + H_2O$$

Each of these four hydrogen atoms is temporarily dissociated from its external electron when water is split in System II. Each of the four dissociated electrons is boosted to a higher energy state by a quantum of light, transferred through chlorophyll, in System II and again in System I. Each direct photochemical event requires only one quantum (as ought to be true on photodynamic princi-

ples), and a total of eight quanta is required to incorporate one molecule of carbon dioxide into carbohydrate.

SUGGESTED READING

Asimov, I., *Photosynthesis*, Basic Books, New York, 1968. A brilliant exposition by a master popularizer, with emphasis on the dark reactions.

Bolin, B., The carbon cycle, *Sci. Amer.* **223**(3):124–135, September 1970.

Cloud, P., and A. Gibor, The oxygen cycle, *Sci. Amer.* **223**(3):110–123, September 1970.

Govindjee, and R. Govindjee, The primary events of photosynthesis, *Sci. Amer.* **231**(6):68–82, December 1974.

Haupt, W., Role of light in chloroplast movement, *BioScience* **23**:289–296, 1973.

Krogman, D. W., *The biochemistry of green plants*, Foundations of Modern Biochemistry Series, Prentice-Hall, Englewood Cliffs, N. J., 1973.

Miller, K. R., The photosynthetic membrane, *Sci. Amer.* **241**(4):102–113, October 1979.

Chapter 8

Respiration

Western skunk cabbage, *Lysichiton kamtschatkensis*. This species and a number of its relatives respire so rapidly as they come up in the spring that the internal temperature of the inflorescence may reach 40° or 45°C, several degrees higher than the normal human body temperature (37°C). Skunk cabbage blooms at a time when the air temperature often drops below freezing, and the high rate of respiration may be of some importance in helping the plant to withstand low temperatures. [Photo by W. H. Hodge.]

INTRODUCTORY SUMMARY

We have seen that light energy is captured by green plants and transformed into chemical energy that is bound up in foods such as glucose that have been made from raw materials. This anabolic process is called photosynthesis. Foods are also broken down within living cells, releasing chemical energy (and heat) into the metabolic system. This catabolic process is called **respiration.** Respiration is always, in a chemical sense, an oxidation-reduction reaction, in which food is oxidized and some other substance—typically oxygen—is reduced. Photosynthesis and respiration collectively form the principal mechanism by which light energy is transformed into metabolic energy.

Respiratory energy is transmitted into the metabolic system chiefly in two forms, (1) the high-energy bond linking the third phosphate group into adenosine triphosphate (ATP) and (2) the reducing power of reduced pyridine nucleotides (NADH and NADPH) (Fig. 8.1). NAD and NADP are not chemical formulas, but conventional abbreviations for complex substances more formally known as nicotinamide-adenine dinucleotide and nicotinamide-adenine dinucleotide phosphate.

ATP transfers energy by giving up its third phosphate group to some other compound. The bond by which this third phosphate group is attached to the ATP

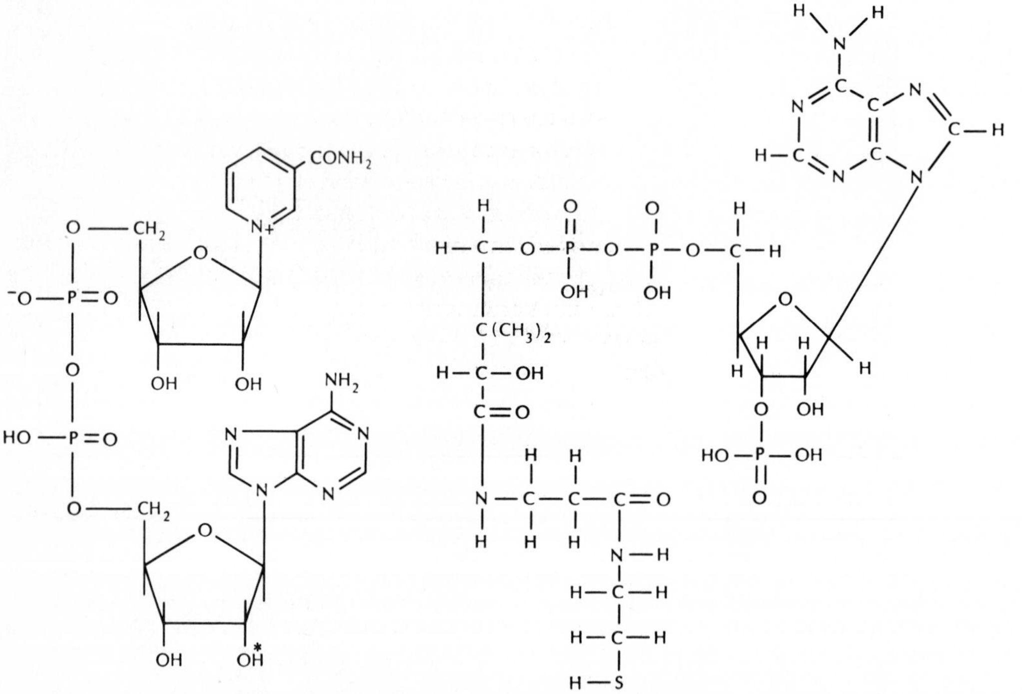

Fig. 8.1   Structure of NAD (A) and Coenzyme A (CoA) (B).

molecule is a high-energy bond, often represented by the symbol ~.

NADH and NADPH transfer energy by passing a pair of reactive hydrogen atoms to a hydrogen acceptor, which is thereby reduced. In the mitochondria NADH is coupled to the cytochrome system for the efficient production of ATP, but elsewhere in the cytoplasm NADH and NADPH transfer some of their energy (in the form of reactive pairs of hydrogen atoms) more directly into anabolic processes.

Respiration also provides carbon-containing compounds which can be used as building blocks for other protoplasmic constituents. For example, acetyl-CoA (a 2-carbon compound attached to Coenzyme A) is an intermediate product in typical respiration of both glucose and fats. The acetyl units of acetyl-CoA can be further broken down (oxidized), with further release of respiratory energy, or some of them can be shunted into anabolic processes.

As here defined, respiration includes both **aerobic respiration,** in which free oxygen is used and foods are completely broken down to such products as water and carbon dioxide, and **anaerobic respiration,** in which free oxygen is not used and foods are only partially broken down, leaving at least one end product that can be further respired aerobically. Anaerobic respiration is often also called **fermentation.** A familiar anaerobic respiratory process, carried out by yeasts, uses glucose to form ethyl alcohol and carbon dioxide. Ethyl alcohol can be respired aerobically by other organisms (including man) to form water and carbon dioxide.

Readers who have previously been exposed to some biological chemistry may note that the term respiration is often used in a narrower sense, to include only what is here called aerobic respiration. For our purposes the broader definition is more useful, inasmuch as no other term is available for the necessary concept of respiration as a means of converting stored energy into metabolic energy. We may also note that the principal steps in some of the most familiar kinds of anaerobic respiration are identical to the early steps in typical aerobic respiration.

The most typical respiratory substrates (foods used in respiration) are hexose monosaccharides, such as glucose. Many other substances, such as pentose monosaccharides, glycerol, fatty acids, amino acids, alcohol, and acetic acid, can also be respired, usually by being converted into respiratory intermediates.

With regard to substances used and products formed, typical photosynthesis (leading to the formation of glucose) and typical respiration (in which glucose is aerobically respired) may be regarded as exact opposites. The equations are:

Photosynthesis:

$$6CO_2 + 6H_2O + \text{light energy} \rightarrow$$
$$C_6H_{12}O_6 + 6O_2$$

Respiration:

$$C_6H_{12}O_6 + 6O_2 \rightarrow$$
$$6CO_2 + 6H_2O + \text{chemical energy and heat}$$

These simplified equations mask a set of essential differences in the actual processes of photosynthesis and respiration. The series of steps by which glucose is respired is very different from the series of steps by which it is made, although cytochromes, ATP, and pyridine nucleotides are involved in both series.

The most common pathway for respiration of glucose, in both plants and animals, may be considered as consisting of two major parts, which occur in sequence in different parts of the cell, using different sets of enzymes and chemical reactions. The *first part*, carried on in the soluble part of the cytoplasm, consists of a series of reactions known as the **EMP pathway.** It liberates a relatively small amount of energy, which is largely passed into NADH and ATP. The EMP pathway leads to the production of a 3-carbon compound called pyruvate (pyruvic acid) (Fig. 8.2); free oxygen is not used in the process.

The *second part*, carried on chiefly in the mitochondria, consists of a series of reactions known as the **TCA cycle,** closely linked to the cytochrome system. The TCA cycle breaks down pyruvate into carbon dioxide and pairs of energy-bearing, reactive hydrogen atoms. These pairs of hydrogen atoms are passed into the cytochrome system and eventually unite with free oxygen to form water. On their way through the cytochrome system they govern the formation of ATP from ADP and phosphate ion.

Intermediates in the TCA cycle may be diverted at any of several steps in the cycle, entering other metabolic processes. The TCA cycle may also be short-circuited into a related cycle called the **glyoxylate cycle,** which releases a smaller amount of energy and produces some 4-carbon compounds that can be used as metabolic building blocks.

Glucose can also be respired by a completely different route, called the pentose phosphate cycle, with an energy yield comparable to that of the combined EMP pathway and TCA cycle. Here we see a good example of the principle of alternative or **backup systems,** which permeates both plant and animal physiology. An organism that can perform some necessary function in more than one way has a competitive advantage under environmental conditions which inhibit one of those ways but not the other. Possible parallels in human society should not escape attention.

THE EMP PATHWAY

The respiratory breakdown of glucose to pyruvate follows a course now generally known as the EMP, or Embden-Meyerhof-Parnas, pathway, after three of the many biochemists who contributed to its elucidation, Gustave Georg Embden (1874–1933), Otto Meyerhof (1884–1951), and Jakob Karol Parnas (1884–1949). Within a period of about 10 years following 1914 the essential features of the EMP pathway were established. The conversion of glucose to pyruvate via the EMP pathway is often referred to as **glycolysis.**

The EMP pathway (Fig. 8.3) is common to the majority of living organisms, both plant and animal. It occurs chiefly in the soluble part of the cytoplasm, rather than being associated with mitochondria or other organelles.

In the simplest terms, the EMP series of reactions may be summarized:

Fig. 8.2   Structure of pyruvate (A) and glucose-6-phosphate (B).

$$C_6H_{12}O_6 \rightarrow 2CH_3COCOOH + 4H$$

glucose   pyruvate

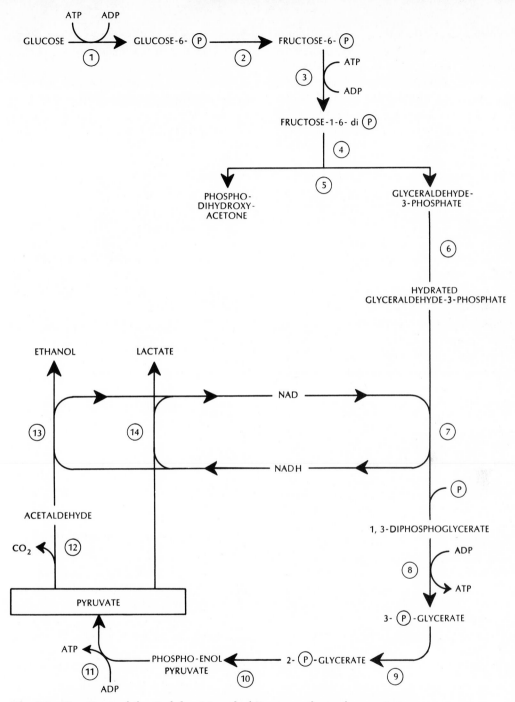

Fig. 8.3 Reactions of the Embden-Meyerhof-Parnas pathway for respiratory breakdown of glucose to pyruvate, showing also the conversion of pyruvate to ethyl alcohol or acetic acid in anaerobic respiration. [Adapted from Beevers.]

The hydrogen atoms shown at the right in this reaction are not released as gaseous hydrogen. Instead they react in pairs with NAD to form NADH. Two molecules of ATP are also formed from ADP and phosphate ion in this set of reactions. Taking account of the immediately related energy-converting reactions, the respiratory breakdown of glucose to pyruvate may therefore be summarized:

$$C_6H_{12}O_6 + 2ADP + 2^-H_2PO_3 + 2NAD \rightarrow$$
glucose   phosphate ion

$$2CH_3COCOOH + 2ATP + 2NADH$$
pyruvate

The reader should remember that in this summary the abbreviations ADP, ATP, NAD, and NADH are used for complex molecules, rather than being chemical formulas as such. There is also a further complexity in the nature of NADH discussed under the cytochrome system on a subsequent page. The eleven steps in the EMP pathway are shown in Fig. 8.3.

THE TCA CYCLE

In most eukaryotes, including all the higher plants (Embryobionta) and higher animals, respiratory breakdown of pyruvate occurs chiefly within the mitochondria by a cyclic pathway variously known as the **citric acid cycle,** the tricarboxylic acid cycle, the **TCA cycle,** or the **Krebs cycle,** after Sir Hans Adolf Krebs (1900– , Fig. 8.4), a German (later British) biochemist who proposed its basic features in 1937. Many prokaryotes (bacteria and blue-green algae) also have the TCA cycle, even though they lack mitochondria, but some others do not have all of the necessary enzymes.

The TCA cycle, in combination with the closely linked cytochrome system, causes the complete oxidation of pyruvate

Fig. 8.4 Sir Hans Adolf Krebs (1900–), who elucidated the TCA cycle.

to form water and carbon dioxide. Free oxygen is required, and energy is liberated. Much of the energy is transferred into ATP, but some of it is lost as heat. Fifteen molecules of ATP can be produced from ADP and phosphate ion through the respiration of one molecule of pyruvate by the TCA cycle and the associated cytochrome system.

The successive reactions of the TCA cycle (Fig. 8.5) may be outlined as follows.

(1) Pyruvate, a 3-carbon compound, is broken down to a 2-carbon (acetyl) compound that is attached to Coenzyme A (Fig. 8.1), forming acetyl-CoA. In the process a molecule of CO_2 is released, a molecule of water is taken on, a phosphate ion is attached to ADP (forming ATP), and a pair of reactive hydrogen atoms are transferred to NAD, forming NADH. Coenzyme A is a complex substance chemically related to the nucleotides, which takes

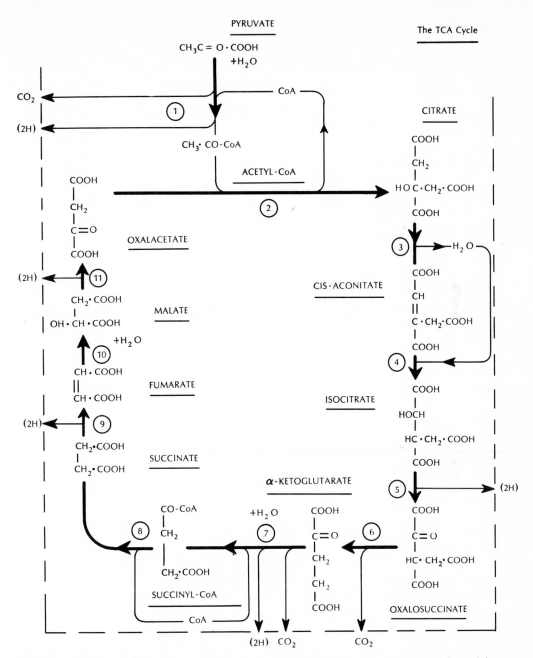

Fig. 8.5 Reactions of the TCA cycle for respiratory breakdown of pyruvate. [Adapted from Beevers.]

part in a number of metabolic reactions, especially some of those involving acetate. This "first step" in the TCA cycle is actually a series of several reactions forming two small cycles that can collectively be considered as one unit in the larger TCA cycle. Several enzymes, forming what is called an oxidative decarboxylation system, are required.

(2) Acetyl-CoA combines with oxalacetate (a 4-carbon compound) forming citrate (a 6-carbon compound) and releasing CoA. The CoA is immediately available to be used again in step 1 or in some other reaction. One ATP supplies the necessary energy for step 2, being converted into ADP and phosphate ion in the process. The requisite enzyme is called citrate synthetase.

(3) Citrate gives up a molecule of water to form cis-aconitate, another 6-carbon compound. No significant energy change occurs in this step. The enzyme aconitase catalyzes this reaction and also the next.

(4) Cis-aconitate takes on a molecule of water to form isocitrate, a 6-carbon compound that has the same components as citrate, but with the atoms differently arranged. No significant energy change occurs in step 4.

(5) Isocitrate is changed to oxalosuccinate, another 6-carbon compound, and another pair of reactive, energy-bearing atoms are transferred to NAD, forming NADH. The requisite enzyme is called isocitrate dehydrogenase.

(6) Oxalosuccinate is converted to α-ketoglutarate, a 5-carbon compound, and a molecule of CO_2 is released in the process. The responsible enzyme is oxalosuccinate decarboxylase, acting in the presence of manganese ions.

(7) α-Ketoglutarate reacts with CoA to form succinyl-CoA, the succinyl part of which is a 4-carbon compound. A molecule of CO_2 is produced in the formation of succinyl-CoA, a molecule of water is taken on, and another pair of reactive hydrogen atoms are transferred to NAD, forming NADH. An oxidative decarboxylation similar to that in step 1 governs the set of reactions in step 7.

(8) Succinyl-CoA releases succinate, another 4-carbon compound, and CoA. The CoA is immediately available again for either of its two reactions in the TCA cycle or for other uses. A molecule of ATP is formed from ADP and phosphate ion in step 8, through the intervention of another phosphate carrier called guanosine diphosphate (GDP). The breakdown of succinyl-CoA to succinate and CoA releases energy which is used to attach a phosphate ion to GDP, forming GTP. The phosphate group is then transferred from GTP to ADP, forming ATP. The enzyme governing step 8 is called succinyl CoA synthetase.

(9) Succinate forms fumarate, another 4-carbon compound, and releases another pair of reactive hydrogen atoms, which are fed directly into the cytochrome system (discussed in subsequent paragraphs) instead of being transferred to NAD. The requisite enzyme is called succinate dehydrogenase.

(10) Fumarate forms malate, another 4-carbon compound, taking on a molecule of water in the process. The governing enzyme is called fumarase. No significant energy change occurs in this step.

(11) Malate gives rise to oxalacetate and releases two more hydrogen atoms (to form more NADH) in the process. The oxalacetate, in turn, is ready to react with another molecule of acetyl-CoA (fed in from the first step in the breakdown of pyruvate), completing the cycle. The requisite enzyme is called malate dehydrogenase.

It will be seen that the oxidation of one molecule of pyruvate, which contains only 4 hydrogen atoms, results in the release of 5 pairs (10 atoms in all) of reactive, energy-bearing hydrogen atoms. The 6 additional atoms of hydrogen (4 + 6 = 10) come from the 3 additional molecules of water that are taken on in steps 1, 7, and 10 of the cycle. The 5 pairs of hydrogen atoms released in one round of the TCA cycle react eventually (after going through the cytochrome system) with 5 atoms of oxygen, forming 5 molecules of water. Thus with 3 molecules of water used and 5 produced, the oxidation of a molecule of pyruvate results in the net production of 2 molecules of water. We have seen that 3 molecules of carbon dioxide are also produced in the TCA cycle. The series of reactions might thus be summarized as

$$CH_3COCOOH + 3H_2O + 2\tfrac{1}{2}O_2 \rightarrow$$

pyruvate

$$3CO_2 + 5H_2O + \text{energy (heat and ATP)}$$

Altogether the energy thus released from one molecule of pyruvate can account for the formation of 15 ATP from ADP and phosphate ion.

Although the primary function of the TCA cycle is evidently the respiratory breakdown of pyruvate, a number of other substances can also be fed into the cycle at appropriate points and respired. Most importantly, acetyl-CoA formed in the preliminary breakdown of fatty acids (the fatty acid spiral, discussed on a subsequent page) can be fed in at step 2. Several amino acids can also be introduced into the TCA cycle. Thus glutamic acid can be converted to α-ketoglutarate, which is formed in step 6 and used in step 7 of the cycle, and aspartic acid is interconvertible with oxalacetate, which is formed in step 11 and used in step 2 of the cycle.

Intermediate products of the TCA cycle may also be diverted into other metabolic processes. Glutamic acid, one of the necessary amino acids, is made from α-ketoglutarate. Isocitrate can be enzymatically split to yield succinate and glyoxylate. The necessary amino acids glycine and serine, as well as some other important metabolites, are made from glyoxylate. Acetyl-CoA is a source of acetate, the building block for fats and some other protoplasmic constituents. Succinate is a necessary substrate for the formation of pyrrolic compounds, including cytochrome and chlorophyll. Citrate, malate, and fumarate also have other metabolic uses. Thus the fate of any particular molecule of pyruvate entering the TCA cycle depends on the circumstances. It is not irretrievably consigned to being converted then and there into water and carbon dioxide.

Diversion of TCA cycle products to other uses will, of course, break the cycle and prevent further inflow of pyruvate, unless the supply of one or another of the requisite acids is replenished from some other source. One such possible source in some lower organisms and in some tissues of higher plants is the glyoxylate cycle, discussed on a subsequent page, which leads to the accumulation of malate. More importantly, there are several mechanisms by which free carbon dioxide is incorporated into compounds convertible into one or another of the TCA cycle acids. For example, phospho-enol-pyruvate, one of the late products in the EMP pathway, may react with carbon dioxide to form oxalacetate, governing the formation of one ATP (or releasing one phosphate ion into solution) in the process. This conversion of phospho-enol-pyruvate to oxalacetate may be the principal source of replenishment for the TCA cycle acids. Pyruvate may also take on carbon dioxide to form malate, using NADH as the source of the necessary hydrogen. All such reactions, of course, require the participation of appropriate enzymes.

The TCA cycle, leading to the production of NADH as a primary carrier of energy, is obligately linked (in eukaryotes) to the cytochrome system, which transfers much of that energy into the high-energy phosphate bond of ATP. The rate of the series of reactions depends heavily on the availability of ADP and phosphate ion. Thus there is a built-in tendency toward self-regulation of the rate of respiration. When ATP is being used rapidly, the supply of ADP builds up, encouraging the oxidation of more pyruvate and the production of more NADH, which (through the action of the cytochrome system) replenishes the supply of ATP at the expense of ADP and phosphate ion. When the use of ATP slackens, the rate of ADP formation diminishes, and respiration slows down. Metabolic self-regulation by such **feedback** mechanisms is a pervasive feature of both plant and animal physiology.

The linkage of the TCA cycle to the cytochrome system is less than perfect, and the self-regulation of the rate of respiration is therefore more complex than the foregoing paragraph might suggest. We shall see in the discussion of the pentose phosphate cycle that there is a mechanism for transferring NADH into and out of the mitochondria. In this as in other aspects of metabolism, the cross-connecting influences, checks, and balances are wondrously complex.

THE CYTOCHROME SYSTEM

The pairs of reactive hydrogen atoms released in the TCA cycle bear the bulk of the energy released in the process. With the exception of the pair released in the conversion of succinate to fumarate, these hydrogen atoms react immediately with NAD. Through the activity of the cytochrome system of enzymes these pairs of energized atoms are then passed from one compound to another until they finally unite with oxygen to form water. At specified reaction points along the way, portions of their energy are used to govern the formation of ATP from ADP and phosphate ion. Each such pair of hydrogen atoms accounts for the formation of three molecules of ATP in passing through the cytochrome system.

We can now explore some of the complexities masked in the relatively simple summary given above. Before reacting with hydrogen, NAD bears a single positive charge. It is sometimes convenient to indicate this charge with a plus sign, NAD^+. When it reacts with a pair of hydrogen atoms, NAD^+ accepts an electron from each hydrogen atom, but only one of the protons (hydrogen nuclei). The other proton is temporarily turned loose into the surrounding medium. After accepting one hydrogen atom and an additional electron, the NAD molecule may be called NADH. NADP, mentioned on earlier pages as an acceptor of hydrogen atoms in some related processes, is similar to NAD in actually accepting two electrons, but only one proton from a pair of reactive hydrogen atoms. During their passage from NAD into and through the cytochrome system, the hydrogen atoms are sometimes intact and sometimes dissociated into protons and electrons. This fact is probably of critical importance in the series of reactions and the formation of ATP. The dissociated proton and electron of a particular hydrogen atom do not necessarily find each other again to form the identical atom, but the same number of hydrogen atoms leave the system as enter it. Thus it is convenient and not misleading to think of the hydrogen molecules as if they maintained their identity throughout the process, even when dissociated into an electron and a proton for each molecule.

The successive carriers of the hydrogen atoms or their electrons in the cytochrome system appear to be as follows: NADH → flavoprotein → iron-sulfur protein (perhaps of more than one kind, in a series) → ubiquinone → cytochrome b → cytochrome c_1 → cytochrome c → cytochrome a → cytochrome a_3 → water. The sequence may actually be more complex, with ubiquinone and cytochrome b forming a sort of internal cycle. Each successive step in the series involves a reduction of the energy of the traveling electrons (or hydrogen atoms), according to standard thermodynamic principles. At each of three steps in the series the energy is released as a "package" large enough to be captured in the formation of ATP. In the other steps the energy is released in packages too small to be captured metabolically and is dissipated as heat.

The precise way in which the traveling hydrogen atoms promote the formation of ATP is still (1981) not entirely clear, and efforts to elucidate the mechanism necessarily invoke physicochemical principles and mechanisms that are beyond the scope of an elementary textbook. Current thinking, embodied in what is called the **chemiosmotic hypothesis,** visualizes the repeated transport of pairs of electrons and their more or less closely associated protons back and forth across a membrane within the mitochondrion. It is thought that pairs of atoms are transported on carrier molecules from within a membrane-enclosed space to an outer region (still within the mitochondrion). There

the protons and electrons of the hydrogen atoms are dissociated and make their way separately back across the membrane into the enclosed space, and the protons govern the formation of ATP on the return journey (one molecule of ATP for each pair of protons on each such journey). Each pair of energized hydrogen atoms is thought to make the journey back and forth across the membrane three times (more or less intact on the outward trip, but as separated protons and electrons on the return trip) before finally combining with oxygen to form water.

The well-known deadly effect of cyanide relates in large part to the fact that it prevents the transfer of hydrogen from cytochrome a_3 to oxygen into water. All the cytochromes become reduced, and the TCA cycle is blocked, in both plants and animals. In animals cyanide also conjugates with the iron in the hemoglobin of the blood, stopping the distribution of oxygen.

Carbon monoxide likewise combines with the iron in the terminal cytochromes of respiration, stopping their transfer of hydrogen and electrons. A concentration of 2 percent of carbon monoxide in the air—a level frequently attained on busy streets—is enough to depress respiration significantly in sugar maples and many other kinds of trees.

About 60 percent of the energy associated with the pair of hydrogen atoms of NADH is transferred into ATP via the cytochrome system. The remaining 40 percent is lost as heat during the transfer.

We noted in an earlier paragraph that the pair of hydrogen atoms released in the formation of fumarate from succinate are fed more directly into the cytochrome system, rather than being delivered to NAD. Ubiquinone appears to be the primary acceptor. Only two molecules of

ATP are produced by the passage of this pair of hydrogen atoms through the cytochrome system. Presumably the production of 2 ATP instead of 3 reflects the fact that the early part of the cytochrome system has been by-passed, and one of the ATP-yielding steps therefore is missed.

ENERGY RELATIONS IN THE EMP PATHWAY AND TCA CYCLE

A little more than 1/6 of the energy stored in glucose is released in breaking it down to pyruvate. Nearly all of this energy is transferred into ATP and NADH with very little being lost in the first instance as heat. Each molecule of glucose can yield a net of 2 ATP and 2 NADH in being broken down to pyruvate. About half of the energy stored in NADH is potentially transferable into ATP (3 ATP for each NADH) via the activity of the cytochrome system. Thus there is a net potential release of 8 ATP in the breakdown of one molecule of glucose to pyruvate.

In more formal thermodynamic terms, each mole (gram molecular weight) of NADH has about 52,000 calories of stored energy associated with the pair of reactive hydrogen atoms. In each mole of ATP the bond attaching the third phosphate ion represents about 8,000 calories.* Thus in the formation of 3 moles of ATP through the action of 1 mole of NADH, about 24,000 calories are transferred into ATP and about 28,000 calories escape as heat.

The free (chemically stored) energy of glucose is calculated at about 674,000 calories per mole. About 120,000 calories

are transferred into NADH (2 moles at 52,000 each) and ATP (2 moles at 8,000 each) in the breakdown of glucose to pyruvate via the EMP pathway. With the ultimate formation of 3 ATP by each molecule of NADH, we get 8 moles of ATP in all, representing a potential 64,000 calories of useful metabolic energy from the 120,000 calories released.

Breakdown of pyruvate in the TCA cycle yields another 15 ATP per molecule. Each of the 5 pairs of hydrogen atoms released in one turn of the cycle results eventually in the formation of 3 ATP, except the pair released in the transformation of succinate to fumarate, which is delivered to ubiquinone and governs the formation of only 2 ATP. That accounts for 14 ATP. The fifteenth is formed via the intervention of GDP (guanosine diphosphate) in the transformation of succinyl-CoA to succinate. One ATP is used in the early part of the TCA cycle, and one is formed in the next step, so that these two cancel out of the balance.

Eight ATP from the breakdown of glucose to pyruvate, plus 15 from each of the two pyruvate molecules, gives a total of 38 ATP for each molecule of glucose that is completely respired to CO_2 and H_2O. Experimental efforts to measure ATP production in respiration have yielded figures as high as 36 ATP per molecule of glucose, thus approaching the theoretically proper number. Inasmuch as some ATP is almost certainly lost in the course of the experiment before it can be measured, the experimental results are in essential harmony with the theoretical calculations.

Thirty-eight ATP, at 8,000 calories per mole, represent 304,000 calories for each mole of glucose. We have noted that the free energy of glucose is about 674,000 calories per mole. Thus a little less than half (about 45 percent) of the free energy

*One calorie is the amount of energy required to raise the temperature of 1 gram of water by 1 degree centigrade, under certain specified conditions. The calories referred to by human nutritionists represent 1000 of these unitary calories and are properly called kilocalories.

of glucose may be converted into ATP during typical respiration. The rest is lost as heat.

These figures do not take into account some minor complicating factors and are thus a bit oversimplified. The NADH produced in the formation of pyruvate from glucose may not all go through the cytochrome system to produce ATP. Some of its energy may be more directly transferred into various metabolic processes, and at least under some conditions a good deal of it is probably lost as heat. Even some of the NADH produced in the TCA cycle may under some circumstances escape indirectly into the soluble part of the cytoplasm. The figure of 8,000 calories per mole of ATP is itself only a rough approximation. It is therefore probably safer to say that about half of the free energy of glucose is transferred into ATP and similar compounds that provide energy for various metabolic processes.

RESPIRATION OF FATS

Fats can be partly or completely respired, yielding metabolic energy in the form of NADH and eventually ATP. The first step is the digestion of fat to form glycerol and fatty acids. This is a hydrolytic process that uses up three molecules of water per molecule of fat (see Chapter 3). A small amount of energy is liberated in this process, but not enough to generate any ATP; instead the energy is lost as heat.

Glycerol and fatty acids follow separate paths of respiratory breakdown, each leading to the formation of compounds that can be either fed into the TCA cycle or (in plants) converted into sucrose.

Glycerol is successively converted into glycerol phosphate and triose phosphate. The first of these steps requires the energy

from 1 ATP. The second step releases one pair of reactive hydrogen atoms to form NADH. Since each NADH can govern the formation of 3 ATP, there is a net potential yield of 2 ATP in the conversion of glycerol to triose phosphate. The triose phosphate can either be synthesized into sucrose or converted into pyruvate and fed into the TCA cycle.

Fatty acids are progressively broken down, in a complex series of reactions known as the **fatty acid spiral,** to yield acetyl-CoA. The acetyl-CoA can either be fed directly into the TCA cycle or used as building blocks for other protoplasmic materials. In fatty seeds, and perhaps in other plant tissues, it can also be converted in considerable part into sucrose through a series of processes beginning with the glyoxylate cycle, discussed on a later page. Animals, on the other hand, or at least the higher animals, lack an efficient means of converting acetyl-CoA into sucrose.

In the fatty acid spiral carbon is detached from the fatty acid chain in 2-carbon units and combined with CoA, forming acetyl-CoA. CoA is the necessary primer for the first step in the spiral, and a separate enzyme is required for each subsequent step. After the first pair of carbon atoms has been detached from the fatty acid, the remainder of the fatty acid molecule, now bearing also its own newly acquired CoA, is fed back into the early part of the spiral. Another pair of carbon atoms is detached (and other molecule of acetyl-CoA formed) each time around, until the fatty acid has been completely converted into the acetyl part of acetyl-CoA. Attachment of the CoA to the fatty acid chain requires the energy from one ATP, but the removal of each acetyl unit from the chain governs the formation of 2 NADH (convertible into 6 ATP), so that there is a net release of metabolically useful energy into the system.

The respiratory breakdown of fats to form triose phosphate and acetyl-CoA characteristically occurs in the mitochondria. The machinery of the TCA cycle is there immediately available to carry the respiratory process through to completion.

THE GLYOXYLATE CYCLE

We have seen that acetyl-CoA is a key product in the early part of the TCA cycle, as well as being produced in the fatty acid spiral. Acetyl-CoA can also be passed into a series of reactions that yield some energy and lead to the accumulation of malate. Malate is a 4-carbon compound that is an intermediate in the TCA cycle and is also used as building blocks for some more complex substances in the cell. As we shall see in the section on the pentose phosphate cycle, it also plays a key role in transporting the energy of NADH out of and into the mitochondria. The series of reactions leading from acetyl-CoA to the accumulation of malate is known as the **glyoxylate cycle.** The glyoxylate cycle occurs in a number of micro-organisms and at least in certain tissues of higher plants, notably in germinating seeds that have a considerable supply of reserve fat. In these seeds the glyoxylate cycle is linked to the subsequent conversion of malate into sucrose, which is readily transportable to the embryo. The glyoxylate cycle typically takes place in minute organelles called glyoxysomes or peroxisomes, which are further discussed in a subsequent paragraph under the heading photorespiration.

The glyoxylate cycle has some of the same steps as the TCA cycle, using the same enzymes. The steps from acetyl-CoA

to isocitrate and from succinate to oxalacetate are the same in both cycles. The principal difference lies in what happens to the isocitrate, a 6-carbon compound. In the TCA cycle the isocitrate is broken down (decarboxylated) in a series of energy-yielding steps to give succinate, a 4-carbon compound. In the glyoxylate cycle the isocitrate is broken down more directly to yield succinate and glyoxylate (a 2-carbon compound). Glyoxylate combines with additional acetate (as acetyl-CoA) to form malate. Malate is also formed from succinate via fumarate in the glyoxylate cycle, just as in the TCA cycle. In the glyoxylate cycle the malate formed from succinate and fumarate is enough to keep the cycle going, and the malate formed from glyoxylate and acetyl-CoA is available for other uses. According to conditions in the cell at the time, it may be used directly or it may first be converted into oxalacetate in an energy-yielding process comparable to that of the TCA cycle.

Here as elsewhere it should of course be understood that molecules of a given kind are not branded and separately herded into different corrals. In order to keep track of balances, we distinguish in the glyoxylate cycle between malate formed from succinate and malate formed from glyoxylate and acetyl-CoA. The plant makes no such distinction. The important thing here is that the glyoxylate cycle leads to the accumulation of malate and compounds formed from it.

In the glyoxylate cycle, as in the TCA cycle, the steps from succinate to fumarate and from malate to oxalacetate yield energy in the form of pairs of reactive hydrogen atoms which are transferred to NAD.

In the fatty seeds mentioned above, malate is converted into oxalacetate, and oxalacetate is converted into phospho-enol-pyruvate. Then, by a reversal of the steps of the EMP glycolytic pathway, phospho-enol-pyruvate is converted into glucose and finally sucrose. The conversion of oxalacetate to phospho-enol-pyruvate requires one ATP, and the conversion of phospho-enol-pyruvate into glucose requires energy in the amounts earlier indicated as being produced in the EMP pathway. Thus the energy released in the conversion of acetyl-CoA to malate and then oxalacetate is largely used up again in the conversion of oxalacetate into sugar. The primary function of the glyoxylate cycle here is as a bridge in the conversion of fat to sugar, rather than as a source of energy or TCA cycle acids.

PHOTORESPIRATION

It has been known for many years that under conditions of very rapid photosynthesis much of the photosynthate is immediately reconverted into carbon dioxide and water, without a comparable output of metabolically useful energy. This process of destruction of photosynthate has been called **photorespiration.** Oxygen is used, and carbon dioxide and water are produced. The nature and significance of photorespiration are only now beginning to become clear.

It is now established that photorespiration occurs in the peroxisomes and that its principal substrate is glycolate ($CH_2OHCOOH$), a 2-carbon organic acid. The photosynthetic bacteria do not produce glycolate, but the vast majority of plants with System II photosynthesis divert some part of their photosynthetic energy into the production of glycolate.

Peroxisomes (Fig. 8.6) have several interrelated functions, notably the glyoxylate cycle and the destruction of hydrogen peroxide, in addition to photorespiration.

The very name peroxisome refers to the destruction of peroxide, which is fostered by the enzyme catalase within these bodies. As a result of a complex series of steps in which glyoxalate is an important intermediate product, about half of the glycolate that enters the peroxisome is oxidized to form carbon dioxide and water, taking up the oxygen that is released in the breakdown of hydrogen peroxide. No ATP is formed in this process; the energy is merely dissipated as heat. The other half of the glycolate is converted through successive reactions into metabolically useful products such as glycine. Glycine is a simple amino acid which differs from glycolate only in that an amine (NH_2) group is substituted for the OH group on one of the carbon atoms, as shown in Fig. 8.7.

The balance among the several different activities within peroxisomes differs in different kinds of plants, in different parts of the same plant, and in the same peroxisome according to the circumstances. The glyoxylate cycle can apparently proceed in the absence of photorespiration and catalase activity, and the term glyoxysome is sometimes used for some of these organelles instead of peroxisome.

THE PENTOSE PHOSPHATE CYCLE

Another way of respiring glucose, which by-passes both the EMP pathway and the TCA cycle, is variously known as the **pentose phosphate cycle,** the **pentose phosphate pathway,** or the **pentose shunt.** This system, which operates in the soluble part of the cytoplasm, is roughly comparable to the TCA cycle in the yield of useful energy per unit of glucose respired. Pentose phosphate (i.e., a 5-carbon, phosphorylated sugar) is a key

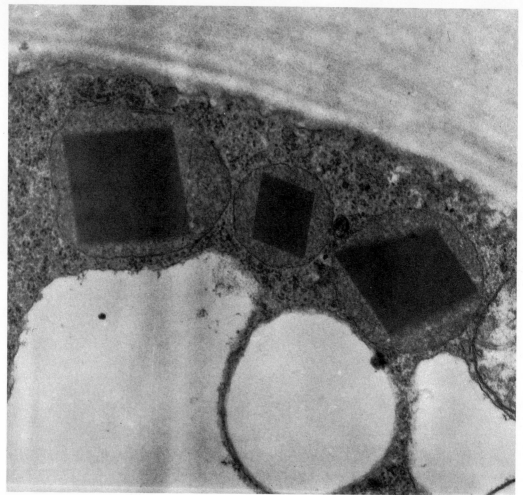

Fig. 8.6 Peroxisomes in tobacco, containing catalase in crystalline form. Catalase governs the breakdown of hydrogen peroxide to water and oxygen. (× 60,000.) [Electron micrograph courtesy of T. E. Jensen and J. G. Valdovinos, from *Planta* **77**:306, 1967; copyright by Springer-Verlag, Heidelberg.]

Fig. 8.7 Structure of glycolate (*A*) and glycine (*B*).

intermediate product which gives the cycle its name. The primary energy-bearing products of the pentose phosphate cycle are NADPH and NADH. Much of this product is probably used directly in various reductive processes in the cell, rather than governing the formation of ATP.

The proportion of glucose that is respired by the pentose phosphate cycle, as compared to the EMP-TCA route, varies with the kind of organism, the specific tissue, and the circumstances. A number of anabolic processes in the cell require ample supplies of NADPH or NADH. The reducing power of these compounds, residing in their highly labile pair of hydrogen atoms, permits them to foster some reactions that cannot obtain the requisite energy from ATP alone. For example, the synthesis of fatty acids from acetate requires the removal of some of the oxygen from the acetate. This removal is accomplished by a complex series of reactions that leads to the transfer of oxygen atoms into newly formed molecules of water and carbon dioxide and the incorporation of hydrogen from NADPH and NADH into the fatty acid. The conversion of nitrate ($^-NO_3$) ions into amino ($^-NH_2$) groups for incorporation into amino acids likewise requires NADPH and NADH. In some plant tissues under some circumstances as much as a third of the hexose that is oxidized follows the pentose phosphate cycle. In some animal tissues, such as actively secreting mammary glands, the proportion is even higher, amounting to well over one half.

The fact that the pentose phosphate cycle operates in the soluble part of the cytoplasm may be correlated with several other facts. Although ADP and ATP pass readily through the mitochondrial membrane, NADH and NADPH do not; they must transfer their energy to other compounds to pass it through. The TCA cycle produces much more NADH than NADPH, whereas the pentose phosphate cycle produces more NADPH. The NADPH is essential for necessary anabolic processes in the soluble part of the cytoplasm. Thus it is metabolically efficient to isolate the TCA cycle in specialized organelles (the mitochondria), where it can be closely linked to the cytochrome system for producing ATP, and to maintain the pentose phosphate cycle in the soluble part of the cytoplasm, where the NADPH and NADH that are produced are readily available for anabolic activities.

By this time the reader should not be surprised to find that there is also a cross-linkage between the pentose phosphate cycle and the TCA cycle, so that each can to a considerable extent perform the functions of the other. NADP can be formed from NAD either inside or outside the mitochondria by transferring a phosphate group from ATP. The energy of NADH can pass into or out of the mitochondria by a system involving malate and oxalacetate; NADH reacts with oxalacetate to form malate and NAD. Malate passes readily through the mitochondrial membrane. The reverse reaction, forming oxalacetate and NADH, can then be carried out. The oxalacetate passes readily back and forth through the mitochondrial membrane and can react again with NADH to form malate. The malate oxalacetate system acts as a two-way carrier of energy from NADH into and out of the mitochondria, the direction of net movement depending on the supply and demand. The only known way to convert the energy of NADH into the energy of ATP requires the cytochrome system. The malate-oxalacetate carrier system can deliver NADH produced in the soluble part of the cytoplasm to the place within the mitochondria where it can negotiate the cytochrome system to produce ATP. The same malate-oxalacetate carrier system can deliver NADH from the TCA cycle to any necessary place in the soluble part of the cytoplasm. We have already noted that malate and oxalacetate are also key products in the glyoxylate cycle. Thus these several respiratory processes, occurring in different parts of the cytoplasm, are bound into a common respiratory system.

The EMP pathway and the pentose phosphate cycle are also cross-linked by a triose phosphate called phosphoglyceraldehyde, which is an intermediate product in both. Such cross-linkage between different metabolic processes, sometimes even between processes that are not closely related from the standpoint of the human observer, are the rule rather than the exception. For purposes of understanding we find it necessary to consider certain sequences of metabolic reactions as distinctive systems, but the plant is not bound by our mental organization. The subcellular organization of certain enzymes into package units in the mitochondria, chloroplasts, and other organelles produces a statistical likelihood that some kinds of molecules will follow certain chemical courses under a given set of circumstances, but the circumstances are always changing and the fate of any individual molecule cannot be predicted in advance.

ANAEROBIC RESPIRATION

Respiration that does not use free oxygen is called **anaerobic respiration,** in contrast to aerobic respiration. When carried on by bacteria and other micro-organisms, it is also called **fermentation,** a term that antedates any understanding of the essential similarity of fermentation to typical respiration in metabolic function and in many of the individual chemical steps. Anaerobic respiration is always an incomplete respiration. It releases only part of the free energy of the substrate, and it has among its end products something that can be further oxidized (at least in the

laboratory) to yield additional energy. (The term fermentation is also commonly extended to cover certain bacterial aerobic respiratory processes, such as the formation of acetic acid from alcohol, in which there is an end product that can be further respired.)

Most anaerobic respiratory processes follow the EMP pathway up to the production of pyruvate. In this respect anaerobic respiration is similar to the most common kind of aerobic respiration. The differences lie in what happens to the pyruvate. In aerobic respiration it is fed into the TCA cycle, in which free oxygen is used. In anaerobic respiration the pyruvate is disposed of in various ways, some of which yield a little more energy, but others of which use up some of the energy that had been liberated during the formation of the pyruvate.

One of the most important anaerobic respiratory processes is the fermentation of sugar by yeast to form ethyl alcohol (C_2H_5OH) and carbon dioxide. Bread, wine, and beer are familiar products for which alcoholic fermentation plays an essential part. Alcoholic fermentation follows exactly the EMP pathway from glucose to pyruvate, but it does not move the pyruvate into the TCA cycle. Instead the pyruvate is converted first to acetaldehyde and then to ethyl alcohol, as shown in steps 12 and 13 of Fig. 8.3. A molecule of carbon dioxide is liberated in the conversion of pyruvate to acetaldehyde. This step is similar to the first step in the TCA cycle, in that pyruvate is converted into an acetyl compound by the extraction of a molecule of carbon dioxide, but it uses a different enzyme and does not use CoA. Two hydrogen atoms are then taken up from NADH, forming alcohol (ethanol) and liberating NAD. This step requires an input of energy (from NADH), balancing the earlier step in the EMP pathway in which NADH is produced. The net yield of metabolically useful energy in the anaerobic respiration of glucose to alcohol and carbon dioxide is thus only 2 ATP. This is barely more than 5 percent of the amount (38 ATP) that can be obtained by coupling the TCA cycle to the EMP pathway. Yet fermentation by yeast proceeds so rapidly that it provides enough ATP for the organism.

The lactic acid bacteria also follow the EMP pathway in respiration, producing pyruvate. The pyruvate is then converted directly to lactic acid ($CH_3CHOHCOOH$), under the influence of the enzyme lactic dehydrogenase. Two atoms of hydrogen are taken on from NADH in this process, just as in the formation of alcohol from acetaldehyde (see step 14 in Fig. 8.3). The net yield of energy is again 2 ATP per molecule of glucose.

Under conditions of oxygen shortage plants that ordinarily respire pyruvate by the TCA cycle may temporarily convert it to ethyl alcohol or lactic acid, which can be accumulated in the system with less metabolic disruption than pyruvic acid. With the return of more normal supplies of oxygen, these storage products are converted back to pyruvate, which can be fed into the TCA cycle. Lactic acid is likewise formed as a temporary product in mammalian muscles during hard exercise, when the immediately available supply of oxygen is inadequate to carry all of the pyruvate through the TCA cycle.

Under appropriate circumstances both alcohol and lactic acid can be fed into the TCA cycle by organisms that have the proper enzymes, but typical yeasts and the lactic acid bacteria do not have all the requisite enzymes of the TCA system. In humans ethyl alcohol is readily converted into acetate and sent into the TCA cycle. Methyl alcohol (wood alcohol), on the other hand, has only a single carbon atom and cannot be effectively respired by humans or other eukaryotes.

CAVEAT

The foregoing survey of some of the more important respiratory processes does not exhaust the possibilities. The bacteria, in particular, have a wide array of both aerobic and anaerobic respiratory processes beyond those discussed here. In the eukaryotes, on the other hand, respiration is largely channeled into a relatively few pathways—those that have proven most efficient in the oxygen-rich milieu in which most eukaryotes live.

SUGGESTED READING

Dickerson, R. E., Cytochrome c and the evolution of energy metabolism, *Sci. Amer.* **242**(3):136–153, March 1980.

Hinckle, P. C., and R. E. McCarty, How cells make ATP, *Sci. Amer.* **238**(3):104–123, March 1978.

Slater, E. C., Z. Keniuga, and L. Wojtczak, (eds.), *Biochemistry of mitochondria*, Academic Press, New York, 1967. One of the papers, by H. A. Krebs, expounds the role of the malate-oxalacetate system in transporting reducing power (as NADH) into and out of the mitochondria. This mechanism was discovered only shortly before.

Chapter 9

Molecular Genetics:
From Genes to Characters

Space-filling model of portion of DNA macromolecule.
[Courtesy of Professor M. H. F. Wilkins, Medical Research
Council, Biophysics Unit, King's College, London.]

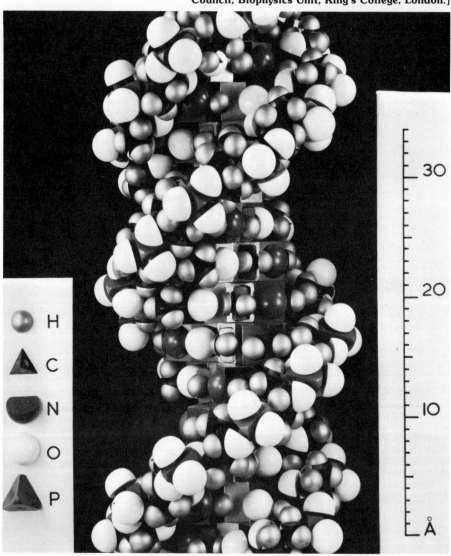

Oh chromosomes, my chromosomes,
We've learned to know you better.
We know the code of DNA,
We can translate each letter.
Our thymine must have adenine,
Our guanine mates with cytosine;
Their messenger, pure RNA,
Puts our proteins together.

(Sung to the tune of "Maryland, My Maryland,"
on the occasion of the XI International Botanical
Congress, in 1969; courtesy of G. L. Stebbins.)

SOME HISTORICAL BACKGROUND

We have noted in Chapter 4 that the individual genes, borne on the chromosomes, govern most of the hereditary features of the cell and that the critical constituent of chromatin (and thus of genes) is a class of compounds called DNA (deoxyribonucleic acid). Present knowledge, although still incomplete, permits a fairly good working knowledge of the chemical and structural nature of genes and how they function. This knowledge did not come into being overnight. Like other scientific concepts, present concepts of molecular genetics are based on a somewhat disorderly accumulation of information, punctuated by occasional brilliant syntheses and also by the sometimes gradual emergence of ideas for which no one person or team is wholly responsible.

A proper understanding of molecular genetics must begin with nucleic acid. The existence of nucleic acid was discovered in 1869 by the Swiss biochemist Friedrich Miescher (1844–1895), who found that a persistent residue remained in cell nuclei after the protein had been hydrolyzed with pepsin. Neither biologists nor biochemists were overwhelmed with the significance of this discovery. Biologists were busy classifying organisms and debating the significance of evolution, and biochemists were more interested in proteins.

Research on nucleic acids picked up somewhat after the turn of the century. By the late 1920s their chemical composition had been worked out, and it had become clear that there are indeed two types of such acids. Both types have a common essential structure of a series of pentose sugar molecules connected to each other by phosphate groups; to each sugar molecule is also attached one or another of several kinds of molecules that are collectively known as bases (Fig. 9.1). The phosphate and sugar groups are identical and repetitive throughout the chain, but there are four different kinds of bases, which may occur in any sequence. In one set of nucleic acids the pentose sugar is ribose; these are called **ribonucleic acids, or RNA** for short. In the other set the pentose sugar differs from ribose in having one less atom of oxygen; these are called **deoxyribose nucleic acids, or DNA.**

DNA is heavily (though not exclusively) concentrated in the nucleus, but RNA is more abundant in the cytoplasm. DNA in eukaryotes occurs in conjugation with proteins, notably proteins of the group called **histones,** forming what is called **chromatin,** or **nucleoprotein** complexes. The relatively simple and repetitive structure of DNA contrasts sharply with the tremendous diversity of proteins. Therefore, it should perhaps not be surprising that even after DNA was found to be largely localized in the chromosomes, attention was at first concentrated on the protein component of nucleoprotein complexes as the most important constituent of genes.

By the mid 1930s the suspicion began to grow that DNA rather than protein is the critical constituent of genes. Experiments with bacteria reported in 1944 by

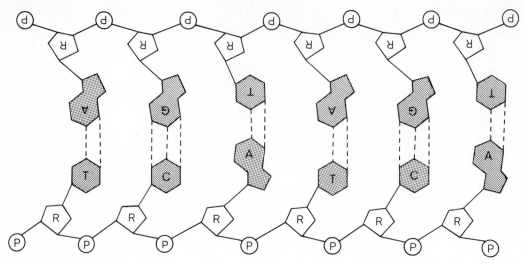

Fig. 9.1   Diagrammatic structure of DNA: A, adenine; C, cytosine; G, guanine; P, phosphate group; R, deoxyribose; T, thymine.

the American biologist Oswald T. Avery (1877–1955) and his associates strongly tilted the balance toward DNA. In retrospect Avery's work seems decisive, but the matter remained controversial until 1952, when some convincing experiments on viruses were reported by another group. It is of course significant that the DNA of bacteria and viruses differs from that of eukaryotes in not being conjugated with proteins.

Concurrently, but approaching the problem from another angle, the American biologists George W. Beadle and E. L. Tatum and their associates developed the concept during the 1940s that each gene governs the formation of a single enzyme. As we shall see, their *one gene–one enzyme* concept must now be refined to *one gene–one polypeptide strand,* but their idea was basically sound. The Beadle and Tatum experiments, chiefly using the fungus *Neurospora,* are further discussed in Chapter 29.

In 1953 James D. Watson and F. H. C. Crick brilliantly synthesized existing in-

formation to produce a structural model for DNA that proved to be essentially correct. Watson, a young American (1928–   , Fig. 9.2), was then a postdoctoral student working with Francis Crick (1916–   ) in Cambridge, England. It is virtually certain that the same solution would have occurred before long to others, perhaps to members of a rival group headed by Maurice Wilkins, who produced much of the information used by Watson and Crick. The personal rivalries are still echoing, but Watson, Crick, and Wilkins shared the Nobel prize in 1962 for their combined work.

THE STRUCTURE OF DNA

The Watson and Crick model of DNA (Fig. 9.3) shows it to be a double helix with regular cross-connections between the two members. The phosphate-ribose chain forms the main strand of each helix, and the associated bases form the cross-connections.

The bases in a DNA molecule are of

Fig. 9.2 James D. Watson (1928–), who, together with Francis Crick, produced a structural model of DNA. [Photo courtesy of Harvard University News Office.]

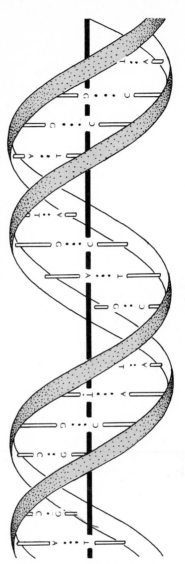

Fig. 9.3 Diagrammatic structure of DNA: the double helix. The spiral bands represent the strings of ribose and phosphate groups. The central rod is inserted only to show the symmetry.

four kinds, known chemically as **adenine, guanine, cytosine,** and **thymine** (Fig. 9.4). RNA likewise has four kinds of bases, three being the same as those of DNA, the fourth being **uracil** instead of thymine. Thymine and cytosine belong to a group called **pyrimidines,** whereas adenine and guanine belong to another group called **purines.** Pyrimidines consist of a ring of four carbon and two nitrogen atoms, to some of which are attached other atoms or submolecular groups of atoms. Purines have a double ring structure, with the two rings sharing one common side. One of the rings is similar to a pyrimidine ring, and the other has three carbon and two nitrogen atoms.

The cross-linkage between the two strands of a DNA double helix always involves one purine and one pyrimidine base. This is in part a reflection of the size of the two kinds of base with respect to

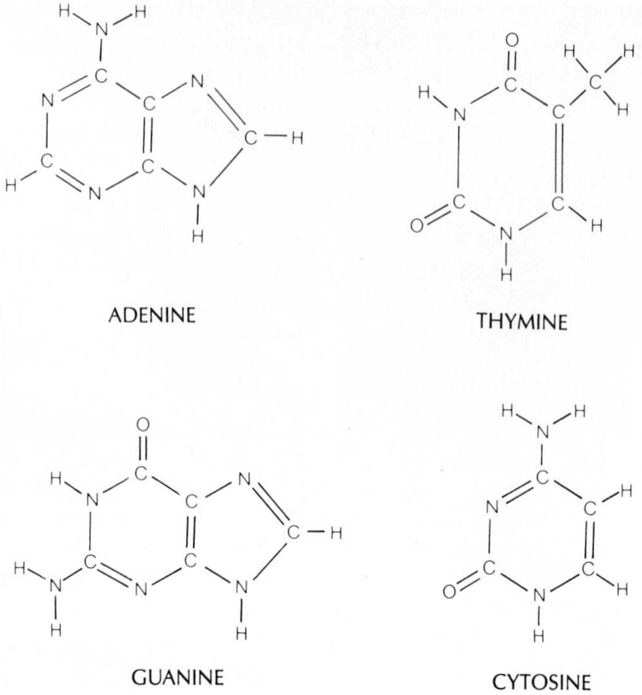

Fig. 9.4   Structural formulas of adenine, thymine, guanine, and cytosine.

the diameter of the helix. Two pyrimidines reaching toward each other from opposite points on their respective helices would fail to meet, whereas there would not be enough room for two purines in the same situation. A purine and a pyrimidine meet at just the right distance to permit the formation of hydrogen bonds between them. More than mere size is involved, however, because adenine always links up with thymine, and cytosine always links up with guanine. If one helix has adenine attached at a particular point, the opposite helix always (at least normally) has thymine, and vice versa; and if one helix has cytosine attached at a particular point, the opposite helix has guanine, and vice versa. If each of these bases is represented by its first letter as a convenient symbol, then the linkage is always A—T (or T—A) or G—C (or C—G). This sort of

pairing is said to be **complementary.** Within these limitations, the bases may be distributed along a particular helix in any sequence, and the same base may appear in two or more successive nucleotides. (A **nucleotide** is a unit of a nucleic acid chain, consisting of a phosphate group, a pentose sugar, and its attached base.)

REPLICATION OF DNA

The two strands that make up the double helix of DNA have complementary pairs of bases throughout their length; thus they form complementary strands. Often the two strands of a given double helix are for convenience designated as + and −. Under proper conditions the two strands can unwind and act as templates against which new complementary strands form. Each + strand is a template against which a new complementary − strand is formed, and each − strand is a template against which a new complementary + strand is formed. Formation of the new complementary strands proceeds apace with the unwinding. In eukaryotes the unwinding and replication begin simultaneously at many sites along the length of the very long double helix and proceed in both directions from each site until the process is completed and all the segments of the newly formed complementary strand have been joined together. In bacteria the double helical strand has no ends, and is said to be circular. In these organisms (or at least in *Escherichia coli*, which has been intensively studied) separation begins at a specific site and proceeds in both directions until the whole strand has been replicated. The double helix of DNA in bacteria is not nearly as long as that in eukaryotes, and one site for the initiation of replication is functionally adequate.

At the completion of replication we have two double helices of DNA, each consisting of a + strand and a − strand, which are complementary to each other. The + strand of the original double helix has been incorporated into a new double helix, in which the − strand has been formed as a complement to the +. The − strand of the original double helix has likewise been incorporated into a new double helix, in which the + strand has been formed as a complement to the − Replication of DNA is therefore said to be *semiconservative*, reflecting the fact that half of each double helix has been conserved from the immediately previous one.

TRANSCRIPTION: FROM DNA TO RNA

Genetic information encoded in DNA is transmitted to the rest of the cell through RNA, which in turn governs the formation of polypeptide strands (and thus eventually proteins). The critical initial step in this transfer of information is the formation of a complementary strand of RNA against a small portion of one of the strands of the double helix of DNA. Here, just as in DNA replication, the DNA strand acts as a template against which a complementary strand is formed, but the complementary strand is RNA instead of DNA. We have noted that RNA contains uracil instead of thymine. Uracil, like thymine, is a pyrimidine base, and like thymine it is complementary to the purine base adenine in template reactions.

Obviously the individual strand of DNA cannot serve as a template for the formation of RNA when it is joined to its complementary strand of DNA in the double helix. Segments of the two strands of the double helix must dissociate over distances of at least the length of the RNA molecules to be formed. Thus the two strands of the double helix are at any one time in complementary association along parts of their length and separated at others, these others being the active sites at which RNA is being formed. The RNA strands, although long in terms of number of nucleotides (sometimes in the thousands), are very short in relation to the tremendously elongate strands of DNA, in which the nucleotides must be numbered in the millions.

Both of the strands of DNA in the double helix govern the formation of molecules of RNA along portions of their length, but not in complementary positions. If a particular portion of one of the two strands governs the formation of a molecule of RNA, then the other strand does not ordinarily govern the formation of anything at all over that portion of its length. This fact is perfectly logical when one considers the alternative. We shall see that it is the sequence of bases on the strands of RNA formed against the DNA strands that determines the sequence of amino acids in proteins subsequently formed. If every protein-governing RNA strand were matched by another such strand governing the formation of a different (but specifically determined) protein, then a tremendous and unnecessary burden would be placed on the metabolic system and on evolutionary change. No matter how useful a particular protein might be, its necessary analogue would likely be useless at best and disastrous at worst.

The many specific kinds of RNA transcribed against DNA strands in the nucleus may be grouped into three general types, known as **messenger RNA** (mRNA), **transfer RNA** (tRNA), and **ribosomal RNA.** Messenger RNA consists of long to very long chains, containing from a few hundred to several thousand (sometimes as many as 12,000) nucleotides. It is the messenger RNA that directly governs the formation of various kinds of polypeptide chains in the cytoplasm. A molecule of transfer RNA commonly consists of some 75 to 85 nucleotides. At this level of study we may consider that there are 20 principal kinds of transfer RNA, and we shall see that each of these kinds has an affinity for a particular one of the 20 amino acids commonly found in proteins.

Ribosomal RNA of eukaryotes is made on the part of the chromosome (or chromosomes) that lies within the nucleolus. It comes in three size classes, with about 120, about 2,000, and 4,200–4,500 nucleotides, respectively. (This compares to about 120, about 1,540, and about 3,200 in bacteria.) Ribosomal RNA and some specific proteins are organized to form ribosomes, further discussed on subsequent pages. The beginning of organization of the ribosomes occurs in the nucleolus, but final assembly takes place in the cytoplasm. The ribosomal proteins and ribosomal RNA together form a structural-functional complex.

We are now in position to give a molecular definition of a typical gene. It is a particular segment of a DNA helix that governs the formation of a particular kind of molecule of RNA. As we shall see, it is by governing the kinds of messenger RNA that the genes indirectly govern the kinds of polypeptide chains that are formed, and it is these individual polypeptide chains that alone or with others form the essential proteins of the cell. This definition of a gene does not include the so-called operator genes, discussed on a subsequent page.

TRANSLATION: FROM RNA TO POLYPEPTIDES

A more detailed account of the function of ribosomes and RNA requires a consideration of the base sequences on RNA. As soon as the Watson-Crick model of DNA structure was accepted, it seemed obvious that the sequence of bases in DNA (and consequently in RNA) formed some sort of genetic code. It was easy to speculate that the code would consist of triplets—three consecutive bases—since a smaller number would not provide the 20 combinations necessary to govern the 20 amino acids and a larger number would be unnecessarily redundant. There are 64 different possible triplet sequences of bases, four different bases being available. Using letters to represent the four RNA bases, these sequences for messenger RNA are AAA, AAC, AAG, AAU, ACA, AGA, AUA, CAA, GAA, ACG, ACU, AGU, etc. In 1961, much sooner than any such discovery had been anticipated, the American biochemist Marshall Nirenberg

Fig. 9.5 Marshall Nirenberg (1927–   ), American biochemist, who with Heinrich Matthaei cracked the genetic code. [Courtesy of National Institutes of Health.]

(1927–   , Fig. 9.5) together with the German biochemist Heinrich Matthaei (1929–   ), demonstrated that UUU was the codon for the amino acid phenylalanine. The genetic code was cracked and soon wholly broken (Fig. 9.6). Using artificial (synthetic) RNA of known constitution, Nirenberg and his colleagues soon determined the codon for each amino acid. A given kind of amino acid can be specified by anywhere from one to six different codons. For example, there are four codons for valine: GUU, GUA, GUC, and GUG. Some other possible codons, such as UAG, do not govern any amino acid but instead signal STOP, END OF LINE.

The strand of messenger RNA that is transcribed against the DNA template may be roughly compared to a very long, flexible comb with many teeth. The row of ribose and phosphate units compares to the back or solid part of the comb, and the

alanine	GCA	GCC	GCG	GCU		
arginine	AGA	AGG	CGA	CGC	CGG	CGU
asparagine	AAC	AAU				
aspartic acid	GAC	GAU				
cysteine	UGC	UGU				
glutamine	CAA	CAG				
glutamic acid	GAA	GAG				
glycine	GGA	GGC	GGG	GGU		
histidine	CAC	CAU				
isoleucine	AUA	AUC	AUU			
leucine	CUA	CUC	CUG	CUU	UUA	UUG
lysine	AAA	AAG				
methionine	AUG					
phenylalanine	UUC	UUU				
proline	CCA	CCC	CCG	CCU		
serine	AGC	AGU	UCA	UCC	UCG	UCU
threonine	ACA	ACC	ACG	ACU		
tryptophan	UGG					
tyrosine	UAC	UAU				
valine	GUA	GUC	GUG	GUU		
Terminus	UAA	UAG	UGA			

Fig. 9.6 The RNA code for amino acids.

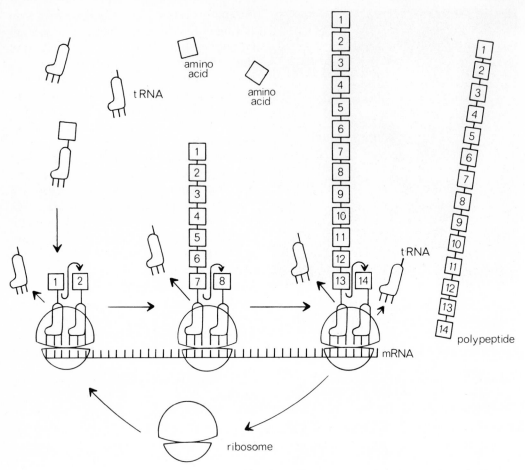

Fig. 9.7   Diagram of function of messenger RNA, transfer RNA, and ribosomes in making polypeptide strands.

attached bases compare to the teeth. Beginning at a particular end of the comb, each successive set of three teeth (bases) constitutes a signal (**codon**) for a particular amino acid.

The amino acids are assembled into a polypeptide chain, in the sequence specified by the messenger RNA, through the intervention of ribosomes and transfer RNA. The messenger RNA provides the code, the transfer RNA brings in the proper amino acids one at a time, and the ribosome mediates the assembly of these amino acids in a simple, progressively elongating chain (Fig. 9.7).

Each ribosome (Fig. 9.8) consists of two globular subunits of unequal size, associated to form a figure that might be compared to a compact, asymmetrical dumbbell or to an ellipsoid that is constricted around a transverse line somewhat away from the middle. The weight ratio of the two subunits is about 3:2 in bacteria and about 2:1 in eukaryotes. Ribosomes of

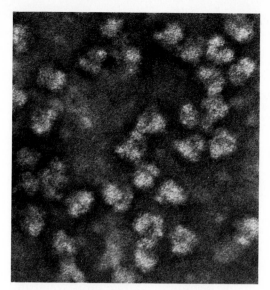

Fig. 9.8 Ribosomes from pea seedlings. (× 240,000.) [Electron micrograph courtesy of S. T. Bayley; from *J. Mol. Biol.* **8**:Fig. 8e, following p. 236, 1964.]

eukaryotes are commonly about 16–20 nm in diameter and those of bacteria a little smaller, mostly 15–17 nm. Within each of these two groups, differences in measurements may more nearly reflect differences in methods of preparation for study than actual differences *in vivo*.

Ribosomes in the chloroplasts and mitochondria of eukaryotes may well be distinctive in function, as compared to other ribosomes in the same cell. Otherwise the ribosomes of a given cell appear to be essentially alike. Probably any one can perform the same function as any other.

Each molecule of transfer RNA has a complex, three-dimensional shape that has been variously compared to a crumpled clover-leaf or a distorted L. At one end it has a sort of molecular hook for a particular amino acid, and at the other end it has a triplet of bases (an **anticodon**)

complementary to one or another codon of the messenger RNA. Typically there are 20 classes of transfer RNA, one for each of the 20 amino acids. Most of these 20 classes of transfer RNA have subclasses, reflecting the fact that there is more than one way to code for a particular amino acid. In the example of valine that we used earlier, there are four different messenger RNA codons, each complementary to its own subclass of valine-carrying transfer RNA. The number of kinds of transfer RNA, based on their triplet code component, approaches the theoretical limit of 64. At this level of study we need not be concerned with the possible significance of differences within any one of the 20 general classes of transfer RNA.

In polypeptide synthesis a ribosome becomes attached to a strand of messenger RNA. The attachment is to a specific site on the smaller subunit of the ribosome, near its junction with the larger subunit. In the simplest case the messenger RNA is initially attached near a particular end of the strand. As so attached, the messenger RNA presents two adjacent codons for possible chemical reaction.

The ribosome–messenger RNA complex now attracts a particular kind of transfer RNA molecule, which apparently spans the junction line of the two subunits of the ribosome. The triplet anticodon of the transfer RNA must complement the codon of the messenger RNA to which it is attached. Thus a messenger RNA codon GUG can accept only a molecule of transfer RNA bearing the complementary anticodon CAC. A molecule of transfer RNA bearing the CAC anticodon will bring with it a molecule of the amino acid valine. A molecule of transfer RNA with its attached amino acid has been compared with a tug towing a barge.

While this first molecule of transfer

RNA is still attached, another molecule of transfer RNA comes in and binds temporarily in the same way to the next codon on the messenger RNA molecule. The first molecule of transfer RNA now gives up its attached amino acid, which links to the exposed end of the amino acid that is attached to the second molecule of transfer RNA. The ribosome now rolls for a distance of one codon along the strand of messenger RNA. The codon that accepted the original molecule of transfer RNA is thus removed from the reaction site, and its connection with the ribosome is severed. The first molecule of transfer RNA, having given up its amino acid, dissociates from the ribosome-messenger RNA complex and is cast loose into the cytoplasm. It can then pick up another molecule of amino acid and again bind to the ribosome-messenger RNA complex, with the proper codon-anticodon relationship to the messenger RNA.

The motion of the ribosome brings the next (third) codon of the messenger RNA into the reaction center, alongside the second codon, which still bears its attached molecule of transfer RNA with its short chain of two amino acids. The newly available codon of the messenger RNA now attracts a suitable molecule of transfer RNA, which has the proper anticodon and carries its own amino acid. The two amino acids attached to the second molecule of transfer RNA are passed on to the third transfer RNA, forming a chain of three amino acids. The ribosome again rolls for a distance of one codon along the messenger RNA, freeing the second molecule of transfer RNA from the reaction. Thus a constantly lengthening chain of amino acids (making up a polypeptide strand) is formed as successive molecules of amino acid are carried in by successive molecules of transfer RNA. Always the most recently arrived amino acid in the chain is next to the transfer RNA, and the earlier ones are progressively farther away in the chain.

The ribosome continues to roll along the strand of messenger RNA, adding the proper amino acid to its growing chain at each stop. Finally it reaches either the end of the molecule, or a codon (UAG, UAA, or UGA) for which there is no complementary transfer RNA molecule and which therefore specifies the end of the line. The messenger RNA then dissociates from the ribosome and releases its polypeptide chain into the cytoplasm. The ribosome is now ready to start all over again to mediate the formation of a new polypeptide chain, beginning at the proper starting place on the same or another strand of messenger RNA.

As soon as a ribosome has moved away from its starting place on the strand of messenger RNA, so as to be physically and chemically out of the way, the messenger RNA strand is ready to receive another ribosome at the starting point. During active protein synthesis a single strand of messenger RNA may thus have several or many ribosomes scattered along its length.

All of the steps in transcription and translation, as here described, are of course mediated by the proper enzymes. A more detailed presentation of the story would divide some of these steps into smaller steps, each with its own enzyme or set of enzymes.

FROM POLYPEPTIDES TO PROTEINS

Once the completed polypeptide chain has been released from the ribosome–messenger RNA complex on which it was formed, it may associate with one or more similar or different polypeptides to form a protein molecule, or it may be a complete protein molecule by itself. There are also sometimes some post-transcriptional alterations of the polypeptide chain before it is incorporated into a protein molecule. In any case, the protein molecule has a predetermined, complex, three-dimensional structure resulting from the folding and twisting of the individual polypeptide chains that were assembled link by link through the agency of the transfer RNA–messenger RNA–ribosome mechanism.

The newly formed molecule of protein may immediately conjugate with some nonproteinaceous molecule in the cell, or it may remain as a simple protein. In simple or conjugated form, as the case may be, it usually acts as an enzyme, mediating some particular chemical change among particular kinds of molecules in the protoplasm. A small proportion of protein is not enzymatic. Some of this nonenzymatic protein is, for example, incorporated into cell walls as structural protein. Whether the histone proteins that contribute to the structure of chromosomes are enzymatic or not is perhaps a matter of definition.

REGULATION OF GENE ACTIVITY

Our discussion so far might imply that every kind of gene and every strand of messenger RNA is as active as any other, and for the same period of time. The situation is, in fact, quite otherwise. At any one time only certain genes in any given cell are producing RNA, and different genes may be active at the same time in different cells of a multicellular organism. Furthermore, different kinds of messenger RNA have different lengths of functional life. Some can govern the formation of only a few polypeptide strands

before they disintegrate, whereas others can function much longer.

The complex mechanism that governs the time of activity of particular genes is perhaps best regarded as one of differential repression. Current conceptual models (derived from studies of bacterial systems, and not yet demonstrated in eukaryotes) divide the DNA of the chromosome into three kinds of units, called **regulator genes, operator genes,** and **structural genes.** The regulatory genes transcribe RNA in the way already described, but the proteins eventually produced, called **repressors,** have a very specific function. They conjugate with the operator genes. In this conjugated form the operator genes repress the activity of adjacent structural genes. Various kinds of smaller, nonprotein molecules that are not an integral part of the chromosome can interact with the repressors (acting as inducers or derepressors), blocking their effect on the operator genes. Freed from the influence of the repressors, the operators activate the structural genes, which function in the ordinary way to govern the formation of RNA. Most of the proteins eventually governed by the structural genes become the ordinary metabolic enzymes of the cytoplasm. The question of what regulates the regulators has as yet scarcely been opened.

Because the operator genes, as now conceived, do not function directly as templates to produce new RNA, it might be argued that they do not fit the definition of genes given on p. 142, and that therefore either the definition should be changed or the operators should not be called genes. The argument eventually becomes a semantic one, because from a functional standpoint the operator might be regarded as an integral part of the DNA strand that produces RNA along a part of its length. Definitions that are useful and sufficient at one level of study frequently need to be refined in order to be useful at another level. We shall see that there is a similar problem regarding what structures should and should not be called leaves.

We must visualize a complex, interacting system of control of gene action, with built-in reciprocal influences. Given a particular gene that eventually governs the formation of a particular kind of polypeptide strand, a difference in the repressor protein, affecting the time and extent of activity of the gene, is just as important to the organism as the substitution of a slightly different gene that would cause a slightly different kind of polypeptide strand to be formed. The nature of the repressor protein is in turn governed by messenger RNA produced by regulatory genes, which thus have their principal effect in controlling the activity of other genes. There is good reason to believe that a great deal of evolutionary change depends on changes in these regulatory genes, with resultant effects on the proportions of cytoplasmic constituents at different times or in different cells, rather than on changes in the *kinds* of molecules that are supplied to the cytoplasm.

The internal environment of the cell also plays a role in determining the amounts and proportions of different proteins that are formed. The proteins can be formed only by the template mechanism we have described, but the template in turn can act only with the amino acids that are provided to it. If there is a shortage of a particular amino acid in the cell, the formation of those proteins requiring that amino acid will necessarily be inhibited. The protoplast of a living cell is in a highly complex state of multiple actions and interactions, and any change in any one of the chemical reac-

tions is likely to have a branching and interlocking chain of consequences.

The external environment also influences the activity of the cell's protein-forming mechanism, through its influence on the internal conditions in the cell. The action of van't Hoff's law (relating to the influence of temperature on the speed of chemical reactions) is strongly modified by differences in the stability and lability of different enzymes at different temperatures, and therefore even a simple change in temperature will favor some metabolic actions and repress others. The rate of respiration, for example, may continue to increase at increasing temperatures that inhibit photosynthesis. Differences in the supply of mineral nutrients obtained from the outside environment will necessarily have differential effects on the rates of various reactions within the cell. Even the amount of available water can have an important effect on the reactions and chemical balances in the cell.

FROM PROTEINS TO CHARACTERS

At one level of study, the proteins formed by the DNA–RNA–ribosome mechanism may be considered as the characters of the organism, but these are not the characters that must be chiefly considered in an organismic approach. Students of the whole organism must be concerned with structure, form, and function. Structure, form, and function are of course eventually controlled by the interaction of heredity and environment, but the control is highly complex, involving many intermediate steps and interactions.

Most proteins are either enzymes or serve as necessary carriers of catalytically active prosthetic groups. Looked at from the other end, enzymes are simple or conjugated proteins. The important catalyst chlorophyll, for example, is a porphyrin rather than a protein, but it acts only when conjugated to a particular protein. The enzyme consists of the protein plus its attached chlorophyll (the prosthetic group). The sequential action of a particular set of enzymes is necessary for the formation of chlorophyll; and the chlorophyll, once formed, can carry on its own enzymatic function only when conjugated to the proper protein. The enzymatic action of chlorophyll further requires the proper physical and chemical milieu, plus the coordinated action of a whole set of other enzymes involved in the photosynthetic process (see Chapter 6). The same sort of complex system of factors governs any physiological process or ultimate cell product that one might wish to consider (Fig. 9.9).

Characters of cells or organisms, at the level of structure, form, and function, are thus seen to result from a complex series of interactions. Each character is influenced by a number of genes, and each gene influences a number of characters. It has even been said that every gene has at least some influence over every character, and conversely that every character is influenced by every gene. This may be an overstatement, but it properly emphasizes the complexity of the hereditary control mechanism.

It is equally true, however, that in the development of any character some genes are more important than others. Certain genes may govern the formation of enzymes without which the character in question absolutely cannot develop. They govern essential links in the **biosynthetic chain** leading from gene to character. Other genes may only indirectly affect the particular biosynthetic chain under consideration. They may govern enzymes which through competition or interaction may depress or promote the activity of essential enzymes in that biosynthetic chain. Other genes may be involved only to the extent of contributing to the general milieu in which the particular biosynthetic chain is constructed.

A corollary of the concept of the biosynthetic chain is that the chain may be broken at any link. In simple presence or absence characters (as, for example, green versus albino corn plants, or red versus white flowers in snapdragons) it may therefore appear that the character is governed by a single gene or pair of genes. It is more accurate in such cases to speak of single gene differences. A difference in a single gene may determine the presence or absence of a character, if all the other necessary genes are present. But a difference in some other gene may also determine the presence or absence of the same character, if that is the only weak link in the chain.

Another fact to be considered in connection with gene interaction is that a particular metabolic product may have several different possible uses. For example, a substance called protoporphyrin IX (Fig. 9.10) appears to be a necessary precursor to both the chlorophylls and the heme pigments. Anything that influences the series of steps leading to the formation of protoporphyrin IX will necessarily influence the formation of both chlorophyll and heme.

Individual molecules of protoporphyrin IX have two principal possible fates (Fig. 9.11). Some of them, under the influence of the proper enzyme, take on a magnesium atom, becoming Mg protoporphyrin IX. Mg protoporphyrin IX (if broadly defined) is apparently the precursor to all chlorophylls. Other molecules of protoporphyrin IX take on an atom of iron, becoming Fe protoporphyrin IX, the

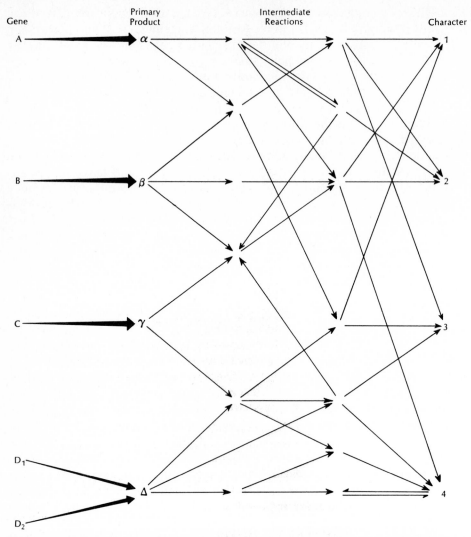

Fig. 9.9 Diagram showing some actions and interactions in the genetic control of characters. [From G. L. Stebbins, *Processes of organic evolution*, copyright 1966. Reprinted by permission of Prentice-Hall, Englewood Cliffs, N. J.]

apparent precursor to all heme pigments. There are further branch points in each of these two main pathways for protoporphyrin IX, leading to the formation of the final individual kinds of hemes and chlorophylls. If through some enzymatic failure the production of protoporphyrin IX is prevented, then neither heme nor chlorophyll can be formed. On the other hand, if the enzyme governing the formation of Fe protoporphyrin IX from protoporphyrin IX fails to act, then only the production of heme pigments is prevented, but the chlorophylls can continue to be pro-

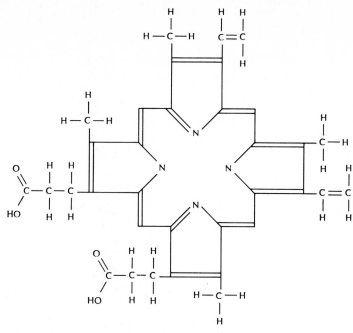

Fig. 9.10 Structure of protoporphyrin IX.

duced until the absence of heme pigments disrupts the metabolic activity of the cell.

In the heme-chlorophyll example here given, the organism would never get beyond the one-celled stage in reproduction if it completely lacked the ability to produce heme, because the heme pigments are essential for respiration. It would be quite possible, however, to have a heme-governing enzyme that acted only under certain conditions. It might, for example, act essentially normally at temperatures of up to 20°C, but fail to act at temperatures substantially higher than that.

THE CHEMICAL NATURE OF GENE MUTATIONS

We have seen that an ordinary gene may now be considered to be a particular segment of a DNA strand which eventually governs the formation of a particular

kind of polypeptide strand. We have also seen that DNA is self-replicating, each double helical strand producing under proper conditions another double helical strand just like itself. This fidelity of chemical replication of DNA is not absolutely perfect, however. Accidents happen, and there are even conditions that increase the frequency of accidents in general (see Chapter 29), and particular genes that increase the frequency of particular sorts of accidents in other genes.

An error in replication immediately sets up an imbalance in the pairing of the two strands that make up the double helix. They are no longer fully complementary. There is a delicate repair mechanism that can frequently restore the status quo ante. The offending new item is chemically snipped out and replaced by a piece of proper structure. One may surmise that the repair mechanism itself

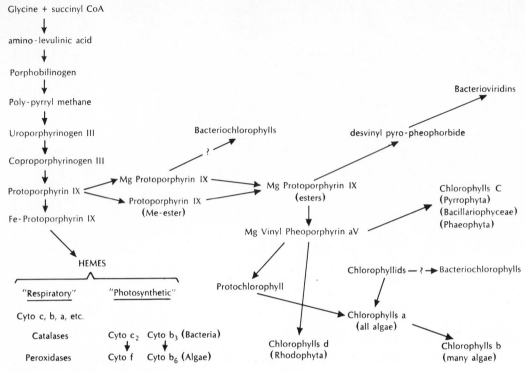

Fig. 9.11 Postulated routes of biosynthesis of porphyrins. [After R. M. Klein and A. Cronquist, *Quart. Rev. Biol.* **42**:119, 1967.]

might sometimes go awry, snipping out the original rather than the changed DNA.

If an error in DNA replication escapes repair before the next replication, it tends to be perpetuated in subsequent replications of the changed strand. The newly modified DNA has essentially the same degree of stability in its new form as it did before the accident; the likelihood (or unlikelihood) of subsequent accident is about the same as the likelihood of the first one. The original gene has undergone **mutation** to produce a different gene.

Several different sorts of accidents may occur in the replication of a given gene. The most obvious kinds are the deletion, insertion, and substitution of particular nucleotides. Either the insertion or the deletion of a single nucleotide in the sequence would of course upset the triplet code for amino acids, changing every single codon from the point of insertion or deletion to the end of the line. Obviously an insertion or deletion that was not itself very close to the terminus of the gene would cause the gene to specify a very different protein. Switching, for convenience, to a consideration of the strand of messenger RNA that would be formed against the gene, we can see that the insertion of a single extra A in the first codon might change a previous series of codons from AAU GUG CCC CAA GGG GAC to AAA UGU GCC CCA AGG GGA. The specified amino acids would thus be changed from asparagine–valine–

proline–glutamine–glycine–aspartic acid, to lysine–cysteine–alanine–proline–arginine–glycine. Deletion of a single nucleotide would have a similarly drastic effect on the codons. The likelihood that such a drastically changed protein would fit properly into the metabolic system of the cell is minimal. It might simply be a useless dud, or it might actively interfere with the functions of other proteins in the cell.

Substitution of one nucleotide for another in a particular codon during replication of the DNA would have a much smaller chemical effect than insertion or deletion. It would merely change the amino acid specified by that particular codon, without affecting other codons. Some substitutions, in fact, might have no presently recognizable effect because they would merely provide alternative ways of specifying the same amino acid. Thus the amino acid glutamine is specified by both CAA and CAG on the messenger RNA (complementary to GTT and GTC on the gene governing the messenger); the substitution of G for the final A in CAA has no effect, according to present information, on the resulting protein. Conceivably some subtle differences might eventually be found to result from different ways of coding for the same amino acid, but we cannot yet specify what they might be.

Obviously a mutation resulting from nucleotide substitution should have a better chance of fitting into the metabolic pattern of the cell than one resulting from insertion or deletion. Its possible effects might range all the way from complete loss of function of the enzyme to no significant effect at all. It might merely change the optimum conditions for operation of the slightly changed enzyme, increasing the rate under some conditions and decreasing it under others, without changing the *nature* of the enzymatic action. Because of the natural selection of favorable mutations that have occurred in the past, the chances are that any particular new mutation from the standard will decrease rather than increase the fitness of the organism in its present habitat, but some few of them must turn out to be at least potentially beneficial. Substitution mutants must provide a major share of the raw material for evolution.

EPILOGUE

There are many complexities and variations on the fundamental pattern of molecular genetics presented on the foregoing pages. For example, the codon-anticodon pairing of transfer RNA with messenger RNA is sometimes less than perfect. The first two members of the triplet must pair, but some variation is possible in the third one. For further example, two consecutive segments of DNA on a chromosome may function individually as genes and collectively as another gene. The stop signal between them can sometimes be overridden. One is tempted to write a biological version of Murphy's Law: Everything is more complicated than you thought it was. The tremendous diversity of life, and the adaptability of individuals to environmental hazards and stress, reflect these chemical and eventually morphological and functional differences. Yet it is only by seeing through the variations to the fundamental repetitive patterns that we can have any hope of understanding nature.

SUGGESTED READING

Beadle, G., and M. Beadle, *The language of life,* Doubleday, Garden City, N.Y., 1966. Science by George, writing by Muriel; good symbiosis.

Crick, F. H. C., The genetic code: III, *Sci. Amer.* **215**(4):55–62, October 1966.

Judson, H. F., *The eighth day of creation. Makers of the revolution in biology,* Simon & Schuster, New York, 1979.

Kornberg, R. D., and A. Klug, *The nucleosome, Sci. Amer.* **244**(2):52–64, February 1981. The elementary subunit of chromosome structure is a DNA superhelix wound on a spool made of histone proteins.

Maniatis, T., and M. Ptashne, A DNA operator-repressor system, *Sci. Amer.* **234**(1):64–76, January 1976.

Portugal, F. H., and J. S. Cohen, *A century of DNA: A history of the discovery of the structure and function of genetic substance,* MIT Press, Cambridge, Mass., 1977.

Sayre, A., *Rosalind Franklin and DNA,* Norton, New York, 1975. A different view of the events leading to the discovery of the double helix.

Stein, G. S., J. S. Stein, and L. J. Kleinsmith, Chromosomal proteins and gene regulation, *Sci. Amer.* **232**(2):47–57, February 1975.

Temin, H. M., RNA-directed DNA synthesis, *Sci. Amer.* **226**(1):24–33, January 1972.

Watson, J. D., *The double helix,* Atheneum, New York, 1968. A controversial personal account of the events leading up to the classic Watson-Crick paper; may remind some readers of a Browning poem.

Watson, J. D., *Molecular biology of the gene,* 3rd ed., Benjamin-Cummings, Reading, Mass., 1976. A clear, step-by-step presentation in a classic textbook.

Watson, J. D., and F. H. C. Crick, Molecular structure of nucleic acids, *Nature* **171**:737–738, 1953. Nobel-prize–winning paper.

Chapter 10

Sex and the Meiotic Cycle

Artificially induced crossing over, following X-ray treatment.
[Photomicrograph courtesy of C. C. Moh.]

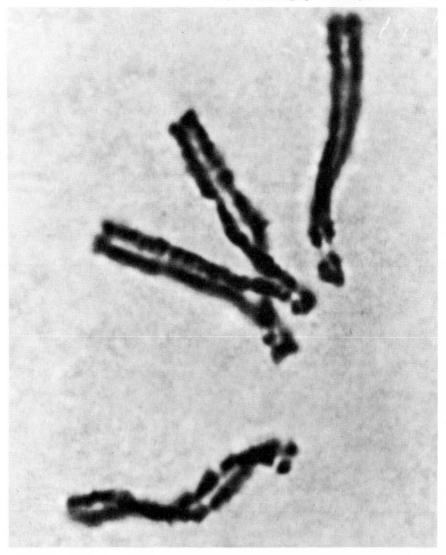

THE NATURE OF SEXUALITY

Reproduction, that is, the formation of new individuals, occurs in a number of different ways. Among unicellular organisms, cell division is also reproduction, since the two daughter cells are physiologically independent of each other and do not remain attached. Multicellular organisms have various means of reproduction. One of the simplest means is the formation of special reproductive cells, called **spores,** which are liberated from the organism and develop directly into new individuals. Among the algae, spores often have one or more slender, whiplike structures called **flagella,** which enable them to swim freely. Motile spores such as these are called **zoospores** because they possess the animal-like feature of motility.

Most kinds of organisms, whether they have any other means of reproduction or not, reproduce by the fusion of two different cells into a single cell that develops into a new individual. Any cell capable of fusing with another cell to form a new individual organism is called a **gamete,** and the cell formed by fusion of gametes is called a **zygote.**

Among some algae, the two gametes that fuse are similar in appearance and are therefore called **isogametes.** Some isogametes are so much like zoospores that they can be distinguished from zoospores only by their behavior (i.e., by the fact that two gametes fuse before a new individual develops, whereas a zoospore develops directly into a new individual). Even when the gametes that fuse are similar in appearance, however, they are often chemically unlike, and there are all transitions between typical isogamy and the more common condition in which the two gametes that fuse are distinctly unlike (heterogamy).

Heterogametes (morphologically unlike gametes that can fuse to form a zygote) are commonly of two types: a small, motile gamete called the **sperm** and a larger, nonmotile gamete called the **egg.** The zygote formed by the fusion of a sperm and an egg is called a **fertilized egg.** The sperm is the male gamete, and the egg is the female gamete. An individual that produces male gametes but not female is called a male; an individual that produces female gametes but not male is called a female; an individual that produces both male and female gametes is said to be a hermaphrodite.

Fusion of isogametes is called **conjugation,** and fusion of heterogametes is called **fertilization.** The term **syngamy** covers both conjugation and fertilization. Reproduction by fusion of gametes (either conjugation or fertilization) is called **sexual reproduction.** Sexual reproduction among the more familiar kinds of plants and animals always involves fertilization rather than conjugation.

When two gametes fuse, their nuclei generally also fuse into a single nucleus, but the individual chromonemata (and eventual chromosomes) retain their identity. The number of chromosomes present in subsequent mitotic divisions is equal to the sum of the numbers present in the two gametes. For example, if each of the 2 gametes has 5 chromosomes, the number of chromosomes formed during the prophase of the next division of the zygotic nucleus is 10. It is convenient to refer to the gametic number of chromosomes by the symbol n and to the zygotic number by the symbol $2n$. The number represented by n varies, according to the kind of organism, from 1 to more than 500, but numbers of 3 to 50 are more common. One complete set of chromosomes is called a chromosome complement, or a **genome.** Ordinarily a gamete has one chromosome complement, and a zygote has two chromosome complements. A cell or nucleus that has one chromosome complement is said to be **haploid** (Greek *haploos,* "simple"), and a cell with two chromosome complements is said to be **diploid** (Greek *diploos,* "double").

In any diploid nucleus, each chromosome from one complete haploid set is ordinarily matched by a similar chromosome from the other haploid set. Any two such matching chromosomes are called **homologous chromosomes.** Each of the chromosomes of any homologous pair carries genes affecting the same set of characters. For every gene on a given chromosome, there is ordinarily a similar gene in the same position on the homologous chromosome. Two similar genes occupying corresponding positions on homologous chromosomes may be exactly identical, or, due to past mutation, they may differ slightly in their chemical structure and resulting effect on the organism. The hereditary differences among individual organisms are largely due to such genic differences which have arisen through mutation. If, in a given individual, the two comparable genes from a pair of homologous chromosomes in a cell are exactly identical, the individual is said to be **homozygous** for that pair of genes; if the two comparable genes are not exactly identical, the individual is said to be **heterozygous** for that pair of genes. In most kinds of plants and animals, an individual diploid organism is usually homozygous for many genes and heterozygous for many others.

HISTORICAL SUMMARY

Any historical summary of scientific development requires oversimplification. Scientific knowledge grows by the efforts of many different workers in different countries, and a scientist trying to solve some problem must frequently refer to papers published in several languages. Fusion of gametic nuclei in plants was first reported by Eduard Strasburger (1844–1912) in 1877, but Strasburger thought that the fusion nucleus then dissolved. It was another German botanist, C. J. Friedrich Schmitz (1850–1895), who first showed clearly, in 1879, that the essential feature of the sexual process in plants is ordinarily the union of two nuclei to form the primary nucleus of the new individual. The cell theory, which is generally attributed to Schleiden and Schwann (as in Chapter 4), was foreseen by the French naturalist R. J. H. Dutrochet (1776–1847), who stated in 1824, "The globular corpuscles which make up all the tissues of animals are really globular cells of extreme smallness. . . . Growth results from the increase in volume of the cells, and from the addition of new little cells." Dutrochet did not distinguish between true cells and other globules visible in tissues, however, and he was unaware of the existence of the nucleus. It is clear that when sufficient data become available to permit the formation of a new concept, the concept will be formed. If one person does not do it, another soon will, and similar conclusions are often reached independently and nearly simultaneously by two or more scientists.

The importance of sperms in stimulating the development of the egg was suspected soon after the first observation of human sperm by Ludwig Hamm, a pupil of Leeuwenhoek. The way the sperm performs its function was not fully understood until two centuries later. The actual penetration of the sperm into the egg was first seen by the English zoologist Alfred Newport (1829–1907), who in 1854 reported his observations on the fertilization of frog eggs. Two years later the German botanist Nathanael Pringsheim (1823–1894) reported the penetration of the sperm into the egg in *Oedogonium*, a green alga. The fact that the penetration of the sperm into the egg is followed by fusion of the sperm nucleus with the egg nucleus was reported in 1876 by the German zoologist Oskar Hertwig (1849–1922); and fusion of gametic nuclei in plants was reported in 1877 by Eduard Strasburger, working on *Spirogyra*, a green alga.

It can readily be seen that if each repetition of the cycle of sexual reproduction led to a doubling of the number of chromosomes in the nucleus of each cell, the number of chromosomes would soon increase beyond bounds. In 1883 the Belgian zoologist Eduard van Beneden (1846–1910) reported that the nuclei of the sperm and egg in the parasitic worm *Ascaris* contained only half as many chromosomes as the nuclei of ordinary body cells of the worm, and his results were soon duplicated in observations of other kinds of animals. In 1888 Strasburger demonstrated the existence of a similar reduction in flowering plants, and in 1894 he set forth the doctrine that reduction in the number of chromosomes is a normal and regular part of the life cycle of all sexually reproducing organisms, balancing the increase in number brought about by gametic fusion. This principle had been foreseen by the German zoologist August Weismann (1834–1914, Fig. 10.1),

Fig. 10.1 August Weismann (1834–1914), German biologist, who made important contributions to the theory of heredity, and who foresaw the necessity and the mechanism of reduction division.

who in 1887 even predicted the mechanism of reduction. Actual discovery of this mechanism was made almost simultaneously by several different investigators working independently in different countries, and it was elucidated in a series of papers published in 1890–1893. There is an interesting legend, perhaps true, that a graduate student at a well-known American university discovered reduction division in plants a year or so before Strasburger, but was dissuaded from publicly reporting it by his major professor. The student left the university soon after without obtaining his doctorate.

MEIOSIS

The Typical Meiotic Process

The process by which the number of chromosomes in a cell is reduced from $2n$ to n is called **reduction division,** or **meiosis.** Meiosis may occur at any time in the life cycle, depending on the kind of organism. In some algae the first division of the zygote after fertilization is a reduction division, and the zygote itself is the only cell in the life cycle that has $2n$ chromosomes. In many other plants, as in most animals, a diploid body of some size is formed from the zygote by repeated mitotic divisions, and reduction division affects only certain special reproductive cells or tissues of the mature diploid organism. Among animals the gametes are usually the immediate products of meiosis, but among most plants the cells formed in meiosis produce (by a series of mitotic divisions) a haploid body of two to many cells, on or in which the gametes are eventually borne. The essential features of meiosis are the same in both plants and animals, regardless of the point in the life cycle at which it occurs.

Meiosis (Fig. 10.2) typically consists of two successive nuclear divisions (with or without the formation of cell walls) during which the individual chromosomes are divided only once, so that four haploid nuclei are produced from an original diploid nucleus. The actual reduction occurs during the first division, but the second division is also an integral part of the process. Any cell capable of undergoing meiosis is called a **meiocyte.** Among plants the immediate products of meiosis are usually spores, and a meiocyte is therefore often called a **spore mother cell.** Spores formed as a result of reduction division are sometimes called **sexual spores** to distinguish them from other kinds of spores.

The prophase of the first division of meiosis is essentially similar to the prophase of mitosis, except that the homologous chromosomes become associated in pairs. Each chromosome becomes visibly double late in the prophase, just as in ordinary mitosis, and a pair of homologous chromosomes therefore consists of four chromatids. The chromosomes remain associated in pairs through the metaphase. At anaphase the members of the pairs are separated; one chromosome of each pair moves to one pole, and the other moves to the other pole. Each of the daughter nuclei formed during the first division has the n rather than the $2n$ number of chromosomes. The first division of meiosis is followed immediately by the second division, in which the two chromatids of each chromosome are separated, just as in ordinary mitosis. As a result of these two divisions, which together comprise meiosis, the original $2n$ nucleus gives rise to four nuclei each with n chromosomes.

The actual reduction in the number of chromosomes per nucleus takes place in the first of the two divisions of meiosis, inasmuch as whole chromosomes rather than chromatids become separated at that time and move to the poles. The second division is essentially mitotic as regards the process itself, but the results differ from those of mitosis for this reason: When the two chromosomes of a pair become separated at anaphase of the first division of meiosis, they often exchange parts rather than separating cleanly. This **crossing over,** as the exchange is called, ordinarily involves only one of the two chromatids from each of the two chromosomes of a pair; of the two chromatids that

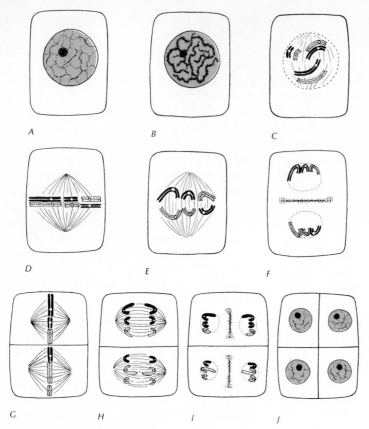

Fig. 10.2   Diagram of meiosis, with n = 3. The chromosomes of one complement are blacked in, and those of the other are stippled. *A*, resting stage, before meiosis; *B*, early prophase; *C*, late prophase, with paired chromosomes; *D*, first metaphase, with paired chromosomes, *E*, early anaphase, showing one possibility of segregation; *F*, telophase; *G*, second metaphase; *H*, second anaphase; *I*, second telophase; *J*, haploid cells after completion of meiosis.

make up a chromosome, one exchanges some parts with a chromatid from the homologous chromosome whereas the other remains unaffected. As a result, the two chromatids that make up a chromosome are no longer identical in all respects. An example for a single pair of chromosomes is shown diagrammatically in Fig. 10.3.

Each of the two daughter cells formed in the first division of meiosis may contain some chromosomes whose two component chromatids are not identical in all respects because one of the two chromatids of the chromosome has been affected by crossing over whereas the other has not. The separation of chromatids during the second division of meiosis may result,

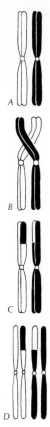

Fig. 10.3 Diagram of crossing over. *A*, a pair of homologous chromosomes, each consisting of two chromatids; *B*, chromosomes twisted together; *C*, chromosomes separate again, after exchanging parts of chromatids; *D*, the four ultimate chromosomes after completion of meiosis.

therefore, in the formation of daughter nuclei that are not genetically identical, and this second division differs in that respect from ordinary mitosis. The amount of crossing over varies according to the size and number of chromosomes, and according to other factors. In general, the fewer and longer the chromosomes, the more abundant the crossing over.

Variations in the Meiotic Process

Reduction division leading directly or eventually to the formation of either iso-gametes or male gametes ordinarily follows the pattern just described, and the four cells produced from each original meiocyte are similar in size and function. Reduction division leading directly or eventually to the formation of female gametes sometimes follows a slightly modified pattern, in which most of the cytoplasm from the original meiocyte is retained with one of the nuclei produced, and the other nuclei degenerate and disappear. The first division in such instances is similar to any other primary meiotic division, except that the cytoplasm is very unequally divided. The nucleus that gets only a small amount of cytoplasm may or may not divide again; in any case, it or its daughter nuclei soon degenerate. The nucleus that gets the larger share of cytoplasm during the first division divides again, just as in any other second meiotic division, except that most of the cytoplasm is again retained with one of the daughter nuclei, and the other daughter nucleus degenerates. The original 2n nucleus thus gives rise to only one functional n nucleus, and this one functional nucleus with the reduced number of chromosomes retains most of the cytoplasm of the original 2n cell; otherwise the meiotic processes leading to the formation of female gametes are identical to those leading to the formation of male gametes.

The production of female gametes in higher animals regularly follows the meiotic pattern in which only a single functional haploid cell is produced from an original diploid meiocyte. Among plants this type of meiosis is less common, but tendencies toward it are shown by most of the plants in which some of the spores produced in reduction division are of the special type known as megaspores.

The Results of Meiosis

There are two direct important results of meiosis. One is the obvious fact that the chromosome number is reduced from $2n$ to n. The other is that the resulting nuclei, with the reduced number of chromosomes, may carry different combinations of hereditary factors and thus be genetically unlike. Without allowing for the complicating effects of crossing over, a relatively simple case in which $n = 2$ may be considered. The two chromosomes in the diploid nucleus that are derived eventually from one of the ancestral gametes may be called A_1 and B_1, and their homologues, derived eventually from the other ancestral gamete, may be called A_2 and B_2. At the metaphase of the first division of meiosis, these chromosomes are arranged in pairs: A_1–A_2 and B_1–B_2. In the ensuing separation, one of the A chromosomes and one of the B chromosomes go to one pole, and the others go to the other pole. One possibility is that chromosomes A_1 and B_1 will migrate to one of the poles and A_2 and B_2 to the other. The resulting nuclei will be similar to those of the parental gametes from which the zygote was formed (except for the effects of crossing over). The other possibility is that chromosomes A_1 and B_2 will migrate to one pole and chromosomes A_2 and B_1 to the other. These combinations are different from those of the parental gametes from which the zygote was formed. Each of the two chromosome complements, A_1–B_2 and A_2–B_1, has a complete set of genes governing the hereditary characters

of the organism, but the two sets may not be alike in all their effects.

The larger the number of chromosomes making up the chromosome complement, the larger the number of different possible chromosome combinations as a result of meiosis. If $n = 2$, the number of possible combinations as a result of reduction division is 4, these being A_1–B_1, A_1–B_2, A_2–B_1, and A_2–B_2 (crossovers still ignored). If $n = 3$, then the number of possible combinations is doubled: Each of the combinations already seen for the condition in which $n = 2$ may exist with the addition of chromosome C_1 or C_2. If $n = 4$, the number of possible combinations is again doubled, and so on. Mathematically, the number of possible combinations of chromosomes obtainable from normal meiosis may be represented as 2^n (i.e., 2 to the power of n) if n represents the number of chromosomes in the cell after reduction. In human beings, for which n typically equals 23, the number of possible combinations of chromosomes obtainable by normal reduction division in the reproductive cells of a single individual is 2^{23}, which is 8,388,608. Inasmuch as crossing over results in the formation of chromatids (and eventually chromosomes) having different combinations of genes from those that previously existed, the effect of crossing over is to increase still further the number of possible chromosome combinations produced by reduction division.

THE BIOLOGICAL SIGNIFICANCE OF SEXUALITY

The biological significance of sexuality is clear; it produces new and different combinations of hereditary characteristics. Mutation and sexual reproduction are the raw materials of evolution.

Consider two individual organisms of the same species, each of which has one beneficial characteristic not possessed by the other. Let one such beneficial character be designated as a_1, and its less beneficial state as a_2, while the other beneficial character is designated as b_1, and its less beneficial state as b_2. By sexual reproduction these two beneficial characters may be combined in the same organism. The principle is the same whether the mature organism is diploid or haploid but is perhaps best introduced at the haploid level. One of the two original haploid organisms will have the constitution a_1b_2, and the other will have the constitution a_2b_1. The zygote formed by fusion of gametes of these two organisms will have the constitution $a_1a_2b_1b_2$. If the factors governing a and b are borne on different chromosomes, then reduction division from the zygote gives us four possible combinations of these factors: a_1b_1, a_1b_2, a_2b_1, and a_2b_2. Two of these nuclei repeat the parental types, and the other two represent combinations different from the parental types. One of the two combinations different from the parental types embraces both the favorable characteristics from the two parents, whereas the other is burdened with both the unfavorable characteristics. If three pairs of characters are considered instead of two, and if each of these characters is governed by a single gene, then the number of possible combinations of these characters would be eight, and so on.

We have seen in Chapter 9 that many characters are controlled by groups of genes, rather than by single genes, but this is merely an added complexity which does not change the basic situation. By gametic union and reduction division, new and different gene combinations are obtained, with correspondingly different kinds of individuals as a result. Most organisms produce many more offspring than can possibly survive, and the resulting competition for survival gives an opportunity for the more suitable gene combinations to be perpetuated. The mechanism of evolution is further discussed in Chapter 30.

SUMMARY

Among the higher animals, including humans, sexual reproduction is associated with the differentiation of distinctly male and female individuals, and in common usage this differentiation is considered the essential feature of sexuality. When some of the simpler kinds of organisms are also considered, however, an expanded concept is useful and proper. The most nearly characteristic feature of sexuality, in the broad sense, is the regular, alternating doubling and halving of the number of chromosomes in the nuclei. The doubling results from fusion of gametes; the halving is brought about by meiosis. Reduction division (meiosis) is just as much a part of the sexual cycle as is gametic fusion. Sexual reproduction results in the formation of new and different combinations of existing genes. The appearance of new characters (or the modification of existing ones) by gene mutation, and the formation of new and different combinations of existing characters by sexual reproduction, together constitute one of the most important forces in the march of organic evolution.

SUGGESTED READING

Crow, J. F., Genes that violate Mendel's rules, *Sci. Amer.* **240**(2):134–146, February 1979.

PART III
ORGANISMS

Moccasin flower, *Cypripedium acaule*, a common orchid of the northeastern United States.

Chapter 11

The Prokaryotes: Bacteria, Viruses, and Blue-Green Algae

Escherichia coli, the human colon bacillus, being attacked by bacteriophage. (x 70,000.)
[Electron micrograph courtesy of T. F. Anderson.]

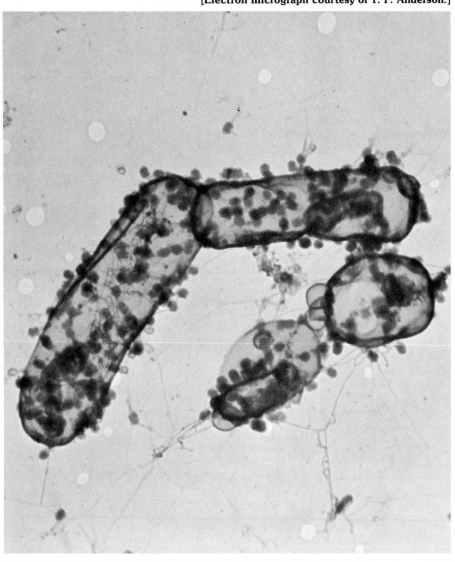

In Chapter 4 we noted that prokaryotic cells, characteristic of the bacteria and blue-green algae, have a less complex organization than the eukaryotic cells found in other plants and animals. The difference between the prokaryotes and eukaryotes is indeed so profound that some biologists consider them to constitute distinct kingdoms of organisms. On the basis of comparisons of modern organisms, plus the fossil record, biologists are agreed that the prokaryotes are more primitive than and ancestral to the eukaryotes. Before proceeding to a comparison of prokaryotes and eukaryotes, we should note that it is customary and convenient to consider the viruses in connection with the bacteria, but the ancestry and proper taxonomic status of the viruses are still highly debatable. The viruses are ignored in the following table of some of the more obvious differences between prokaryotes and eukaryotes.

In spite of these several features in

Prokaryotes	Eukaryotes
Protoplasm a relatively rigid, usually nonvacuolate gel, relatively resistant to desiccation, osmotic shock, and thermal denaturation	Protoplasm typically more fluid, generally vacuolate, and more sensitive to desiccation, osmotic shock, and thermal denaturation
Membrane-bounded protoplasmic organelles lacking	Membrane-bounded protoplasmic organelles of various sorts present in the cytoplasm
Nucleus, when present, relatively simple in structure, and without a bounding membrane	Nucleus with complex internal structure, and bounded by a definite membrane
Chromosomes without histone proteins, and with little or no protein of any sort	Chromosomes usually with histone proteins as an important structural component
Cell division amitotic	Cell division typically mitotic
Typical sexual processes lacking, but genetic material sometimes exchanged by other "parasexual" processes	Typical sexual processes (with alternating nuclear fusion and reduction division) often present
Flagella, when present, of relatively simple structure	Flagella, when present, compound
Many members can use atmospheric nitrogen	Cannot use atmospheric nitrogen
Always of small size, ranging from single celled and microscopic to multicellular but without differentiated tissues or organs	Variously unicellular and microscopic to large and complex

common, the bacteria and blue-green algae play very different roles in the balance of nature. With minor exceptions, the blue-green algae are food producers, whereas the bacteria are consumers. The role of the blue-green algae foreshadows that of eukaryotic plants, which are indeed much more important as food producers than are the blue-green algae. The world could get along handily (after some upset) without the blue-green algae, but the bacteria are vitally important to the downswing of the recycling process, providing the essential raw materials to fuel the upswing. The bacteria share this role with the fungi, but the fungi would be hard put to fill it alone, and we would have a very different sort of world indeed if all the bacteria were eliminated in some unforeseeable catastrophe.

Bacteria

HISTORY AND DEFINITION

Bacteria were first observed by Leeuwenhoek in pepper infusions which he prepared in 1676. His drawings of bacteria from his own teeth, made in 1683, include examples of the three common shapes currently recognized in bacterial classification (spherical, rod shaped, and spiral). Early microbiologists did not usually distinguish the bacteria as a group from other microorganisms, and the term *animalcule* (or its equivalent in other languages) was rather generally applied to all unicellular organisms and some very small multicellular ones. The bacteria were first recognized and defined as a group by the German botanist Karl Wilhelm von Naegeli (1817–1891), who proposed the name Schizomycetes (fission fungi) for them in 1857.

Bacteria may be defined as prokaryotes

that do not contain phycobilin pigments, and that either do not contain chlorophyll or (in a relatively few species) do contain chlorophyll but do not release oxygen as a result of photosynthesis. The photosynthetic bacteria are of great biologic interest, but very little present-day importance. About a third of the kinds of bacteria are visibly pigmented, and the remainder are essentially colorless. Most bacteria are unicellular, but some form simple or branched filaments, or even nonfilamentous colonies of coherent cells. Most kinds have a definite cell wall, which is usually nitrogenous. Bacteria store food reserves in a number of forms, including droplets of fat, granules of protein, and especially granules of various kinds of polysaccharides. Both starch and glycogen are included in the wide range of polysaccharides found in diverse bacteria. Bacteria ordinarily reproduce by cell division, which is amitotic. Some forms produce spores. Transfer of genetic material from one cell to another has been demonstrated in a number of kinds of bacteria, but in all carefully studied examples the process is parasexual rather than truly sexual, that is, genetic material is exchanged without the formation of a zygote.

DISTRIBUTION

Bacteria are widely distributed, on and beneath the surface of the earth, in fresh water and in the sea, on and in other organisms, and on the dust particles that float in the air. They do not generally occur inside normal, healthy cells of other organisms, but otherwise they are usually found wherever food is available to them. Almost all naturally occurring organic compounds can be used as food by one or another kind of bacteria, and some can derive energy through oxidation of inorganic substrates. Many bacteria require free oxygen; others can use it or grow anaerobically without it; others are obligate anaerobes. Some bacteria live and grow at temperatures as high as 90°C.

CLASSIFICATION

In sexually reproducing organisms, species are usually interbreeding or potentially interbreeding populations that can be distinguished morphologically from other such populations. Neither the concept of interbreeding populations nor the concept of morphologically recognizable species is readily applicable to the bacteria. Bacteria which are morphologically similar may be physiologically so different that they must be regarded as separate species in any generally useful classification. Among the pathogenic bacteria, a single physiological difference, caused by a single gene mutation, may be of critical importance in determining the effect of the bacterium on the host and vice versa. The fact that bacterial populations consist of very large numbers of individuals increases the likelihood that mutants may be produced. Because of these problems, the definition of species in bacteria is especially difficult, and there are wide differences of opinion as to the total number of species in the group. More than 1000 species of bacteria are usually accepted, and many others have not yet been characterized.

It is customary to recognize about 10 orders of bacteria. One of these, the Eubacteriales, makes up more than half the class. Another large order, the Pseudomonadales, is similar to the Eubacteriales in most respects: The Pseudomonadales have rod-shaped to spiral cells with terminal flagella or no flagella; the Eubacteriales have spherical to rod-shaped cells with flagella more generally scattered over the surface. It is these two orders that are usually intended when reference is made to bacteria without further qualification. Other orders have special features that set them apart from the main group of bacteria. For example, the Actinomycetales are filamentous, the Spirochaetales are spiral and lack a cell wall, and the Mycoplasmatales lack a cell wall and are unstable in shape.

GENERAL MORPHOLOGY AND CYTOLOGY

Most bacteria are unicellular, the cells having one of three more or less definite shapes: (1) spherical, (2) rod shaped, and (3) spiral or corkscrew shaped. Spherical bacteria are called **cocci** (sing., coccus); rod-shaped bacteria (Fig. 11.1) are called **bacilli** (sing., bacillus), or merely rods; and corkscrew-shaped bacteria (Fig. 11.2), except the spirochaetes, are called **spirilla** (sing., spirillum). These differences are not absolute, nor do they define proper taxonomic groups, but the descriptive

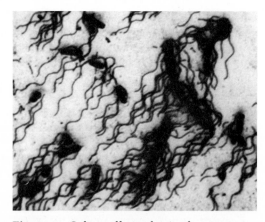

Fig. 11.1 *Salmonella typhosis*, the cause of typhoid fever. Note the long flagella. (× about 6000.) [Courtesy CCM: General Biological, Inc., Chicago.]

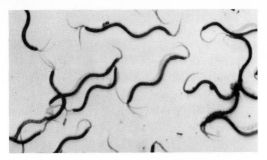

Fig. 11.2 *Spirillum* sp. (× about 1250.) [Courtesy CCM: General Biological, Inc., Chicago.]

terms are useful. Cocci vary from about 0.2 to 4 nanometers in diameter; bacilli and spirilla usually resemble cocci in diameter, but are longer, sometimes as much as 40 nanometers. Cells of some of the sulfur bacteria are as much as 60 nanometers long and 25 nanometers thick. Some of the spirochaetes are even longer, up to 500 nanometers, but are still very slender.

The Cell Wall

The cell wall of bacteria generally consists of two layers, the outer layer commonly being the thicker. Amino acid polymers (proteins and shorter-chain polypeptides) are usually the principal constituents. A small proportion of lipid is usually associated with the proteins and polypeptides, perhaps as a cementing substance, and the resulting wall layers have no evident interior structure. Various polysaccharides are also frequently associated with the other wall constituents, and in some few bacteria the wall is composed largely of cellulose arranged into micelles, just as in higher plants.

Hexose amines are also commonly found in bacterial cell walls. They differ from hexose sugars, such as glucose, in having an amine group (i.e., a small nitrogen-containing group, such as NH_2) substituted for one or more of the hydroxyl (OH) groups of the sugar.

It is sometimes convenient to refer to the amino acid polymers and associated hexose amines and polysaccharides of bacterial cell walls collectively as **peptidoglycan.**

The cell wall is often surrounded by a thick gelatinous sheath. This may form a discrete capsule surrounding the cell or may dissolve into the medium as slime or mucus. Typically the gelatinous sheath is composed of polysaccharide, with different polysaccharides being produced by different kinds of bacteria, but in some bacteria the capsule is proteinaceous. Phospholipids, iron compounds, and polypeptides may also occur in the sheath.

Flagella

Many of the bacteria are nonmotile. Many others swim by means of one, two, or several slender, threadlike or whiplike projections, known as **flagella,** which are generally longer than the breadth of the cell. The flagella commonly have a rotary motion, and rates as high as 40 revolutions per second have been calculated for some species.

A bacterial flagellum consists of a single proteinaceous fibril, with relatively small protein molecules helically wound to form a hollow core. The flagellum passes through the cell wall and often terminates in a basal granule in the outer part of the cytoplasm. In some bacteria several flagella are aggregated or even loosely twisted together and function as one, but no bacteria are known to have the complex type of flagellum, with two central and nine peripheral strands, that is typical of flagellated cells in eukaryotes.

Cytoplasm

The cytoplasm of bacteria is less highly organized than that of eukaryotes. It contains ribosomes and various kinds of granules, including polysaccharide food reserves, and RNA is regularly present. Mitochondria, plastids, golgi bodies, and a central vacuole are absent.

The cytoplasm is bounded by a unit membrane similar to the plasma membrane of eukaryotic cells. Often there is another unit membrane outside the peptidoglycan layer of the cell wall. Bacteria that have a cell wall but lack this outer membrane adsorb and retain a stain devised by the Danish physician Hans C. J. Gram (1853–1938), and are said to be Gram-positive. Bacteria that do not take Gram's stain are said to be Gram-negative. Some other chemical characteristics of various kinds of bacteria appear to be correlated with the reaction to Gram's stain.

The Hereditary Material

The protoplast of a bacterial cell characteristically includes a DNA-bearing organelle that usually divides at about the same time as the cell. Sometimes this organelle divides well before the division of the cell, so that the cell contains two or more such organelles of apparently identical composition.

This DNA-bearing organelle has often been called a nucleus or regarded as the evolutionary precursor of the eukaryotic cell nucleus, but it might better be compared to the eukaryotic chromosome in its extended (interphase) state. It consists of an elongate double helix—a great macromolecule—of DNA. The bacterial strand of DNA is conventionally said to be circular, or to form a ring. This is true in the sense that it is continuous, forming a closed structure without free ends, but it does not form a geometrical circle. Its shape might better be compared to that of a long piece of string, thrown loosely onto the floor after the two ends have been joined. This organelle in bacteria differs from the chromosome of eukaryotes in its "circular" form, and in not having protein as a part of the fundamental structure. Bacteria have nothing at all comparable to the nuclear membrane, nucleolus, and nuclear sap that are familiar components of the eukaryotic nucleus. The trend of current usage is to call this major DNA-bearing organelle of bacteria a chromosome.

In addition to the circular chromosome, many bacterial cells have one or more also circular but very much shorter strands of DNA, called **plasmids.** Plasmid DNA has a double helical structure, just as in bacterial and eukaryotic chromosomes. Plasmids are further discussed on p. 176.

SPORES

Many bacteria produce thick-walled resting cells which are highly resistant to such unfavorable environmental conditions as drought and extremes of heat and cold. Although they may not be really comparable to the spores of other kinds of organisms, these resting cells are generally called **spores,** or more specifically, **endospores.** Sporulation is commonly induced by conditions unfavorable for growth, but the physiological mechanism of the response is complex. An insufficient supply of glucose in the cell is often the trigger.

During the formation of a bacterial spore from a vegetative cell, there is a change in the nature of some of the proteins of the protoplasm, and most of the protoplasmic constituents other than water become concentrated into a rela-

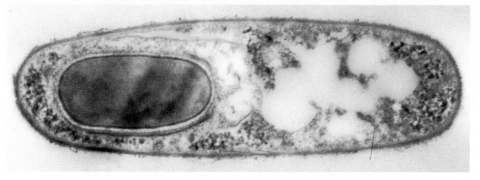

Fig. 11.3 *Bacillus cereus*. Electron micrograph of thin section, showing spore formation. (× about 30,000.) [Courtesy of George B. Chapman.]

tively small mass around which a thick new wall is formed (Fig. 11.3). The volume of the spore is commonly about 1/10–1/4 that of the vegetative cell from which it is formed. Most of the water remaining in the cell is probably bound (by hydrophilic colloids or otherwise) so that it is not available as a medium for chemical reactions. Metabolic activities slow down to a rate at which they cannot be detected, so that from a practical standpoint the spore is in a state of suspended animation. Spores of bacteria have been sealed in containers at a high vacuum at the temperature of liquid helium (—269° to —271°C) without losing their power to give rise to normal, vegetatively active cells. Resistance of spores to drought is doubtless similar to their resistance to cold, relating to dehydration and suspension of normal metabolic activities.

Many bacteria can also withstand temperatures far below freezing while still in the vegetative state. In such cases, most of the bacterial population dies at or soon after the time of freezing, but those individual cells not immediately killed may persist more or less indefinitely, although they are not quite so cold resistant as spores.

When suitable conditions for growth

are re-established, the spore germinates, either immediately or after a necessary period of rest. **Germination** is a general term applied to the resumption of growth by any spore, seed, or other propagule after a period of dormancy. The spore swells, the old wall splits, and the vegetatively active cell, with its own new wall, is released (Fig. 11.4). A single vegetative cell (representing the complete organism) usually gives rise to only a single spore, and the germination of the spore gives rise to only a single vegetative cell. Therefore spores of unicellular bacteria are not reproductive bodies in the usual sense. In this respect ordinary bacterial spores differ from most other bodies generally called spores.

Several other kinds of spores are produced by one or another special group of bacteria, such as actinomycetes and sheathed bacteria.

REPRODUCTION

Bacteria ordinarily reproduce by means of cell division. The cell appears merely to become constricted into two cells, and the process is commonly referred to as **fission,** binary fission, or simple fission. "Simple fission" is more complex than

Fig. 11.4 *Bacillus cereus.* Electron micrograph of thin section, showing a germinating spore. (× about 22,000.) [Courtesy of George B. Chapman.]

casual observation would suggest, however. Very often a transverse double membrane develops, from the margins toward the center, before the visible fission occurs (Fig. 11.5). The wall is then deposited between the two layers of the membrane, progressing from the outside to the center. We noted earlier that chromosomal division in bacteria may precede or be associated with cell division.

Under favorable conditions, bacterial cells may divide as often as once every 20 minutes. Starting with a single cell, such a rate of reproduction would give 2 cells in 20 minutes, 4 cells in 40 minutes, 8 cells in 1 hour, 64 cells in 2 hours, 512 cells in 3 hours, 4,096 cells in 4 hours, 32,768 cells in 5 hours, 1,073,741,824 cells in 10 hours, and more than 4,000,000,000,000,000,000,000 in 24 hours. This figure may be expressed in a more convenient mathematical form as 4×10^{21}. If the same rate of increase could be maintained for about 3 days, the volume of bacteria would be greater than that of the earth. Competition for space or nutrients (or both) soon limits the rate of bacterial reproduction in any particular instance, even if other factors remain favorable, but the potentiality for rapid increase to exploit a food supply is obviously enormous.

SEXUALITY

Typical sexual processes, involving fusion of gametes (and their nuclei) to form a zygote, followed eventually by reduction division, have not been demonstrated

Fig. 11.5 *Bacillus cereus.* Electron micrograph of thin section of dividing cell. (× about 35,000.) Note the partition developing from the margins toward the center. The attenuated dumbbell in the middle is a "nucleus" in process of division. The cell evidently has other nuclei as well. [Courtesy of George B. Chapman and James Hillier.]

in bacteria and probably do not occur. It is abundantly clear, however, that many kinds of bacteria do transfer genetic information from one individual to another. Processes such as those in bacteria, which transfer genetic material without the formation of a zygote, are sometimes called **parasexual processes.**

Parasexual processes in bacteria fall into three general categories: transformation, conjugation, and transduction. In all of these, genes are transferred unidirectionally from a donor cell to a recipient; there is no regular reciprocal transfer.

Transformation is the transfer of small bits of DNA (sometimes only a single gene) from one bacterial cell to another, without the formation of a direct passageway and without the use of a virus as a

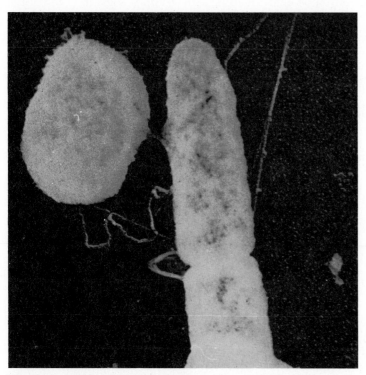

Fig. 11.6 *Escherichia coli.* Electron micrograph showing two strains, with different characteristic shapes, in conjugation. ($\times$ 7000.) The more elongate individual at the right is also in process of fission. In this instance the genetic material passes only one way. The cell at the right continues to divide, without showing any influence of the other cell. The progeny of the cell at the left show a variety of types combining the characters of the two cells in conjugation; and new recombinant types continue to be formed for a number of generations before each line becomes stabilized. [Courtesy of T. F. Anderson, E. L. Wollman, and François Jacob.]

carrier. The DNA is released into the surrounding medium and is taken into another cell by a complicated physico-chemical process. Cell-free extracts from one strain of a bacterial species can be used in the laboratory to transfer genetic information to another strain, and the same sort of transfer evidently occurs in nature.

Bacterial **conjugation** involves the formation of a tube between two adjacent cells. Both plasmid and chromosomal DNA can pass through the conjugation tube, always in one direction only. Only one of the two complementary strands of a double helix is transferred from one cell to another. Plasmid DNA makes the trip in the form of a complete ring that can replicate to restore the original double helical structure; both the donor and the recipient cell thereafter carry a reconstituted plasmid. Transfer of chromosomal DNA through a conjugation tube is much less common. It requires that one of the strands of the double helix break to produce free ends. Part (or even all) of the broken strand can then pass into the recipient cell. There parts of the transferred strand can be incorporated into the chromosome during a subsequent replication.

Plasmids can often be transferred between rather widely different species. They typically carry genes that are helpful under some conditions but not vitally necessary for ordinary metabolism. Factors for resistance to antibiotic drugs are often borne on plasmids. Thus a mutation for drug resistance may spread far beyond the bacterial species in which it originated. The potential hazards to human health are evident and are a justified cause of serious concern.

Transduction is the transfer of genetic material from one bacterial cell to another via a bacteriophage (a virus that infects bacteria). It is discussed on p. 190.

ABSORPTION OF FOOD

Some bacteria require only inorganic nutrients, but the majority require organic compounds of one sort or another. Of these, some require only an organic energy source, such as glucose, and can manufacture all the other organic compounds they need. Others require one or more organic compounds in addition to those used as a source of energy. A few kinds of bacteria require such a complex organic environment that it has not yet been possible to culture them outside the living cells that they parasitize. Some bacteria that manufacture their own carbohydrates from inorganic materials still require some organic compounds, especially nitrogenous compounds, from an external source.

The more complex organic compounds, such as proteins and polysaccharides, which are used by bacteria, must be digested before they can be absorbed. Digestion is accomplished by enzymes that the bacteria release into the surrounding medium. Such enzymes are sometimes called **exoenzymes,** to distinguish them from **endoenzymes,** which are retained in the cell. All living cells contain endoenzymes, but among plants the production of exoenzymes is chiefly confined to the bacteria and fungi, which absorb food from their surroundings. The ability of some bacteria to cause certain diseases rests on their capacity to digest living tissues. Some of the more virulent toxins produced by bacteria, however, are by-products of their metabolism, rather than digestive enzymes.

SOME TERMS DESCRIBING THE NUTRITION AND ECOLOGY OF ORGANISMS

An organism that makes its own food from raw materials is said to be **autotrophic.** Autotrophic organisms either are plants or stand on the borderline between plants and animals, and their nutrition is therefore said to be **holophytic.** Organisms that do not make their own food are said to be **heterotrophic.** Heterotrophic organisms may obtain their food by eating (ingesting) it or by absorbing it. Organisms that eat their food either are animals or are on the borderline between plants and animals, and their mode of nutrition is therefore said to be **holozoic.** Some heterotrophic plants that absorb their food use dead organic matter and are said to be **saprophytes,** or **saprobes.** Others obtain their food from living organisms and are said to be **parasites.** The organism that provides food for a parasite is called the **host.** The term parasite is also applied to animals which are so much smaller than their host that the depredations of any one individual seldom cause serious harm. The term host is also commonly extended to cover any organism that provides "lodging," that is, a place of habitation, for another organism. No sharp line can be drawn between parasites and saprophytes. Some of the "saprophytic" wood-rotting fungi utilize only the nonliving parts of their host and grow just as well on a log as on a living tree; yet they may cause the death of living cells of the host, so that they might with some reason be called parasites.

Some kinds of organisms, including many bacteria, are either parasitic or saprophytic, according to the circumstances. An organism that is usually parasitic but has the power (the faculty) of becoming a saprophyte under some circumstances may be called a **facultative saprophyte.** Likewise an organism that is usually saprophytic but has the power of becoming a parasite under some circumstances may be called a **facultative parasite.** A parasite that cannot live in any other way is called an **obligate parasite,** and a saprophyte that cannot live in any other way is called an **obligate saprophyte.** Most parasitic bacteria can be grown on nonliving media and, from a laboratory standpoint, are therefore facultative saprophytes. The term facultative saprophyte is also applied to autotrophs that can live as saprophytes.

When two different kinds of organisms spend much or all of their lives in close physical association, deriving mutual benefit, they are said to have a **symbiotic** relationship, and each of the two organisms is a symbiont. The term **symbiosis** is sometimes extended to include all prolonged close associations of two different kinds of organisms regardless of benefit or harm, but in this broad definition the word loses much of its conceptual utility. A close physical association between two different kinds of organisms, in which one kind is benefited and the other is neither benefited nor harmed, is called **commensalism.** A plant that grows attached to some larger plant, without deriving nourishment from its host, is called an **epiphyte.**

All these different terms, except holozoic, may be used to describe some bacteria, or the relationships between some bacteria and other organisms.

THE NITROGEN CYCLE

The elements used by plants go through a cycle of chemical changes as they are absorbed, used, and then restored to a form in which they can again be absorbed. We are beginning to learn that similar principles apply to materials used in human societies. The earth, like a spaceship, has limited resources that must be recycled. The nitrogen cycle (Fig. 11.7), in which bacteria play several important roles, is an interesting and delicately balanced cycle that is commonly studied in beginning botany courses.

Most plants absorb nitrogen from the soil principally in the form of nitrate (NO_3^-) ions, although they can also absorb ammonium (NH_4^+) ions. The plants use the nitrogen in the formation of proteins and other compounds, such as nucleic acids. Animals that eat the plants digest and reassemble these proteins to form their own proteins.

Since nitrate ions are continually being removed from the soil by green plants, the supply of nitrates would eventually be exhausted if there were no way to replenish it. Neither the elaborated nitrogenous compounds formed by plants and animals nor the nitrogenous wastes excreted by

animals are directly available for reuse by most green plants. The conversion of nitrogen from proteins and other nitrogenous compounds into the form of nitrates involves several steps and is due in large part to the activities of bacteria, although some of the earlier steps are also taken by the fungi.

Ammonification

The nitrogenous compounds of both plant and animal bodies and the nitrogenous wastes excreted by animals serve as a food supply for many different kinds of bacteria (and fungi) that cause decay. Ammonia (NH_3) is produced as a result of the decay. Some of the ammonia escapes into the air and is lost to the cycle, but most of it reacts immediately with water to form ammonium hydroxide (NH_4OH), and various ammonium salts are derived from the ammonium hydroxide without the intervention of bacteria.

Remains of higher plants are usually relatively poor in nitrogen and rich in carbohydrates such as cellulose. When

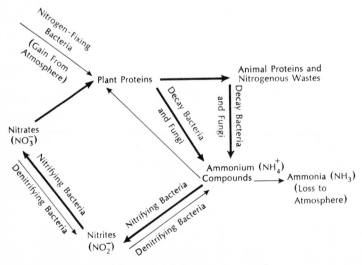

Fig. 11.7   Principal features of the nitrogen cycle.

these plant remains are used as food by the bacteria, all the nitrogen is needed for the production of microbial cell material. The carbohydrates supply enough energy so that all the nitrogen can be assimilated, but since some of the carbohydrate is oxidized in the process, the ratio of carbohydrate to nitrogen is decreased. Another organism may then feed on the first organism and also on the remaining carbohydrate. It assimilates all the nitrogen and oxidizes more carbohydrate, again decreasing the ratio. This continues until the energy supply from carbohydrate is insufficient to permit the assimilation of all the nitrogen, and the protein is used for energy. Proteins consist essentially of carbohydrate derivatives with amine groups attached. When protein is used for energy, the amine group is split off as ammonia, and the carbohydrate derivative is then oxidized. Thus ammonification occurs when there is more organic nitrogen than can be assimilated.

Nitrification

Ammonium ions can be absorbed directly by most plants as a source of nitrogen. However, when dissolved in water, ammonium compounds readily produce some gaseous ammonia (NH_3) which tends to escape into the air and be lost to the cycle. Oxidation of ammonium compounds to the more stable nitrates is brought about exclusively by the activities of several kinds of bacteria. The first step, in which ammonium compounds are used and nitrites (NO_2^- compounds) are produced, is carried on especially by *Nitrosomonas*, a very short, flagellated, aerobic rod. The further oxidation of nitrites to nitrates is due chiefly to *Nitrobacter*, a very short, nonmotile, aerobic rod. *Nitrobacter*, *Nitrosomonas*, and other bacteria

that oxidize inorganic nitrogen compounds are called **nitrifying bacteria.** Energy is released in the oxidation of these nitrogen compounds, and some of this energy is used by the nitrifying bacteria in the manufacture of their own food from raw materials.

Denitrification

There are several factors that increase the complexity of the nitrogen cycle. We have noted that nitrogen may be lost to the cycle by the escape of gaseous ammonia into the air. A number of kinds of bacteria cause denitrification, a change essentially opposite to nitrification. Sometimes the denitrification proceeds to the formation of nitrous oxide (N_2O) or free nitrogen, which escapes into the air and is lost to the cycle.

The denitrifying bacteria are facultative or obligate anaerobes; they extract oxygen from nitrates or nitrites, rather than using atmospheric oxygen. Essentially what happens is that the oxygen from the nitrates or nitrites is used to oxidize the hydrogen of the organic foods, forming water as one of the end products. More energy is released in the oxidation of the hydrogen than is used in separating the oxygen from the nitrates or nitrites, so that there is a net transfer of energy into the metabolic system of the bacteria. Most denitrifying bacteria use atmospheric oxygen, if it is available in sufficient quantity, and carry on denitrification only under conditions of oxygen shortage. Denitrification is favored by poor aeration of the soil, and nitrates may eventually disappear from water-logged soils.

Nitrogen Fixation

Some of the bacteria and some of the blue-green algae differ from all other organisms in their ability to use atmospheric nitrogen. Some of the **nitrogen-fixing bacteria,** as they are called, are saprophytes which live in the soil; others are symbionts, most of which live in the roots of some higher plants, especially legumes. The two most important genera of free-living (i.e., nonsymbiotic) nitrogen-fixing bacteria are *Azotobacter* and *Clostridium*. *Clostridium* is an anaerobic, spore-forming rod. *Azotobacter* is a relatively large, aerobic, non–spore-forming short rod.

Many plants of the legume family (Fabaceae, also called Leguminosae), including clover, beans, peas, lupines, vetch, and alfalfa, develop characteristic swellings, called **nodules** (Fig. 11.8), on their roots. Some of the cells of these nodules are inhabited by bacteria belonging to the genus *Rhizobium*, an aerobic, non–spore-forming rod (in culture), the cells of which usually exhibit bizarre, irregular shapes when growing in the nodule. The development of the nodule is apparently stimulated by growth promoting substances produced by the bacteria, which invade near the root tip.

Fig. 11.8 Roots of soybean, showing bacterial nodules. [Courtesy of the Nitragin Co., Inc.]

Some nitrogen compounds produced by the nodule bacteria become available to the host plant for the formation of protein and other nitrogenous compounds, but the details of the process are not yet understood.

Leguminous crops are frequently grown in rotation with other crops as a means of restoring nitrogen compounds to the soil. If the crop itself is harvested and carried away, there may be little or no benefit to the soil, because the amount of nitrogen left in the roots may not exceed the amount of nitrogen absorbed from the soil and removed with the crop. Therefore the crop is often plowed under; it is then spoken of as green manure. This practice improves soil texture by adding humus as well as adding nitrogen.

A sprinkling of other kinds of flowering plants, in several different families, also harbor nitrogen-fixing bacteria in root nodules. Most of these latter bacteria are filamentous and belong to a group called actinomycetes.

It has recently been discovered that several kinds of nitrogen-fixing bacteria (including *Azotobacter*) grow in the soil around the roots of rice. Rice has a well-developed internal aerating system, and some of the oxygen and gaseous nitrogen delivered to the roots escapes into the water-logged soil, providing a milieu suitable for the bacteria. The relationship between the rice and the bacteria has been described as associative symbiosis. It appears that different **cultivars** (cultivated varieties) of rice differ in the degree to which they tolerate or promote the growth of the bacteria. A breeding program is being undertaken to exploit the agricultural possibilities. Maize, sugar cane, and some other kinds of plants foster a similar sort of associative nitrogen fixation, at least under some conditions.

The free-living nitrogen-fixing bacteria are widespread in the soil, being absent mainly from highly acid soils, and it is probably a safe assumption that they are already present in any natural soil capable of supporting them. The symbiotic nitrogen-fixing bacteria are also widely distributed in nature, but no one strain can inhabit all different kinds of legumes, and it is often advisable to dust legume seeds with the proper bacteria before planting, especially if the particular crop has not previously been grown in the area. The legume can grow in the absence of the bacteria, but it does not do as well, and has no special value then as a soil builder.

Summary of the Nitrogen Cycle

The principal steps in the nitrogen cycle are these:

(1) Nitrates are absorbed by plants and used in the manufacture of proteins and other organic nitrogen compounds. Some of the proteins produced by the plants are consumed by animals and used in the formation of their own proteins.

(2) Plant and animal proteins (and other nitrogenous compounds) and nitrogenous wastes are used as food by decay bacteria and fungi, resulting in the formation of ammonia and ammonium compounds.

(3) Ammonium compounds are oxidized to nitrites by *Nitrosomonas* and other bacteria.

(4) Nitrites are oxidized to nitrates by *Nitrobacter* and other bacteria.

The bacterial steps in the nitrogen cycle can be bypassed by plants that harbor symbiotic fungi in their roots, forming a fungus-root combination known as a mycorhiza (see Chapter 14). The fungus absorbs complex organic nutrients from decaying organic remains such as leaves and stems, and some of the nitrogen is passed on to the root of the host plant. In moist, tropical, forested regions most of the nitrogen is probably cycled through mycorhizae rather than through nitrifying bacteria.

BACTERIA AND DISEASE

Before about 1850 bacteria were not generally suspected of causing disease. In 1850 two French workers noted large numbers of microscopic rods in the blood of cattle that had died from anthrax. Later it was shown that these rods, now known as *Bacillus anthracis*, are living organisms and are regularly present in infected animals. Final proof of the bacterial cause of anthrax was obtained in 1876 by the German physician and bacteriologist Robert Koch (1843–1910, Fig. 11.9), who was able to obtain the bacteria in essentially pure culture and to cause the disease by injecting these bacteria into the bloodstream of cattle. Several years earlier Pasteur had for practical purposes demonstrated the bacterial nature of the pebrine disease of silkworms, but his experiments did not rigidly eliminate the possibility of some other interpretation. The first definite proof that a disease may be caused by bacteria is therefore generally attributed to Koch.

Some of the diseases caused by bacteria are diphtheria, gonorrhea, leprosy, pneumonia, plague, scarlet fever, syphilis, tetanus, tuberculosis, typhoid fever, fire blight of pears, crown gall of many plants, and soft rot of potatoes and fruits. Botulism is caused by one of the most powerful known poisons, produced by *Clostridium botulinum*, a spore-forming anaerobic bacillus. The organism thrives on meat and other high-protein foods (such as peas

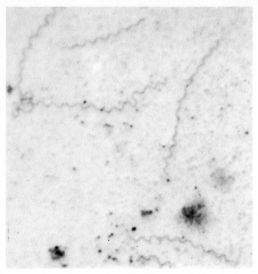

Fig. 11.10 *Treponema pallidum*, the cause of syphilis. (× about 2500.) [Courtesy CCM: General Biological, Inc., Chicago.]

Fig. 11.9 Robert Koch (1843–1910), German bacteriologist and physician, who first proved that a disease can be caused by bacteria. [Courtesy of the New York Academy of Medicine.]

and beans) in the absence of air, and the spores often survive the temperature used in home canning. Fortunately, the poison itself is readily destroyed by heat, and it is therefore generally recommended that all home-canned high-protein foods be boiled for 10 or 15 minutes before they are used or even tasted. The temperatures used in commercial canning of high-protein foods are normally high enough to destroy the spores of *C. botulinum*.

AUTOTROPHIC BACTERIA

Some kinds of bacteria can manufacture their own food from raw materials; that is, they are autotrophic. An external source of energy is of course necessary. Auto-

tropic bacteria that use light as the source of the requisite energy for food manufacture are said to be **photoautotrophic**, or **photosynthetic**. Autotrophic bacteria that use the energy released in the oxidation of certain inorganic compounds are said to be **chemoautotrophic**, or **chemosynthetic**.

Chemoautotrophic (Chemosynthetic) Bacteria

We have noted that the nitrifying bacteria use, in the formation of their own food, some of the energy released by the oxidation of nitrogen compounds. The iron bacteria are another group of chemosynthetic autotrophs. They facilitate the oxidation of iron, and use in food production some of the energy that is thereby released. One such oxidation may be represented as follows:

$$4FeCO_3 + O_2 + 6H_2O \rightarrow$$

ferrous carbonate

$$4Fe(OH)_3 + 4CO_2 + energy$$

ferric hydroxide

Some iron bacteria can oxidize manganese as well as iron as a source of energy.

Some of the iron bacteria, such as *Gallionella*, a unicellular, curved rod belonging to the Eubacteriales, may be fully autotrophic. Others, such as *Sphaerotilus*, a multicellular, filamentous bacterium belonging to the order Chlamydobacteriales (sheathed bacteria), require an organic source of nitrogen. *Gallionella* is widely distributed in iron-bearing waters. *Sphaerotilus* is often found in sewage-disposal plants and in polluted streams.

There are two groups of autotrophic sulfur bacteria, one chemosynthetic, the other photosynthetic. The chemosynthetic sulfur bacteria are colorless and utilize in the manufacture of their food some of the energy released in the oxidation of sulfur. Some of the chemical processes relating to the sulfur may be represented as follows:

$$2H_2S + O_2 \rightarrow 2S + 2H_2O + energy$$

hydrogen sulfide

$$2S + 2H_2O + 3O_2 \rightarrow 2H_2SO_4 + energy$$

sulfuric acid

Photosynthetic Bacteria

Several related genera of bacteria contain chlorophyll and associated carotenoid pigments and carry on photosynthesis. They vary in actual color from green (green bacteria) to more often purple, red, or brownish (all usually called purple bacteria), according to the concentration of the various pigments.

Bacterial photosynthesis has been dis-cussed and compared to typical photosynthesis in Chapter 7. The photosynthetic bacteria apparently all (except *Halobacterium* contain a specific chlorophyll, called bacteriochlorophyll, and they often contain other chlorophylls as well. Bacterial photosynthesis does not result in the release of oxygen.

Many of the photosynthetic bacteria use sulfur compounds such as H_2S in the photosynthetic process. If carbohydrate is, for convenience, represented by the generalized formula (CH_2O), two of the more common bacterial photosynthetic processes may be represented as follows:

$$2H_2S + CO_2 + light\ energy \rightarrow$$

hydrogen sulfide

$$(CH_2O) + 2S + H_2O$$

$$2S + 3CO_2 + 5H_2O + light\ energy \rightarrow$$
$$3(CH_2O) + 2H_2SO_4$$

sulfuric acid

The genus *Halobacterium*, which inhabits highly salty water, as in drying pools along the seacoast, has a unique photosynthetic apparatus that lacks chlorophyll. Its light-absorbing pigment is rhodopsin, the *visual purple* of animal physiology, which is a carotenoid conjugated with a protein. Photosynthetic generation of ATP by *Halobacterium* via rhodopsin appears to be a membrane-related process, similar in that respect to the respiratory generation of ATP in the mitochondria of eukaryotes. Beyond that the details are still obscure.

ECONOMIC IMPORTANCE

Bacteria are an integral part of the balance of nature and play a particularly important role in the nitrogen cycle. The rotting of meat and other foods, which is caused chiefly by bacteria and which, of course, represents an economic loss, is merely one aspect of the necessary nitrogen cycle by which organically bound nitrogen is restored to a form in which it can be used by green plants. Bacteria are normal inhabitants of the digestive tracts of animals. No mammal is known to produce an enzyme that attacks cellulose, and the ability of horses, cattle, and other herbivores to digest the cellulose that makes up so much of their diet is due to the presence in their digestive tract of certain bacteria and protozoa. Certain bacteria, notably *Escherichia coli,* are symbionts in the human lower intestine, where they produce significant amounts of some of the B vitamins. Bacteria carry on a number of fermentative processes, including lactic acid fermentation (which results in the souring of milk). Bacteria cause many diseases of plants, humans, and other animals, but from the standpoint of human life and economics the enormous damage they cause through disease and decay is far outweighed by their value in the nitrogen cycle and as intestinal symbionts.

EVOLUTIONARY ORIGIN AND RELATIONSHIPS

The bacteria very probably represent the most primitive group of living organisms, a group from which all other kinds of organisms are eventually descended. The origin of the bacteria themselves is obscure. According to a widely accepted hypothesis, organic matter was first formed by happenstance chemical reactions promoted by ultraviolet light and taking place in the fresh (not yet salty) water of the sea. In the absence of any destroying or consuming agent, the organic molecules very gradually accumulated, and reactions among them led to the formation of larger molecules. Electrochemical attractions might be expected to hold some of these molecules together as colloidal particles. When such an aggregation of molecules came to include the proper chemical compounds to promote the formation of additional such compounds from the less complex organic substances in the surrounding medium, the first organism came into being. Under this concept, the first organisms were heterotrophic.

As more and more of the previously formed organic molecules became incorporated into the molecule systems that represented the first organisms, a premium must have been placed on the ability to use less and less complex substances as raw materials, and finally on the ability to tap some new source of energy. If, as most students of the subject believe, the atmosphere at that time contained little or no free oxygen, these organisms must have been anaerobic. Any new source of energy, therefore, would not involve free oxygen and thus would not resemble the chemosynthetic processes now carried on by the iron bacteria and some of the sulfur bacteria. The first new source of energy may have been the anaerobic manipulation of sulfur. Development of a primitive, anaerobic cytochrome system of respiratory enzymes probably accompanied or preceded this step.

Another major step in evolutionary progress was the development of primitive, anaerobic photosynthesis, with the first chlorophyll originating from cytochrome by the substitution of magnesium for iron in the center of the porphyrin ring. These first photosynthetic organisms probably used sulfur in the photosynthetic process, much as some of the bacteria do today. Bacterial photosynthesis, which does not release free oxygen, is now regarded as the evolutionary precursor of algal and higher plant photosynthesis, which does. According to a now widely accepted view, it was not until a considerable supply of atmospheric oxygen had been built up by the newer type of photosynthesis that aerobic respiration, using a new and more efficient form of cytochrome, became possible.

Microscopic, ellipsoid-cylindric objects from deposits 3.2 billion years old in South Africa have been taken to be ancient bacteria and have been given the name *Eobacterium isolatum.* Some other South African fossils of comparable age appear to show stages in binary fission and are thought to represent prokaryotes of either bacterial or blue-green algal nature. These South African prokaryotes may be the oldest fossils of any kind.

Viruses

HISTORY AND DEFINITION

During the latter part of the nineteenth century, after Koch's and Pasteur's demonstrations that some bacteria cause disease, a great deal of attention was devoted to determining the causative agents of particular diseases. In 1892 the Russian botanist Dmitri Ivanovski (1864–1920) reported the transmission of tobacco mosaic disease by means of sap that had been passed through filters that were supposed to retain all bacteria. Six years later hoof-and-mouth disease of cattle was shown also to be caused by an agent that could pass through bacteriological filters, and similar reports for other diseases soon followed. Disease-causing agents that could pass through filters came to be called filterable viruses, the word virus being taken from Latin *virus*, a poison. As knowledge of the nature of these disease-causing agents has increased, it has become customary to speak of them merely as viruses, and it has come to be recognized that some of them may exist in the host without causing disease.

More definite knowledge of the nature of viruses began with the publication in 1935 of a paper by the American virologist W. M. Stanley (1904–1971) entitled *Isolation of a crystalline protein possessing the properties of tobacco mosaic virus* (see Fig. 11.11). Following on the improvement of the electron microscope during the 1940s, EM studies showed that the crystals of viruses are actually composed of many individual complex units, which have come to be called **virions.** The virion is now regarded as the basic structural unit of a virus, the form in which it is transferred from one host (or cell) to another.

Experimental biochemical studies of viruses during the 1950s and 1960s, largely under the intellectual leadership of the American virologist Salvador E. Luria (1912–   , Fig. 11.12), demonstrated that viruses resemble typical organisms in their basic biochemical features, notably in having nucleic acids that govern the formation of particular proteins. Indeed viruses have been used as the essential experimental organisms for much of the important work in molecular biology, inasmuch as they present some of the essential biological processes without the overlay of secondary products and self-regulation found in more complex organisms.

Viruses may now be defined as ultramicroscopic organisms whose genetic material is an element of nucleic acid (either DNA or RNA) that reproduces inside living cells and uses the internal machinery of those cells to direct the synthesis of specialized particles, the virions, which

 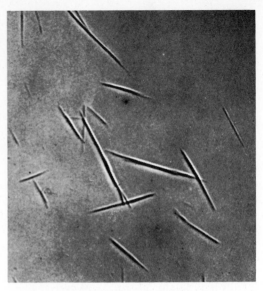

Fig. 11.11 (*Left*) W. M. Stanley (1904–1971), American virologist, who was the first to "crystallize" a virus. (*Right*) "Crystals" of tobacco mosaic virus. (× 800.) [Courtesy of W. M. Stanley.]

Fig. 11.12 Salvador E. Luria (1912–   ), American virologist. [Courtesy of S. E. Luria.]

contain the viral nucleic acid and transfer it to other cells.

STRUCTURE

A **virion** consists of a core of nucleic acid (either DNA or RNA, but not both) surrounded by a sheath of protein (the **capsid**), which in some viruses is further enclosed by an **envelope.** The nucleic acid may occur as single or double strands; DNA is more often in double strands, and RNA more often in single ones. The envelope is a unit membrane comparable to the plasma membrane of eukaryotic cells. Often it is in fact a portion of the plasma membrane of the host cell, acquired by the virion during its exit from the cell. In the pox (and some other) viruses, on the other hand, the envelope is an integral part of the virion, unrelated to the plasma membrane of the host. Some virions have one or more additional proteins, polypeptides, or other substances enclosed by the envelope but outside the capsid, or contained within the capsid. None of them, however, has a set of enzymes anywhere nearly complete enough to permit normal metabolism distinct from the metabolism of the host cell. They depend on the host cell not only for nutrients, but also for essential respiratory and other enzymes.

Virions (Fig. 11.13) are in general smaller than bacteria, although there is some overlap in size. Both the limit of resolving power of conventional microscopes and the size of the pores in the finest ordinary bacterial filters nearly coincide with the size of the smallest organisms that carry on the vital metabolic processes within their own bodies and can be cultivated on nonliving media. The smallest bacteria are coccoid types belonging to the *Rickettsia* group. They are only about 200 nm in diameter, at the

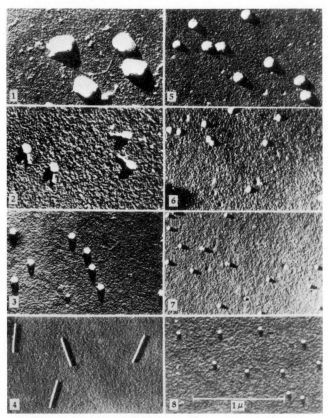

Fig. 11.13 Virions of various types: (1) vaccinia; (2) T-2 bacteriophage, with tail; (3) T-3 bacteriophage; (4) tobacco mosaic; (5) influenza; (6) rabbit papilloma; (7) bushy stunt of tomato; (8) polio. (All to scale indicated at lower right.) [Electron micrograph courtesy of R. C. Williams and other members of the virus laboratory, University of California, Berkeley.]

ultimate limits of resolving power of conventional microscopes. Some of the largest virions, such as those of the pox viruses, are up to about 300 nm long and nearly as thick. The pox viruses have the most complex structure and system of enzymes of all viruses, but they are still clearly viruses, not bacteria. Tobacco mosaic virus (TMV) virions are about 300 nm long but only 15 nm thick, long enough but not thick enough to be seen with the light microscope. Some of the smallest virions, such as the one that causes hoof-and-mouth disease, are only about 10 nm long and thick.

The proteinaceous sheath of a virion is called the **capsid.** It contains one or several kinds of proteins, according to the kind of virion, in one layer or in successive layers. Capsids come in two general styles: elongate and with the protein molecules helically arranged, or more or less isometric. The helical capsids, such as those of TMV, are typically open ended and hollow (Fig. 11.14), with the nucleic acid wound in a tight spiral around the inside of the capsid. Isometric capsids completely enclose the nucleic acid core, instead of being open ended. Often they are polyhedral, with all the faces forming equilateral triangles. Brick-shaped virions, such as those of the pox viruses, have the nucleic acid wholly enclosed and are generally grouped with the isometric type, as are also the tadpole-shaped viruses that parasitize some bacteria.

CLASSIFICATION

Knowledge of the essential nature of viruses is so recent, and the information is still restricted to such a small proportion of the known kinds, that no generally accepted classification has yet been de-

Fig. 11.14 Schematic representation of the structure of tobacco mosaic virus. The six-turn segment shown here corresponds to about 1/20 the length of the full virion. The central hole is about 40 Å in diameter. [Courtesy of D. Caspar; from D. Caspar, in Horsfall and Tamm, *Viral and rickettsial infections of man,* Lippincott, Philadelphia, 1965.]

vised. It has been customary merely to name each virus for the disease it causes, or even to use arbitrary symbols. Efforts to establish a binomial nomenclature comparable to that used for other organisms have not been enthusiastically received. It has been customary for some decades to recognize three general groups, according to the host: animal viruses, plant viruses, and bacterial viruses, or **bacteriophages** (the name often shortened to **phage**). It should be noted, however, that some of the "plant viruses" are equally or more at home in the insects that transmit the disease and might therefore with some justification be considered to be animal viruses.

Recent attempts at a more formal classification rely heavily on three features that occur in various combinations among different viruses: a core of DNA or RNA; a helical or isometric capsid; and the presence or absence of an envelope. The proper taxonomic weighting of these characters has not yet been agreed upon. It may be noted that almost all known plant viruses have an RNA core, whereas animal viruses have either DNA or RNA, according to the kind of virus. The tailed phages (Fig. 11.15) form a distinctive group of isometric DNA viruses without an envelope, but other phages lack a tail, and some of them have an RNA core.

Fig. 11.15   Virions of T-4 phage. (× about 70,000.) [Electron micrograph courtesy of R. C. Williams.]

It is even debatable whether a natural classification, in the sense of a classification in accord with probable phylogeny, is theoretically possible for viruses. If most or all of them are derived from bacteria by progressive simplification, as some biologists believe, then a natural classification should be possible. If many or all of them are derived eventually from normal intracellular particles of their own or other hosts, as others believe, then a phylogenetic approach is scarcely possible.

DISPERSAL AND REPRODUCTION

Virions are the static, chemically quiescent form of viruses, having no metabolism and no direct reproduction. Their only function is the dispersal of the virus from one cell (or host) to another. New virions do not arise by growth and division of existing virions, but are formed directly by aggregation and organization of newly formed viral molecules within the protoplasm of the host cell. The nucleic acid of the virus interacts with the enzyme system of the host to bring about the formation of molecules that become organized *de novo* into new virions; it subverts the metabolic system of the host into the production of virus.

The essential feature of virus dispersal is the introduction of viral nucleic acid into a host cell of a susceptible organism. This introduction occurs in four general ways, according to the kind of virus. Bacteriophages inject their nucleic acid into the host cell, leaving the empty capsid as "ghost" outside the bacterial cell wall. In some other viruses the envelope merges with the plasma membrane of the host cell. The capsid is then introduced into the cytoplasm intact. The whole virion of vaccinia is engulfed into a vacuole within the infected cell. Finally, virions may be introduced into the cell by or within a parasite. Many plant diseases are carried by sucking insects, a few by nematodes, and some even by intracellular parasites of the chytrid group of fungi. The arboviruses (i.e., arthropod-borne viruses) of humans and other vertebrates are also carried by sucking parasites, but these must make their way into the individual host cell after being introduced into the bloodstream by the carrier parasite.

Once the virion is introduced into the cell, the capsid protein dissociates from the nucleic acid core, and the nucleic acid goes to work. It subverts the host protoplasm into the production of molecules that can later be assembled into virions. Thus the infecting virions cease to exist as organized units. This initial dissolution of the virions, known as the **eclipse phenomenon,** is an essential part of the life cycle of all viruses; indeed it has come to be considered the most important definitive characteristic of the group. Bacteriophages, in which only the nucleic acid enters the host cell, are otherwise functionally similar to viruses in which these components dissociate after entry.

The viral nucleic acid often separates into individual genes or RNA segments during the eclipse phase. These are recombined eventually when new virions are produced. This behavior permits the exchange of genetic material between related viruses in the same cell. The new virions that are eventually produced show various combinations of the characters of the original types, as well as reconstituting the originals. This sort of genetic exchange serves the same function as sexuality in more complex organisms.

If the reproductive cycle is to be completed, the new viral components that a host cell has been subverted to produce must sooner or later be organized into new virions. Formation of new virions is initiated by some sort of interaction between viral and host protoplasm. This generally does not begin until enough material is available to form a considerable number of virions. Once initiated, the process is self-driven and self-regulated by the chemical attractions among the specific viral molecules. Capsid formation appears to be a terminal process. The nucleic acid that is incorporated into a capsid is not usually released again into the same cell.

Bacteriophages sometimes carry bacterial genetic material from one cell to another, in a process known as **transduction.** Some of the phage virions are misassembled to contain some bacterial DNA instead of part or all of the viral DNA. After transmission to a new host cell, the bacterial DNA may persist as a plasmid or may be incorporated into the chromosome. Inasmuch as the same phage may attack a number of different kinds of bacteria, there is an opportunity for transfer of genetic material between bacteria that are not closely related. Here, as in the transfer of plasmids through a conjugation tube, there is a possibility to spread individual features such as drug resistance far beyond the point of origin.

ECONOMIC IMPORTANCE

The plant and animal viruses are economically important because of the diseases that they cause. Some of the diseases caused by plant viruses are potato leaf roll, rice dwarf, sugar beet curly top, tomato bushy stunt, and a series of different **mosaic diseases** of such plants as beans, cucumbers, peas, peaches, sugar beets, sugar cane, tobacco, turnips, and wheat. The mosaic diseases are so named because they tend to inhibit chlorophyll production in a more or less definite pattern (often along the main veins) in the leaves of infected plants (Fig. 11.16). Many plant viruses are transmitted by leaf hoppers, aphids, and other insects, in which they are often quite at home, multiplying with or without apparent damage to the insect host. Other virus diseases of plants are mechanically transmitted, usually through some point of injury. Plant viruses can also be transmitted by grafting.

Some of the diseases caused by animal viruses are dengue fever, several types of human and equine encephalitis, influenza, measles, mumps, polio, chickenpox, smallpox, the common cold, hoof-and-mouth disease of cattle, rabies, hog cholera, distemper, myxomatosis of rabbits, and granulosis of butterflies and moths. Cold sores and warts are also caused by viruses, and it has become increasingly evident that a virus or viruslike element is involved in some human cancers. A large proportion of animal viruses are transmitted by contact, but some, such as the yellow fever virus and some of the encephalitis viruses, are transmitted by mosquitoes or other biting insects.

The degree of importance of phages in limiting the numbers of bacteria remains to be determined. Many of the phages ("temperate phages") are so well adapted to their hosts that they are carried on from generation to generation in morphologically undifferentiated form, only occasionally maturing into definite virions that are liberated by death and rupture of the host cell. Like other viruses, individual kinds of bacteriophage often have a number of different hosts, being highly virulent to some, but causing little damage to others.

Fig. 11.16 A single leaf of tobacco, showing characteristic symptoms of the mosaic disease. [Photo by H. A. Allard, courtesy of the U.S. Agricultural Research Service.]

Blue-Green Algae

HISTORY AND DEFINITION

The term **algae** is applied to most of the chlorophyll-bearing plants other than those in which the zygote normally begins its growth and development as a parasite on the parent plant. A few colorless plants are included in the algae because of their similarity to undoubted algae, as are also a few plants in which the zygote does begin its growth as a parasite on the parent plant. The chlorophyllous bacteria are excluded from the algae by their unusual type of photosynthesis.

As so defined, the algae are mostly water plants, and mostly relatively simple in structure, although some of the marine algae are large and structurally complex. Nearly all plants that float free in the ocean are algae; many of the plants inhabiting fresh waters and ocean shores are algae, but others are basically land plants that have returned to an aquatic habitat. Some algae do occur on land, especially in moist places, but these are always small, simple organisms. Algae never have specialized water-conducting tissues, as do many land plants, and only a few of the larger marine algae have tissues comparable to the phloem (a food-conducting tissue) of many land plants. Any complete plant body, such as that of an alga, that lacks specialized conducting tissue is called a **thallus,** particularly if it is multicellular but relatively simple in external form, not being divided into parts that resemble roots, stems, and leaves.

Gametes of algae are generally borne in unicellular gametangia, and never in structures resembling the multicellular gametangia of land plants. A **gametangium** is any structure in which gametes are borne.

The algae are now considered to be a convenient artificial group rather than a natural taxon. The definition of algae in approximately the modern sense dates from a "natural" system of plants proposed in 1789 by the French botanist A. L. de Jussieu (1748–1836). In 1836 the algae were divided by the Irish botanist W. H. Harvey (1811–1866) into three groups which correspond fairly well to the red algae, brown algae, and green plus blue-green algae, as those groups are now understood. The blue-green algae were recognized as a group distinct from the green algae in 1853 by the Swiss botanist Karl Wilhelm von Naegeli, although he did not propose a formal name. The oldest (1860) formal name for the blue-green algae is Myxophyceae, but the more descriptive name Cyanophyceae has been more generally adopted. The characteristic ending for names of algal classes is -*phyceae*.

The blue-green algae (Cyanophyceae) may be defined as prokaryotes (Schizophyta) that possess chlorophyll *a* and phycobilin pigments and that release oxygen as a result of photosynthesis. The prokaryotic protoplasmic organization distinguishes them from all other algae, and the photosynthetic system distinguishes them from the bacteria. Those bacteria that do carry on photosynthesis have neither chlorophyll *a* nor phycobilins, and they do not release oxygen.

DISTRIBUTION

Blue-green algae occur in fresh and salt water and in moist subaerial habitats throughout the world. They are more abundant in fresh water than in salt water, and more abundant near the surface than at depths of more than a few feet. Sunny habitats that are covered by a thin film of water for prolonged periods usually harbor blue-green algae that become dormant during dry periods. Temporary pools are likely to contain more blue-green algae than any other sort. Many blue-greens occur at or just beneath the surface of the soil. They are also commonly present in and about hot springs, as for example in Yellowstone Park, where they contribute to the brilliant color of Mammoth Hot Springs and others. The Red Sea is said to have been so named because of the abundance of *Trichodesmium*, a blue-green alga that is red.

Many blue-green algae occur in association with other organisms as epiphytes, commensals, symbionts, or parasites. Some of them are enslaved within the cells of other organisms, such as various fungi, protozoa, and other algae, providing food for the host cell while deriving no apparent benefit from the relationship. Such a condition is called **helotism.** The endosymbiotic hypothesis of the origin of eukaryotes, discussed in Chapter 12, is founded on the existence of these helotic intracellular blue-green algae.

CHARACTERISTICS

External Morphology

Blue-green algae occur as single cells, as small cell colonies, and as multicellular filaments which are themselves sometimes aggregated into colonies. Cell colonies may be flat and only one cell thick, or spherical and hollow, or elongate and quasi-filamentous, or massive and cubical, or formless. The gelatinous sheaths of the individual cells may or may not remain readily distinguishable in the colonial forms. In truly filamentous blue-green algae the walls of most adjacent cells are in contact, rather than being separated by sheath material. The individual chain of cells in a filament is called a **trichome;** usually the trichome is provided with a common gelatinous sheath. The cells of a trichome may or may not be all alike.

Blue-green algae have no evident means of locomotion, but some filamentous ones move nonetheless. The whole filament may move slowly forward and backward, either in a straight line or in a spiral, or the end of the filament may wave slowly back and forth. Students are not agreed about the mechanism of locomotion.

Pigmentation

All photosynthetic organisms (except *Halobacterium*) contain chlorophyll and associated carotenoid pigments. The chlorophyll of the blue-green algae is always and only chlorophyll *a*, which is the essential chlorophyll of all photosynthetic organisms other than bacteria. The principal carotenoid pigment of blue-green algae is β-carotene, just as in most other algae and in higher plants. Two other carotenoid pigments, myxoxanthine and myxoxanthophyll, are usually also present in some quantity. Myxoxanthophyll is apparently unique to the blue-green algae. Some blue-green algae also have other carotenoid pigments.

In addition to chlorophyll and carotenoid pigments, blue-green algae consistently have another class of pigments called **phycobilins.** These are important accessory photosynthetic pigments in the blue-green algae, the red algae, and the cryptomonads.

As we noted in Chapter 7, phycobilin pigments are of two general types: phycocyanins, which absorb green, yellow, and red light while transmitting blue; and phycoerythrins, which absorb blue-green, green, and yellow light while transmitting red. Most Cyanophyceae contain the blue pigment phycocyanin, and some also have the red pigment phycoerythrin; a few have phycoerythrin but lack phycocyanin. The phycocyanins and phycoerythrins of the blue-green algae are mostly not identical with those of red algae and cryptomonads, differing in small features of the protein constituent.

The typical color of the blue-green algae is just what the name suggests. Variations in the kinds and proportions of pigments in the different species also produce many other colors, including diverse shades of purple, red, yellow, brown, and blackish.

Cytology

An individual cell of a blue-green alga (Fig. 11.17) consists of a protoplast, surrounded by a wall that is usually enclosed in a gelatinous sheath. The principal wall component is one or more of a group of mucopolymers composed of glucosamine, amino acids, and muramic acid, with some glucopyranose, galactose, and pentoses. Most of these substances belong to the general group known as hemicellulos-

es. As in the bacteria, these substances may collectively be considered to form a peptidoglycan. The walls of many blue-green algae also contain cellulose, which (according to the genus) may or may not be organized into the sort of fibrils commonly seen in eukaryotes. The gelatinous sheath surrounding the cell wall is generally composed of pectic substances.

The protoplast consists of two more or less well-defined portions, a colorless **central body** and a pigmented outer region called the **chromoplasm.** The central body contains most of the DNA and is evidently of nuclear nature, but it is not bounded by a nuclear membrane, and there is no nucleolus. The DNA of blue-green algae has not been as intensively studied as that of bacteria. It appears that there are several DNA strands in the central body, and it is clear that these resemble the bacterial chromosome in being "circular" and in lacking attached proteins. Ribosomes are scattered throughout the protoplast, but are most numerous in the central body. There is no central vacuole.

The chlorophyll, carotenoids, and

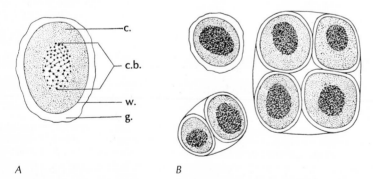

A   *B*

Fig. 11.17   (*A*) A single cell of a blue-green alga, showing parts: c., chromoplasm; c.b., central body; g., gelatinous sheath; w., cell wall. (*B*) *Gloeocapsa*, single cell and colonies. (× about 1000.)

phycobilins of blue-green algae occur together in organized units of diverse size and shape, which are scattered throughout the chromoplasm and give it its characteristic color. In some genera these photosynthetic units are tiny granules on the order of 0.3–0.5 microns in diameter. In others they are flattened sacs or disks (thylakoids) which may be derived (phylogenetically) by fusion of smaller granules. The disks, when present, are well separated, and not organized into the definite stacks found in many eukaryotic algae.

Cell division (Fig. 11.18) is strictly amitotic in the blue-green algae, with the central body merely seeming to be constricted into two equal parts. Cytokinesis is accomplished by the centripetal growth of a transverse membrane, sometimes accompanied by constriction or furrowing of the protoplast.

Mitochondria, organized chloroplasts, endoplasmic reticulum, golgi bodies, and double-membrane structures in general are unknown among the blue-green algae. These organelles are eukaryotic structures, whereas the blue-green algae are prokaryotic. Even the structure and composition of the cell wall recall the bacteria rather than typical eukaryotes.

Blue-green algae commonly store food as small granules of carbohydrate in chemical combination with protein. The carbohydrate fraction of these granules is called **cyanophycean starch.** It is a branching polyglucan (i.e., a polymer formed by linking glucose molecules together), often containing small amounts of pentose and some other hexose sugars in addition to glucose. In composition and type of linkage and branching it is very much like the amylopectin fraction of the starch of higher plants, but it usually has only about 20–30 glucose residues per molecule, in contrast to the 2000 or more in amylopectin.

NITROGEN FIXATION

Many blue-green algae, including *Gloeocapsa, Nostoc,* and *Anabaena,* resemble some of the bacteria in being able to use atmospheric nitrogen. In filamentous genera nitrogen fixation typically occurs in the **heterocysts,** which are scattered throughout the filament or borne at the tip. Heterocysts are enlarged, thick-walled, specialized cells that lack phycobilins. Their contents appear clear under the light microscope, but in EM preparations it can be seen that they have a complex internal membrane system. Nitrogen fixation tends to be inhibited by free oxygen, and these nonphotosynthesizing cells furnish a better milieu for the process. The blue-green algae are less important than the bacteria in nitrogen fixation, but it is clear that they play a major role (along with some nitrogen-fixing bacteria) in maintaining the fertility of rice paddies. The ability to use free nitrogen is restricted to prokaryotes.

REPRODUCTION

Reproduction in blue-green algae is accomplished by cell division, by fragmentation of colonies or filaments, and by spores.

There are two types of spores. One type, called an **akinete,** is formed by enlargement of a single vegetative cell, accompanied by an increase in thickness of the wall. An akinete is a resting cell that carries the organism over a period of unfavorable conditions, and it usually germinates as soon as favorable conditions for growth are restored. The other type of spore, called an **endospore,** is formed by repeated division of a proto-

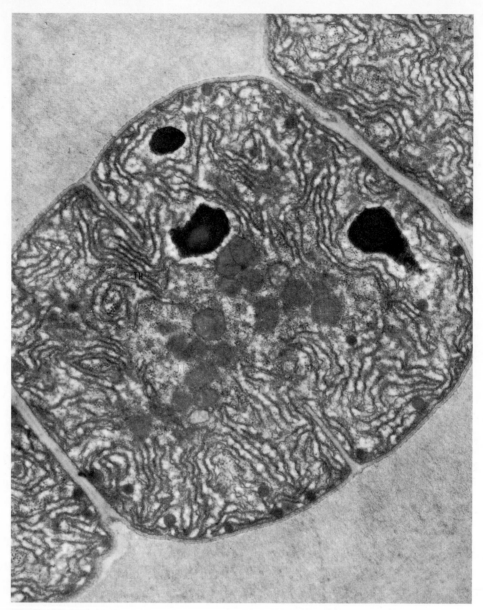

Fig. 11.18 Electron micrograph of thin section of *Nostoc pruniforme*, a blue-green alga, showing a cell in process of division. (× about 27,000.) The developing partition intrudes into the cell, as here seen, from the upper left and lower right. The contorted lines permeating the chromoplasm are the photosynthetic lamellae. The region without the lamellae is the central body. [Electron micrograph courtesy of Thomas E. Jensen and C. C. Bowen.]

plast within a cell wall. Both akinetes and endospores are nonmotile, and no motile reproductive structures of any sort are known in the blue-green algae.

In some kinds of blue-green algae fragmentation of the filament occurs at special points where two vegetative cells are separated by a double-concave disk of gelatinous material, called a **separation disk.** In blue-green algae that have heterocysts, fragmentation of the filament often occurs between two heterocysts or between a heterocyst and an ordinary cell.

No sexual processes of any sort are known among the blue-green algae.

CARBONATE DEPOSITION

There is a complex chemical balance between dissolved carbon dioxide, carbonic acid (H_2CO_3), and dissolved bicarbonates in natural waters. Depletion of the dissolved carbon dioxide tends to cause the formation of insoluble carbonates from dissolved bicarbonates, as, for example, in the equation

$$Ca(HCO_3)_2 \rightarrow CaCO_3 + H_2O + CO_2$$

The use of dissolved carbon dioxide for photosynthesis by algae and other water plants, therefore, tends to cause precipitation of carbonates. The fact that some algae are much more prone than others in the same habitat to accumulate lime on or within their bodies suggests, however, that these calcareous algae also play a more direct role in the process. Many of these algae, especially certain blue-green, green, and red algae, probably extract carbon dioxide more or less directly from dissolved bicarbonates, leading to the deposition of such insoluble carbonates as calcium carbonate, magnesium carbonate, and calcium magnesium carbonate. The **marl** that accumulates at the bottom of

some lakes and bogs is composed largely of calcium carbonate formed through the activities of green and blue-green algae, and the precipitation of calcium carbonate through chemical activities of algae in the geologic past is believed to have been an important factor in the formation of many or most limestone deposits.

CLASSIFICATION AND REPRESENTATIVE GENERA

The blue-green algae are grouped into three, four, or five orders, by different algologists, on the basis of the structure of the thallus and mode of reproduction. It has been customary to recognize about 1500 species, but comprehensive studies in progress by the American algologist Francis Drouet (1907–   ) suggest that many of these represent mere growth responses to differing environmental conditions. It may therefore eventually become necessary to admit a much smaller number of species, perhaps fewer than a hundred, but the matter is still highly controversial.

Gloeocapsa

Gloeocapsa (Fig. 11.17) commonly occurs on damp rocks. The individual cells are more or less spherical or ellipsoid and are aggregated into amorphous colonies of less than 50 cells. The sheath is often colored red, blue, violet, yellow, or brown by one or more pigments called gloeocapsin, and the sheaths of the individual cells retain their identity. Reproduction is by cell division and happenstance fragmentation of the colony.

Oscillatoria

Oscillatoria (Fig. 11.19) is one of the most common genera of algae and occurs in a

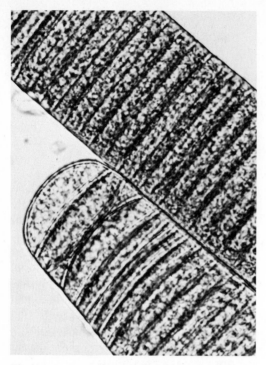

Fig. 11.19 *Oscillatoria* filaments. (× about 700.) [Photomicrograph courtesy of George Schwartz.]

wide variety of freshwater and subaerial habitats, including hot springs. It forms well-defined, unbranched, cylindrical, uniseriate filaments that occur singly or irregularly interwoven in layers of indefinite extent. The gelatinous sheath, if present at all, is extremely thin. Reproduction is by fragmentation at separation disks. The filaments of *Oscillatoria* and some related blue-green algae wave slowly from side to side and sometimes also move very slowly from place to place.

Nostoc

Nostoc (Fig. 11.20) occurs on bare soil and in fresh water, either free floating or attached to submerged vegetation. The cells form trichomes which resemble the trichomes of ordinary filamentous blue-green algae except for being much contorted and twisted. These trichomes have individual sheaths and are aggregated into a rather firm, amorphous, or globular colony with a common sheath. Mature colonies are readily visible to the naked eye, often several centimeters thick. There are scattered heterocysts. The trichomes often separate between a heterocyst and a vegetative cell, and thus a mature colony contains many trichomes. When the colony becomes mature, many of the cells develop into akinetes.

ECONOMIC IMPORTANCE

We noted in Chapter 7 that the food chain begins with plants. All classes of algae are ultimate sources of food for fish, but the blue-green algae are less important than some other groups, such as the diatoms and green algae. The significance of blue-green algae in nitrogen fixation and carbonate deposition has already been noted.

If allowed to grow unchecked, algae often become so abundant in reservoirs as to give the water a bad taste. Some of the blue-green algae are particularly offensive. Most algae are much more susceptible to poisoning by copper than are human beings and most other organisms, and small quantities of copper sulfate are therefore often added to public water supplies during the summer. The role of blue-green and other algae in eutrophication is discussed in Chapter 31.

EVOLUTIONARY ORIGIN AND RELATIONSHIPS

The blue-green algae are undoubtedly derived from the photosynthetic bacteria. The presence of System II photosynthesis (which releases oxygen as a by-product) in the blue-green algae marks a definite

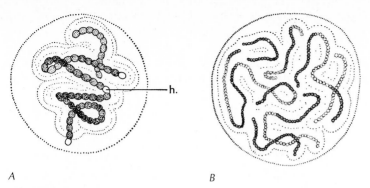

Fig. 11.20   *Nostoc*, two colonies: h., heterocyst. [(A) × 300; (B) × 100.]

advance over the bacteria, which have only System I. The blue-green algae must therefore be derived from the photosynthetic bacteria (though not necessarily from any existing genus). The evolutionary divergence of the two groups evidently occurred so long ago that most of the connecting forms have disappeared. Even the photosynthetic bacteria are a mere evolutionary remnant of what was probably once a much larger and more varied group.

Fossils that almost certainly represent blue-green algae are known from as far back as the Ordovician period, more than 400 million years ago. *Gloeocapsomorpha*, an Ordovician fossil named for its resemblance to *Gloeocapsa*, has definite cells aggregated into colonies held together by a matrix. Fossils that resemble *Nostoc* have been found in Cambrian rocks, but the identity of these is less certain.

Some Precambrian fossils also very probably represent blue-green algae. Slender, branching filaments thought to be of this group are associated with the presumed bacterial remains in Precambrian iron-ore deposits in Michigan and elsewhere; and blue-green algae are believed to have been involved in the formation of some Precambrian limestones in Montana, among other places. Other fossils believed to represent blue-green algae come from Australian deposits about a billion years old. More doubtful specimens, named *Archaeosphaeroides barbertonensis*, occur with *Eobacterium* in South African rocks dated at 3.2 billion years.

SUGGESTED READING

Barghoorn, E. S., The oldest fossils, *Sci. Amer.* **224**(5):30–42, May 1971.

Berg, H. C., How bacteria swim, *Sci. Amer.* **233**(2):36–44, August 1975.

Brill, W. J., Biological nitrogen fixation, *Sci. Amer.* **236**(3):68–81, March 1978.

Burris, R. H. (arranger), Future of biological N_2 fixation, *BioScience* **28**:563–592, 1978. A symposium of five articles by different authors.

Butler, J. G., and A. Klug, The assembly of a virus, *Sci. Amer.* **239**(5):62–69, November 1978.

Cairns, J., The bacterial chromosome, *Sci. Amer.* **214**(1):36–44, January 1966.

Campbell, A. M., How viruses insert their DNA into the DNA of the host cell, *Sci. Amer.* **235**(6):103–113, December 1976.

Davey, R. B., and D. C. Reanney, Extrachromosomal genetic elements and the adaptive evolution of bacteria, *Evolutionary Biology* **13**:113–147, 1980.

Delwiche, C. C., The nitrogen cycle, *Sci. Amer.* **223**(3):136–146, September 1970.

Echlin, P., The blue-green algae, *Sci. Amer.* **214**(6):74–80, June 1966. A scientifically accurate popular account.

Rossmoore, H. W., *The microbes, Our unseen friends*, Wayne State University Press, Detroit, 1976. Readable, nontechnical but informative.

Stanley, W. M., Isolation of a crystalline protein possessing the properties of tobacco-mosaic virus, *Science* **81**:644–645, June 28, 1935. The classic paper that opened the way to an understanding of the nature of the viruses.

Stoeckenius, W., The purple membrane of salt-loving bacteria. *Sci. Amer.* **234**(6):38–57, June 1976.

Walsby, A. E., The gas vacuoles of blue-green algae, *Sci. Amer.* **237**(2):90–97, August 1977. They regulate buoyancy.

Williams, G., *Virus hunters*, Knopf, New York, 1959. An interesting popular treatment.

Wood, W. B., and R. S. Edgar, Building a bacterial virus, *Sci. Amer.* **217**(1):60–74, July 1967.

Chapter 12

Green Algae

Hydrodictyon reticulatum, a green alga. (× 500.)
[Photomicrograph courtesy of Elsa O'Donnell-Alvelda.]

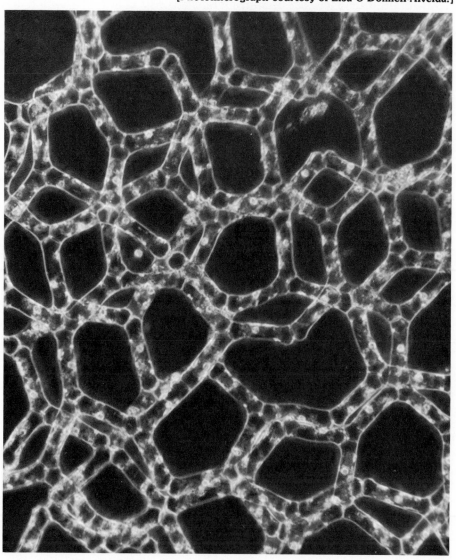

ORIGIN OF THE EUKARYOTES

There is such a profound difference in protoplasmic organization of prokaryotes and eukaryotes that some biologists have felt the need for an extraordinary explanation. The traditional concept has been that the eukaryotic organization evolved from the prokaryotic type by gradual increase in specialization and complexity. As the nature of the differences has become more clear through EM and biochemical studies, many molecular biologists, in particular, have found this explanation insufficient. Inspired especially by the presence of DNA in such cytoplasmic organelles as chloroplasts and mitochondria, and considering DNA to be essentially nuclear in nature, they have revived and modified the **endosymbiotic hypothesis** of the origin of eukaryotes, which in various forms has been floating around for many years but has not until recently attracted much support.

The essential feature of all versions of the endosymbiotic hypothesis is that chloroplasts are, in evolutionary origin, endosymbiotic prokaryotes that were taken into heterotrophic unicells of some other kind. Other eukaryotic organelles are considered either to be modified chloroplasts or to be additional endosymbionts. The characteristic double membrane of chloroplasts is considered to be derived by union of the plasma membrane (a unit membrane) of the endosymbiont and that of the host cell, resulting from the original enclosure of the endosymbiont after invagination. Other double membranes of the eukaryotic cell, including the nuclear membrane and endoplasmic reticulum, are considered to have evolved from this original double membrane.

In purporting to answer the question of the origin of double membranes and euka-

ryotic organelles in general, the endosymbiotic hypothesis raises a host of other questions that cannot be explored in the space available here. These difficulties have led one recent proponent of the concept (Raven, in 1970) to envision three separate origins of the eukaryotes, invoking as chloroplastid endosymbionts not only the blue-green algae but also two other hypothetical groups of ancient prokaryotic autotrophs which have left no other trace of their existence. Another proponent of the endosymbiotic concept (Margulis, 1969) considers not only chloroplasts but also mitochondria and flagella to be separately acquired endosymbionts. Peroxisomes have also been suggested to be endosymbionts.

Opponents of the endosymbiotic hypothesis tend to consider it as fantastically improbable, although they do not deny the existence of some modern endosymbiotic blue-green algae that have no relation to the origin of the eukaryotic condition. Members of this school of thought see the blue-green algae, or their immediate evolutionary forebears, as being directly ancestral to the eukaryotes, and some of them see the photosynthetic lamellae of blue-green algal cells as the possible precursors of other double-membrane structures. They see no problem in the presence of DNA in various cytoplasmic organelles, inasmuch as DNA is well scattered through the chromoplasm of the blue-green algae as well as being concentrated in the central body. In essence, they use Occam's Razor (the philosophical principle that explanations should not be unnecessarily complicated) to cut the throat of the endosymbiotic hypothesis. A satisfactory resolution of the controversy is not yet in sight.

The student should become aware that such controversy is inherent in scientific progress. Any new idea that has some initial plausibility is examined from various points of view by different scientists, to see how well it fits with other things we know or think we know. Ideas come easily to some workers, who may indeed come to be known to their colleagues as "idea men." Often these idea men perceive scientific truths before they become obvious to others, and a period of controversy, commonly involving the need to acquire more evidence, ensues before the idea is generally accepted. At least as often, these initially plausible ideas do not survive careful scrutiny. But like a student who fails a course, a rejected idea may come back again, sometimes to make the grade, sometimes not. The concept of organic evolution was proposed many times before Darwin's massive work made it scientifically respectable. On the other hand, the venerable idea of the inheritance of acquired characters (characters acquired by the individual in direct and immediate adaptation to the environment) still shows no signs of becoming acceptable to the scientific community. Almost any important new idea, good or bad, is initially resisted. Not many of those that do not make the grade find their way into general textbooks, but the scientific literature is full of them. Initial rejection is no proof of merit.

DEFINITION, TAXONOMY, AND DISTRIBUTION

The green algae are eukaryotic algae with grass-green chloroplasts that have chlorophyll *a*, chlorophyll *b*, and yellow carotenoid pigments in about the same proportions as in the higher plants (Embryobionta). The cell wall usually contains cellulose or cellulose-like polysaccharides. The most common food reserve is starch.

The flagella, when present, are generally of the whiplash type, described below, without the lateral, hairlike appendages (mastigonemes) found on some other kinds of flagella.

The green algae make up a division Chlorophyta, with two classes, the Chlorophyceae and Charophyceae. The Chlorophyceae are by far the larger of the two classes, consisting of about 6500 species, and it is usually this class that is meant when "green algae" are spoken of without further qualification. The Chlorophyceae were first recognized as a group, with approximately the present taxonomic limits, in 1863 by the German botanist Gottlob Ludwig von Rabenhorst (1806–1881), one of the most eminent algologists of his time. The Charophyceae (stoneworts) are structurally more complex than most of the Chlorophyceae, and are not further discussed here.

Green algae occur in fresh and salt water, and in many (mostly moist) subaerial habitats, including snowbanks. About 10 percent of the species are marine; these occur chiefly in shallow water along the seashore, often attached to rocks where they are exposed at low tide. A few grow in deeper water, up to as much as 100 m in warm seas. The larger nonfilamentous forms are chiefly marine. The most common algal components of lichens are green algae, especially the genus *Trebouxia*. A few green algae are colorless; some of these are saprophytes, but others are internal parasites of plants and animals.

STRUCTURE

General Morphology

Green algae may be unicellular, colonial, or multicellular, or they may have the nuclei scattered in a continuous protoplast. Multicellular green algae may be filamentous, may form thin, leaflike thalli up to several inches across, or may have various other shapes. Often the cells are not all alike, but the thallus lacks the complex differentiation of tissues that characterizes the vascular plants.

Unicellular flagellates such as *Chlamydomonas* (Fig. 12.1) are believed to represent one of the most primitive types of body organization among the algae. Evolutionary development of the algae from the motile unicell follows four principal lines:

1. Loss of the cell wall and flagella and development of an amoeboid organization
2. Organization of the cells into a colony without loss of motility
3. Loss of both motility and the capacity for vegetative cell division, although the capacity for nuclear division may be retained, so that a multinucleate filamentous protoplast (said to be *coenocytic*) is formed
4. Loss of motility (except for reproductive

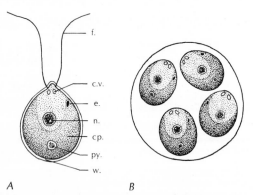

A   *B*

Fig. 12.1 *Chlamydomonas.* (*A*) Vegetative cell: c.p., chloroplast; c.v., contractile vacuole; e., eyespot; f., flagellum; n., nucleus; py., pyrenoid; w., cell wall. (× 1400.) (*B*) Four cells produced by reduction division within the zygote wall.

cells) but retention of the capacity for vegetative cell division

The first of these tendencies is often associated with the loss of chlorophyll and further evolution along protozoan rather than algal lines. The second tendency has inherently limited possibilities, and the third had only moderate success. The fourth tendency has infinite possibilities for variation. All of these evolutionary tendencies occur among the Chlorophyceae, and to some extent in other groups of algae as well.

Cell Wall

The cell wall of green algae generally consists of two layers. The inner layer is firm and usually composed largely of one or another kind of cellulose, or of cellulose-like polysaccharides based on monosaccharides such as xylose or mannose instead of glucose. The outer layer is softer and consists of pectic substances, forming a gelatinous sheath that usually tends to dissolve continuously into the surrounding water as new material filters out through the micelles of the cellulose layer. The cellulose layer of the cell wall may itself be divided into two layers, corresponding more or less to the primary and secondary walls of the cells of higher plants. The cell walls of some green algae (and other algae) may also become impregnated with calcium carbonate, so much so that in some marine forms the whole plant is greenish white.

Chloroplasts and Food Reserves

Green algae may have one, two, several, or many chloroplasts in a cell, the shape differing in different taxa. The single, massive, cup-shaped chloroplast of *Chlamydomonas* and many other flagellated green algae is believed to be a primitive type.

Chlorophyll *a* is the principal chlorophyll in the green algae, as it is in all other groups of algae and in higher plants. β-carotene is usually the principal carotene, and lutein the principal xanthophyll. Except for some specialized marine members, the pigmentation of green algal chloroplasts is essentially like that of the Embryobionta.

In most green algae, and in many other algae, each chloroplast contains one or more specialized proteinaceous bodies called **pyrenoids,** on or around which starch accumulates. Pyrenoids are also present in the chloroplasts of many other algae, but in these other groups they seldom accumulate starch. The pyrenoid is a local center of concentration of the enzymes used to transform primary photosynthetic products into food reserves. It may be thought of as the first evolutionary answer to the need to remove photosynthate (mainly glucose) from the reaction system of the chloroplast. Some of the larger green algae lack pyrenoids and accumulate starch in leucoplasts, much as in the higher plants. Many green algae produce oils as well as starch, and a few produce oils *instead* of starch.

Flagella

All flagella of eukaryotes have in common certain structural features that distinguish them collectively from the always much simpler flagella of bacteria. Indeed this uniformity of structure is one of the important factors bolstering the present belief in the basic evolutionary unity of the eukaryotes, since there is no obvious reason why equally effective flagella might not be organized in any of a number of quite different ways.

In contrast to the bacterial flagellum, consisting of a single proteinaceous fibril, the eukaryotic flagellum is compound, consisting of several proteinaceous fibrils, chemically different from bacterial flagella and arranged in a particular pattern. The all but universal pattern for these compound flagella is a circle of nine fibrils surrounding a core of two (Fig. 12.2). Usually the central fibrils are longer than the outer ones and extend out beyond them. Each of the nine peripheral fibrils is itself compound, typically consisting mainly of two longitudinal strands that are homologous with the microtubules of the cytoplasm. The two central fibrils consist of a single strand each, and they are chemically somewhat different from the nine peripheral double fibrils.

At the base of each flagellum, just

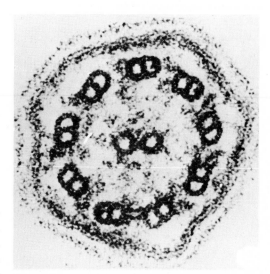

Fig. 12.2 Electron micrograph of cross section of flagellum of *Chlamydomonas reinhardii,* a green alga. (× 110,000.) [Electron micrograph courtesy of Cell Research Institute, University of Texas, published by David Ringo in *J. Ultrastructure Res.* **17**:273, 1967.]

inside the plasma membrane, is a granule called a **blepharoplast,** or basal body. A cross section through the basal body generally shows nine peripheral units in a ring. These are comparable to the nine peripheral fibrils of the flagellum, except that each consists of a linear triplet instead of a doublet. The pair of flanges often seen (in EM preparations) on the peripheral fibrils of the flagellum represents an imperfect development of the third subunit (microtubule) of the blepharoplast marginal units. The blepharoplast is homologous with the centriole, an organelle that plays a role in mitosis of some flagellated algae and most animals. Indeed the same granule serves as centriole and blepharoplast in some flagellates.

There is some diversity in details of flagellar structure within the general pattern just described (Fig. 12.3). Flagella that are more or less round in cross section, without lateral appendages, are called the **whiplash** type. With few exceptions, green algal flagella are of the whiplash type. Flagella with slender lateral fibrils (technically called mastigonemes) extending out from the central shaft are called the **tinsel** type. The mastigonemes can be individually distinguished only in EM preparations, and tinsel flagella often appear to be flattened or band shaped when studied with the light microscope. Many flagellates have two dissimilar flagella, one whiplash, one tinsel.

Tinsel flagella may be further classified according to whether they have one or two rows of lateral fibrils. Those with only one row, like the teeth of a comb, may be called the **pectinate** type (from Latin *pecten*, a comb). Those with two rows, one on each side, like the two rows of barbs on the shaft of a feather, may be called the **pinnate** type (from Latin *pinna*, a feather). Both the pectinate and the pinnate types are probably derived (in an evolutionary sense) from the whiplash type. Algologists find it useful to describe and name still other variations in flagellar structure.

Flagella of green algae are attached at or near one end of the cell, when present, and are usually two or four in number, although uniflagellate and multiflagellate types are also known. The flagella of any one cell are alike.

Flagellated cells of green algae usually have a small, orange-red, light-sensitive **eyespot** in the cytoplasm, typically near the base of the flagellum, and often embedded in the margin of the chloroplast. It is regarded as a modified part of the plastid.

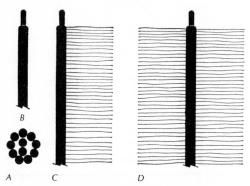

Fig. 12.3 Flagellar structure: *A*, cross section, showing two central and nine peripheral strands; *B*, whiplash type, without lateral appendages; *C*, pectinate type, with one row of lateral appendages; *D*, pinnate type, with two rows of lateral appendages. The duplex nature of the peripheral strands, not shown in this set of diagrams, can be seen in the electron micrograph in Fig. 12.2.

Vacuoles

Flagellated green algae generally have two (seldom more) small contractile vacuoles, commonly near the base of the flagella.

The vacuolar contents are largely water, but probably include waste materials as well. Each vacuole undergoes a recurrent cycle of slow distension and forcible contraction, followed by another slow distension, and so on. It is believed that the contractile vacuoles are excretory structures, and that they help regulate osmotic balances within the cell.

The nonflagellated genera of green algae usually lack contractile vacuoles and have a central vacuole similar to that described in Chapter 4. Sometimes there are several large vacuoles instead of only one, or the vacuole may be traversed by strands of cytoplasm. Some green algae, notably some of those that are subaerial rather than aquatic, have no very evident vacuoles of any sort.

REPRODUCTION

Some green algae are wholly asexual, some have only sexual reproduction, and others reproduce both sexually and asexually. Among the unicellular members of the group, cell division is also reproduction. Several types of spores are produced by different members of the group.

Spores

The most common type of spore in the green algae is the **zoospore,** a motile, flagellated cell that usually lacks a cell wall. The sporangium in which the zoospores are borne is usually derived from an ordinary vegetative cell; sometimes it is a specialized cell morphologically unlike other cells. The entire protoplast of the sporangium may be transformed into a single zoospore; more often the protoplast undergoes one or more mitotic divisions, leading to the formation of two to many (seldom more than 32) spores within the original cell wall.

Zoospores are commonly released through a pore that develops in the wall of the sporangium. The zoospore swims about for a period of a few minutes to two to three days; one or two hours is a common time for many species. It then retracts or loses its flagella and secretes a wall, becoming a vegetative cell. In multicellular or colonial types, the vegetative cell then undergoes a series of mitotic divisions to produce a new colony or organism.

Zoospores are haploid or sometimes diploid. In addition to being borne in sporangia, they are sometimes formed by reduction division from a zygote, or by mitotic divisions from thick-walled resting cells (akinetes). Zoospores of many genera of green algae are very similar to vegetative cells of *Chlamydomonas*, except for lacking a cell wall.

Zygospores are discussed under the sexual cycle.

Sexual Cycle

The sexual cycle in green algae, as in other truly sexual organisms, involves alternating stages with n and 2n chromosomes. The 2n stage is brought about by fusion of gametic nuclei, and the n condition is restored by reduction division.

In the simplest situation, as in some species of *Chlamydomonas*, ordinary flagellated vegetative cells act as gametes and fuse to produce a zygote. The zygote then undergoes reduction division to produce four spores that develop directly into unicellular vegetative plants (Fig. 12.1) In some other species of *Chlamydomonas* a vegetative cell gives rise by one or more mitotic divisions to 2, 4, 8, or 16 smaller daughter cells that are then released from the wall of the parent cell and function as gametes.

One type of evolutionary advance from the situation in *Chlamydomonas* leads to the formation of a multicellular thallus from the zygote or from zoospores by a series of mitotic divisions. In many green algae mitotic division occurs only during the haploid stage, so that a multicellular haploid thallus may be produced, while the zygote remains the only diploid cell. In other green algae mitotic divisions may take place during both the haploid and the diploid stages of the life cycle, so that the complete sexual cycle consists of:

1. Fusion of gametes to form a diploid zygote
2. Development of a diploid thallus by mitotic divisions
3. Production of haploid cells (typically zoospores) by meiosis affecting some or all of the cells of the diploid thallus
4. Development of a haploid thallus from each zoospore by a series of mitotic divisions
5. Formation of gametes within some or all of the cells of the haploid thallus

The diploid and haploid thalli may be alike or different.

Alternation of haploid and diploid (or *n* and *2n*) **generations,** such as in the foregoing cycle, is characteristic of the higher plants (subkingdom Embryobionta) as well as of many thallophytes. Because the reproductive body of the haploid (or *n* chromosome) generation is the gamete, it is customary to call this generation the **gametophyte.** Because the reproductive body of the diploid (or *2n*) generation is the spore, it is customary to call this generation the **sporophyte.** It should be noted that the spore itself, being the product of reduction division, is typically haploid and therefore represents the first step in the gametophyte generation.

The green algae show all conditions from typical isogamy (see Chapter 10) to pronounced heterogamy. It is sometimes useful to divide the heterogamous condition into **anisogamy** and **oögamy.** In anisogamy the two gametes that fuse are both motile but differ in size. In oögamy one gamete (the sperm) is small and motile, whereas the other (the egg) is larger and nonmotile. Anisogamy may be regarded as an evolutionary link between isogamy and oögamy. Isogamous and anisogamous green algae usually have gametangia that are morphologically undifferentiated from vegetative cells, and **syngamy** (fusion of gametes) occurs in the water, external to the parent plants. Oögamous green algae generally have gametangia that are morphologically different from each other and from vegetative cells, and syngamy occurs in the female gametangium. A male gametangium is called an **antheridium** regardless of whether it is unicellular (as among the algae) or multicellular (as among the embryophytes). A female gametangium is called an **oögonium** when it is unicellular (as among the algae) and an **archegonium** when it is multicellular (as among the embryophytes).

Zygotes of many green algae develop a thick wall and go into a resting stage that may last from a few days to several months. Such thick-walled zygotes are called **zygospores.** Zygospores ordinarily undergo reduction division immediately on germination, or even before, so that the zygospore itself is the only diploid cell in the life cycle.

Several groups of green algae, in contrast, have thin-walled zygotes that germinate within a day or so after being formed; these develop mitotically into diploid thalli that eventually produce zoospores or gametes as a result of reduction division. Most of the green algae that produce zygospores occur in fresh water, and most of those in which the zygote does not go into a pronounced resting stage are marine.

REPRESENTATIVE GENERA

Chlamydomonas

Chlamydomonas (Fig. 12.1) is a large genus of unicellular flagellates occurring in standing fresh water, on damp soil, and on persistent snowbanks. Waters rich in ammonium compounds, such as barnyard pools, provide an especially favorable habitat for some species. The cells are spherical to subcylindric or pear shaped and have a definite cellulose-like polysaccharide wall that in some species is enclosed by a gelatinous pectic sheath. There are two flagella and a well-developed eyespot. The flagella are usually said to be of the whiplash type, as in other flagellated green algae, but very inconspicuous mastigonemes have been demonstrated in some species. Most species have a pair of contractile vacuoles near the base of the flagella. There is no central vacuole, and the nucleus is often more or less central in position. Most species have a single cup-shaped chloroplast, which is often so large that it occupies most of the protoplast, usually at the opposite end from the flagella. Usually there is only one pyrenoid.

Asexual reproduction occurs by one or a series of mitotic divisions of the protoplast to form 2, 4, 8, or 16 (typically 4) daughter protoplasts within the wall of the parent cell. After the daughter protoplasts have each developed flagella and a wall, the wall of the parent cell ruptures or softens and the daughter cells are released. These daughter cells grow directly into mature vegetative cells. They have sometimes been called zoospores, but they differ from ordinary zoospores of other kinds of algae in having a cell wall.

When the plant grows on damp soil instead of in water, the daughter protoplasts sometimes remain in the gelatinous matrix formed by dissolution of the old cell wall and reproduce for generation after generation without forming flagella. Amorphous colonies of hundreds or thousands of cells in a common gelatinous matrix may be formed in this fashion, but as soon as the colony is flooded with water the individual cells develop flagella and are released from the matrix. These temporary colonies resemble normal colonies of the related genus *Palmella*, which produce *Chlamydomonas*-like flagellated cells only as reproductive bodies. Many different kinds of algal flagellates produce *Palmella*-like temporary colonies of nonmotile cells under certain conditions.

Most species of *Chlamydomonas* are isogamous, although the gametes are often chemically differentiated. Ordinary vegetative cells of many species can function as gametes under proper conditions. The protoplasts of the fusing cells may or may not escape from the enclosing wall before fusion.

The zygote has four flagella, and remains motile for several hours, or in some species for as much as two weeks, before losing the flagella and developing a thick wall. Often the zygote continues to manufacture food and grow during the "resting" stage. In most species the starch reserves of the zygote are eventually changed into oils, and considerable quantities of a diffuse red carotenoid pigment (hematochrome) are produced. *Chlamydomonas* is the most common cause of the red coloring sometimes seen at the surface of snowbanks ("red snow") at high altitudes or latitudes.

Eventually the protoplast of the zygote undergoes reduction division, producing four haploid protoplasts within the zygote wall. Usually these immediate products of reduction division acquire walls and are released by rupture of the zygote wall, developing flagella immediately after liberation. Each then matures directly into a vegetative cell. Sometimes the immediate products of reduction division undergo further mitotic divisions before liberation, so that 8, 16, or 32 "zoospores" are produced within the wall of the zygote.

Chlamydomonas is evidently a relatively primitive type within the green algae. All of the other sorts of green algae might reasonably have evolved from something similar to Chlamydomonas, although presumably not from any existing species. Some botanists have thought that all of the other eukaryotic algae, except the red

algae, might also have evolved from something like Chlamydomonas, but this concept is more debatable.

Ulothrix

Ulothrix (Fig. 12.4) is a genus of green algae occurring mainly in fresh water, often in flowing streams. The cells are cylindrical, longer or shorter than thick, and are united end to end in uniseriate, unbranched filaments of indefinite length. In some species the filament has a gelatinous sheath. All cells are alike except that the basal one is slightly modified into a holdfast. Each cell has a large central vacuole and a single large, thin chloroplast that forms a broad band partly or

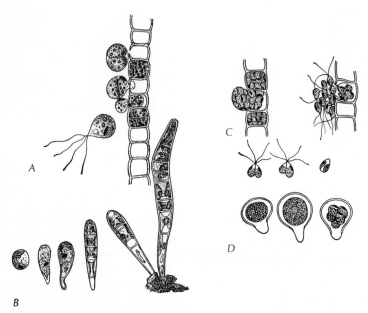

Fig. 12.4 *Ulothrix. A*, Formation of zoospores; *B*, germination of zoospore to form a new filament; *C*, formation and escape of gametes; *D*, germination of zygote with formation of zoospores. [From W. H. Brown, *The plant kingdom*, Ginn, Boston, 1935, 1963. Courtesy of M. A. Brown.]

wholly encircling the protoplast. The chloroplast has one or several pyrenoids.

Asexual reproduction usually occurs by the formation of quadriflagellate zoospores. The whole protoplast of a cell may be changed into a zoospore, but more often 2 to 32 zoospores are formed by mitotic divisions of a protoplast within the original cell wall. The zoospores usually escape through a pore that develops in the cell wall. After swimming about for a period of several hours to several days, the zoospore settles down, loses its flagella, and becomes a holdfast cell from which a new filament is produced by a series of mitotic cell divisions.

Asexual reproduction may also occur by the formation of nonmotile spores called **aplanospores,** comparable to the endospores of blue-green algae. The whole protoplast of a cell may round up and secrete a new wall, or zoospores that fail to be liberated from the parent cell may lose their flagella and secrete a wall. Each aplanospore gives rise to a new filament by a series of mitotic cell divisions.

Sexual reproduction is isogamous. The gametes are biflagellate and have an eyespot. Although the gametes all look alike, those produced by a single filament do not fuse with each other. The zygote soon loses its flagella, secretes a thick wall, and goes into a "resting stage" during which it continues for some time to carry on photosynthesis and accumulate food reserves. The zygote is the only diploid cell in the life cycle, its first nuclear division being meiotic. The zygote gives rise to 4 to 16 zoospores or aplanospores, each of which develops into a new filament.

The free-swimming, unicellular gametes and zoospores of *Ulothrix*, complete with flagella and eyespot, suggest that *Ulothrix* evolved from a unicellular flagellate ancestry. Here we have one of the many examples that illustrate the oversimplified aphorism (see Chapter 30) that *ontogeny* (the life history of the individual) *is a recapitulation of phylogeny* (the evolutionary history of the group).

Pleurococcus

Pleurococcus (Fig. 12.5) is one of the most common genera of green algae. It forms a thin green coating on tree trunks, stone walls, flower pots, and so on, usually in shaded or protected sites, and it also occurs in fresh water. It has been shown to be able to absorb water from moist air. The side of the tree that is most protected from

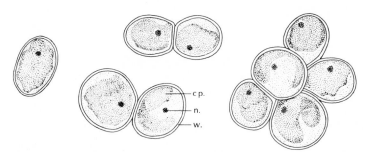

Fig. 12.5 *Pleurococcus:* cp., chloroplast; n., nucleus; w., cell wall. (× 1700.)

evaporation usually has the thickest coating of *Pleurococcus*.

Pleurococcus is nonmotile and typically unicellular. There is a well-developed cell wall and no gelatinous sheath. Solitary cells are spherical to ellipsoid. Sometimes the cells do not separate after division, and small colonies are formed. When growing in water, *Pleurococcus* sometimes forms irregularly branched filaments of as many as 50 or more cells. Each cell has a single large, thin chloroplast near the wall, usually without pyrenoids. There is no evident vacuole. Reproduction is exclusively by cell division. *Pleurococcus* is usually considered to be a reduced and modified derivative of *Ulothrix*-like ancestors.

Oedogonium

Oedogonium (Fig. 12.6) occurs in small permanent ponds and slow streams, often as an epiphyte on larger aquatic plants. The cells are cylindric and are joined end to end in uniseriate, unbranched filaments of indefinite length. The basal cell is differentiated into a holdfast, and the gametangia are visibly different from vegetative cells. Each vegetative cell has a large central vacuole and a single reticulate (netlike) chloroplast that extends the whole length and circumference of the cell. The chloroplast has many pyrenoids. Due to a complex process that accompanies mitosis, many cells of a filament have one or more narrow concentric girdles of

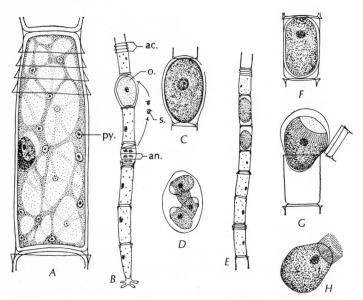

Fig. 12.6 *Oedogonium*. A, Vegetative cell of filament (× 500); B, part of filament, with reproductive structures (× 85); C, oögonium, with included zygote (oöspore) (× 170); D, zygote germinating to produce four zoospores (× 170); E, part of filament, with zoosporangia (× 85); F, zoosporangium; G, zoospore escaping from zoosporangium; H, zoospore: a.c., apical cap; an., antheridium; o., oögonium, with egg included; py., pyrenoid; s., sperm.

hemicellulose at one end, forming a characteristic apical cap.

Asexual reproduction occurs by the formation of either zoospores or nonmotile spores. Zoospores are borne only in cells that have an apical cap. The entire protoplast of the cell becomes a single zoospore that bears a ring of rather short flagella near one end and seldom has an eyespot. The cell wall breaks open about at the junction of the apical cap with the rest of the wall, liberating the zoospore. After an hour or so, the zoospore settles down, retracts its flagella, and secretes a wall. A new filament is then formed by a series of mitotic divisions.

Sexual reproduction in *Oedogonium* is oögamous, the different sexes of gametes being produced by the same or different filaments, according to the species. Some species have male and female filaments of about equal size, but in others the male filaments are very small, few-celled, and usually epiphytic on the larger female filaments. Antheridia are shorter than vegetative cells and are often borne in series. One or more often two sperms are borne in an antheridium. The sperms somewhat resemble zoospores but are smaller and have fewer and often larger flagella, these still borne in a ring near one end. Oögonia are obviously swollen and larger than vegetative cells. Each oögonium contains a single large egg with abundant food reserves and no large vacuole. The sperm is attracted to the egg by a chemical stimulus and enters the oögonium through a pore near one end.

The fertilized egg develops into a thick-walled zygote that is eventually liberated by decay of the oögonial wall. The zygote often remains in a resting condition for a year or more. Eventually the zygote undergoes reduction division, producing four zoospores. Each zoospore gives rise to a new filament by a series of mitotic divisions.

Although *Oedogonium* resembles *Ulothrix* in being filamentous with a basal holdfast, it is more advanced in several ways. The most important evolutionary advance, from our standpoint here, is that *Oedogonium* has well-differentiated male and female gametes, in contrast to the isogametes of *Ulothrix*.

Ulva

Ulva (Fig. 12.7), the sea lettuce, is a common alga along ocean shores between high and low tide and at depths down to about 10 m. The thallus consists of a holdfast and a blade. Plants are often broken loose by the tide, and may be washed up on shore. The blade is an irregular sheet two cells thick and several inches or even a foot or more long, superficially suggestive of a lettuce leaf. The holdfast is a cluster of rhizoids at one end of the blade. The rhizoids are filaments that are divided by occasional cross walls into multinucleate segments. (The term **rhizoid** is applied to any structure that resembles a root in position and function but lacks the complex tissues of the roots of higher plants.) The cells of a blade are all about alike, and their gelatinous sheaths are confluent to form a tough, gelatinous matrix. Each cell has a single large, often cup-shaped chloroplast with a single pyrenoid. There is no large vacuole.

The sexual cycle includes an alternation of similar diploid (sporophytic) and haploid (gametophytic) generations. Many of the cells near the margin of a haploid thallus become gametangia in which 32 or 64 biflagellated gametes are produced by a series of mitotic divisions. The gametes are liberated through a terminal pore in a nipplelike protrusion that

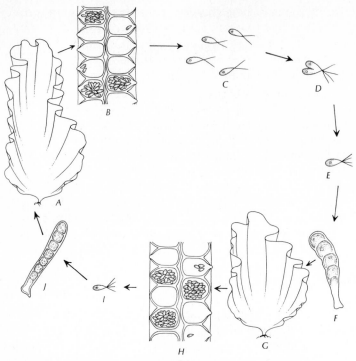

Fig. 12.7   Life cycle of *Ulva*, the sea lettuce. *A*, mature gametophyte, about a third natural size; *B*, enlarged segment of margin of thallus, showing gametangia, some already emptied, some ready to empty; *C*, isogametes; *D*, fusion of isogametes; *E*, zygote; *F*, young, filamentous sporophyte; *G*, mature sporophyte; *H*, enlarged margin of thallus, showing sporangia, some already emptied, some ready to empty; *I*, zoospore; *J*, young, filamentous gametophyte.

develops at the surface of the thallus. Syngamy occurs in the water. Some species are isogamous, others anisogamous. Gametes produced on one thallus do not fuse with each other, in the species that have been investigated. In some species a gamete that fails to fuse may function as a zoospore and develop into a haploid thallus.

The zygote formed by fusion of biflagellate gametes has four flagella. It soon settles down, loses its flagella, and se-

cretes a wall. It germinates within a day or so thereafter and eventually produces a diploid thallus by a series of mitotic divisions. At an early stage in its development the blade resembles an ordinary filamentous green alga, illustrating the recapitulation principle.

The mature diploid thallus is similar to the mature haploid thallus. Many of the cells near the margin of the diploid thallus become sporangia that resemble the gametangia of the haploid thallus. The first division of the developing sporangium is meiotic; 32 or 64 quadriflagellate haploid zoospores are produced by a series of subsequent mitotic divisions. Each zoospore develops into a haploid thallus in the same way that a zygote develops into a diploid thallus.

Ulva and *Oedogonium* show different sets of evolutionary advances over filamentous green algae such as *Ulothrix*, which have isogamous sexual reproduction and only a single diploid cell in the life cycle. *Ulva* has largely retained the isogamous reproduction of genera such as *Ulothrix*, but it has a much larger and more complex thallus, and unlike *Ulothrix* it has similar haploid (gametophytic) and diploid (sporophytic) thalli. In more technical jargon, *Ulva* has *isomorphic alternation of generations*. We shall see that the subkingdom Embryobionta probably originated from oögamous green algae with isomorphic alternation of generations, that is, from algae which combined the two different sorts of advances shown by *Oedogonium* and *Ulva*.

FOSSIL FORMS

The green algae have a long fossil history, dating back at least to the Precambrian Bitter Springs deposits in Australia. This formation, thought to be about a billion years old, contains both filamentous and coccoid types of presumed green algae. Fossils that probably represent some of the larger green algae are known from as far back as the Cambrian period. Forms clearly identifiable with the modern order Dasycladales (a group of complex marine species) occur in rocks as old as the late Precambrian, and are abundantly represented in calcareous deposits of most of the geologic periods from the Ordovician to the present. *Rhabdoporella,* a small, slender, unbranched form from Ordovician deposits, may be near the ancestral prototype of the order.

ECONOMIC IMPORTANCE

Like the blue-green algae, the green algae are important as an ultimate source of food for aquatic animals (especially in fresh water), in the formation of limestone deposits, and as contaminants that give a bad flavor to water supplies. Some of the green algae are involved in the formation of coral reefs, but the coralline red algae are more important in this regard. Green algae are also important inhabitants of oxidation ponds in sewage-treatment plants, providing through photosynthesis the oxygen necessary for rapid decomposition of the sewage by bacteria. The **water bloom** that sometimes forms on the surface of reservoirs, lakes, and ponds in summer (often reflecting accelerated eutrophication) is composed largely of planktonic green and blue-green algae. The term **plankton** applies to microscopic or barely megascopic algae and aquatic animals that float freely in the water, being carried about by the currents and waves.

EVOLUTIONARY RELATIONSHIPS

The green algae are thought to be derived eventually from the photosynthetic bacteria, as are also the blue-green algae. Possibly a single group of photosynthetic bacteria developed the capacity to use water as a source of hydrogen and release free oxygen in photosynthesis, and the Chlorophyceae and Cyanophyceae diverged from this proto-algal stock that contained chlorophyll *a* as the principal chlorophyll and β-carotene as the principal carotene. This possibly oversimplified view does not account for the fact that the usual nuclear structure of the Pyrrophyta (see Chapter 13) may prove to be more primitive than that of other eukaryotes such as the green algae, and further study is in order. It is generally agreed that the whole subkingdom Embryobionta is derived eventually from the green algae. The green algae also appear to be directly ancestral to some of the protozoa; indeed each of the larger groups of algal flagellates appears to have given rise to protozoan types. Some proponents of the endosymbiotic hypothesis of eukaryotic evolution would turn this story around and derive various groups of algal flagellates from protozoans, but both views accept evolutionary connections at several points between algal and protozoan flagellates.

SUGGESTED READING

Allsopp, A., Phylogenetic relationships of the procaryota and the origin of the eucaryotic cell, *New Phytol.* **68**:591–612, July 1969.

Dodge, J. D., *The fine structure of algal cells,* Academic Press, London and New York, 1973. Beautiful EMs.

Gibor, A., *Acetabularia:* A useful giant cell, *Sci. Amer.* **215**(5):118–124, November 1966. A physiologically oriented discussion.

Pickett-Heaps, J. D., *Green algae. Structure, reproduction and evolution in selected genera,* Sinauer, Sunderland, Mass., 1975. Beautiful EMs.

Preston, R. D., Plants without cellulose, *Sci. Amer.* **218**(6):102–108, June 1968. Some marine green algae such as *Acetabularia, Caulerpa,* and *Bryopsis.*

Satir, P., How cilia move, *Sci. Amer.* **231**(4): 45–52, October 1974. "The cilia bend when the microtubules, powered by ATP, slide past one another."

Tiffany, L. H., *Algae, the grass of many waters,* 2nd ed., C. C Thomas, Springfield, Ill., 1958. A readable popular account by a noted authority, emphasizing ecology, economic importance, and interesting and unusual algae.

Chapter 13

Other Algae

Fucus on the rocks, coast of Maine.

The green algae and blue-green algae were discussed at some length in previous chapters because of their evolutionary significance, as well as because of their importance as primary food producers especially in fresh water. The blue-green algae provide a sort of evolutionary link between the bacteria and the eukaryotes, and the green algae are the ancestors of the land plants. The other groups of algae find their place in this text for various reasons. The dinoflagellates and chrysophytes are important as primary food producers, especially in the oceans. A reasonable estimate is that about a third of the photosynthesis in the world is carried on by marine algae. The brown algae and red algae are conspicuous in tidal and shallow-water marine habitats, and the euglenoids and cryptomonads are of considerable evolutionary interest.

These remaining groups of algae are here considered to comprise 13 classes, organized into 6 divisions, as indicated in Chapter 2. The red algae (Rhodophyta) are so different from all the others that they are recognized in nearly all modern systems as a distinct division with one or two classes. The others all appear to be more or less related and are grouped by different algologists into one to seven or more divisions. The difference in number of recognized divisions does not so much reflect disagreements about relationships among the classes as it does differences of opinion about how best to express the relationships in a formal taxonomic system.

An ecological classification of these several groups of algae comes out somewhat differently. The red algae (Rhodophyta) and brown algae (Phaeophyta) are chiefly multicellular, often large and conspicuous, and they mostly grow attached to the bottom in shallow water (rarely deeper than 100 m) or in the intertidal zone. The Euglenophyta, Pyrrophyta, Cryptophyta, and Chrysophyta are mostly planktonic and unicellular, too small to be seen individually with the naked eye, although some few are filamentous or form small cell colonies. The red algae and brown algae, the familiar seaweeds, are abundant and conspicuous along shore, but it is the plankton of the open ocean that begin the food chain for most of the great variety of marine animal life. It is appropriate to discuss the primarily planktonic groups first.

The **phytoplankters** (plankton algae) are largely concentrated in the upper few meters of the seawater. Below that most of the light, especially photosynthetically useful light, is screened out. Some of the red algae can grow attached to the ocean bottom at depths of as much as 100 or even 175 m in very clear water, but such extreme records are of little significance in the general pattern of marine life.

The different groups of phytoplankters are not equally abundant in different regions. The diatoms (class Bacillariophyceae of the division Chrysophyta) are more abundant in cooler regions, and are well represented in fresh as well as salt water. The other chrysophytes and the dinoflagellates are mostly more abundant in tropical and subtropical marine waters.

Phytoplankters are eaten by zooplankters, which in turn become the food for larger kinds of marine animals. Dead or damaged phytoplankton cells that are not immediately eaten tend to sink gradually toward the bottom. It has been estimated that as much as 20 percent of the phytoplankton cells in surface waters eventually sink to some deeper level, where they nourish deep-water organisms. Nitrogen, phosphorus, and other important nutrient elements thus tend to be progressively removed from surface waters, and thermal stratification of the water hinders their return. Upwelling of water from the ocean deeps occurs consistently in certain areas, however, especially along the west coasts of continents in association with offshore currents. In such areas the sea is highly productive, and fishing fleets from many nations congregate.

A practical political result of these facts, with which some readers may be familiar, is that Ecuador claims a 200-mile limit of jurisdiction in the Pacific Ocean, embracing an area of upwelling associated with the Humboldt Current. The refusal of some other nations to recognize this claim has led to the seizure of tuna boats, and much international friction. Similar claims are now being made by some other maritime nations, and a revision of international law of the sea is under consideration.

EUGLENOIDS

The euglenoids form a coherent group of some 500 species poised on the borderline between the plant and animal kingdoms. First recognized as a natural group in 1838 by the German botanist Christian G. Ehrenberg (1795–1876), they are here treated as a division Euglenophyta, with a single class, Euglenophyceae. They are not so abundant and important in the food chain as some other flagellates, but they are interesting as easily available organisms that demonstrate the lack of a sharp boundary between the plant and animal kingdoms. Nearly all of them are unicellular flagellates that lack a cell wall. Many of them have chloroplasts and carry on photosynthesis. They all have a gullet at one end, and some of the ones without chlorophyll ingest small particles of food through this gullet.

The cytoplasm of *Euglena* is bounded externally by a plasma membrane. Just within the plasma membrane is a fairly firm but usually flexible, proteinaceous layer, called the **pellicle** or **periplast.** One or more flagella, usually of the pectinate type, emerge from the base of the gullet. Often there is an eyespot near the gullet, but it is not associated with the chloroplast as it is in the green algal flagellates. Near the base of the gullet are one or more contractile vacuoles that eventually discharge their contents into the gullet and disappear. New contractile vacuoles are formed by fusion of smaller vacuoles.

Nutrition of euglenoids varies from autotrophic to saprobic to holozoic. Some of the chlorophyllous forms are wholly autotrophic, but many of them require vitamin B_{12} or an organic source of nitrogen. Many or all of the chlorophyllous species are facultatively saprobic, losing their chlorophyll when deprived of light. The chloroplasts, when present, are grass green and have the same chlorophylls (*a* and *b*) as the Chlorophyceae, plus β-carotene and several xanthophylls. The principal xanthophyll is diadinoxanthin, which also occurs as a minor pigment in dinoflagellates. Studies now under way suggest a chemical similarity among the major xanthophylls of the euglenoids, dinoflagellates, cryptomonads, chrysophytes, and brown algae, in contrast to those of the blue-green, green, and red algae. The principal food reserve of euglenoids is paramylon, an insoluble polyglucan chemically similar to the laminarins of the Chrysophyta and Phaeophyta.

Euglenoids ordinarily (always?) reproduce by cell division. The chromosomes do not elongate during interphase, and mitosis follows a unique, perhaps relatively primitive pattern. Microtubules are present in a spindlelike arrangement within the nucleus, but there is no centriole and there are no tractile fibers. The nucleolus retains its identity throughout, pinching in two as the chromosomes separate. Only the Pyrrophyta, which differ notably in that the chromosomes lack associated protein, can be compared to the euglenoids in these respects, and even so there are some further differences in detail between the mitotic processes of the two groups.

The most common and familiar genus of euglenoids is *Euglena* (Figs. 13.1, 13.2), which occurs in both fresh and salt water, but is most common in "fresh" water that is rich in organic matter. The cells have a flexible pellicle and change shape freely while they swim, varying from slenderly cigar shaped to pear shaped. There are usually numerous discoid to band-shaped chloroplasts. There is only one normally developed flagellum, and a second very short one contained within the gullet. Thick-walled resting cells are frequently produced.

Euglena (and presumably other euglenoids) make lysine (an essential amino acid) in an unusual way, following what is known as the amino-adipic pathway. The majority of the fungi also follow this same pathway in making lysine, but no other organisms are yet known to do so.

Euglena presents perhaps the best case for an endosymbiotic origin of undoubted chloroplasts (see p. 202). The chloroplasts of *Euglena* are more nearly independent, physiologically, than those of other algae, less thoroughly integrated into the economy of the cell. An endosymbiotic interpretation would also help to explain the presence in *Euglena* of an evident but nonfunctional gullet; it seems unlikely that a gullet would evolve in advance of the need.

DINOFLAGELLATES

The dinoflagellates (class Dinophyceae in the division Pyrrophyta), diatoms (class Bacillariophyceae in the division Chrysophyta), and several other groups of Chrysophyta make up the great bulk of the marine plankton algae. These several groups of phytoplankters differ collectively from the green algae and euglenoids in having a relatively high ratio of carot-

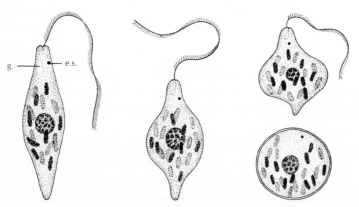

Fig. 13.1 *Euglena.* Varying shapes assumed by a single cell: e.s., eyespot; g., gullet. The cell at lower right is in a resting stage, with a definite wall and without a flagellum. (× 500.)

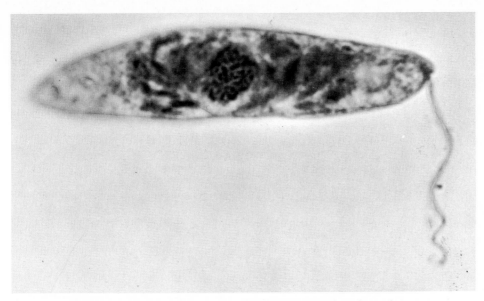

Fig. 13.2 *Euglena.* (× about 1000.) [Courtesy CCM: General Biological, Inc., Chicago.]

enoids to chlorophyll. Furthermore, their xanthophylls are often brownish rather than yellow, and other, nonplastid pigments are sometimes scattered through the cytoplasm (especially in the dinoflagellates). The cells therefore do not have the bright grass-green color characteristic of the green algae and euglenoids. These planktonic groups (other than the euglenoids and green algae) all lack chlorophyll *b*, but most of them have chlorophyll *c* instead. All of them of course contain chlorophyll *a*, which is the essential chlorophyll of all photosynthetic organisms other than bacteria.

The dinoflagellates (Fig. 13.3) are a coherent group of about 1000 species of unicellular flagellates that make up the bulk of the class Dinophyceae in the division Pyrrophyta. Some forms (the armored dinoflagellates) have a definite, specialized cell wall; others have the cell bounded by a firm periplast that usually

has a fixed shape. Some few members of the class are filamentous, or amoeboid (with neither cell wall nor flagella), or form small cell colonies, and the term dinoflagellate does not strictly apply to these forms. The dinoflagellates were first recognized as a natural group by Ehrenberg in 1838.

Like the euglenoids, the dinoflagellates are poised on the border between the plant and animal kingdoms. Most of them are photosynthetic autotrophs (therefore plants), but others are saprobic (absorbing their food), or even holozoic organisms that ingest small particles of food (thus animals). Some individuals are at once autotrophic and holozoic, both making and eating food.

Most dinoflagellates occur in the ocean, especially in warm regions. Some genera and species have abundant (often red) pigments distributed through the cytoplasm, masking the plastids. Dinoflagel-

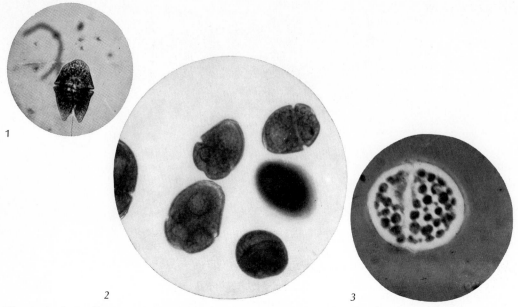

Fig. 13.3   Miscellaneous dinoflagellates: (1) *Gymnodinium* (× 450), note the trailing flagellum; (2) *Katodinium* (× 350); (3) *Glenodinium*, showing chloroplasts (× 900). [Photomicrographs courtesy of John Lee and John J. A. McLaughlin.]

lates often become so numerous that they color the seawater bright yellow to red. Some species of *Noctiluca*, *Gymnodinium*, *Peridinium*, and other genera are luminous in the dark, and when the individuals are very numerous, the sea glows at night. This phosphorescence of seawater can be caused by a wide range of plankton organisms, both plant and animal, but dinoflagellates are by far the most common cause.

The sporadic red tides on the coast of Florida are caused mainly by *Gymnodinium breve*, and other dinoflagellates of this and other genera cause similar phenomena elsewhere. *Gonyaulax polyhedra* has been identified as a species responsible for red tides along the California coast. The red tides caused by *Gymnodinium breve* kill large numbers of fish and other marine animals, and some other species of

this and other genera of dinoflagellates have similar effects, although still others are apparently harmless, even when sufficiently abundant to color the water. *Gymnodinium breve* reached an abundance of more than 20,000 individuals per cubic centimeter of water during the great red tides along the Florida coast in the summer of 1971. A concentration of 250 individuals per cubic centimeter is enough to begin to poison fish. The effect is due to a specific neurotoxin released into the water as a result of metabolic activities of the flagellate.

The poisonous red tide along the coast of New England in the fall of 1972 was caused by *Gonyaulax tamarensis*, which is regularly present in the area but became extraordinarily abundant for some weeks. Clams characteristically feed by filtering plankton from the water, and at this time they developed such great concentrations of the dinoflagellate toxin as to be highly poisonous to any vertebrate that ate them, although the clams themselves were not obviously affected. New England shellfish had to be temporarily withdrawn from the market, with attendant economic distress in the area.

Red tides are not a new phenomenon in the world, but they may have become more frequent and intense in recent years. It is believed that inadvertent fertilization of the seawater near shore with nitrate and phosphate wastes from human activities may now be an important causal factor.

The chloroplasts of dinoflagellates, when present, are mostly golden brown to somewhat yellowish green. They contain chlorophyll *a*, chlorophyll *c*, β-carotene, and several characteristic xanthophylls, generally including peridinin (the principal xanthophyll, nearly unique to the division Pyrrophyta), diadinoxanthin (the major xanthophyll of the euglenoids), and dinoxanthin.

Dinoflagellates have two dissimilar flagella, inserted in a transverse or spiral groove (the girdle), with one flagellum trailing and the other lying in the groove (Fig. 13.4). Usually the groove more or less completely encircles the cell. The trailing flagellum is at least sometimes of the pinnate type. The transverse flagellum is pectinate and undulate along an axis to which it is attached, as if the axis had been shortened like a drawstring. This axis is a structure in addition to the 11-strand (9 + 2) flagellum that is attached to it. The transverse flagellum appears to be responsible for the forward motion of the cell as well as its characteristic twirling.

The division Pyrrophyta, of which the dinoflagellates are the principal members,

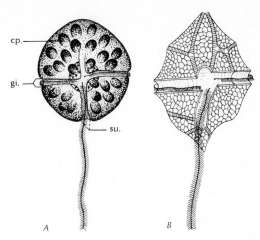

Fig. 13.4 (A) *Gymnodinium:* cp., chloroplast; gi., girdle; su., sulcus. (× 1500.) (B) *Peridinium.* (× 500.)

is unique among eukaryotic organisms in that the chromosomes lack proteins. Each chromonema apparently consists largely or wholly of a number of reduplicated parallel strands of DNA. The chromosomes do not elongate during interphase, and they have often been described as beaded in appearance. They do not have a centromere, there are no tractile fibers during mitosis, and there is no spindle. These features, especially the absence of proteins from the chromosomes, have often been interpreted as archaic, and some biologists would even establish a separate kingdom, Mesokaryota, intermediate between the Prokaryota and Eukaryota, to accommodate the pyrrophytes. On the other hand, the cytoplasm appears to be fully eukaryotic, and the pyrrophytes are linked to other algal flagellates by a series of features, including pigmentation, flagellar structure, food reserves, wall structure, and the frequent presence of an eyespot. Even the unusual mitosis might be compared in some respects with that of the euglenoids, although the euglenoids do have histone proteins associated with the DNA of the chromosomes, like other eukaryotes.

Gymnodinium (Fig. 13.4) may be taken as representative of the dinoflagellates that lack a wall. The cell is bounded by a firm periplast with a fixed shape. As in the euglenoids, there is a plasma membrane external to the periplast. The girdle encircles the cell in a loose spiral, and the separated ends of the spiral are joined by a longitudinal furrow called the sulcus. Most species have numerous small, golden-brown chloroplasts, and as we have noted some species have a diffuse red pigment as well. According to the species, the organisms are autotrophic, saprobic, holozoic, or at once autotrophic and holozoic. Reproduction is usually by cell division. Sometimes the cell goes into a resting stage and produces a cellulose wall. The protoplast of the cell may emerge from its wall and return directly to the vegetative state, or several zoospores may first be formed within the wall.

The armored dinoflagellates, such as *Peridinium* (Fig. 13.4), have a more or less *Gymnodinium*-like protoplast surrounded by a wall that usually contains cellulose, consisting of a definite number of plates arranged in a definite pattern. One of these plates encircles the cell at the girdle, dividing the remainder of the wall into two parts, each composed of several plates. The flagella emerge through small pores in the wall. The wall in armored dinoflagellates is a special structure, not really the same as the cell wall in other plants. It lies just within the plasma membrane, rather than being external to it. Depending on the genus, there may or may not be a pellicle just within the wall of armored dinoflagellates.

Nearly all of the armored dinoflagellates are photosynthetic, forming starch as a reserve food. Some also evidently ingest food particles by an undetermined means, which may perhaps involve extrusion of part of the protoplast between the plates of the cell wall.

CRYPTOMONADS

The cryptomonads, a group of less than 100 seldom-seen species of mostly unicellular flagellates making up the division Cryptophyta, are economically insignificant. They are of great interest to students of broadscale relationships among the algae, but their combination of shared and distinctive features does more to raise questions than to provide answers. They are here mentioned mainly as a teaser, to emphasize the fact that many botanical problems remain to be solved.

CHRYSOPHYTES

The chrysophytes (division Chrysophyta) are unicellular, colonial, or sometimes filamentous algae, with or without a cell wall. The cell wall, when present, seldom contains appreciable amounts of cellulose, and it is often silicified; very often it consists of two overlapping halves, a condition that reaches its extreme development in the diatoms. The flagella, when present, are typically two and dissimilar, one pinnate and one whiplash; or the whiplash flagellum may be reduced or wanting. Less often there are two equal whiplash flagella. The chloroplasts are mostly yellowish green to golden brown, with a high proportion of carotenoids to chlorophyll, the carotenoids commonly including diatoxanthin and often fucoxanthin and sometimes diadinoxanthin, but seldom peridinin. Chlorophyll *b* is lacking, but chlorophyll *c* (or sometimes *e*) is often present in addition to the omnipresent chlorophyll *a*. The food reserves are generally chrysolaminarin (a β-1,3-linked polyglucan allied to laminarin and paramylon) or oils or both. The Chrysophyta were first recognized as a natural group in 1914 by the German botanist Adolf Pascher (1887–1945, Fig. 13.5), whose overall scheme for algal classification is the foundation for most modern systems.

The major taxonomy of the Chrysophyta is in a state of flux. The diatoms form a well-marked group, the class Bacillariophyceae, sharply distinct from the rest of the division. Traditionally most of the other members of the division have been referred to two additional classes, the Xanthophyceae and Chrysophyceae; another small group of more doubtful position, the Chloromonadophyceae, has often also been included. Electron micrograph-

ic, biochemical, and cultural studies in recent years have led to the segregation by some algologists of two or more additional classes, most notably the Prasinophyceae and Haptophyceae. Many of the genera traditionally referred to the Xanthophyceae and Chrysophyceae have not yet been studied in the detail necessary to permit their certain assignment in such classifications. Among these nondiatomic chrysophytes we shall content ourselves with a discussion of a common and widespread genus, *Halosphaera*, and a large and abundant group known informally as coccolithophores.

Halosphaera

Halosphaera (Fig. 13.6), a plankton alga common in warm seas, is one of the genera now referred by some algologists to the class Prasinophyceae. It has large, spherical, nonmotile individual cells with numerous, small, discoid chloroplasts. The nucleus is suspended in the middle of the large central vacuole by cytoplasmic strands, and the peripheral cytoplasm contains the chloroplasts. The cell wall is composed of two overlapping halves, which fit together like the halves of a pillbox or medicinal capsule, but this construction can be demonstrated only after careful chemical treatment. Reproduction is by repeated division of the protoplast within the cell wall to form a number of heart-shaped zoospores, which are then liberated. The zoospores lack a cell wall and are much like ordinary vegetative cells of another genus, *Pyramimonas*. Electron micrographic examination discloses a covering of tiny scales on the pellicle. Four flagella arise from between the lobes at the front end of the zoospore. These flagella are covered with tiny scales that are somewhat different

Fig. 13.5 Adolf Pascher (1887–1945), German botanist, who made fundamental contributions to the classification of algae. [Photo courtesy of Einar Teiling.]

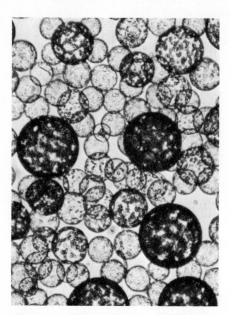

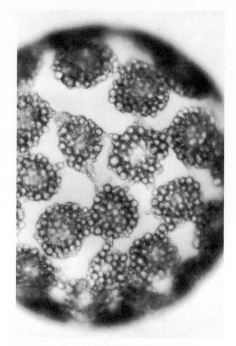

Fig. 13.6 *Halosphaera minor,* one of the smaller species; living cells in surface view. (*Left*) Cells grown in culture. (× 130.) (*Right*) A single mature cell, ready to sporulate, showing clusters of developing zoospores. (× 375.) [Photos courtesy of Mary Parke.]

from the scales of the pellicle, and they are also provided with numerous minute hairs, apparently unlike the mastigonemes found on pectinate and pinnate flagella in other groups. There is a single, large, somewhat cup-shaped chloroplast.

Coccolithophores

An abundant group of chrysophytan marine flagellates is characterized by the presence of numerous small, calcified disks embedded in the more or less gelatinous surface of the pellicle. These disks, which are readily visible microscopically, are called **coccoliths,** and the cells that bear them are called **coccolithophores**

(Fig. 13.7). It has been suggested that the coccoliths are a way of getting rid of excess calcium after carbon dioxide has been extracted from dissolved calcium bicarbonate [$Ca(HCO_3)_2$] for photosynthesis, but this interpretation is controversial. The coccoliths persist after the death of the cell, and their accumulation on the sea bottom has fostered the development of extensive chalk beds in warm regions. Large Mesozoic chalk deposits, in particular, are attributed to coccolithophores. Not all chalk beds have been formed by coccolithophores, however.

Coccolithophores have two whiplash flagella, and at least sometimes a third, flagellumlike structure called a **haptonema.** The haptonema appears to consist

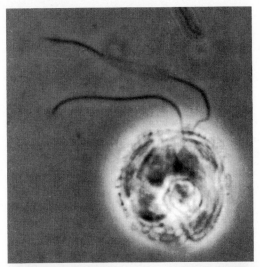

Fig. 13.7 (*Above*) *Hymenomonas carterae*, a common coccolithophore. (× 2000.) [Photo courtesy of Barry S. C. Leadbeater and P. J. Merrick.] (*Below*) Coccoliths of *Coccolithus huxleyi* (× 7000.) [Photo courtesy of Barry S. C. Leadbeater.]

in part of microtubules, but these are not arranged in the characteristic 9 + 2 structure of true flagella. It has been thought that the haptonema may serve for temporary anchorage of the cell to some other surface. The flagellar structure, the presence of the haptonema, and some other features indicate that the coccolithophores belong with the chrysophytan group that many algologists now consider to constitute a distinct class, the Haptophyceae.

Diatoms

The Bacillariophyceae, or diatoms, are unicellular or loosely colonial algae with a silicified wall composed of two halves that fit together like the halves of a petri dish or pillbox. Unlike that of some of the other chrysophytan algae with pillbox walls, the wall structure of the diatoms is evident to the light microscope. The diatoms further differ from all other Chrysophyta (so far as known) in that the vegetative cells are diploid rather than haploid. The only flagellated cells are the sperms of some genera that have a single, pinnate flagellum.

Diatoms are common in both fresh and salt water, particularly in temperate or cool regions. In freshwater lakes and ponds they tend to be more numerous in the spring and fall than during the hot summer months. They make up a large proportion of the phytoplankton in the cooler parts of the ocean. Many of them also occur attached to rocks or aquatic plants, and some are found in moist, subaerial habitats. The centric diatoms (Fig. 13.8) are chiefly marine, whereas the pennate diatoms (see Fig. 2.5) are more common in fresh water. An ample supply of dissolved silicates tends to favor the growth of diatoms, but at least some species can be cultured in a silicon-free medium, where they produce unsilicified cell walls.

The diatom cell has a well-developed central vacuole and one or more variously shaped chloroplasts. Sometimes the nucleus is suspended in a broad band of cytoplasm that passes through the vacuole. The chloroplasts are typically golden brown or somewhat yellowish and bear chlorophylls *a* and *c*, β-carotene, and ε-carotene, fucoxanthin, diatoxanthin,

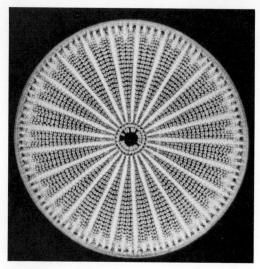

Fig. 13.8 *Arachnoidiscus*, a centric diatom. [Courtesy CCM: General Biological, Inc., Chicago.]

and diadinoxanthin. A few species lack chloroplasts and live as saprophytes. Food reserves of diatoms are chiefly oils and often also chrysolaminarin.

The cell wall of diatoms is very complex and has received a great deal of study because of the taxonomic usefulness of small differences in its structure. The outer of the two overlapping halves of the wall is called the **epitheca,** and the inner the **hypotheca.** The matrix of the wall consists largely of pectin or a pectinlike substance, without cellulose. On this is deposited a layer of hydrated silica, chemically similar to opal. Often the whole cell, or colonies of cells, are embedded in a watery gelatinous substance, which is at least sometimes composed largely of pectic acid.

The siliceous material of the wall is deposited as a continuous sheet, often with spiny processes or appendages, and regularly arranged with thin spots or perforations, so that a distinct pattern is formed. The pattern follows one of two general types. In the **centric diatoms** (Fig. 13.8) it tends to be radially symmetrical from a central point. In the **pennate diatoms** (Figs. 13.9, 13.10) there is a longitudinal (commonly bilaterally symmetrical) pattern arranged on both sides of a longitudinal strip called the axial field. Often the axial field is traversed by a complex longitudinal slot, the *raphe.* The raphe is a passageway through the wall, often V shaped in cross section, with the separated parts of the wall being aligned like the tongue and groove of floor boards. The raphe is usually interrupted at midlength by a channeled thickening of the wall, the central nodule, and sometimes also has a similar thickening at each end, the polar nodules.

Pennate diatoms with a raphe are often motile, the movement being alternately forward and backward in the direction of the long axis. The mechanism of movement is still controversial, but it appears to involve cytoplasmic flow in direct contact with the water along the outer fissure (the outer arm of the V, in cross section) of the raphe.

Asexual reproduction of diatoms occurs by cell division. Division is mitotic,

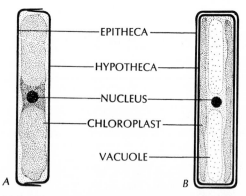

EPITHECA

HYPOTHECA

NUCLEUS

CHLOROPLAST

VACUOLE

A   *B*

Fig. 13.9 A typical pennate diatom: *A,* side view; *B,* top view.

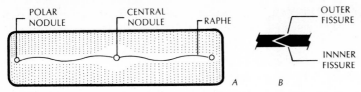

Fig. 13.10 Wall structure of a pennate diatom: *A*, top view; *B*, sectional view of raphe.

and each of the two daughter cells retains one of the two halves of the original cell wall. Each cell then produces a new half that fits into the older half. Thus of the two daughter cells, one is the same size as the parent and the other is slightly smaller. Repeated cell division therefore leads to a gradual reduction in the average size of the descendants of the original cell.

Restoration of diatom cells to the maximum size for the species follows on sexual reproduction. Vegetative cells of diatoms are diploid, and the first step in sexual reproduction is reduction division. Thereafter the process varies in detail according to the kind of diatom, but it commonly involves release of the protoplast from the cell wall, restoration of the diploid condition (usually by syngamy), enlargement of the naked diploid protoplast, and finally the formation of a new wall around the zygote. The zygote wall is silicified, but differs in detail from the wall of vegetative cells. The germinating zygote usually divides mitotically into two vegetative cells, each of which produces a complete new wall composed of the usual two parts.

In the centric diatoms syngamy is commonly oögamous. From 4 to 128 uniflagellate (pinnate) sperms are produced by some cells, whereas other cells produce each a single egg. In the pennate diatoms the gametes that fuse are more often amoeboid, and frequently alike; in any case, they lack flagella.

Many centric diatoms also produce resting cells called **statospores** (Fig. 13.11); these occur only in the Chrysophyta. The wall is composed of two usually very unequal overlapping halves, the smaller half forming a silicified or unsilicified plug closing a pore in the highly silicified larger half of the wall. The wall of the statospore is distinctly different in detail from that of the vegetative cell in which it is produced.

The silicified part of the cell wall of diatoms is scarcely affected by the death and degeneration of the rest of the cell,

Fig. 13.11 Statospores of various sorts.

and diatom shells therefore tend to accumulate at the bottom of any lake or sea in which diatoms live. When other sediments are slight, deposits consisting almost wholly of diatom shells may be formed. Such deposits, especially those that have been preserved from the geologic past, are called **diatomaceous earth.** A soupy diatomaceous ooze is now accumulating at the bottom of Klamath Lake, in southern Oregon, and in some ·other freshwater lakes, but the most important deposits of diatomaceous earth are of marine origin. Oil wells in the Santa Maria oil fields of California pass through a layer of marine diatomaceous earth about 3000 feet thick. Near Lompoc, California, surface beds of marine diatomaceous earth several miles across and over 700 feet thick are being exploited by quarrying (Fig. 13.12).

The oldest fossil diatoms of thoroughly authenticated date come from the early part of the Cretaceous period, but most diatomaceous earth formations are of late Cretaceous or later age. All late Cretaceous genera are still represented by living species, and some late Cretaceous species are apparently identical with modern ones. All of the older fossils of diatoms appear to be marine. Freshwater types make their first appearance in the Oligocene.

The most important commercial use of diatomaceous earth is in filtration of liquids, particularly in sugar refineries.

Fig. 13.12   Diatomaceous earth quarry near Lompoc, California. [Photo courtesy of Celite Division, Johns-Manville Corporation.]

Small quantities of powdered diatomaceous earth added to a sugar solution will retain the suspended impurities from the solution when the mixture is forced through a filter. Some other uses of diatomaceous earth are as a fine abrasive (as in silver polish and, at least in the past, in toothpaste), as an additive to paint to increase the night visibility of signs, and as an insulating material, especially in blast furnaces and other high-temperature installations. At temperatures of over 1000°F, diatomaceous earth is a more effective insulator than asbestos.

BROWN ALGAE

The brown algae are a well-defined group of about 1500 species, nearly all of them marine, which collectively make up the division Phaeophyta. Some of them are microscopic, branching filaments, but most of them have a larger, more complex thallus, ranging from a few centimeters to as much as 60 m long. The larger forms are usually complex in structure, with several different kinds of cells. There are no unicellular brown algae.

Brown algae are especially common along and near the seashore in the cooler parts of the world (see chapter-opening figure), both in the intertidal zone and wholly submerged. Most of them are attached to the substrate (sometimes even to other algae) by a basal holdfast; a few, such as some species of *Sargassum*, are free-floating in the open ocean.

The cell wall of brown algae has a cellulosic inner layer and a gelatinous, pectic outer layer usually composed largely of algin. Algin is merely a convenient name to cover a varying mixture of polymers of several organic acids and their calcium salts. Mannuronic, gulonic, and glucoronic acids are among the principal components of these algin polymers.

Cells of brown algae contain one to more often several or many variously shaped, often discoid chloroplasts. The chloroplasts bear chlorophylls *a* and *c*, β-carotene, and large amounts of fucoxanthin and violaxanthin, plus other xanthophylls. Fucoxanthin gets its name from *Fucus* (Fig. 13.13), one of the more common genera of brown algae. It is also an important xanthophyll in diatoms and some of the other Chrysophyta. Fucoxanthin is largely responsible for the typical brownish color of the brown algae.

The nucleus of brown algae contains one or more nucleoli and divides mitotically. A typical spindle is formed, originating at the two centrioles. A number of clear vesicles containing a substance called **fucosan** are usually aggregated

Fig. 13.13   *Fucus.* (× 2/3.) [Photo by W. H. Hodge.]

around the nucleus. Fucosan has many of the properties of tannin; it is variously thought to be a waste product or to have some as yet undiscovered metabolic function. Like the tannins of higher plants, it may play a defensive role in making the plant unpalatable to potential predators. Vegetative cells of brown algae usually contain many small vacuoles, or occasionally a large central vacuole.

The most common food reserve of the brown algae is laminarin, a soluble polysaccharide composed of about 15 to 30 glucose residues primarily in β-1,3 linkages, and with mannitol as a terminal unit. Mannitol (a 6-hydrogen alcohol that can be oxidized to yield the hexose sugar mannose) and minute globules of fat are also frequently found in the cytoplasm as storage products. Laminarin is chemically allied to the chrysolaminarin of the Chrysophyta and the paramylon of the euglenoids.

The brown algae have diverse sorts of life cycles. Genera such as *Ectocarpus* have similar diploid (sporophytic) and haploid (gametophytic) generations. *Laminaria* (see Fig. 2.5) and other kelps have a relatively large and conspicuous sporophyte and a microscopic, filamentous gametophyte. In *Fucus* and *Sargassum* fertilization follows immediately on reduction division, so that the sperms and eggs are the only haploid cells. Sporophytes of many brown algae also produce unreduced (diploid) zoospores that develop directly into new sporophytes, bypassing the sexual cycle. The free-floating species of *Sargassum* reproduce asexually by the detachment of fragments that float away and develop into new plants, never producing sexual structures.

Motile cells of brown algae generally lack a cell wall and have two dissimilar flagella, one whiplash and one pinnate.

Sometimes the whiplash flagellum is much reduced or even wanting.

Visitors to rocky coasts, seeing kelp, bladder-wrack (*Fucus*), and other brown algae in abundance, may easily get an exaggerated idea of their importance in the oceanic food chain. With the notable exception of some species of *Sargassum*, however, they are primarily littoral or sublittoral, and absent from the open sea. *Sargassum* (Fig. 13.14) is chiefly a tropical and subtropical genus with a complex sporophyte, commonly a meter or more long, and divided into stemlike and leaflike parts. The Sargasso Sea, a large area in the Atlantic Ocean between the West Indies and Africa that is relatively little affected by ocean currents, gets its name from the abundance of *Sargassum*.

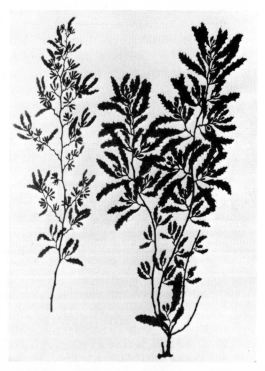

Fig. 13.14 *Sargassum platycarpum.* [New York Botanical Garden photo.]

The larger species of brown algae, particularly the kelps (*Laminaria* and other members of the order Laminariales), are harvested as a major source of food by the Japanese and as a source of the colloidal gel algin in Europe and America. Algin is used as a stabilizer or a moisture retainer in a wide variety of commercial products, including ice cream, cake frosting, paint, and various pharmaceuticals; it is also used in the processing of natural and synthetic rubber latex.

The brown algae have a long fossil history, extending back nearly to the base of the Palaeozoic era, more than half a billion years ago. Some of the Devonian members of the group, such as *Protosalvinia,* were partially adapted to life on land, occurring above as well as below the high tide line.

RED ALGAE

The red algae (Fig. 13.15) are a sharply marked group of about 3500 species, the vast majority of them marine, which collectively make up the division Rhodophyta. They were first clearly recognized as a distinct group in 1836 by the Irish botanist W. H. Harvey (1811–1866). Nearly all of them are multicellular, forming well-developed, often compactly branched thalli that are typically composed of compacted or distinct filaments, without much differentiation of tissues. They occur in all oceans, but are most common in tropical and warm-temperate regions, especially in the Southern Hemisphere. The marine species are almost always attached rather than free-floating. Some species occur in the intertidal zone; many others are wholly submerged, at depths of up to 175 m. Red algae occur at greater depths than any other photosynthetic organisms.

The cell wall of red algae consists of a firm inner layer and a gelatinous or mucilaginous outer layer, both composed largely of carbohydrate (or modified carbohydrate) polymers. The inner layer commonly but not always contains some cellulose, usually with some carbohydrate molecules other than glucose incorporated into the cellulose. The polymers of the soft outer layer can be considered to be pectic compounds, in a broad sense. They are commonly based in considerable part on sulfated galactose. Galactose is a hexose, not very different from glucose in structure. Agar and carageenin are such galactosulfate polymers that are abundant wall components of many red algae.

Plasmodesmata in the ordinary sense are lacking from the red algae. In most of them, however, the cross wall formed between two sister cells in mitosis is imperfect, with a large central pore. It was at one time thought that the cytoplasm was continuous through these pit connections, but EM studies indicate that the pit is typically (always?) closed off by a thin membrane or plug. Similar pits often develop by dissolution of the wall (or the major part of the wall) between adjacent cells that were derived from different mitotic divisions. The pit connections have been thought to function in the transfer of food and other substances from cell to cell, but their possible function must be considerably restricted by the fact that the cytoplasm is not actually continuous from one cell to another.

Some species of red algae have uninucleate cells, whereas others are multinucleate. The nucleus, although usually small and inconspicuous, has a well-developed nucleolus and a nuclear membrane. There has been some question as to whether the nucleus of red algae is fully comparable in all respects to that of other

Fig. 13.15   Some common red algae. (*Top*) *Phycodrys rubens*, from the Atlantic coast of North America and Europe. (*Bottom left*) *Microcladia coulteri*, from the Pacific coast of the United States. (*Bottom right*) *Polysiphonia harveyi*, from the northeast coast of the United States. (All × about 1/2.) [New York Botanical Garden photos.]

eukaryotes, but no real differences have been disclosed by the limited number of studies so far undertaken. Nuclear division is mitotic.

There are no flagella, nor any motile structures of any sort, among red algae. Neither is there a centriole, so far as present studies show. Current opinion is that these structures have never been present in the ancestry of the red algae (aside from the much simpler flagella of eventual bacterial ancestors), rather than that they have been lost in the course of evolution. Chloroplasts, mitochondria, golgi bodies, and microtubules are present, as in other eukaryotes.

Red algae have one to many chloroplasts in a cell. These contain chlorophyll a and usually d, α- and β-carotene, lutein, and sometimes other xanthophylls, phycoerythrin, and often phycocyanin. Often there is only a small amount of chlorophyll as compared to the amount of phycobilin pigments.

Phycoerythrin and phycocyanin are phycobilin pigments, which have been discussed in connection with the blue-green algae in Chapter 11. Like the carotenoids, they absorb light energy and transmit some of it into the photosynthetic process. Phycoerythrin absorbs reasonably well throughout the green part of the spectrum, with its absorption peak near the boundary of green and yellow. Only green light penetrates very far through water, the other wavelengths being practically all screened out within a few meters. Red algae that grow in the lower part of the intertidal zone or wholly submerged in the ocean commonly have abundant phycoerythrin and little or no phycocyanin, but those from the upper intertidal zone and from fresh water commonly have a considerable amount of phycocyanin in addition to phycoerythrin and thus do not have the characteristic red color of the deep-water species.

The carbohydrate reserves of red algae usually accumulate as small cytoplasmic granules of floridean starch, a branching polyglucan very much like the amylopectin fraction of the starch of higher plants. Red algae also accumulate greater or lesser amounts of floridoside, which is unique to this group. A molecule of floridoside consists of one molecule of galactose combined with one molecule of glycerol. Galactose is thus seen to be a major product of the red algae, appearing in food reserves as well as in the cell wall. Galactose also occurs in several other groups of plants, but seldom as such a major product.

Life cycles of red algae are extremely diverse and often very complicated, so much so that different stages have often unwittingly been given names as completely different genera. *Porphyra*, for example, is well known on rocks in the intertidal zone in both the Old and the New World, forming a thallus consisting of a flat or convoluted sheet, one or two cells thick and several centimeters to a meter or more across. A shell-boring, filamentous alga, long known as *Conchocelis rosea*, has been demonstrated to be a stage in the life cycle of the alga otherwise known as *Porphyra umbilicalis*, and other species of *Porphyra* also have a *Conchocelis* stage in the life cycle (Fig. 13.16). A complicated special terminology has been developed for the reproductive structures of red algae, but is not presented here.

The red algae are so different from all other algae that they are taxonomically rather isolated. They have most often been compared to the blue-green algae, which they resemble in having phycobilins as important accessory photosynthetic pig-

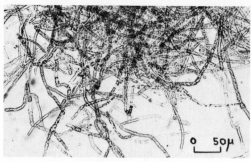

Fig. 13.16 *Porphyra leucosticta*, the leaflike thallus at the left, the *Conchocelis* phase at the right. [Photos courtesy of Peter Edwards, from his Illustrated Guide to the Seaweeds and Sea Grasses in the Vicinity of Port Aransas, Texas, published in *Contrib. Marine Science*, suppl. to Vol. 15, 1970.]

ments, in the complete absence of flagellated cells, and in a number of detailed chemical and ultrastructural features. On the other hand, the red algae are distinctly eukaryotic, in contrast to the prokaryotic organization of the blue-green algae. If the eukaryotic condition evolved only once, as most evolutionary biologists believe, then the red algae must be related to the other eukaryotic algae, and in a sense more closely related to them than to the blue-green algae. One possible answer to this problem is to consider that there was an early evolutionary dichotomy among the most primitive eukaryotes (derived from blue-green algae), and that the red algae retained a number of primitive chemical and ultrastructural features that have been lost by other eukaryotes. On the other hand, this thought does not fit well with the idea that the euglenoids and especially the dinoflagellates may be among the most primitive eukaryotes, based on mitotic behavior and (in the dinoflagellates) chromosome structure. Concepts of evolutionary relationships among the algae are still in a state of flux, but algologists agree that the red algae stand well apart from all other groups.

Red algae are economically important as the source of the colloids agar and carageenin. Agar is widely used as a culture medium for bacteria, as a bulk producer in laxatives, and as a stabilizer or filler in various commercial foods. *Gelidium* and various other genera are the source of agar. Carageenin, which has

Fig. 13.17 *Chondrus crispus*, the Irish moss. (About natural size.) [Courtesy CCM: General Biological, Inc., Chicago.]

properties somewhat different from those of agar and algin, has traditionally been obtained from *Chondrus crispus* (Fig. 13.17), the "Irish moss," which is harvested on the coasts of northern Europe, New England, and the Maritimes. Colloids similar to carageenin are now obtained from other red algae elsewhere in the world. *Porphyra* is used for food in the Orient, both as a prime ingredient of soup and as a flavoring for other foods. The plant is cultivated on tidelands in Japan and is harvested annually.

The coralline algae (Fig. 13.18), constituting the family Corallinaceae, are marine red algae in which the thallus becomes strongly calcified. Such genera as *Lithothamnion* and *Porolithon* are important contributors to the growth of coral

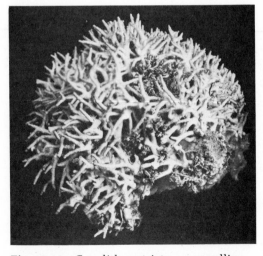

Fig. 13.18 *Gonolithon strictum*, a coralline red alga from the Bahama Islands. (Slightly less than natural size.) [New York Botanical Garden photo.]

reefs. The algae commonly grow closely intermingled with coral animals, which are also heavily calcified. The coralline algae have been active reef-builders in the geologic past as well as in recent times, and fossils that are identified as representing the modern genus *Lithothamnion* are known from as far back as the Triassic period. Other fossils that very probably represent calcareous red algae are abundantly represented in strata of Ordovician to Jurassic age.

SUGGESTED READING

Boney, A. D., *A biology of marine algae,* Hutchinson, London, 1966. Strong on brown and red algae.

Boney, A. D., *Phytoplankton,* Edward Arnold, London, 1975.

Chapman, A. R. O., *Biology of seaweeds,* University Park Press, Baltimore, Md., 1979.

Dawson, E. Y., *Marine botany,* Holt, Rinehart & Winston, New York, 1966. Useful for additional information on algae.

Duddington, C. L., *Seaweeds and other algae,* Faber & Faber, London, 1966; republished as *Flora of the sea,* Crowell, New York, 1967.

Frederick, J. F., and R. M. Klein (eds.), Phylogenesis and morphogenesis in the algae, *Ann. N.Y. Acad. Sci.* **175**(2):413–781, 1970. A symposium reporting current research and thought.

Gantt, E., Phycobilisomes: Light-harvesting pigment complexes, *BioScience* **25**:781–788, 1975.

Isaacs, J. D., The nature of oceanic life, *Sci. Amer.* **221**(2):146–162, September 1969.

McElroy, W. D., and H. H. Seliger, Biological luminescence, *Sci. Amer.* **207**(6):76–89, December 1962.

Werner, D. (ed.), *The biology of diatoms,* Botanical Monographs 13, University of California Press, Berkeley, 1977. A symposium.

Chapter 14

Fungi

Lepiota americana, a common mushroom. (About twice natural size.) This genus contains both edible and poisonous species, as well as species that are poisonous to some people and harmless to others. [Photo by Hugh Spencer, from National Audubon Society Collection/Photo Researchers, Inc.]

Fungus (pl., **fungi**) is a Latin word, originally applied to mushrooms, the meaning of which has been gradually extended to cover molds, yeasts, and all other organisms that seem to be related to mushrooms. It has been estimated that there are more than 40,000 known species of fungi, and that at least as many more, perhaps twice as many, remain to be recognized.

One way to consider the ecological significance of the great groups of organisms is to think of the food makers (plants), food eaters (animals), and food absorbers (bacteria and fungi). In a formal taxonomic sense, the bacteria and the fungi are usually associated with the plant kingdom, but they play a very different role in the unending cycle of nutrients and energy in the organic world. Some biologists consider the fungi as a third kingdom, coordinate with the plant and animal kingdoms, but this classification also has problems. In recent years it has even become progressively more doubtful that the fungi are a proper natural group, especially inasmuch as some of the subgroups within the traditional fungal class Phycomycetes may have evolved from very different flagellated ancestors.

As food absorbers, rather than food eaters, the fungi do not need and have not developed the syndrome of features associated with motility and the capture of food that characterize the animal kingdom. Instead they retain the simpler, less rigidly ordained and less tightly integrated structures of plants. In any two-kingdom system, the fungi must go with the plants.

The fungi are unicellular to more often multicellular or acellular eukaryotic plants, without chlorophyll, and typically more or less filamentous in structure. Often some of the filaments are compacted into bodies of definite macroscopic form but without much differentiation of tissues. The walls of the cells or filaments are most often chitinous, less commonly of cellulose. Gametangia, when present, are "unicellular," although they often have many nuclei. Some of the simpler kinds of fungi are unicellular, with or without flagella; and the slime molds have a special type of multinucleate, noncellular protoplast called a **plasmodium.** Fungi have diverse sorts of life cycles, often very complex, that generally involve the formation of one or more sorts of spores as reproductive bodies.

The fungi are here considered to constitute a taxonomic division (Fungi) of plants, with two subdivisions, the Myxomycotina and Eumycotina. This familiar arrangement has some weaknesses, but perhaps no more than any of the other current schemes. The Myxomycotina, with a single class Myxomycetes (slime molds), are so different from the other fungi that they must be treated separately.

MYXOMYCOTINA: MYXOMYCETES (SLIME MOLDS)

The Myxomycetes, or slime molds, are fungi that in part or all of the "vegetative" state are amoeboid and usually holozoic, consisting of a mass of protoplasm in which the nuclei are not separated by cell walls. The characteristic "vegetative" or assimilative body of most slime molds is called a **plasmodium** (Fig. 14.1). The plasmodium flows slowly across the substrate, engulfing and ingesting bacteria and other small organic particles in an amoeboid manner.

Myxomycetes are common but inconspicuous inhabitants of dead wood and other vegetable remains, often spending most of their lives within the substrate and emerging only when about to produce

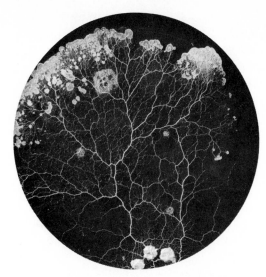

Fig. 14.1 Plasmodium of a slime mold, *Physarum*. (Much enlarged.) [Courtesy CCM: General Biological, Inc., Chicago.]

sporangia. There are about 500 species in all.

There has been continuing controversy as to whether the slime molds are better associated with the plant or the animal kingdom. They are certainly animal-like in nutrition, but their reproductive structures (Fig. 14.2) are more plantlike. Furthermore, some unusual slime molds such as *Plasmodiophora* are food absorbers rather than eaters and have often been considered to form a link to the true fungi.

EUMYCOTINA

General Features

The Eumycotina, or true fungi, have definite walls and are always saprobic or parasitic; they absorb their food rather than eating it. Except for the class Oömycetes, the walls of Eumycotina are generally chitinous. Chitin is a polymer of glucose-like units that contain some nitrogen. Chemically, a molecule of fungus chitin typically consists of eight N-acetyl-glucosamine residues arranged in a pair of cross-linked four-unit chains, in contrast to the long, simple chains of insect and crustacean chitin.

Fig. 14.2 Fruiting bodies of slime molds, on dead wood. (*Left*) *Lycogala epidendrum*. (About half natural size.) [New York Botanical Garden photo.] (*Right*) *Stemonitis* (× 3.) [Courtesy CCM: General Biological, Inc., Chicago.]

Most of the Eumycotina are more or less filamentous; a small proportion are unicellular and uninucleate or have a multinucleate nonfilamentous protoplast. A single fungus filament is called a **hypha.** The hypha may or may not have cross-partitions (**septa**); nonseptate hyphae have the nuclei scattered in a continuous protoplast and are said to be **coenocytic.** Most hyphae are branched, and a mass of branching hyphae is called a **mycelium.** Many parasitic fungi also produce special short hyphae, called haustoria, that enter the cells of the host and absorb nutriment. The term **haustorium** is also applied to organs of similar function but different structure produced by some other kinds of parasitic plants. Many fungi produce some relatively short hyphae that penetrate the substrate and anchor the external mycelium. Such hyphae are called **rhizoids,** a general term applied to any structure that performs some of the functions of a root (especially anchorage) but is anatomically rather simple, consisting of only one or a few cells, or part of a cell.

The Eumycotina, like the bacteria, commonly release exoenzymes into the surrounding substrate. Different kinds of fungi produce different kinds of exoenzymes, and a large proportion of the naturally occurring organic substances can be used as food by one or another kind of fungus. Food materials digested by the action of these enzymes are absorbed by the fungus. Absorbed food that is not immediately used is stored in the cytoplasm in a variety of different chemical forms, the form often being influenced by the original nature of the food. Reserve carbohydrates of a wide range of types occur in different fungi. The alcohol mannitol and minute droplets of oil are also common food reserves.

The sexual cycle is extremely varied among different fungi, with nuclear fusion (**karyogamy**) often being delayed until long after the gametic nuclei have been brought together in the same protoplast (**plasmogamy**). The first division of a fusion nucleus is usually meiotic, but there are exceptions.

The structures associated with nuclear fusion and reduction division provide some of the most consistent guides to systematic relationships within the fungi; the stage in which these structures are produced is called the **perfect stage.** Often a **fruiting body** of definite macroscopic size and form is produced at the perfect stage. A fruiting body is, theoretically, composed of closely compacted hyphae, but the compaction is often so thorough that the filamentous nature is more theoretical than actual.

Reduction usually leads directly, or after one or a few intervening mitotic divisions, to the formation of spores. Many fungi have life cycles in which several different kinds of spores are produced. Spores produced immediately or very soon after meiosis are, in this text, designated as **sexual spores,** in contrast to **asexual spores,** which are not associated with reduction division. Some botanists prefer the terms **meiospore** and **mitospore,** respectively.

Spores of the Eumycotina are usually nonmotile and have a definite wall, except in the small classes Chytridiomycetes and Oömycetes, which have naked spores with one or two flagella. When two flagella are present, one is whiplash and the other pinnate.

The term **conidium** (or gonidium) has often been used for any asexual spore in the fungi or elsewhere. Among the fungi, conidia are usually produced by the differentiation and abscission of a hyphal tip, and often the same hypha cuts off a series of conidia one after another (see Fig. 14.22). Alternatively, an inner wall is extruded through the outer wall at the hyphal tip, and a conidium is cut off. Any hypha on which conidia are produced may be called a **conidiophore.** Many fungi produce more than one kind of conidium, or they produce conidia of such distinctive type as to warrant a distinctive name. Mycologists do not generally use the term conidia for spores produced in sporangia, even when these are strictly asexual.

The Eumycotina are here considered to consist of five classes, the **Chytridiomycetes, Oömycetes, Zygomycetes, Ascomycetes,** and **Basidiomycetes,** plus an artificial group, the **Fungi Imperfecti,** which lack sexual stages and cannot be clearly referred to one of the other groups.

The Chytridiomycetes, Oömycetes, and Zygomycetes have some features in common and have often been considered collectively to constitute a class **Phycomycetes.** They are typically coenocytic and more or less filamentous, with the only cross walls in the mycelium being those that delimit the reproductive structures. The perfect stage, usually consisting of a zygospore or an oöspore, is not borne on a definite fruiting body. Most "Phycomycetes" produce numerous spores in a sporangium, unlike nearly all the Ascomycetes and Basidiomycetes. On the other hand, the Chytridiomycetes and Oömycetes differ collectively from all other true fungi in having flagellated gametes or zoospores, and the Oömycetes differ from all other fungi in a series of features, as noted below.

Fig. 14.3 (*Left*) Elias Magnus Fries (1794–1878), Swedish mycologist, the father of systematic mycology. (*Right*) Heinrich Anton DeBary (1831–1888), German mycologist, who worked out the life cycle of *Puccinia graminis* and made many other important mycological contributions. [Photos courtesy of Department of Plant Pathology, Cornell University.]

The Chytridiomycetes, commonly called **chytrids,** are of interest to mycologists especially because some of them are among the simplest of all fungi, consisting of a multinucleate cell that eventually produces flagellated zoospores or isogametes. Some other chytrids show transitional stages toward the typical mycelium of more advanced fungi. The chytrids are of little economic interest, although one species, *Synchytrium endobioticum,* causes the wart disease of potatoes. Many mycologists recognize an additional class, Hyphochytridiomycetes, differing from the Chytridiomycetes in some important technical features.

Oömycetes

General Features. The Oömycetes are filamentous or unicellular, coenocytic fungi that generally have cellulose as an important wall constituent; rarely they also have chitin. They produce motile cells at some stage(s) in the life cycle. These motile cells are biflagellate, with one whiplash and one pinnate flagellum. The few members that have been tested make lysine by the diaminopimelic pathway, unlike all other groups of fungi, which so far as known follow the aminoadipic pathway. These features collectively set the Oömycetes off from the other Eumycotina.

Some of the simpler members of the Oömycetes are not filamentous, and are similar in gross structure to some of the simpler chytrids. These are of interest mainly to mycologists and will not be discussed here. Two other orders, the Saprolegniales and Peronosporales, merit our attention.

Saprolegniales. The Saprolegniales, or water molds, occur most commonly in fresh water, but many of them inhabit moist soil. They are saprophytes living on dead plant or animal remains, or they are facultative or obligate parasites of algae, fish (see Fig. 2.6), and various small aquatic animals, or occasionally they are parasites of the roots of vascular plants. They have a well-developed mycelium which develops chiefly within the substrate, the external mycelium being chiefly reproductive in function. Numerous biflagellated zoospores are produced in sporangia at the hyphal tips, and some members of the group also produce thick-walled conidia in chains at the ends of individual hyphae.

Sexual reproduction involves the formation of antheridia and oögonia, with the sperm nucleus being transferred to the oögonium through a conjugation tube. The fertilized egg secretes a thick wall and goes into a resting stage, at which time it is called an **oöspore.** Fusion of the sperm and egg nuclei in the oöspore is delayed for some time, often until shortly before germination. The first nuclear division of the germinating oöspore is believed to be meiotic.

Peronosporales. The Peronosporales are closely related to the Saprolegniales, and are evidently derived from them. They differ from the Saprolegniales in several ways. They are mostly terrestrial rather than aquatic, and mostly parasitic rather than saprophytic. Furthermore, the structure that in the Saprolegniales remains attached to the mycelium and develops into a zoosporangium is in the Peronosporales generally released from the tip of the hypha as a conidium. A conidium of the Peronosporales often germinates directly to produce a new mycelium, as do the conidia of most other fungi, but if it falls into the water, it may produce zoospores instead. The conidium of the Peronosporales is thus evidently a modified sporangium. The sexual cycle of the Peronosporales is basically similar to that of the Saprolegniales.

Phytophthora infestans (Fig. 14.4), a member of the Peronosporales, causes the late blight of potatoes. Originally native to the Andes of South America, as is the potato, the fungus has now become generally distributed wherever potatoes are grown. The first severe outbreak of the disease in the cultivated crop occurred in 1845 and 1846 and resulted in crop failure over much of Europe and eastern North America. The great Irish Famine of 1845–1847 resulted from these two years of crop failure, for potatoes were then the principal staple food of Ireland. Students of history may recall that the Irish Famine set off a great wave of emigration from Ireland to the United States.

The conidia of *Phytophthora infestans* are borne on specialized hyphae (conidiophores) that emerge from the leaf of the diseased potato plant through the stomates (see Chapter 21). The conidium is released into the air and is wind borne.

Subsequently, according to the conditions, it may function as a sporangium and produce zoospores, or it may germinate directly to form a new mycelium, thus immediately spreading the disease.

There are many parasitic Peronosporales which, like *Phytophthora infestans,* produce conidia apically on conidiophores that typically extend out into the air through the stomates of the leaf of the host. These external hyphae are often so numerous as to make the leaf surface finely downy. Such members of the Peronosporales, and the diseases they cause, are known as **downy mildews.**

Zygomycetes

The Zygomycetes are filamentous, coenocytic fungi that have chitinous walls and lack flagellated cells of any kind. Most of them produce sporangia with numerous spores. Sexual reproduction involves the fusion of multinucleate gametangia at the hyphal tips to form a multinucleate zygospore. The most widespread and important order in the class is the **Mucorales.**

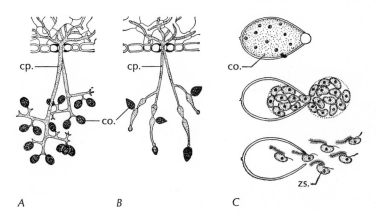

A   *B*   *C*

Fig. 14.4 Downy mildews. *A, Plasmopara viticola,* the downy mildew of grapes; *B, C, Phytophthora infestans,* cause of the late blight of potatoes: co., conidium; cp., conidiophore; zs., zoospore.

The Mucorales are a common group of terrestrial (rather than aquatic) saprophytes or facultative (seldom obligate) parasites that are vegetatively rather similar to the Saprolegniales. They usually have an extensive coenocytic mycelium within the substrate, and a smaller aerial reproductive mycelium. The wall is usually or perhaps always chitinous. Sporangia borne on the aerial mycelium produce numerous spores, each with its own wall and without flagella.

Rhizopus stolonifer (R. nigricans) (Figs. 14.5, 14.6), the common bread mold, may be taken as typical of the Mucorales and of the Zygomycetes as a whole. Although it is generally saprophytic, the species is a rather weak facultative parasite and will grow, for example, on strawberry fruits. Bread mold has become much less abundant in the United States since bakers began to use calcium propionate "to retard spoilage."

Ascomycetes

Definition and History. The Ascomycetes are fungi in which the sexual spores (**ascospores**) are usually few and definite in number (typically eight) and are produced within a sporangium called an **ascus.** The term ascus is derived from the Greek word *askos*, meaning "bladder, sac, or container," and the Ascomycetes are often called **sac fungi.**

The Ascomycetes are mostly filamentous, with chitinous walls, only rarely with appreciable amounts of cellulose. The filaments are generally septate (provided with cross-partitions), but the segments delimited by the septa often contain several or many nuclei, and each septum commonly has a central opening through which cytoplasm and often nuclei may pass.

Many Ascomycetes reproduce asexually by conidia that are successively cut off from the ends of conidiophores, and only seldom carry on sexual reproduction. A conidium germinates directly to form a new hypha and eventually a mycelium, instead of giving rise to zoospores as in many Oömycetes.

The Ascomycetes were first recognized as a coherent group by the Czech mycolo-

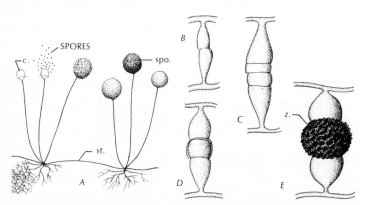

Fig. 14.5 *Rhizopus stolonifer*, the common bread mold. *A*, Aerial mycelium, with a part of the internal mycelium (× 35); *B–E*, stages in sexual reproduction [(B) × 125, (C–E) × 85]: c., columella; spo., sporangium; st., stolon; z., zygospore.

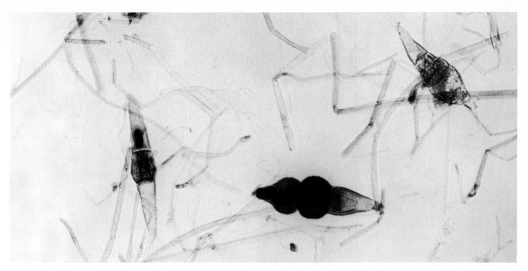

Fig. 14.6 *Rhizopus stolonifer,* showing stages in zygospore formation. [Courtesy CCM: General Biological, Inc., Chicago.]

gist August Corda (1809–1849) in 1842, some 26 years after the ascus itself had been recognized as a significant structure by the German botanist Christian Gottfried Daniel Nees von Esenbeck (1776–1858) in 1816. The greatest contribution to our knowledge of the morphology and taxonomy of the class was made by two amateur French mycologists, the brothers Louis René Tulasne (1815–1885) and Charles Tulasne (1816–1884), in a series of papers published from 1861 to 1865.

Sexual Cycle. Except for some relatively simple forms, such as yeasts, the sexual cycle in Ascomycetes (Fig. 14.7) usually conforms more or less closely to the pattern described below.

One or several adjacent cells or multinucleate segments at the end of a hypha of the vegetative mycelium become differentiated as an **ascogonium** that is usually somewhat broader than an ordinary vegetative hypha. Usually a slender prolonga-

tion, called a **trichogyne,** grows out from the end of the ascogonium. Specialized apical segments of the hyphae of the same, or more usually a different, mycelium act as antheridia. The trichogyne grows to the antheridium, and the walls between them dissolve. The nuclei of the antheridium then pass through the trichogyne into the body of the ascogonium, where they become paired with the ascogonial nuclei, without fusion. Sometimes antheridia are not produced or do not function fully, and the nuclei of the ascogonium become paired without the introduction of any foreign nuclei. Or foreign nuclei can be introduced in other ways, such as by the transfer of specialized conidia.

After one or more sets of nuclei are paired in the ascogonium, several filaments, called **ascogenous hyphae,** usually grow out of the ascogonium. Some or all of the nuclei of the ascogonium move into the ascogenous hyphae, and usually a number of mitotic divisions take place,

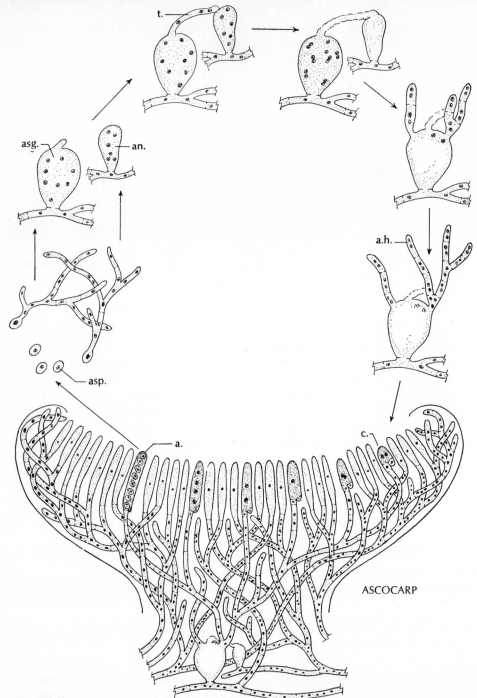

Fig. 14.7   Life cycle of a typical Ascomycete: a., ascus; an., antheridium; asg., ascogonium; asp., ascospore; a.h., ascogenous hypha; c., crozier; t., trichogyne.

but the nuclei remain paired. The ascogenous hyphae usually branch repeatedly and sooner or later develop septa. The segments nearer the ascogonium are often multinucleate, but each terminal segment contains only a single pair of nuclei. Each of these terminal, binucleate hyphal segments is the forerunner of one or more asci. Often the young, developing asci proliferate by a complex process that maintains their binucleate condition.

Coincident with the growth of the ascogenous hyphae there is usually developed a very compact mycelium originating directly from the vegetative mycelium, which comes to surround the group of ascogenous hyphae; and slender hyphae, originating from or in the same way as this compact mycelium, become intermingled with the ascogenous hyphae. This compact mycelium usually has a definite form and structure, according to the genus and species. Together with the included associated ascogenous hyphae and eventual asci, it is known as the **fruiting body.** The fruiting body of an Ascomycete may also be called an **ascocarp.**

The asci are often borne in a distinct layer, intermingled with sterile hyphae, covering part of the external or internal surface of the fruiting body. Such a layer of asci and sterile hyphae is called a **hymenium.**

The two nuclei in the developing ascus fuse to form a diploid nucleus. The diploid nucleus then undergoes reduction division, producing four haploid nuclei. Usually each of these undergoes a single further mitotic division, so that there are eight nuclei (and ultimately eight ascospores) in the ascus. In due time the ascus breaks open, liberating the spores. Each spore can develop into a new vegetative mycelium.

Taxonomy and Distribution.   It has been estimated that there are about 15,000 known species of Ascomycetes. They occur as saprophytes and parasites (chiefly of plants and insects) in terrestrial habitats throughout the world. Only a few are aquatic, but these include both freshwater and marine species.

It is customary to segregate the yeasts and some other Ascomycetes that do not have definite fruiting bodies as a small subclass Hemiascomycetes. The remainder of the class, in which definite fruiting bodies are formed, is now generally divided into two subclasses, the Euascomycetes and Loculoascomycetes, on the structure of the asci and fruiting bodies.

Yeasts.   Ascomycetes that are more or less distinctly unicellular and produce solitary asci are called **yeasts** (Fig. 14.8). They make up a large part of the order Saccharomycetales, also called Endomycetales. Asexual reproduction occurs by cell division. Most yeasts divide into two initially unequal cells by a process known as **budding.** A small projection or bud develops on a cell, the nucleus divides mitotically, and one of the daughter nuclei moves into the bud. The bud continues to enlarge, although the opening remains small. Eventually the opening is closed by a double partition, and the two cells separate. Under conditions favorable for rapid growth, the initiation of new buds may so far precede the maturation and separation of previously formed buds that branching, temporary colonies are formed.

The typical sexual cycle in yeasts involves fusion of two equal or unequal uninucleate cells, followed almost immediately by nuclear fusion. The resulting cell, which may properly be called a zygote, enlarges and becomes an ascus.

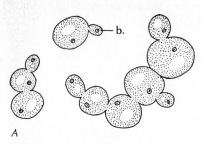

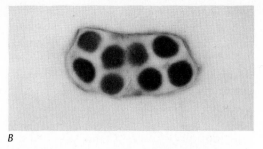

A   *B*

Fig. 14.8 Yeast. (*A*) Baker's yeast, *Saccharomyces cerevisiae:* b., bud. (× 2000.) (*B*) Ascus and ascospores of *Schizosaccharomyces octosporus.* [Courtesy CCM: General Biological, Inc., Chicago.]

The diploid nucleus undergoes meiosis, after which the four resulting haploid nuclei may mature directly into ascospores, or they may first divide mitotically so that eight ascospores are produced in the ascus. There are several minor variations on this general pattern, and some yeasts lack the sexual cycle entirely. These **asporogenous yeasts** (in contrast to the typical **sporogenous yeasts**) are in a technical sense Fungi Imperfecti, but they are commonly grouped with the yeasts because of their obvious relationship.

Yeasts are economically important chiefly because of the fermentation that many of them carry on, releasing alcohol and carbon dioxide as end products. The equation for alcoholic fermentation of glucose is

$$C_6H_{12}O_6 \rightarrow 2C_2H_5OH + 2CO_2 + energy$$

Only a relatively small proportion of the energy bound up in glucose is released in this process, and many kinds of organisms (including humans) can use alcohol as a source of respiratory energy. The economically important alcoholic fermentations are carried on almost exclusively by yeasts. The chemistry of alcoholic fermentation has been discussed in Chapter 8.

The common cultivated yeast, *Saccharomyces cerevisiae,* exists in a number of races that vary in physiological characteristics. The baker's yeasts are strains in which further growth is inhibited when alcohol reaches a concentration of 4–5 percent. The carbon dioxide produced by the yeast causes the dough to rise, and the texture of the finished product (e.g., bread) is due to the numerous small bubbles of carbon dioxide formed by fermentation. Alcohol is merely a by-product in the making of bread, and evaporates during baking. The common brewer's yeasts are strains of *S. cerevisiae* in which growth continues until the alcohol reaches a concentration of 14–17 percent. Higher concentrations of alcohol are obtained only by distillation. Other species of *Saccharomyces* and related genera are used in the brewing of such beverages as sake, ginger beer, Jamaica rum, pulque, and some kinds of wine.

Neurospora. *Neurospora* is a saprophytic mold that characteristically produces numerous pink or orange conidia; it is sometimes called pink bread mold. The fruiting bodies are oval or pear shaped and 1/4–1/2 mm thick. In the early 1930s the American mycologist Bernard O.

 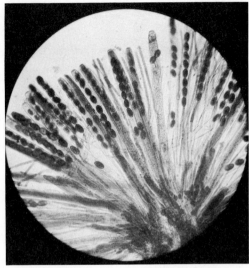

Fig. 14.9   *Peziza.* (*Left*) Ascocarps. (About natural size.) (*Right*) A bit of the hymenium, spread out to show asci and paraphyses. (Much enlarged.) [Courtesy of the Department of Plant Pathology, Cornell University.]

Dodge (1872–1960, see Fig. 29.3) pointed out the usefulness of *Neurospora* as a tool for basic genetical studies, and many students have followed his lead. *Neurospora* is easily cultured and reproduces rapidly. The vegetative mycelium is haploid, as in other Ascomycetes. It produces four, or more often eight, ascospores in each ascus, according to the species, and these are arranged in a row. One of the great advantages of using an Ascomycete for genetic study is that nuclear fusion and reduction division occur within the ascus, leading to the formation of a limited number of spores that utilize all the nuclei formed in meiosis. By careful micromanipulation the spores can be individually removed from the ascus and separately cultured. Such precision is obviously impossible in dealing with diploid organisms such as the higher plants and animals, in which for a number of reasons the nuclei produced in a single

meiosis cannot be followed through to maturity. The use of *Neurospora* in genetic experiments is further discussed in Chapter 29.

Pezizales.   The Pezizales are an order of chiefly saprophytic Ascomycetes in which the developing hymenium is exposed to the air and the ascus opens by a distinct lid (the **operculum**). In *Peziza* (Fig. 14.9) and some other genera the fruiting body is cup shaped, and the hymenium lines the inner surface of the cup. The hymenium consists of numerous cylindric to slightly club-shaped asci, intermingled with sterile hyphae. Each ascus contains eight ascospores in a single row. The ascocarp of *Peziza* ranges from about 2 to as much as 40 cm across.

In *Morchella* (Fig. 14.10), the common **morel,** the fruiting body is commonly about 10 cm long, with a thick stalk and a deeply and characteristically wrinkled

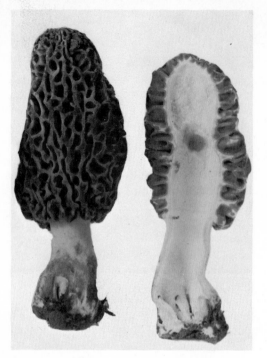

Fig. 14.10 Morel, *Morchella*, fruiting bodies. (Slightly less than natural size.) [Courtesy of the Department of Plant Pathology, Cornell University.]

cap; the hymenium forms the surface layer of the cap. Morels are considered great delicacies by mycophagous gourmets.

Lichens

Many Ascomycetes (especially of the subclass Loculoascomycetes) and a few Basidiomycetes are commonly found in symbiotic association with one or another species of green or blue-green alga. Such fungus-alga combinations are known as lichens. *Trebouxia* is an alga very often found in lichens. It is vegetatively rather similar to *Pleurococcus*, and has often been confused with it. The algae that occur in lichens usually also occur alone, but nearly all the lichen fungi are apparently restricted to their lichen occurrence. It is convenient and customary to regard the name of the lichen fungus as being also the name of the complete lichen.

Lichens commonly form thin thalli up to a decimeter or even a meter long or across. A section through a typical lichen thallus shows a peripheral layer of compact mycelium, and an internal region of more scattered hyphae among which are interspersed cells, filaments, or cell colonies of an alga. Some hyphae of the fungus generally also penetrate the substrate as rhizoids, serving as an anchor and in the absorption of minerals. The fungus sometimes produces haustoria which penetrate and eventually kill some of the enslaved algal cells, but more often a hyphal tip merely becomes closely appressed to the algal cell, without entering it. It is clear enough that the fungus benefits from its association with the alga, which is the source of food. The benefit to the alga is more debatable, but experimental work suggests that some lichen fungi produce substances that stimulate or are necessary to the growth of the included alga. Furthermore, it is clear that some of the places where lichens grow, such as bare rocks, would be inhospitable to either the fungus or the alga by itself. Lichens commonly become dormant during periods of drought and return to vegetative activity under proper conditions of temperature and moisture.

It is convenient and customary to divide lichens into three groups according to the external appearance of the thallus, although there is no sharp distinction between these groups, and this classification bears no relationship to the taxonomic position of the fungi and algae involved. Lichens that form a crust closely appressed to the substrate are called

Fig. 14.11   Crustose lichen, *Parmelia*, on a rock. [Photo by John H. Gerard, from National Audubon Society Collection/Photo Researchers, Inc.]

crustose lichens (Fig. 14.11); lichens with a more leaflike thallus, usually attached to the substrate by a relatively smaller portion, are called **foliose lichens;** and lichens that are more or less bushy-branched are called **fruticose lichens.**

Crustose lichens often occur on bare rocks and are also found as epiphytes on tree trunks and elsewhere. Foliose and fruticose lichens do not occur in such dry habitats as many crustose lichens. The importance of lichens as pioneers on bare rock surfaces is discussed in Chapter 31. Reindeer moss (*Cladonia*, Fig. 14.12), an important forage plant in arctic regions, is a fruticose lichen rather than a true moss. Otherwise lichens are of relatively little direct economic importance.

The alga of a lichen usually reproduces solely by cell division, regardless of whether the alga has any other means of reproduction in the free state. The fungus carries on its normal sexual cycle, leading to the formation of fruiting bodies and spores. Presumably a mycelium formed by germination of one of these spores can sometimes capture the proper alga and reconstitute a lichen, but this is evidently not the usual method of lichen reproduction. Experimental efforts to form lichens from the separate fungal and algal components have only recently begun to succeed. Nutrient conditions that are marginal for the survival of the fungus and alga separately tend to favor formation of the lichen.

Many or most lichens reproduce asexually as lichens (rather than as separated fungi and algae) by means of **soredia.** A soredium is a specialized, small fragment of the lichen thallus, with a peripheral layer of mycelium enclosing a few cells of the alga. The soredium typically originates internally, grows out through the surface of the thallus, and becomes detached as a little ball that may be carried about by the wind or otherwise. On becoming lodged under favorable conditions, the soredium sends out rhizoids and develops directly into a new lichen thallus. Alternatively, small cylindrical projections may grow out from the surface of the thallus and function as soredia.

Basidiomycetes

Definition and History.   The Basidiomycetes are fungi in which the sexual spores (**basidiospores**) are characteristically borne externally on a structure known as a **basidium.** Each basidium bears a small and usually definite number (typically four) of basidiospores, but sometimes ad-

Fig. 14.12   Fruticose lichens. (*Left*) *Cladonia*, fruiting. (Several times natural size.) [Photo by Jack Dermid, from National Audubon Society Collection/Photo Researchers, Inc.] (*Right*) A species of *Usnea* growing draped on a salmonberry in western Washington.

ditional basidiospores are formed after the first set has been released. The basidium is often more or less club shaped, and the Basidiomycetes are frequently called **club fungi.**

The basidium was noted by the French mycologist Joseph Henri Léveillé (1796–1870) in 1837, and the group we now call the subclass **Homobasidiomycetes** (e.g., mushrooms, bracket fungi) was recognized as a natural taxon in 1842 by Corda. The existence of the smuts and rusts (the major part of the present subclass **Heterobasidiomycetes**) as a natural group was recognized as early as 1821 by Elias Magnus Fries (Fig. 14.3). The basidium of the smuts and rusts is somewhat different

from that of the Homobasidiomycetes, however, and it was not until 1887 that all these organisms were brought together as a major group of fungi by Narcisse Patouillard (1853–1926), French pharmacist and mycologist.

Hyphae of Basidiomycetes are usually septate, but the septa are typically incomplete, with a central pore through which cytoplasm (and sometimes nuclei) can pass. In this respect the Basidiomycetes resemble the Ascomycetes.

In addition to the manner in which the spores are borne, most Basidiomycetes differ from most Ascomycetes in developing an extensive mycelium with paired nuclei after nuclear transfer (in the sexual cycle) has taken place. Generally each segment of the hypha, as delimited by successive septa, has two nuclei, just as does the ultimate segment of an ascogenous hypha. One of these two nuclei is derived ultimately from one of the two hyphae or mycelia that were involved in nuclear transfer, and the other from the other. The mycelium is thus physiologically diploid, although the individual nuclei are haploid. Such a mycelium, with two haploid nuclei of different origin in each segment or "cell," is said to be **dikaryotic.** A mycelium with uninucleate segments, or with multinucleate segments in which the nuclei are not associated in pairs, is said to be **monokaryotic.**

Hyphae of a large proportion of the Basidiomycetes are marked by **clamp connections** (Fig. 14.13) which bypass the septa. Usually there is only one clamp connection for each septum. The usual function of the clamp connection is to permit the proper distribution of nuclei so that the dikaryotic condition can be maintained as the filament grows. Growth of the hyphae of a dikaryotic mycelium is largely apical. The nuclei are commonly so large, as compared to the diameter of the hypha, that they do not readily slip by each other. Thus the dikaryotic condition could not be maintained without some such mechanism as the clamp connection. Basidiomycetes with relatively broad hyphae, in comparison to the size of the nuclei, often lack clamp connections. The clamp connection of Basidiomycetes has an evolutionary precursor in the characteristic hook by which young asci of Ascomycetes proliferate.

Like the Ascomycetes, the Basidiomycetes often produce conidia. Some Basidiomycetes, especially the rusts, produce more than one type of conidium, and these are often given distinctive names.

Sexual Cycle.   Like the Ascomycetes, the Basidiomycetes have a variety of ways in which the nuclei may become paired, but they usually follow a more stable pattern in the formation of sexual spores after nuclear fusion. The mycelium produced from a basidiospore or from a uninucleate conidium is usually monokaryotic with uninucleate segments. A common method of nuclear transfer is for two ordinary

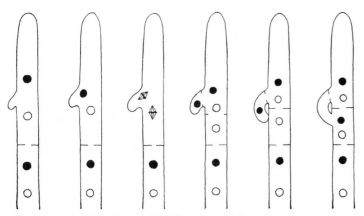

Fig. 14.13   Development of a clamp connection.

monokaryotic hyphae to come into contact, after which the walls between them dissolve and the nucleus from one of the cells passes into the other cell. A dikaryotic mycelium may then be derived directly and solely by growth from this cell, or the existing mycelium may be converted to the dikaryotic condition by migration and mitotic divisions of the proper nuclei.

Many Basidiomycetes produce more or less specialized conidia—variously called pycniospores, spermatia, or oidia—which, after transfer to a hypha of another strain of the same species, donate the single conidial nucleus to a cell of the receptor mycelium, thus initiating the dikaryotic stage. Oidia and spermatia (pycniospores) are often surrounded by or associated with a nectar that attracts insects, and the insects serve as unwitting agents of nuclear transport. Still other ways of transferring nuclei exist.

The fruiting body **(basidiocarp)** of Basidiomycetes is entirely a product of the dikaryotic mycelium. The basidia may or may not be borne in a definite hymenium, comparable to the hymenium of fruiting bodies of Ascomycetes. Some Basidiomycetes produce basidia without the formation of a definite fruiting body.

Nuclear fusion occurs in the young basidium, which at this stage may be called the **probasidium.** In the subclass Heterobasidiomycetes the probasidium is commonly a spore that often has thick walls and goes into a resting period before developing into a mature basidium. Such spores are called **teliospores.** Teliospores sometimes consist of two or more cells, each representing a probasidium. The mature basidium of teliosporic Basidiomycetes generally consist of the hypobasidium, representing the remains of the teliospore cell, and the epibasidium, the

structure produced by germination of the teliospore cell. In the Homobasidiomycetes development of the probasidium generally continues without interruption until the basidiospores are formed.

Meiosis generally occurs in the enlarging basidium. After meiosis there may or may not be a formation of septa that divide the basidium into uninucleate segments. Typically the mature basidium of the Heterobasidiomycetes is septate, whereas that of the Homobasidiomycetes lacks septa. The basidiospores originate as projections from the basidium in a manner that has been roughly compared to budding in yeasts.

A basidiospore typically germinates directly to produce a new monokaryotic mycelium, which eventually becomes or produces a dikaryotic mycelium, as previously described.

Taxonomy and Distribution.   It has been estimated that there are about 15,000 known species of Basidiomycetes. These occur as saprophytes and parasites (chiefly of vascular plants) in terrestrial habitats throughout the world.

The structure of the basidium and of the fruiting body are among the most important characters by which the major groups of Basidiomycetes are distinguished. Some of the differences between the two subclasses, Heterobasidiomycetes and Homobasidiomycetes, have already been noted. The smuts and rusts, which do not have definite fruiting bodies, make up the bulk of the Heterobasidiomycetes. Mushrooms, toadstools, puffballs, stinkhorns, bracket fungi, and other forms with well-developed fruiting bodies are familiar Homobasidiomycetes.

Rusts.   The rust fungi (other than the white rusts, a relatively unimportant group of Oömycetes) constitute the order

Uredinales, a large group of obligate parasites. Individual species of rusts may have a very wide or a very limited range of hosts. Some rusts can complete their life cycle on a single kind of host; these are called **autoecious rusts.** Others require two often quite different kinds of hosts to complete the full sexual cycle; these are called **heteroecious rusts.** Most autoecious rusts are of relatively little economic importance, but the heteroecious rusts cause some of the most destructive of all plant diseases, from an economic standpoint.

Perhaps the most important single plant disease is the **stem rust of wheat,** variously called red rust or black rust, according to the stage in which it is observed. The stem rust of wheat (Fig. 14.14) is caused by *Puccinia graminis*, a heteroecious rust composed of a number of physiological races in which one host is a barberry, typically the cultivated European barberry (*Berberis vulgaris*), and the other host is one or another kind of grain or other grass. The spermatia and aecia (defined below) are produced on the barberry, and the uredospores and teliospores (likewise defined below) are produced on the grass.

The basidiospores of *Puccinia graminis*, formed in the spring, are carried about by the wind and infect barberry leaves, developing a monokaryotic mycelium within the leaf. After a time the mycelium produces a number of specialized conidiophores in pockets (Fig. 14.15) that open to the upper surface of the leaf. Each such pocket is called a **spermogonium** and has also been called a *pycnium*. The spermogonia are barely, if at all, visible to the naked eye. Each conidiophore in the spermogonium cuts off a series of specialized uninucleate conidia, which by different botanists called

spermatia or **pycniospores.** A sugary slime of fragrant nectar produced in the spermogonium attracts insects, which unwittingly transfer the spermatia from one spermogonium to another. Special receptive hyphae extend out through the mouth of the spermogonium, and often also through the stomates or between the epidermal cells of the barberry leaf.

Like many other rusts, *Puccinia graminis* has two physiologically different sexual strains, conveniently called *plus* and *minus*. When a spermatium touches a receptive hypha of the opposite sexual strain, the spermatium fuses with a cell of the hypha. The spermatial nucleus is then apparently passed down through the hypha from one cell to another until it reaches a hyphal cell near the lower surface of the leaf, which becomes the initial cell of the new dikaryotic mycelium. Probably the dikaryotic stage can also be initiated by fusion of hyphae from plus and minus mycelia if these come into contact, as is true in many other rusts.

The initial dikaryotic cell gives rise to a dikaryotic mycelium, and within a few days numerous small, cuplike structures, called **aecia** or **cluster cups,** are formed on the lower surface of the barberry leaf (see Figs. 14.14, 14.15). The individual aecia are mostly less than 1 mm wide, but still readily visible to the naked eye. Each aecium contains a number of conidiophores that cut off specialized conidia, called **aeciospores.**

The aeciospores are binucleate and dikaryotic. They are distributed by the wind, but effective distribution is strictly local, and long-distance transfer of viable aeciospores is seldom if ever accomplished. The aeciospores are infectious only to certain grasses. The race of *Puccinia graminis* that affects wheat also affects some wild grasses.

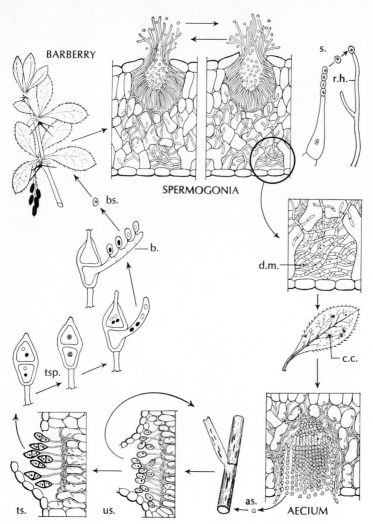

Fig. 14.14 Life cycle of wheat rust, *Puccinia graminis*: as., aeciospore; b., basidium; bs., basidiospore; c.c., cluster cups; d.m., dikaryotic mycelium; r.h., receptive hyphae; s., spermatium; ts., teliosorus; tsp., teliospore; us., uredosorus.

The hypha produced by the germinating aeciospore enters the host through a stomate of the stem or leaf sheath, and gives rise to an extensive dikaryotic mycelium within the host plant. Two kinds of spores are produced, at different times, from this dikaryotic mycelium. Short hyphae arise from a small, compact mycelial mass just beneath the epidermis of the host. These hyphae rupture the epidermis and become visible collectively as a definite sorus. A **sorus** is any cluster of sporangia or externally produced spores, in the fungi or elsewhere, especially if the

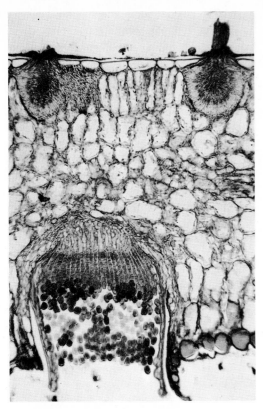

Fig. 14.15 Wheat rust, aecia and spermogonia on barberry leaf. [Courtesy of E. J. Kohl, Lakeside Biological Products, Ripon, Wis.]

the wind and are infectious to the same range of hosts as the aeciospores. In wheat-growing areas the most abundant potential host is wheat, and it is therefore said that the uredospores are produced on wheat and return to wheat. Uredospores are resistant to drying and may retain their viability for some time, with the result that they can be effectively distributed for long distances. The uredospore cycle may be repeated several times during the summer, with no other kind of spore being produced.

Late in the season, when the host plant is nearing maturity, some of the uredosori begin to produce a different kind of spore, called a **teliospore,** thus becoming **teliosori** (Fig. 14.16). New teliosori are also formed directly from the mycelium within the host. Teliosori and teliospores are black.

The teliospores are thick walled and two celled. They generally remain attached to the straw and do not germinate until the following spring. Each cell of the teliospore is a probasidium. The probasidium germinates to form an epibasidium

cluster is large enough to be visible to the naked eye.

The first kind of spore produced by the dikaryotic mycelium is called a **uredospore,** and the sorus in which uredospores are borne is called a **uredosorus** or simply a **uredium.** The uredosori appear in the stem, leaf sheaths, and basal parts of the leaf blades one to two weeks after infection. An individual uredosorus is narrow, about 2–10 mm long, and rusty red from the color of the numerous binucleate, dikaryotic uredospores.

The uredospores are carried about by

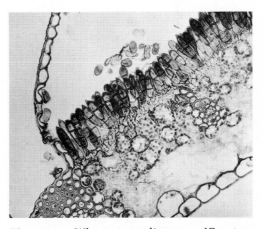

Fig. 14.16 Wheat rust, teliospores. [Courtesy of E. J. Kohl, Lakeside Biological Products, Ripon, Wis.]

that is microscopic, elongate, narrow, and transversely partitioned into four cells. Each cell of the epibasidium gives rise to a single ellipsoid basidiospore. Two of the basidiospores on a given basidium represent the plus strain, and two the minus strain. These strains cannot be called male and female, since each produces spermatia and receptive hyphae, corresponding more or less to male and female structures. The nuclear phenomena associated with the formation of the basidium were discussed earlier.

The basidiospores are forcibly discharged at maturity. They are infectious to barberry, but not to wheat or other grasses.

Puccinia graminis seldom does much damage to the barberry, but it is often very destructive to wheat and other grains. In part of Europe the barberry was commonly believed to be harmful to wheat for hundreds of years before the German mycologist Heinrich Anton DeBary (1831-1888, Fig. 14.3) demonstrated the principle of rust heteroecism by working out the life cycle of *Puccinia graminis* (1864–1865).

Inasmuch as the complete life cycle of the fungus requires the barberry as well as the grass host, it would seem that the disease could be eliminated by eradication of the barberry. This method of control has been highly successful in northern Europe, but much less so in the United States and some other parts of the world. The uredospores, which do not effectively

Fig. 14.17 Harvest scene in a wheat field heavily infested with stinking smut, near Pullman, Washington, in 1956. The clouds of spores will infest fallow fields for miles around. [From G. W. Fischer and C. S. Holton, *Biology and control of the smut fungi,* 1957. Copyright Ronald Press, New York.]

survive the winter in cooler climates, survive very well in the southwestern United States, and carry the infection back to wheat the next year, thus bypassing the barberry stage in the life cycle. New crops of uredospores carry the disease northward through the wheat belt of the Great Plains with the advancing season. The wheat-growing regions of northern Europe, on the other hand, are effectively cut off from the warmer regions where the uredospores might survive, by the Pyrenees, the Alps, and related mountain masses.

Some cultivated varieties of wheat are far more resistant to rust damage than others, and a great deal of effort has been devoted to developing more resistant cultivars by hybridization and selection. The intent is to combine in one **cultivar** (cultivated variety) all the other qualities desired, plus resistance to the rust. The problem is tremendously complicated by the fact that the fungus exists in a number of minor races, and a wheat plant that is resistant to one of these races may be susceptible to others. It is believed that elimination of the barberry is still a proper part of a control program in the United States, in combination with breeding for resistance, because the sexual stage of the life cycle cannot be completed without the barberry. If the fungus reproduces only by uredospores, then new races, which might be able to infect previously resistant cultivars of wheat, can arise only through relatively infrequent mutations, whereas if the full sexual cycle is allowed to take place new races can also be produced by recombination of existing genes.

Smuts. The smuts constitute the order **Ustilaginales,** a large group of fungi which in nature are obligate parasites of flowering plants, commonly affecting particularly the flowers and associated structures. The economically important smuts are mostly parasites of the cereal grasses, such as wheat (Figs. 14.17, 14.18), oats, rye, and maize. The smuts are related to the rusts, which they resemble in their essentially obligate parasitism, lack of definite fruiting bodies, production of teliospores, and often also in having a septate basidium. They differ in their simpler life cycle (always autoecious) and some technical details.

Mushrooms and Toadstools. Basidiomycetes that produce a definite fruiting body with a stalk and a cap are called mushrooms or toadstools (Fig. 14.19). The hymenium is borne on the lower side of the cap, which is more technically called the **pileus.** Typically the lower side of the pileus is divided into many thin, more or less parallel, vertical plates, called **gills,** which radiate from the stalk, called a **stipe,** to the margin of the pileus. The hymenium covers the gills (Fig. 14.20); it consists in large part of basidia, which commonly produce four basidiospores each. Mushrooms and toadstools with gills are often considered to make up a single broadly defined family, the **Agaricaceae;** members of this family are sometimes called **agarics.**

Agarics are mostly saprophytic, occurring especially in soil that is rich in organic matter, or on rotting logs or stumps, etc.; but some are facultative parasites on the roots of trees or elsewhere, or they may be involved in mycorhizal associations. The vegetative mycelium is buried in the substrate and sends up fruiting bodies at irregular intervals of months or years, usually after soaking rains.

Many agarics are edible and are considered great delicacies, but others are

Fig. 14.18 *Ustilago tritici*, the loose smut of wheat. [Courtesy of G. W. Fischer, Department of Plant Pathology, Washington State University.]

chances of error are so great, and the possible results of error so serious, that people who are not competent mycologists are better advised to forego the consumption of wild mushrooms. The common (and edible) field mushroom, *Agaricus campestris*, is probably the ancestor of the common cultivated mushroom (*A. brunnescens*), and at least in the United States only the cultivated specimens are used in commercial canning. These two species are generally regarded as completely safe.

Many nonbotanists apply the term *mushroom* to those agarics and related fungi they believe to be edible, and the term *toadstool* to those they believe to be poisonous. Because edibility is so difficult to determine safely, and because this criterion cannot be used to divide the agarics into taxonomically natural groups, most botanists prefer to use the terms interchangeably, or to use the term mushroom in a broad sense and leave toadstool to popular usage.

Bracket Fungi. The bracket fungi, making up the family **Polyporaceae** of the order Agaricales, are similar in many respects to the Agaricaceae, differing chiefly in the structure of the fruiting body and in the greater tendency to grow on wood. The fruiting body (Fig. 14.21) is usually a firm bracket on the tree or log within which the vegetative mycelium has developed. The lower surface of the bracket is covered with tiny pores that are often so small as to be hardly visible to the naked eye. Each such pore is the external opening of a slender, usually vertical tube in the fruiting body. The hymenium covers the internal surfaces of the tubes. Another family of Agaricales, the Boletaceae, produces fruiting bodies with the form of agarics but the porous structure of the bracket fungi.

extremely poisonous. Some species of *Amanita* (Fig. 2.7) are so virulent that a piece 1 cc in size may be fatal, and there is no known antidote for the poison of these species. Other species of *Amanita* are harmless. Some kinds of mushrooms are poisonous to some people and harmless to others, and a few are poisonous only when consumed more or less concurrently with alcoholic beverages. There is no test, other than actual eating, that will distinguish all poisonous mushrooms from all edible ones, and this test is not conducive to the tester's longevity. It is, of course, possible to learn to recognize individual species that are edible and to learn to recognize also some of the more notorious poisonous species, but the

Fig. 14.19 *Clitocybe brumalis,* a common mushroom. [Courtesy of the Department of Plant Pathology, Cornell University.]

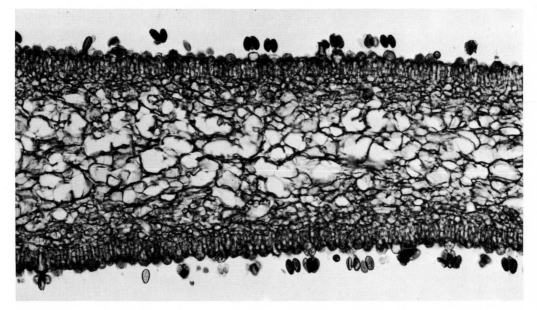

Fig. 14.20 *Coprinus* gill, showing basidiospores. [Courtesy of CCM: General Biological, Inc., Chicago.]

Fig. 14.21 *Polyporus squamosus*, a bracket fungus. [Courtesy of the Department of Plant Pathology, Cornell University.]

Bracket fungi grow on both live and dead trees, or less commonly on organic matter in the soil; a few species are parasitic on other fungi. The species that grow on living trees typically attack directly only the dead cells of which the wood is largely composed, but the enzymes they produce may kill some of the living cells and ultimately contribute to the death of the tree. Bracket fungi often do serious damage to mature standing timber, and also to lumber that is kept moist or in contact with the ground. The vegetative mycelium develops extensively in the substrate before fruiting bodies are produced, and the appearance of the basidiocarps indicates that considerable decay has already occurred.

The bracket fungi, together with certain Ascomycetes (especially the family Xylariaceae) are the principal agents of decay of wood. They thus play a vital role in the recycling of materials so essential to the balance of nature.

Mycorhizae. Many species of fungi enter into a symbiotic association with the roots of vascular plants, forming a structure known as a **mycorhiza.** This term is also extended to cover the products of somewhat similar associations between fungi and the underground gametophytes of *Lycopodium* (see Chapter 16) and some other plants. Many members of the Boletaceae and Agaricaceae are mycorhizal, as are also a number of Ascomycetes and even Zygomycetes. A large proportion of trees are mycorhizal, likewise most members of the heath and orchid families. Maize requires *Endogenes*, a mycorhizal fungus belonging to the Zygomycetes, for normal growth. Well-preserved fossil mycorhizae have been found in coal balls dating from the Lower Carboniferous (Mississippian) period. Mycorhizae are further discussed in Chapter 22.

Fungi Imperfecti

Fungi of which the perfect stage is unknown are called Fungi Imperfecti. When the perfect stage of a fungus previously referred to the Fungi Imperfecti is discovered, the species is transferred to the appropriate class. Often both the perfect and the imperfect (asexual) stage of a fungus are discovered and named long before it is learned that both names apply to the same species. According to the rules of nomenclature, a name based on the perfect stage always takes precedence over one based on the imperfect stage, regardless of which name is the older. One reason for this is that species, genera,

and higher groups that are readily distinguishable in the perfect stage may be wholly indistinguishable in the imperfect stage, so that one name for an imperfect fungus may prove to apply to several different taxa when the perfect stages are discovered.

The imperfect genera *Aspergillus* and *Penicillium* (Figs. 14.22, 14.23) illustrate the situation. Both are common genera, with many species that reproduce abundantly by conidia. In *Aspergillus* the conidiophore is swollen at the tip and covered with very numerous, short, radiating branches, each of which terminates in a chain of conidia. In *Penicillium* the conidiophore divides into two or more branches, each of which terminates in a chain of conidia. The common blue and green molds are mostly species of *Aspergillus* and *Penicillium*. The perfect stages have been found for only a small fraction of the species that have been placed in these two genera. The presently known perfect stages of *Aspergillus* prove to belong to three different genera in the

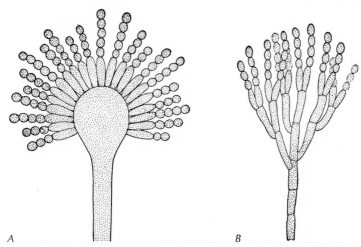

A   *B*

Fig. 14.22   Conidiophores and conidia of *Aspergillus* (*A*) and *Penicillium* (*B*). (× 800.)

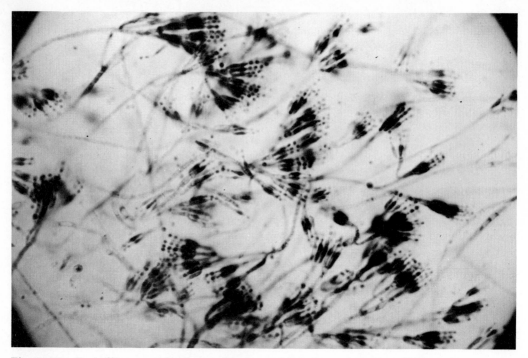

Fig. 14.23 *Penicillium* conidiophores. [Courtesy CCM: General Biological, Inc., Chicago.]

family Eurotiaceae of the order Eurotiales of the Ascomycetes. The presently known perfect stages of *Penicillium* represent two other genera, both likewise belonging to the Eurotiaceae. There is no assurance that perfect stages that may later be found for other species of *Aspergillus* and *Penicillium* might not belong to some entirely different order. Genera of the Fungi Imperfecti are often spoken of as **form genera,** that is, genera which are based on certain morphological characters but which do not necessarily represent natural groups of species.

A large proportion of the species that have been removed from the Fungi Imperfecti on discovery of the perfect stage prove to be Ascomycetes, but some are Basidiomycetes. Conidial stages of Zygomycetes and Oömycetes are more readily recognized and are commonly referred to

the appropriate position in their group rather than to the Fungi Imperfecti. Doubtless many of the approximately 10,000 species now referred to the Fungi Imperfecti will eventually be removed to other classes, but many others have probably lost the perfect stage entirely. It is scarcely to be expected that perfect stages with macroscopic fruiting bodies will be discovered for such dermatophytes (skin-inhabiting fungi) as *Trichophyton gypseum* and *T. purpureum,* which cause athlete's foot.

ANTIBIOTICS

Any substance produced by a living organism, which under natural conditions tends to inhibit the growth of competing organisms, is called an **antibiotic.** Many fungi produce antibiotics that are particularly effective against bacteria. The first antibiotic to attract much attention was **penicillin.** The Scottish medical bacteriologist Alexander Fleming (1881–1955) noticed in 1928 that a species of *Penicillium* that grew as a contaminant in a laboratory culture of *Staphylococcus* inhibited the growth of bacteria for some distance around and beyond the fungal mycelium. Obviously this effect could be achieved only by the release into the culture medium of some substance produced by the fungus. Fleming named this substance penicillin in 1929. Experimental work later demonstrated that penicillin is highly effective in limiting the growth of many Gram-positive bacteria, although relatively harmless to most of the Gram-negative ones.

Penicillin was first used to treat a human illness in 1941, and the dramatically successful results soon led to its widespread use to combat a number of different bacterial diseases and infections, although the original patient suffered a relapse and died after the supply of the drug had been exhausted. Gonorrhea has been particularly amenable to penicillin treatment, although penicillin-fast cases (i.e., cases that do not respond to treatment with penicillin) are now appearing with increasing frequency. The species of *Penicillium* observed by Fleming was *P. notatum,* but commercial penicillin is now obtained largely from a related species, *P. chrysogenum.* Neither of these species is known to have a perfect form. Penicillin is now considered to be a generic name for several chemically closely allied compounds (some of them synthetic) with similar bacteriostatic activity.

The success of penicillin led to an intensive search for other antibiotics. Many fungi and actinomycetes (an order of filamentous bacteria) have been shown to produce antibiotics, but most of these are unsuited to medical use because there is little or no margin between the therapeutic and the toxic dose. Medically useful antibiotics other than penicillin have come chiefly from the actinomycetes, especially the genus *Streptomyces.* The most important of these are the tetracyclines, which are active against a broad range of both Gram-positive and Gram-negative bacteria.

Fungi and bacteria are the only large groups of plants that regularly produce exoenzymes. The antibiotics that they produce are also released into the substrate. It has been suggested that these antibiotics are merely waste products comparable to the toxins produced by some of the pathogenic bacteria, but they also have a definite survival value for the organisms that produce them, in reducing the competition for available food. Although their evolutionary precursors may have been mere waste products, they have become weapons in the constant struggle for existence. We shall see in Chapter 26 that some of the vascular plants also produce antibiotics that play a similar role in reducing competition.

ECONOMIC IMPORTANCE

Fungi cause many serious diseases of plants, and also a few diseases of humans and other animals. They are second only to the bacteria as agents of decay, and are almost alone as the cause of decay of wood. As noted in the discussion of the bacteria, decay is the means by which complex organic compounds are changed into simpler substances that can again be used as raw material for plant growth. Some mushrooms and other fungi such as truffles and morels are used as food, and many people die from mistaking poisonous mushrooms for edible ones. Lichens are significant as pioneers on bare rock surfaces, and many forest trees and even crop plants are dependent on mycorhizal associations for survival. Alcohol is obtained from yeasts, which are also important in the baking industry. A number of other useful chemicals, such as penicillin, are obtained from various kinds of fungi. Although fungi obviously cause enormous damage from a human standpoint, they are also an essential part of the balance of nature. Without fungi our world would be very different indeed.

EVOLUTIONARY RELATIONSHIPS

The relationships of and among the fungi have long been subjects of controversy, and the controversies have not yet been fully resolved. With some reservations about the Ascomycetes and Basidiomycetes, the fungi are probably all eventually descended from algal flagellates that lost the ability to carry on photosynthesis, but they may not all have come from the same group of such flagellates. It is possible that the Myxomycetes have nothing to do with the other fungi. There is a growing conviction that the Oömycetes are derived from Xanthophycean flagellates, but it also seems increasingly likely that the Chytridiomycetes and their descendents the Zygomycetes may have originated from some different group of algae.

Nearly all students of the fungi have agreed that the Basidiomycetes are derived from the Ascomycetes, and that the basidium is a modified ascus. The ancestry of the Ascomycetes is more controversial. Some mycologists seek the origin of the Ascomycetes in the red algae, seeing certain parallels in structures associated with sexual reproduction and in the perforation of the cross-walls of the filaments. Other students (including the author) consider that the Chytridiomycetes, Zygomycetes, Ascomycetes, and Basidiomycetes form an essentially natural group, marked by their consistently chitinous walls, by their unusual method of lysine synthesis, and by the tendency toward progressive delay between plasmogamy and karyogamy. This tendency, which begins in the Zygomycetes, culminates in the dikaryotic condition of many Basidiomycetes.

FOSSIL FORMS

The fossil record provides abundant evidence of the existence of fungi from the Devonian period to the present. Some of the oldest known fossils of vascular plants, from the Rhynie beds of Devonian age in Scotland, bear well-preserved fossil fungi that resemble the Oömycetes and Zygomycetes in having nonseptate hyphae. Fossil leaves of Carboniferous and later ages often have small spots resembling the fruiting bodies of parasitic Ascomycetes, and leaves of Cretaceous and later ages have spots resembling the aecia, uredinia, and telia of rusts. An Upper Carboniferous fossil showing a definite clamp connection has recently been discovered, demonstrating the existence of Basidiomycetes at that time. Fossils of the larger fruiting bodies of fungi are very rarely found. Since the classification of fungi depends largely on the structure of the fruiting bodies and on microscopic details that in any case are not likely to be preserved, the fossil record provides but little evidence as to the phylogeny of the group. The fossil record, so far as it goes, is in harmony with the view, based largely on comparison of living organisms, that the phylogenetic progression was from Chytridiomycetes to Zygomycetes to Ascomycetes to Basidiomycetes.

SUGGESTED READING

Ahmadjian, V., *The lichen symbiosis*, Blaisdell, Waltham, Mass., 1967.

Ainsworth, G. C., *Introduction to the history of mycology*, Cambridge University Press, New York, 1976. Good reading, scientifically accurate.

Alexopoulos, C. J., *Introductory mycology*, 2nd ed., Wiley, New York, 1962. The standard textbook.

Bonner, J. T., How slime molds communicate, *Sci. Amer.* **209**(2):84–93, August 1963.

Christensen, C. M., *The molds and man*, 3rd ed., University of Minnesota Press, Minneapolis, 1965. A popular treatment by a noted mycologist.

Christensen, C. M., *Molds, mushrooms, and mycotoxins*, University of Minnesota Press, Minneapolis, 1975. Informative, entertaining, nontechnical, and authoritative.

Cooke, R., *The biology of symbiotic fungi*, Wiley, New York, 1977. Textbook style, assumes some knowledge of mycology.

Hale, M. E., Jr., *The biology of lichens*, 2nd ed., American Elsevier, New York, 1974.

Hawker, L. E., *Fungi*, Hillary, New York, 1967. Semipopular.

Ramsbottom, J., *Mushrooms and toadstools*, Collins, London, 1953. An interesting, accurate account at the semipopular level written by an outstanding botanist.

Robinson, R. K., *Ecology of fungi*, English Universities Press, London, 1967.

Rose, H., Yeasts, *Sci. Amer.* **202**(2):136–146, February 1960.

Smith, A. H., *The mushroom hunter's field guide*, 2nd ed., University of Michigan Press, Ann Arbor, 1963. An accurate account by an outstanding mycologist, with many color and black-and-white photos.

Chapter 15

Bryophytes
(Mosses and Liverworts)

Hairy-cap moss, *Polytrichum.* (About twice natural size.)
[Photo by Alvin E. Staffan, from National Audubon Society
Collection/Photo Researchers, Inc.]

INTRODUCTION TO EMBRYOBIONTA

The classification of all plants into two primary groups, the Thallophyta and Cormophyta, was proposed by the Austrian botanist Stephan Ladislaus Endlicher (1804–1849) in 1836. The term Embryophyta has often been used, subsequently, in place of Cormophyta, and more recently many botanists have treated Endlicher's groups as subkingdoms, under the names **Thallobionta** and **Embryobionta.** That practice is adopted in this text, but **thallophytes** and **embryophytes** are retained as common names for the groups. The divisions Schizophyta, Chlorophyta, Euglenophyta, Cryptophyta, Pyrrophyta, Chrysophyta, Phaeophyta, Rhodophyta, and Fungi make up the subkingdom Thallobionta. The divisions Bryophyta, Rhyniophyta, Psilotophyta, Lycopodiophyta, Equisetophyta, Polypodiophyta, Pinophyta, and Magnoliophyta, make up the subkingdom Embryobionta. Although there is a fair degree of agreement among botanists as to the groups to be recognized, the nomenclature and the precise ranks at which the various groups should be received are still controversial. The Pinophyta and Magnoliophyta, for example, are often called Gymnospermae and Angiospermae, respectively, and regardless of the technical nomenclature these groups retain the names gymnosperms and angiosperms in informal botanical parlance.

The divisions of Thallobionta are very diverse and have little in common except the absence of some of the features that characterize the Embryobionta. The divisions of Embryobionta, on the contrary, form a well-marked group with many morphological, anatomical, and physiological features in common, and it is clear that the Embryobionta are derived from a particular division (Chlorophyta) of the Thallobionta.

The pigments of the chloroplasts of embryophytes are chlorophylls a and b, β- and usually α-carotene, and several xanthophylls, generally including lutein, cryptoxanthin, and zeaxanthin. Chlorophyll a is the principal chlorophyll, β-carotene is the principal carotene, and lutein is the principal xanthophyll. The carotenes and xanthophylls, collectively called carotenoid pigments, are masked by the more abundant chlorophylls. The most common carbohydrate reserve of the embryophytes is starch. The primary cell wall of embryophytes consists largely of cellulose. Flagellated cells of embryophytes, when present, have only the whiplash type of flagellum. In all these respects the Embryobionta resemble the Chlorophyceae (green algae), and the green algae are generally believed to be ancestral to the Embryobionta.

The embryophytes differ from the green algae and from most thallophytes in that the normal life cycle of the Embryobionta shows a well-developed **alternation of generations** in which the **sporophyte** always begins its development as a parasite on the **gametophyte.** The young sporophyte is called an **embryo.** The only thallophytes with a definite sporophyte attached to the gametophyte are some of the red algae, and these are so different from the embryophytes in other respects that no close relationship seems possible. We have noted that many thallophytes produce one or more kinds of asexual spores in addition to (or instead of) the sexual spores (meiospores) that are formed as a result of reduction division and are part of the normal sexual life cycle. The embryophytes regularly produce sexual spores by reduction division as part of the sexual life cycle, and they never produce any other sort of spore. Asexual reproduction, when it occurs among the Embryobionta, does not involve the formation of spores.

All the less highly modified divisions of the Embryobionta have multicellular antheridia and archegonia in which the male and female gametes (respectively) are borne. With insignificant exceptions, the gametes of the Thallobionta are borne in unicellular gametangia that may or may not be differentiated as antheridia and oögonia.

All divisions of embryophytes except the bryophytes characteristically have specialized conducting tissues called **xylem** and **phloem.** Xylem, a water-conducting tissue, is wholly unknown among the thallophytes; and the only thallophytes that contain a tissue resembling phloem, a food-conducting tissue, are some of the brown algae. Sporophytes of some of the Bryophyta and nearly all the other Embryobionta have characteristic openings in the epidermis, called **stomates,** through which gases are exchanged between the internal tissues and the atmosphere. A stomate is bounded by two specialized cells, called **guard cells,** which regulate the opening. The structure of the stomatal apparatus is discussed in Chapter 21. Stomates are unknown in the Thallobionta. A large proportion of the algae and some fungi are aquatic. The vast majority of the embryophytes are terrestrial, and the Embryobionta are sometimes loosely spoken of as the **Land Plants.** A number of the characteristic differences between the Embryobionta and the Thallobionta merely reflect the adaptation of the Embryobionta to a land habitat.

HISTORY AND CLASSIFICATION

The division Bryophyta is generally (as here) considered to consist of three classes, here called the Bryopsida, Marchantiopsida, and Anthocerotopsida. The Bryopsida are informally called **mosses,** the Marchantiopsida are **liverworts,** and the Anthocerotopsida are **horned liverworts.** The more traditional, but nomenclaturally irregular, name for the Bryopsida is **Musci,** and that for the liverworts is **Hepaticae,** the latter name having been used either for the Marchantiopsida alone or for the Marchantiopsida and Anthocerotopsida collectively.

The name Musci is taken directly from the Latin word for moss (*muscus,* pl., *musci*). The Romans included also under that name some things that are not now considered to be mosses. Linnaeus and his contemporaries did not distinguish between true mosses (in the modern sense) and the club mosses (Lycopodiopsida), which are about as closely related as horses and sea horses. The present delimitation of the class Bryopsida was established (under the name *Musci*) in 1782 by the German botanist Johann Hedwig (1730–1799; Fig. 15.1), who is often called the father of bryology.

The Hepaticae were established as a group in 1789 by the French botanist Antoine Laurent de Jussieu (1748–1836). Jussieu and his successors for more than a century included the horned liverworts with the true liverworts. The horned liverworts (Anthocerotopsida) were first raised to the status of a class coordinate with the true liverworts in 1899 by the American botanist Marshall Avery Howe (1876–1936).

Although the mosses and liverworts had been separately recognized as natural groups before the end of the eighteenth century, it was not until 1821 that these groups were brought together into a more comprehensive group (now called Bryophyta) by the British botanist Samuel Frederick Gray (1766–1828).

Fig. 15.1 Johann Hedwig (1730–1799), German botanist, the "father" of bryology.

CHARACTERISTICS AND LIFE CYCLE

The bryophytes are characterized by their life cycle, in which the sporophyte has no direct connection to the ground and is permanently dependent, wholly or in part, on the gametophyte for its nutrition. They are always small plants, seldom more than 15 cm long and never rising as much as a meter above the surface of the substrate. The gametophyte is always photosynthetic and is generally larger and more conspicuous than the sporophyte, which often appears to be merely a stage in the reproduction of the gametophyte. Gametophytes of all land plants are inherently limited in size. In the bryophytes and related groups, the plant is still dependent on water as a medium in which the sperm swims to the egg, and therefore the sexual structures cannot be raised much above the level of the substrate. The sporophyte of bryophytes, being more or less dependent on the necessarily small gametophyte, is also inherently limited in size. Because of their small size, the bryophytes have little or no need for a specialized conducting system; if their ancestors ever had such a system, the majority of modern bryophytes have lost it. Some of the larger mosses, such as *Polytrichum*, the hairy-cap moss, do have a central strand of elongate, thick-walled cells in the axis of the gametophyte, but these are structurally less modified than typical xylem cells, and their function in water conduction is still debatable.

The sporophyte has the diploid (2n) number of chromosomes. It is typically an unbranched structure, attached to the gametophyte by a basal region called the **foot,** and producing spores in a terminal sporangium called the **capsule.** Usually the capsule is elevated above the foot on a slender stalk called the **seta.** In a few genera the sporophyte consists only of a capsule and its contents, the foot and seta being suppressed.

The spores are the direct products of meiosis. Since they have the reduced (haploid) number of chromosomes, the spores represent the first stage in the gametophyte generation. After being liberated from the capsule, the spores germinate and eventually give rise to green, photosynthetic gametophytes, which may be strictly thallose or may be divided into stemlike, leaflike, and rootlike parts. True roots, stems, and leaves of higher plants are sporophytic rather than gametophytic structures, and they regularly contain xylem and phloem, which are unknown in the bryophytes. The rootlike organs of the gametophytes of bryophytes are called **rhizoids.** Rhizoids of liverworts and horned liverworts function principally or wholly as anchoring structures, but in many of the mosses they appear to be effective also in the absorption of water and minerals. The stemlike and leaflike organs of bryophytes are often loosely called stems and leaves (especially by bryologists), but botanists with broader interests often prefer the technically more accurate terms **caulidia** (sing., caulidium) and **phyllida** (sing., phyllidium), respectively.

The antheridia and archegonia of bryophytes are either microscopic or barely visible to the naked eye. They are borne on the same or different gametophytes, according to the species or genus.

The archegonium (Fig. 15.2) is formed by a series of mitotic divisions from a primary archegonial cell. Just before maturity, the archegonium is a closed, flask-shaped structure, with a swollen base called the **venter,** and a long, slender **neck,** capped by four **cover cells.** The venter is occupied by a large, naked **egg cell.** Immediately above the egg cell is the **ventral canal cell.** Above the ventral canal cell is a row of 4 to 15 or more **neck canal cells,** occupying the interior of the neck of the archegonium. When the archegonium is mature, the cover cells spread apart or fall off, and the neck canal cells and the ventral canal cell disintegrate to form a mucilaginous mass.

The antheridium (Fig. 15.3) is formed by a series of mitotic divisions from an antheridial initial cell. Just before maturity, the antheridium is an ellipsoid or spherical structure with a multicellular jacket one cell thick enclosing a central mass of cells called **androcytes.** Each androcyte changes into a slender, elongate, apically biflagellate sperm. The fla-

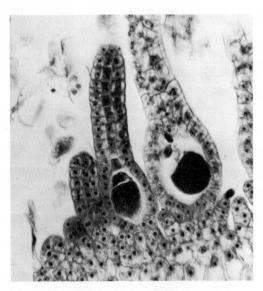

Fig. 15.2 Archegonium of *Marchantia,* a liverwort, in long section. [Courtesy CCM: General Biological, Inc., Chicago.]

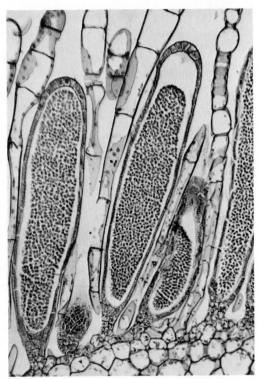

Fig. 15.3 Antheridia of *Mnium affine,* a moss, in long section. [Courtesy of E. J. Kohl, Lakeside Biological Products, Ripon, Wis.]

gella are slightly unequal in length and are both of the whiplash type. The cell walls of the androcytes disappear during the metamorphosis into sperms, and at maturity the sperms lie in a viscous liquid. The antheridium then ruptures and the sperms swim about freely.

Although most bryophytes are terrestrial rather than aquatic, they are dependent on water for the transfer of the sperm to the egg. A film of water from dew or rain is ordinarily sufficient. The sperms are attracted to the archegonia by chemical stimuli. Entering the opening at the end of the neck of the archegonium, the sperm swims through the mucilaginous material of the neck canal until it reaches the egg in the venter of the archegonium.

Fusion of a sperm and an egg constitutes fertilization. Since it receives one full chromosome complement from each of the two gametes, the fertilized egg or zygote has the diploid (2n) chromosome number and constitutes the first step in the sporophyte generation. The zygote grows and undergoes a series of mitotic divisions to form a multicellular embryo that gets its nourishment from the gametophyte. The embryo continues to grow and develops into a mature sporophyte that remains attached to the gametophyte. The developing and mature sporophyte may be wholly dependent on the gametophyte, or it may be green and manufacture some of its own food, according to the genus and species.

ANTHOCEROTOPSIDA (HORNED LIVERWORTS)

The Anthocerotopsida, or horned liverworts, may be defined as bryophytes with thallose, dorsiventral gametophytes with embedded archegonia and antheridia, and with a **meristematic region** (i.e., a region

characterized by continued cell division) near the base of the sporophyte. (The term *dorsiventral* is a descriptive adjective applied to organisms that are more or less flattened, with the two flattened surfaces unlike. Human beings, for example, are dorsiventral.) There are only about 200 species of Anthocerotopsida, occurring in moist places throughout most of the world. *Anthoceros* (Fig. 15.4) is the most familiar genus in the temperate zone, the other genera being chiefly tropical.

The sporophyte of *Anthoceros* consists of a basal foot (embedded in the gametophyte) and a terminal capsule, without an intervening stalk (Fig. 15.5). Just before maturity the capsule is an erect, slender rod commonly 2–5 cm long and about the thickness of a pencil lead. It is green and photosynthetic, with well-developed stomates in the epidermis. The upper part of the capsule matures first and begins to open along lines that extend progressively downward from the top, releasing the spores. The sporophyte continues to grow as long as the gametophyte remains vege-

Fig. 15.4 *Anthoceros* gametophytes. [Photo courtesy of J. Proskauer.]

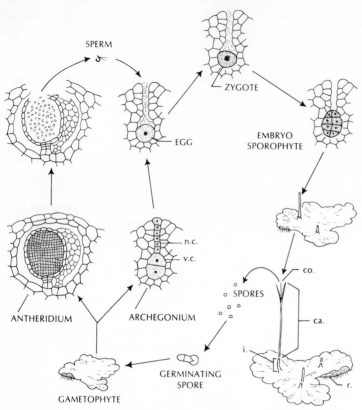

Fig. 15.5   Life cycle of *Anthoceros:* ca., capsule; co., columella; i., indusium; n.c., neck canal cell; r., rhizoid; v.c., ventral canal cell.

tatively active, elongating from a meristematic region above the foot at the base of the capsule, and opening from above.

The Anthocerotopsida are of no economic importance, and have only a minor role as food producers in the play of life. They are of interest botanically because they suggest a possible evolutionary connection of the bryophytes to primitive vascular plants. The presence of stomates and a meristematic region in the sporophyte are considered especially significant in this regard. They are also noteworthy in having a pyrenoid within the chloroplast. Pyrenoids are common

among the algae, but among the Embryobionta they are known only in the Anthocerotopsida and the class Isoëtopsida of the division Lycopodiophyta.

MARCHANTIOPSIDA

The Marchantiopsida, or liverworts, may be defined as bryophytes in which the gametophyte is ordinarily dorsiventral, with superficial (at least in origin) antheridia and archegonia, and in which the sporophyte has neither stomates, nor a meristematic region, nor a **columella** (central sterile column in the capsule). Both the Anthocerotopsida and the Bryopsida usually have a columella. The gametophyte of liverworts may be strictly thallose (**thallose liverworts,** Fig. 15.6) or may be differentiated into a caulidium with attached phyllidia (**leafy liverworts**). In either case the gametophyte is usually attached to the substrate by rhizoids, and these are always unicellular.

The more traditional name for the class Marchantiopsida is Hepaticae, but this name is not in accord with nomenclatural principles that are coming to be widely adopted. Both the name **Hepaticae** and the common name, liverwort, refer to the

Fig. 15.6 *Conocephalum conicum*, a thallose liverwort, belonging to the order Marchantiales. (About twice natural size.) [Photo by John H. Gerard, from National Audubon Society Collection/Photo Researchers, Inc.]

fancied resemblance of the outline of the thallus in some genera to the outline of the human liver. Under the Doctrine of Signatures, popular in Renaissance times, this resemblance led to the belief that the plants were good for ailments of the liver.

Liverworts occur in moist places throughout the world, but are most abundant in the tropics. They grow in swamps and bogs, on wet rocks and cliffs, on soil or rotting logs in dense woods, and as epiphytes on the bark and leaves of trees. A few species are found floating or submerged in water, and some others in sites that are alternately wet (or moist) and dry. There are about 5000 species of Marchantiopsida, distributed among six orders. The vast majority of species belong to a single order (Jungermanniales) of leafy liverworts, but in temperate regions some of the thallose liverworts such as *Marchantia* (order Marchantiales) are more abundant or at least more conspicuous.

Porella

Porella (Figs. 15.7, 15.8) is one of the most common genera of leafy liverworts in North America. Unlike many other Marchantiopsida, it is not limited to moist habitats but often occurs as a pioneer on dry rocks or as an epiphyte on the bark of trees, particularly in the southeastern United States.

The gametophyte is olive or dull green, up to about 10 cm long and 2–4 mm wide, with a very slender, freely branched caulidium, and with well-developed, relatively broad phyllidia ("leaves"), those of the two dorsolateral rows being much larger than those of the single ventral row. Unlike most of the genera that grow in moister places, *Porella* has the phyllidia of the two dorsolateral rows deeply and unequally bilobed, with a comparatively

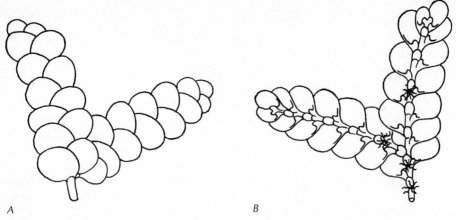

Fig. 15.7   *Porella,* a leafy liverwort. (*A*) Top view. (*B*) Bottom view, showing rhizoids. (× 5.)

large and broad, spreading dorsolateral lobe, and a much smaller, ventrally placed lobe lying parallel and adjacent to the lower side of the caulidium. The forward margin of each dorsolateral phyllidium overlaps the rear margin of the adjacent phyllidium. This arrangement tends to channel drainage water toward the ventral surface of the gametophyte, where some of it is retained in tiny pockets that are bounded on the lower side by the ventral lobes of the dorsolateral phyllidia. Rhizoids are borne in clusters on the caulidia at the bases of the ventral phyllidia. There is little or no differentiation of tissues in the caulidia and phyllidia, and the phyllidia are only one cell thick.

Antheridia of *Porella* are borne singly in the axils of slightly modified phyllidia on special antheridial branches that are otherwise not much different from the sterile branches. The **axil** of a leaf or phyllidium is the position or place at the point of the angle formed by the insertion of the leaf (or phyllidium) on the stem (or caulidium). A structure that is axillary thus lies between the stem (or caulidium) and the base of the leaf (or phyllidium).

The archegonia are borne at the tips of special archegonial branches. Two or three dorsolateral phyllidia nearest the archegonium are laterally fused to form a cup-shaped perianth surrounding the archegonium. The term *perianth* is more generally used for the enclosing outer parts of flowers of the Magnoliophyta (angiosperms) and is applied here only by analogy.

Fertilization takes place in the usual bryophytic fashion. The venter of the archegonium then enlarges with the growth of the young sporophyte, forming a covering called the **calyptra,** but eventually the seta of the sporophyte elongates and the capsule ruptures the calyptra and is exserted beyond the perianth. At maturity the capsule opens along four longitudinal lines, releasing the spores. The spores are distributed by the wind, and under proper conditions they germinate to form new gametophytes.

Marchantia

Marchantia (Fig. 15.9), a common and widely distributed genus in temperate regions, is the liverwort most commonly

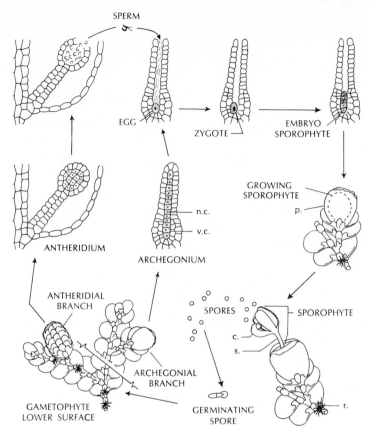

Fig. 15.8   Life cycle of *Porella:* c., capsule; n.c., neck canal cell;
p., perianth; r., rhizoid; s., seta; v.c., ventral canal cell.

studied in general botany courses, although it is a rather highly specialized type, hardly typical of the class as a whole. The gametophyte is a flat, ribbon-shaped, dichotomously branched thallus commonly several centimeters long and about 1.5–2 cm wide, with a thickened midrib along each branch. The lower side of the thallus bears two or more rows of multicellular scales and two kinds of rhizoids. Some of the rhizoids penetrate the soil as agents of anchorage and possibly absorption. Others grow parallel to the lower surface of the thallus on top of the soil. It has been suggested that these tend to channel surface water inward from the margins of the thallus by capillary action, thus helping to distribute the supply of water to all absorptive parts of the thallus.

The thallus has some differentiation of tissues, and is marked by regularly distributed tiny pores, analogous (not homologous) to the stomates of *Anthoceros* and higher plants. The mechanism for constricting or closing the opening under conditions of water stress is both more complicated and less effective than that of true stomates.

Vegetative reproduction in *Marchantia* commonly occurs by death of the older parts of the thallus, leaving the separated branches as independent plants. Adventi-

Fig. 15.9 *Marchantia*, in Utah, assembled to show archegoniophores (upper left), antheridiophores (right), and gemmae cups (left center and lower left). [Photo by Valoie Albrecht.]

tious branches may also be formed on the ventral surface of the thallus, followed by death and decay of the connecting tissues. The term *adventitious* is applied to any structure, in the bryophytes or elsewhere, which arises from mature tissues, or from tissues that would not ordinarily be expected to produce such a structure.

The thallus of *Marchantia* also bears on its upper side some small open cups, each of which contains several reproductive bodies called **gemmae.** The gemmae are multicellular, more or less discoid bodies just large enough to be readily visible to the naked eye. Discharge and swelling of mucilage from club-shaped mucilage cells intermingled with the gemmae break them from their short stalks and force

them out of the cup. Gemmae that lodge in suitable places produce new gametophytes by vegetative growth.

Antheridia and archegonia of *Marchantia* are borne on specialized, ascending or erect branches with a slender, rodlike stalk and an expanded, shallowly eight-lobed cap, arising from the surface on different gametophytes. The **antheridiophore** (male branch) has numerous antheridia sunken in the upper surface of the cap, and the **archegoniophore** (female branch) has a radial row of several archegonia extending toward the tip of each lobe on the upper side. The disk of the archegoniophore is only slightly raised above the thallus at the time of fertilization. After fertilization the archegoniophore continues to grow, commonly reaching a length of 2 or 3 cm. Meanwhile the disk (cap) expands and the margins become inflexed, so that the archegonia are directed downward, and a series of cylindrical processes a few millimeters long develop from the disk. These appendages (**rays**) radiate outward from the center and droop beyond the margins of the disk, like the ribs of an umbrella, giving the mature archegoniophore a very characteristic appearance.

The fertilized egg (zygote) undergoes a series of mitotic divisions to produce a pendant, multicellular sporophyte (Fig. 15.10) a few millimeters long, composed of a small foot, a short, stout seta, and a

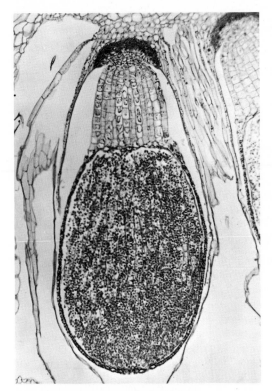

Fig. 15.10 *Marchantia.* Sporophyte in long section. [Courtesy of Triarch Incorporated, Box 98, Ripon, Wis. 54971.]

relatively large capsule. The sporophyte is green and photosynthetic, but it makes only a small proportion of its own food, being otherwise parasitic on the gametophyte. At maturity the capsule splits lengthwise from the tip into an indefinite number of segments, releasing the spores.

BRYOPSIDA

The Bryopsida (mosses) (Fig. 15.11) are bryophytes in which the spore gives rise to a filamentous or thallose **protonema** (from Greek *protos*, first, and *nema*, thread), which in turn gives rise to one or more **gametophores** that have well-developed, mostly equal phyllidia more or less spirally and symmetrically arranged around a caulidium. The protonema is usually a transitional stage that dies after the gametophores have become established. The gametophore is ordinarily the conspicuous and persistent part of the moss plant. It superficially resembles a small, leafy stem, but it lacks xylem and phloem. The phyllidia generally originate in three rows, corresponding to the three cutting faces of the apical meristematic cell, but further growth of the caulidium often obscures the arrangement. Rhizoids of mosses are uniseriate, multicellular filaments with diagonally slanting cross-walls. They are borne on the protonema and usually also (except in *Sphagnum*) on the gametophores, serving as agents of anchorage and absorption. The protonema and gametophores have chlorophyll and make their own food; both have n chromosomes and belong to the gametophyte generation.

The sporophyte may or may not contain chlorophyll, but it is always at least partly dependent on the gametophyte for its nutrition. After fertilization, the archegonium is ruptured by the growth of the

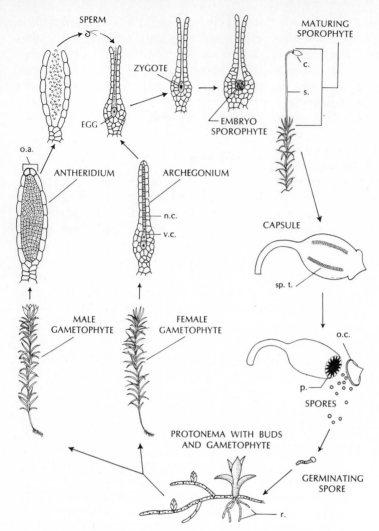

Fig. 15.11 Life cycle of a moss: c., capsule; n.c., neck canal cell; o.a., operculum of the antheridium; o.c., operculum of the capsule; p., peristome; r., rhizoid; s., seta; sp.t., sporogenous tissue; v.c., ventral canal cell.

young sporophyte, and the upper part is carried upward on the developing capsule. This detached upper part of the archegonium continues to grow and enlarge with the sporophyte, forming a covering called a **calyptra,** atop the maturing capsule. The calyptra can often be lifted from the capsule between thumb and forefinger, and it must fall off before the spores can be dispersed.

The capsule of the moss sporophyte is usually more complex than that of other bryophytes, with a larger proportion of sterile tissue, and it often bears more or less functional or vestigial stomates. It usually has a central columella, which may or may not reach the tip of the capsule. In most mosses the capsule has a specialized lid, the **operculum,** which separates in a circular line around the top of the capsule at maturity. The mouth of the capsule is then usually surrounded by a **peristome** (Greek, literally, around the mouth) commonly consisting of two rings of teeth that project inward from the margin and often close the opening. In dry weather the teeth bend outward and permit the discharge of spores; in wet weather they bend inward and retain the spores.

Mosses occur in a wide variety of terrestrial or subaquatic to occasionally fully aquatic habitats, from the tropics to the polar regions, but they are especially common as pioneers on bare rock or earth surfaces in moist regions. Their role as soil builders in the long process by which a bare rock surface becomes covered with soil is discussed in Chapter 31. Many of them grow as epiphytes on the bark of trees, although the "moss" of tree trunks more often consists largely of green algae such as *Pleurococcus.*

There are three major groups of mosses, which may conservatively be treated as orders, under the names Andreaeales, Sphagnales, and Bryales. The Bryales consist of some 14,000 species and make up the bulk of the class, with a large number of families and genera. The Sphagnales consist of a single family, the Sphagnaceae, which has only a single genus, *Sphagnum,* with some 300 species. The Andreaeales, with only about 100 species, differ from the other two groups in having the capsule open along four distinct longitudinal lines, instead of by a lid or irregu-larly, and in some other features. They will not be further discussed here.

Polytrichum

Polytrichum, the hairy-cap moss, is one of the most common genera of mosses in temperate and boreal regions of the Northern Hemisphere. It is somewhat more complex in structure than most mosses, and thus not fully typical, but at this level of study it may represent the Bryales (Fig. 15.11). The gametophores are individually ascending or erect, usually 5–10 cm long, and grow in masses, often as much as several meters wide, covering the ground. There is considerable differentiation of tissues (more than in most mosses), including a central strand of elongate, rather thick-walled empty cells in the caulidium that have been compared to the xylem of vascular plants. The phyllidia are narrow and rather firm, commonly several millimeters long, and (unlike most mosses) have a central band bearing many closely set, thin, longitudinal photosynthetic ridges several cells high and one cell thick arising from the upper surface.

The seta of *Polytrichum* is commonly several centimeters long and rather stiffly erect. The barrel-shaped capsule, commonly 3–5 mm long, is nodding or turned to one side until fully mature, when it becomes erect. It is green and photosynthetic at least when young, and it has some more or less functional stomates much like those of vascular plants. The capsule is covered by a conspicuous, evidently hairy calyptra (whence the name, hairy-cap moss), which falls off at maturity.

Sphagnum

Sphagnum, the peat moss (Fig. 15.12), is common in moist or wet places in cool-

Fig. 15.12   *Sphagnum squarrosum*, showing gametophytes and sporophytes. [Courtesy of Frances Wynne Hillier.]

temperate or arctic regions, often forming extensive **sphagnum bogs.** The chemical activities of *Sphagnum* tend to increase the acidity of the surrounding water, through a colloidal absorption process that results in the breakdown of salts and release of the corresponding acids.

Decay proceeds very slowly under sphagnum bog conditions, and as growth of the gametophores continues from above, the remains of the lower parts of the gametophores and associated plants of other kinds accumulate as **peat.** In Ireland and some other northern countries, peat is used for fuel after being dug and allowed to dry for several months. The record of major vegetational changes over periods of thousands of years can often be deciphered by identification of the pollen grains of other plants preserved at successive depths in a peat bog. Many individual

sphagnum bogs have been in existence since the close of the Pleistocene glaciation, the latest phase of which reached its height some 11,000 years ago.

Unlike the Bryales, the protonema of *Sphagnum* is flat and thallose rather than filamentous. Each protonema gives rise to a single gametophore. Groups of lateral branches are produced at regular, close intervals on the gametophore. Some of these branches droop; others are more or less erect and resemble the main shoot on which they are borne. As the older parts of the gametophore die, the branches become separated from each other and continue to grow as independent, closely massed plants. The gametophores do not have rhizoids.

The phyllidia of *Sphagnum* are one cell thick and are usually composed of two very different kinds of cells in a regular pattern. Some of the cells are long and narrow and contain chloroplasts. Others are somewhat longer and much broader, with spirally thickened walls and without living contents. In cross section the phyllidium shows regularly alternating chlorophyllous and colorless cells. The colorless cells absorb water and release it only gradually under conditions of drought. It is because of this water-retaining capacity that *Sphagnum* has been so commonly used to pack the roots of live plants for shipment.

Sporophytes of *Sphagnum* are relatively small, consisting of a nearly spherical capsule commonly 2–3 mm thick and a basal foot, the whole elevated on a gametophytic branch several millimeters long. The capsule is chlorophyllous when young and has poorly developed, scarcely functional stomates. At maturity it dehisces (opens along a predetermined line) explosively, often shooting the operculum and spores several inches into the air. The sporophyte and associated structures then soon shrivel and degenerate. The gametophores, on the other hand, are perennial and live indefinitely. Because of their early degeneration, the sporophytes of *Sphagnum* are much less often seen than those of other mosses.

FOSSIL FORMS

The oldest fossils that might reasonably be thought to be bryophytes are of Silurian age, but these are too fragmentary for a confident interpretation. The oldest undoubted bryophytes are Devonian antecedents of the leafy liverworts, followed at the base of the Carboniferous by apparent precursors of the Marchantiales. Mosses first appear in the lower part of the Upper Carboniferous, and the beginning of the divergence of the Sphagnales from more typical mosses (Bryales) can be traced to the base of the Permian. Fossil horned liverworts are virtually unknown (or unrecognized).

One of the most interesting and best known fossil genera of bryophytes is *Naiadita*, which is well represented in some Triassic and Jurassic deposits. The sporophyte consists of a small, globular capsule borne on a presumably gametophytic stalk. *Naiadita* is generally regarded as a liverwort, but the arrangement of equal phyllidia in a spiral around the caulidium is also suggestive of mosses. Thus it tends to link the two groups, but it is too recent to tell much about their original divergence.

The fossil record does not tend to connect the bryophytes with any other group. It puts some constraints on the phylogenetic ideas that can be seriously considered, but it still leaves room for some diversity of interpretation based on general theories of the course of evolution.

EVOLUTIONARY RELATIONSHIPS

Most botanists are agreed that the bryophytes are related to the vascular plants, particularly the fossil rhyniophytes (discussed in Chapter 16), and that both the bryophytes and the vascular plants are derived from the green algae. Some of the evidence on which these conclusions are based is summarized in the opening paragraphs of this chapter. It has traditionally been thought that the bryophytes represent a more or less direct link between the green algae and the vascular plants, and that the evolution of land plants from the green algae followed a course of progressive elaboration of the sporophyte accompanied by reduction of the gametophyte. The more general view today is that the bryophytes are derived from the green algae through prerhyniophyte ancestors in which both the gametophyte and the sporophyte were green and physiologically independent. These concepts are further discussed in Chapter 16. One of the most convincing bits of evidence pointing to the derivation of the bryophytes from ancestors with more complex sporophytes is the presence of stomates on many bryophytic sporophytes in which they have little or no functional importance. The stomatal apparatus is a complex evolutionary adaptation to the need of a land plant for access of air to internal photosynthetic tissues, without excessive evaporation during periods of water deficiency. It is not likely that such a mechanism would arise in advance of the need.

None of the three classes of bryophytes can be considered directly ancestral to any of the others. Regardless of which view is taken as to the evolutionary history of the group, each class is advanced in some features and more primitive in others. There has probably been an evolutionary progression from erect gametophytes bearing three rows of equal phyllidia, to dorsiventral gametophytes with the ventral phyllidia reduced or wanting, to strictly thallose gametophytes without phyllidia. The thallose gametophyte of the Anthocerotopsida is thus interpreted as among the most "advanced" in the division, whereas the sporophyte is the least reduced and therefore the most primitive in the division. Even if the older viewpoint is accepted and the sporophyte of the Anthocerotopsida is considered to be an advance toward the condition of the vascular plants, the presence of pyrenoids in the chloroplasts of the Anthocerotopsida is still a primitive character not shared by other bryophytes.

SUGGESTED READING

Conard, H. S., *How to know the mosses and liverworts*, Brown, Dubuque, Iowa, 1956. A profusely illustrated key to the more common bryophytes of North America.

Watson, E. V., *The structure and life of bryophytes*, 3rd ed., Hutchinson, London, 1971. Readable; assumes some botanical knowledge.

Chapter 16

The Vascular Cryptogams (Pteridophytes)

Lady-fern, *Athyrium filix-femina*, showing portion of leaf, with sori partly covered by the pale indusia. (Nearly 10 times natural size.) [Photo by N. E. Beck, Jr., from National Audubon Society Collection/Photo Researchers, Inc.]

INTRODUCTION TO VASCULAR PLANTS

Evolutionary Background

There are two opposing theories as to the nature of the green algal ancestors of the embryophytes and the general course of evolution in the embryophyte subkingdom. According to the **homologous theory** of land plant evolution, the ancestor of the embryophytes was a green alga with isomorphic alternation of generations, that is, the gametophyte and sporophyte were essentially similar. The primitive embryophyte is likewise believed to have had similar gametophyte and sporophyte generations; and from this type evolution proceeded in two directions. In one line the sporophyte became essentially parasitic on the gametophyte, losing much or all of its photosynthetic capacity and coming to be scarcely more than a stage in the reproduction of the gametophyte. The bryophytes represent this line of development. In the other line the sporophyte became the dominant generation, and the gametophyte was reduced, ultimately becoming merely a step in the reproduction of the sporophyte. The vascular plants represent this line of development.

According to the **antithetic theory** of land plant evolution, the ancestor of the embryophytes was a green alga in which the zygote was the only diploid or sporophytic cell. The sporophyte is regarded as an entirely new structure interposed between two gametophyte generations, and evolutionary advance consisted of progressive elaboration of the sporophyte and eventual reduction of the gametophyte. Under the antithetic theory, the bryophytes are regarded as a transitional step from the green algae toward the vascular plants.

At the time the antithetic theory was first proposed, the only green algae in which sexual reproduction was known had only a single diploid cell in the life cycle, but many genera with isomorphic alternation of generations (e.g., *Ulva*) are now known. Evidence from comparative morphology of the bryophytes further suggests that the forms with very small sporophytes have been derived from ancestors with larger sporophytes. The presence of stomates on many moss sporophytes, as mentioned in Chapter 15, is significant in this regard. Although the antithetic theory was once widely accepted, the accumulation of evidence has favored the homologous theory, which has become more and more generally adopted.

Under either theory, there are no fossils connecting the bryophytes to other embryophytes, nor connecting any of the embryophytes to the green algae or any other algae. The interpretation must therefore be based largely on comparison of modern groups. Under either theory, there is a considerable gap between the most archaic known embryophyte and any possible green-algal ancestor. The multicellular antheridia and archegonia of the embryophytes, for example, are quite different from the unicellular antheridia and oögonia of algae.

Fritschiella (Fig. 16.1), a terrestrial green alga with isomorphic alternation of generations and some differentiation of tissues, has attracted some attention as a plant having the potentiality of evolutionary development along the lines of the embryophytes, but no one suggests that *Fritschiella* itself is directly ancestral to the embryophytes. In some cytological features it is unlike the embryophytes and their putative green-algal ancestors.

 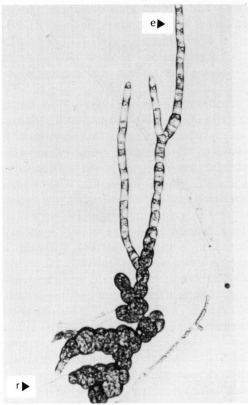

Fig. 16.1 *Fritschiella,* a green alga that may be similar to the ancestors of the land plants. (×
about 175.) Note erect filaments (e), rhizoidal filaments (r), and compact, hardly filamentous
tissue at the base. [Courtesy of Gordon E. McBride; these photos also appeared in *Arch.
Protistenk.* **112:** after p. 374, 1970. Fischer. Jena.]

Nature of Vascular Plants

The divisions Rhyniophyta, Psilotophyta,
Lycopodiophyta, Equisetophyta, Polypodiophyta, Pinophyta and Magnoliophyta
make up a large group characterized by
the presence of specialized conducting
tissues in the sporophyte. These divisions
are often collectively called the **vascular
plants.** The gametophytes of vascular
plants are always small, inconspicuous
bodies which are scarcely more than a
stage in the reproduction of the larger and
more conspicuous sporophyte. The vascular plants clearly constitute a natural
group, and some botanists consider them
to form a single large division, the Tracheophyta. Other botanists prefer to recognize several divisions of vascular
plants, as is done in this text. The two
points of view do not reflect any fundamental conflict in ideas of relationship.
Those who object to placing all vascular
plants in a single division merely believe
that such a classification minimizes the
differences within the group, and that so

large and varied a division is hardly comparable to the other recognized divisions of plants.

The specialized conducting tissues of vascular plants are called **xylem** and **phloem.** Both xylem and phloem are usually complex tissues, with two or more kinds of cells. Xylem, the water-conducting tissue, is characterized by the presence of elongate, thick-walled cells called **tracheids.** Tracheids are dead at maturity, without living contents. They give strength and rigidity to the plant and serve also to conduct water and minerals. They have a well-developed secondary wall with a high proportion of lignin, a complex substance which was described in Chapter 4. The wood of trees and shrubs consists of xylem, but the strands of xylem in the stems of herbaceous (nonwoody) plants are not usually called wood. The structure of xylem is more fully discussed in Chapter 19. The structure of phloem, the food-conducting tissue, is also discussed in Chapter 19.

In the most primitive kinds of vascular plants, such as some of the rhyniophytes (which are known only as fossils), the sporophyte consists only of a dichotomously branched axis that is considered to be a stem. In more advanced kinds of vascular plants, the sporophyte is differentiated into three principal kinds of organs: the **root,** the **stem,** and the **leaf.** In their typical forms, the root is the underground part of the plant, the stem is the aerial axis (including the branches), and the leaves are the flattened, photosynthetic appendages of the stem. The leaves and stem together are often called the **shoot,** as opposed to the root. Roots, stems, and leaves usually differ anatomically from each other, and specialized organs of various sorts can often be identified by

their anatomy as being modified roots, stems, or leaves.

The place on a stem where a leaf is (or has been) attached is called a **node,** and the part of the stem between one node and the next is called an **internode.** Leaves are said to be **alternate** if only one is borne at a node, **opposite** if two are borne (usually on opposite sides of the stem) at a node, and **whorled** if three or more are borne at a node. Leaves of many kinds of plants have a slender stalk, called a **petiole,** and a more or less expanded, flat **blade.** Leaves with the blade attached directly to the stem, without an intervening petiole, are said to be **sessile.** Leaves with the blade all in one piece, as in apple trees and cherry trees, are said to be **simple,** in contrast to **compound** leaves, such as those of many ferns, in which the blade consists of several or many distinct parts or leaflets.

Cryptogams and Phanerogams

One of the most obvious distinctions in the plant kingdom is that between seed-bearing plants, or **phanerogams,** and non–seed-bearing plants, or **cryptogams.** The stamens (microsporophylls) and ovules (young seeds) of seed plants, as described in subsequent chapters, obviously represent male and female structures, whence the name phanerogam (Greek *phaneros,* visible, and *gamos,* marriage). The method of reproduction of cryptogams (Greek *kryptos,* hidden) is not so immediately evident. The phanerogams are here considered to constitute two divisions, the Pinophyta (gymnosperms) and Magnoliophyta (angiosperms, flowering plants). All other divisions of plants are cryptogams. Thus the terms phanerogams and cryptogams conform only in part to the formal taxonomic scheme, just as do the terms vertebrates and invertebrates in the animal kingdom.

Vascular Cryptogams

The vascular cryptogams, considered in this chapter, are those cryptogams that characteristically have a specialized conducting system.

The life cycles of a number of different groups of vascular cryptogams, and also of many mosses, were first worked out by the German botanist Wilhelm Hofmeister (1824–1877, Fig. 16.2). The book presenting his results, published in 1851, is a landmark of botanical history. Hofmeister did not fully understand the processes of gametic union and meiosis, with the associated changes in chromosome number, but he knew that the sperm stimulates the egg to develop into the new sporophyte, and he is responsible for the concept of **alternation of** sporophytic and gametophytic **generations.**

For more than three quarters of a century after Hofmeister's work, it was customary to refer all vascular cryptogams to a single division, the **Pteridophyta,** and some botanists still find this the most useful arrangement. The name comes from *Pteris,* a genus of ferns, plus the Greek ending -*phyta,* meaning plants. *Pteris* is also the classical Greek word for a fern. The ferns make up by far the largest group of pteridophytes (vascular crypto-

Fig. 16.2 (*Left*) Wilhelm Hofmeister (1824–1877), German botanist, who discovered alternation of generations. (*Right*) Walter Zimmermann (1892–1980), German botanist, author of the telome theory.

gams); the other pteridophytes have often been called **fern allies,** or fernlike plants. In this text the ferns are treated as forming a division Polypodiophyta, and the fern allies are considered to represent several small divisions—the Rhyniophyta, Psilotophyta, Lycopodiophyta, and Equisetophyta. The life cycle of ordinary ferns, discussed in subsequent pages, is basically similar to that of most other vascular cryptogams, and these other groups are therefore here discussed mainly in terms of the sporophytes, which are the conspicuous phase of the life cycle. The life cycle of *Selaginella* (division Lycopodiophyta) and a few other vascular cryptogams is significantly different from that of most ferns, in a way that helps to understand the life cycle of seed plants. *Selaginella* is therefore considered separately.

The basic feature of the sexual cycle of all vascular cryptogams is an alternation of sporophytic and gametophytic generations, with each generation being physiologically independent of the other at maturity. The sporophyte generation begins with the fertilized egg and ends with the spore mother cells. The gametophyte generation begins with the spores (produced by reduction division from the spore mother cells) and ends with the male and female gametes. The gametophyte is, in a sense, physiologically independent from the beginning, although whatever food is stored in the spores has necessarily come from the parent sporophyte.

The sporophyte of vascular plants is always the larger and more conspicuous generation, typically provided with vascular tissues and differentiated into roots, stems, and leaves, in contrast to the much smaller and anatomically simpler gametophyte. So dominant is the sporophyte, in comparison to the gametophyte, that botanists and nonbotanists alike think automatically of the sporophyte alone whenever any vascular plant is mentioned.

RHYNIOPHYTA

Nature of Rhyniophytes

The **rhyniophytes** (division Rhyniophyta) are Silurian and Devonian fossil vascular plants of relatively simple structure. More attention is devoted to them here than to most fossil groups because of their importance to any understanding of the evolutionary relationships among the vascular plants. No modern plants belong to this division.

The sporophyte of rhyniophytes consists of a branching axis, with or without tiny leaves. The underground part of the axis is scarcely different from the aboveground part (except in lacking chlorophyll), and it is customary to consider the whole axis to be a stem. Often the underground part of the stem bears rhizoids comparable to the rhizoids of bryophytes. Branching of the stem is variously **dichotomous** (forking into two similar branches) or **sympodial** (i.e., one of the two branches of a pair continues as a segment of the more or less zigzag main axis, whereas the other is smaller and lateral). The sporangia terminate some of the branches or are sometimes borne laterally on the stem.

Rhynia

Perhaps the most important genus of rhyniophyte fossils is *Rhynia* (Fig. 16.3), because it so nicely fulfills the qualifications of an ancestral prototype for all vascular plants. *Rhynia* was described in 1917 from rocks of Middle Devonian age near Aberdeen, Scotland—the now famous Rhynie beds, which have yielded a

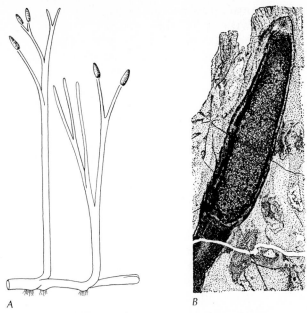

Fig. 16.3 *Rhynia major.* (*A*) Habit. (× 1/3.) [After Kidston and Lang.] (*B*) Fossil sporangium. (× about 5.) [From Kidston and Lang.]

number of different vascular plant fossils showing extraordinary preservation of cellular detail.

The sporophyte of *Rhynia* is a dichotomously branched axis, with a horizontal rhizome (underground stem) and erect aerial branches about 1–3 dm tall. The rhizome is basically similar in structure to the aerial branches, but it differs in bearing clusters of slender, unicellular rhizoids that evidently acted as absorbing structures. Sporangia are borne singly at the tips of some of the aerial branches. There are no leaves.

The aerial stem of *Rhynia* is differentiated into an **epidermis,** a **cortex,** and a **central cylinder** of vascular tissue. The xylem is a small, central strand of tracheids, surrounded by a layer, four or five cells thick, of elongate, thin-walled cells that are considered to be phloem. The phloem is surrounded by a large cortex composed of thin-walled cells, forming a tissue of a relatively unspecialized type, described in Chapter 18 as **parenchyma.** The cortex is differentiated into an inner and an outer zone, and the inner zone is believed to have been photosynthetic. The epidermis, forming the outermost part of the stem, consists of a single layer of cutinized cells. **Cutin,** a waxy substance commonly associated with the epidermis of stems and leaves of vascular plants, impedes the passage of water and water vapor. The only intercellular spaces in the epidermis are the scattered stomates.

The sporangia of *Rhynia* are up to about 1.5 cm long and are nearly cylindrical, tapering slightly toward the tip. The jacket of the sporangium is several cells thick and does not have a special point or line on which to open. The sporangial cavity lacks a columella and contains many spores of about equal size. In some specimens the spores are grouped in tetrads (sets of four), indicating that they are products of reduction division.

The gametophytes of *Rhynia* and other rhyniophytes are unknown. There has been some speculation that specimens without sporangia may be gametophytes. This would be very convenient for botanical evolutionary theory (at least the homologous theory), but the sex organs (antheridia and archegonia) have not been observed in these or any other rhyniophytes.

Cooksonia

The fossil genus *Cooksonia* (Fig. 16.4), from late Silurian and early Devonian rocks, is one of the oldest (perhaps *the* oldest) known genera of fossil vascular plants. It is less well preserved and thus less well known than *Rhynia*, but it appears to be basically similar in having a

Fig. 16.4 *Cooksonia*, one of the oldest known fossils of vascular plants. [After Croft and Lang.]

leafless, dichotomously branching stem with terminal sporangia. The sporangia are globose, rather than elongate as in *Rhynia*. *Cooksonia* is of interest because it carries the rhynioid type far back in time, farther than any of the more complex types of vascular plants. These facts give some support to the thought that all other vascular plants were derived eventually from ancient plants of the rhynioid type.

Evolutionary Relationships

The rhyniophytes are generally considered to be the most primitive known group of vascular plants. All other groups of vascular plants, except possibly some ancient and poorly known fossils, appear to be derived eventually from rhyniophytes. Most students of the evolution of vascular cryptogams conceive of the Lycopodiophyta, Equisetophyta, and Polypodiophyta as separate lines of development from the ancestral Rhyniophyta (although diverse names are used for these same groups by different botanists). The other remaining group of modern vascular cryptogams, the small division Psilotophyta (Fig. 16.5), is variously considered as a set of slightly modified descendants of the Rhyniophyta (still the most widely accepted view), or modified lycopods (Lycopodiophyta), or reduced or relatively primitive ferns (Polypodiophyta).

A comprehensive theory of the origin of other groups of vascular plants from the ancestral Rhyniophyta was presented in 1930 by the German botanist Walter Max Zimmermann (1892–1980, Fig. 16.2). Zimmermann calls the ultimate branches of a dichotomously branching stem **telomes,** and his theory is called the **telome theory.** In Zimmermann's view, all leaves originated (phyletically) as branch stems

Fig. 16.5 *Psilotum nudum,* the whisk-fern, which resembles the rhyniophytes in its dichotomously branched stem, solid stele, absence of true roots, and some other features. (Slightly less than natural size.)

or branch stem systems. Some of the evolutionary series, as suggested by him, are shown in Fig. 16.6. The accumulating evidence from fossils, especially of Devonian age, now gives strong support to Zimmermann's concepts as regards ferns, equisetoids, and early seed plants. The precise evolutionary relationship of the lycopod line to the rhyniophytes, on the other hand, is still uncertain, and many paleobotanists now support the earlier concept that the lycopod leaf is fundamentally different in origin from the leaves of other groups. In this view the lycopod leaf is regarded as an **enation,** a mere outgrowth from the epidermis, which has no evolutionary history of ever being a branch or branch system. The nonvascularized leaf of **Asteroxylon** (Fig. 16.7), a Devonian associate of *Rhynia*, is seen as an evolutionary precursor to the lycopod leaf, with an unbranched midvein. *Asteroxylon* itself is seen as probably descended from plants such as *Zosterophyllum*, which had a naked stem (like *Rhynia*) but lateral rather than terminal sporangia. The telome theory is a good example of the fact that useful new ideas often turn out to have a more limited application than was at first enthusiastically proposed.

POLYPODIOPHYTA (FERNS)

General Features

The ferns (Fig. 16.8) are vascular cryptogams with alternate, generally large and compound leaves that have a branching vein system. The sporangia are borne on ordinary vegetative leaves, or less often on more or less strongly modified leaves, typically in clusters (**sori**) on the lower surface or at the margin of the leaf.

The ferns are here treated as a division, Polypodiophyta, with a single class, Poly-podiopsida, although a case might be made for putting some of the Paleozoic fossils into a separate class. There are about 10,000 living species of ferns, the vast majority of which belong to the order Filicales, also called Polypodiales.

Ferns occur in a wide variety of habitats and climates, but most of them are found in shady rather than sunny places, in moist or wet rather than dry sites and climates, and in tropical or subtropical rather than in temperate or cold regions. No large part of the United States, even the desert regions, is wholly without ferns.

The sporophyte is the conspicuous generation in ferns, as in all vascular plants; the gametophyte is a small thallus seldom over 2 cm long. Both generations are physiologically independent at maturity. The sporophyte is always photosynthetic and the gametophyte is usually so, but the gametophytes of some kinds of ferns are subterranean, colorless, and mycorhizal (that is, they have a symbiotic relationship with a fungus). In some specialized kinds of ferns, not further discussed here, the gametophytes are largely or wholly dependent on the food originally stored in the spores, as they are also in *Selaginella*.

Stem

Some ferns reach a height of more than 20 m, with a trunk up to about 6 dm thick, surmounted by a crown of leaves (Fig. 16.9); but more often the stem is represented only by a simple or branched rhizome (underground stem) from which the leaves arise. Apical growth of the stem is brought about by divisions in **apical initial** cells, or in a group of apical initials forming an **apical meristem** (tissue characterized by cell division).

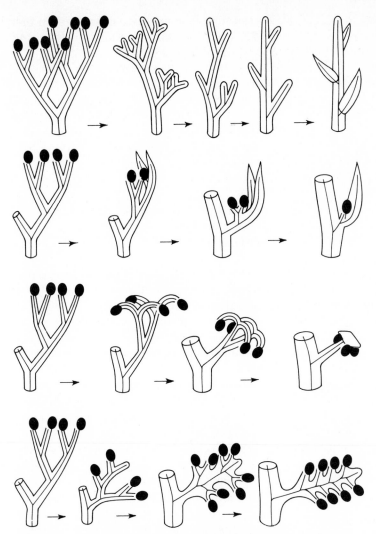

Fig. 16.6 Origin of leaves, sporophylls, and sporangiophores, according to the telome theory. Starting with a dichotomously branched stem with terminal sporangia, we derive: (*top row*) the lycopod leaf; (*second row*) the lycopod sporophyll; (*third row*) the equisetoid sporangiophore; and (*bottom row*) the fern leaf with marginal sporangia. As noted in the text, the sequence shown in the two lower rows is now generally accepted as correct, but the sequence in the two upper rows is probably wrong. The leaves of lycopods are now thought to represent enations rather than telomes.

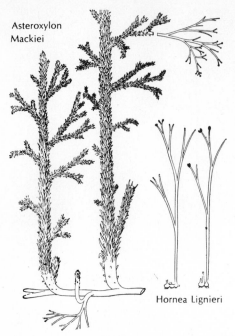

Asteroxylon
Mackiei

Hornea Lignieri

Fig. 16.7 *Asteroxylon* and *Horneophyton* (*Hornea*). The detached fruiting branches associated with the reconstruction of *Asteroxylon* exemplify one of the problems in making reconstructions from associated fossil parts that are not organically united. It is now known that these fruiting branches, as shown in the reconstruction by Kidston and Lang, belong to another kind of rhyniophyte. The sporangia of *Asteroxylon* are borne on the stem, intermingled with the leaves. [From Kidston and Lang.]

Branches of a fern stem characteristically originate by bifurcation of the apical meristem or apical cell, each of the two parts continuing to develop at the apex of a branch. The two branches may be essentially equal, or more often one branch may continue as the main stem, to which the other is attached as a lateral branch. When the branching is lateral, the apical cell or meristem of the lateral branch commonly remains dormant for some time before resuming its meristematic activity, so that the branch actually originates well behind the tip of the main stem.

In cross section the stem of ferns is seen to consist typically of an epidermis, enclosing a cortex, which in turn encloses the central cylinder, or **stele.** In some few ferns the central cylinder consists of a solid core of xylem, surrounded by phloem, as in *Psilotum* and the rhyniophytes. This kind of stele is believed to be primitive, in comparison with other kinds, and is called a **protostele.** More often the central part of the stele consists of parenchyma (thin-walled, relatively unspecialized cells), forming a pith that is surrounded by a cylindrical sheath or network of vascular tissues. Such a stele is called a **siphonostele.** It is regarded as an evolutionary modification of the protostele, with the central cells remaining relatively unspecialized (thus parenchyma) instead of maturing into xylem.

In the most typical kind of siphonostele the vascular tissue forms a cylindrical network of vascular bundles (Fig. 16.10). In a cross section of the stem the vascular bundles form a ring, with the bundles separated by areas of parenchyma (unspecialized, thin-walled cells). Conceptually, and perhaps phyletically, these parenchymatous gaps between the vascular bundles reflect the departure of vascular strands toward leaves or branches, so that **leaf gaps** or **branch gaps** are left in what would otherwise be a continuous cylinder of vascular tissue. The presence of leaf gaps in the stele has been used by some botanists as a conceptually unifying feature for the ferns, gymnosperms, and angiosperms. The leaf gap concept is now under attack, however, and it may not properly apply outside the ferns.

An individual vascular bundle in a fern

Fig. 16.8   Maidenhair fern, *Adiantum pedatum*, in Maryland. [Photo by W. Bryant Tyrrell, from National Audubon Society Collection/Photo Researchers, Inc.]

stem usually has xylem on the inner side (toward the pith), and phloem on the outer side, but sometimes the phloem completely surrounds the xylem. Some ferns, such as the common bracken (*Pteridium aquilinum*, a cosmopolitan species), have a highly complex vascular anatomy which is usually interpreted in terms of a much modified siphonostele.

Root

All divisions of vascular plants except the Rhyniophyta and Psilotophyta ordinarily have the root clearly differentiated from the stem. The typical anatomical differences between roots and stems of vascular plants are discussed in Chapter 20. The primary root, in ferns and other plants, usually originates early in the development of the embryo sporophyte, but in ferns it does not persist until the sporophyte is mature. Roots of the mature sporophyte of ferns originate from internal tissues of the stem, pushing their way through the intervening tissues as they grow. The individual roots are rather small and fibrous, and simple or branched; the branches originate internally.

Fig. 16.9   Tree ferns in the Colombian Andes. [Photo courtesy of José Cuatrecasas.]

Leaf

The leaves of ferns are usually divided into a slender petiole and an expanded blade. Typically the growing young blade unrolls toward the tip, and the leaf is said to be **circinate.** Leaves in which the blade has not yet unrolled, or has only begun to do so, are often called **fiddleheads** (Fig. 16.11).

The leaves of most ferns are compound, but there are also many simple-leaved ferns. In most compound fern leaves, the petiole continues as the main axis or **rachis** of the leaf blade, and the primary segments or leaflets appear as lateral branches from the rachis. Leaves of this type are said to be **pinnately compound** (from Latin *pinna,* feather), and the primary segments are called **pinnae** (sing., pinna). The pinnae may themselves be one or more times compound; the ultimate segments are then called **pinnules.** General features of leaves are further discussed in Chapter 21.

The leaves of most ferns are anatomically more complex than those of most other pteridophytes. They characteristically have a branching vein system, in which the smaller branches are often distinctly dichotomous. The smaller veins sometimes, but not always, form a network. The epidermis is generally cutinized and, except for the guard cells of the

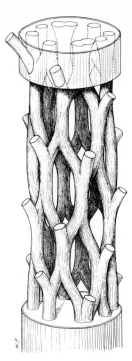

Fig. 16.10 Diagram of stem with dissected stele forming a cylindrical network of vascular bundles.

stomates, usually lacks chloroplasts. The internal tissue is composed chiefly of chlorenchyma (chloroplast-bearing cells) with interspersed veins.

Sporangia

The sporangia of ferns are generally borne on the leaves. The sporophylls (spore-bearing leaves) are in most cases ordinary vegetative leaves, but not infrequently the sporophyll is more or less modified and not photosynthetic. Numerous sporangia are borne on each sporophyll, and in most modern ferns the sporangia are aggregated into distinct groups called sori (sing., **sorus**; Fig. 16.12). A typical sorus is about the size of a pinhead. In most

Fig. 16.11 Fiddleheads of interrupted fern, *Osmunda claytoniana.*

modern ferns the sori are borne on the lower (abaxial) surface of the sporophyll, often at the tips of some of the ultimate veinlets. The brown dots that can fre-

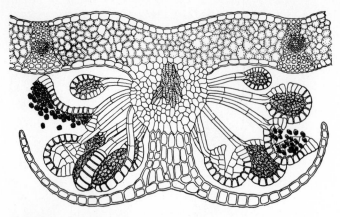

Fig. 16.12   Cross section of part of a fern leaf, showing a sorus with its indusium. [From Transeau, Sampson, and Tiffany, *A textbook of botany*, Harper & Row, New York, 1953.]

quently be seen on the lower side of fern leaves are sori.

The young sorus is usually covered by a thin protective membrane called the **indusium.** The shape and structure of the indusium differ in different kinds of ferns, and the differences furnish useful taxonomic characters. When the sori are borne near the margin of the leaf, the margin is often inrolled to form a single common indusium for a whole row of sori, or individual bits of the margin may be modified to form separate indusia.

There is some variation in the ontogeny and mature structure of fern sporangia. The putatively more primitive pattern is said to be **eusporangiate,** in contrast to the more common **leptosporangiate** pattern. Only the latter type is here described. In leptosporangiate ferns the sporangium (Fig. 16.13) generally develops from a single sporangial initial cell. The jacket layer is only one cell thick at maturity, and dehiscence (opening in a predetermined way) is usually due to moisture changes in a group of thick-walled cells that form a partial or complete ring, the

annulus. Tensions developing within the annulus cause the sporangium to break open at a definite point marked by thin-walled cells called **lip cells.** Usually there is a relatively small and definite number of spores, ranging from 16 to 64 in different species.

The spores have n chromosomes and represent the first step in the gametophyte generation. In most ferns the spores are all alike and give rise to gametophytes that bear both archegonia and antheridia; that is, the plants are **homosporous.** Some few ferns are **heterosporous** instead; that is, they produce two different kinds of spores, giving rise to separate male and female gametophytes, more or less in the manner described on a subsequent page for *Selaginella.*

Gametophyte

The gametophyte of ferns and other pteridophytes is often called a **prothallus.** The prothallus varies in form and structure in different kinds of ferns, but it is always small and inconspicuous. In the

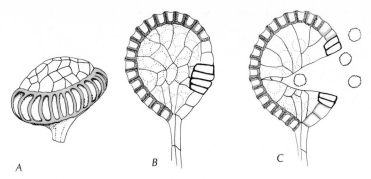

Fig. 16.13   Two types of leptosporangia: (*A*) with annulus complete; (*B, C*) the more common type, with the annulus interrupted by the stalk, and with prominent lip cells.

leptosporangiate ferns the most common type is a thin, flat, more or less heart-shaped, green thallus (Fig. 16.14), seldom as large as the little fingernail. A prothallus of this type is only one cell thick except for a slightly thicker median strip extending back from the notch. Rhizoids, antheridia, and archegonia are borne on the under side of this median strip, the archegonia usually being closer to the notch than are the antheridia. Often the archegonia are produced first, and if an egg is soon fertilized no antheridia are produced.

The archegonium (Figs. 16.15, 16.16) is more or less flask shaped, with a neck that usually protrudes and is often curved toward the rear of the prothallus. Each archegonium contains a single egg. The antheridia (Fig. 16.15) are more or less globose. In the leptosporangiate ferns the antheridium usually projects from the surface of the gametophyte, and an apical cover cell is pushed away from within by expansion of the antheridial contents when the sperms are mature. Leptosporangiate ferns usually have only about 32 sperms in each antheridium. The sperms are naked, coiled, and provided with more or less numerous whiplash flagella

at one end. As in other pteridophytes, the sperms are attracted to the egg in the archegonium by a chemical stimulus, and they require a film of water in which to swim. On reaching the archegonium, the sperm swims through the mucilaginous material formed by the degeneration of the canal cells, until it reaches the egg.

Embryo

Fusion of a sperm and an egg in the venter of the archegonium constitutes fertilization. The fertilized egg has 2n chromosomes and is the first cell of the sporophyte generation. The developing embryo has a foot, a primary root, a primary stem, and a primary leaf. The foot absorbs food from the body of the gametophyte and passes it on to the remainder of the developing embryo. After the new sporophyte has become established as an independent plant, the gametophyte dies and degenerates.

LYCOPODIOPHYTA

General Features

The **lycopods** (i.e., members of the division Lycopodiophyta) are vascular cryp-

Fig. 16.14   Gametophyte of cinnamon fern, *Osmunda cinnamonea*, from beneath, showing rhizoids. (× about 25.) [Photo by Hugh Spencer, from National Audubon Society Collection/Photo Researchers, Inc.]

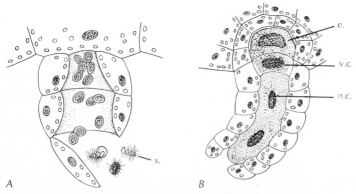

Fig. 16.15   Antheridium (*A*) and archegonium (*B*) of a fern, in long section: e., egg; n.c., neck canal nucleus; s., sperm; v.c., ventral canal nucleus. (× 500.)

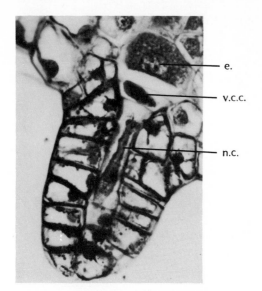

Fig. 16.16 Fern archegonium, in long section: e., egg; n.c., neck canal cell; v.c.c., ventral canal cell. [From H. Bold, *Morphology of plants,* 3rd ed., Harper & Row, New York, 1973.]

togams with simple, narrow, and usually small leaves that have a single, unbranched midvein (rarely two veins) of vascular tissue. The leaves are usually alternate or opposite, seldom whorled. Opposite leaves of lycopods are often modified in position so that both are on the same side of the stem. Many lycopods of the Carboniferous and Permian geologic periods were trees, but all modern (i.e., living) species are relatively small herbs with short or elongate, sometimes vinelike, stems.

Leaves of many lycopods (but not *Lycopodium* itself) have a small appendage on the adaxial side (the side toward the stem, the upper side), a little above the base. This appendage is called a *ligule* (Latin *ligula,* a little tongue), and lycopods are described as **ligulate** or **eligulate** according to whether or not the ligule is present.

The function of the ligule, if any, is dubious. If the leaves of lycopods originated by modification of some branches of a dichotomously branched stem system, as Zimmermann has postulated (Fig. 16.6), the ligule may represent a vestigial branch. It should be emphasized, however, that the fossil transitional forms to support this interpretation have not been found, and many botanists consider the leaves of lycopods to have originated as enations rather than as telomes or telome systems.

The sporangia of lycopods are borne singly in the axils of leaves, or on the adaxial side of the leaf near the base. The sporophylls (sporangium-bearing leaves) of some lycopods are ordinary vegetative leaves (Fig. 16.18), but other lycopods have more or less modified sporophylls that are grouped into a terminal cone called a **strobilus** (Fig. 16.19).

There are about 1200 modern species of lycopods, most of which belong to only two genera, *Lycopodium* and *Selaginella*. Both genera are cosmopolitan, but more abundant in moist, tropical regions than in cooler or drier ones. Both genera have a long fossil history, and *Lycopodium* appears to have survived with relatively little change since the Devonian period. *Baragwanathia,* from Lower Devonian deposits in Australia, is very much like those modern species of *Lycopodium* that do not have a differentiated strobilus.

Lycopodium and *Selaginella* are often called **club mosses.** The name comes from the superficial resemblance of the sporophytes of *Lycopodium* and *Selaginella* to large moss gametophytes, plus the clublike terminal strobilus. Some species of *Lycopodium* are also called ground pine, running pine, or ground cedar, because of a very casual resemblance to pine or cedar twigs.

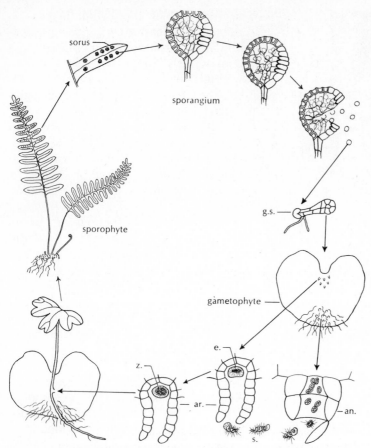

sorus

sporangium

sporophyte

g.s.

gametophyte

e.

z.

ar.

an.

s.

Fig. 16.17 Life cycle of a fern: an., antheridium; ar., archegonium; e., egg; g.s., germinating spore; s., sperm; z., zygote.

Selaginella

Selaginella (Fig. 16.20), with about 500 species, is of especial interest because of its type of life cycle, in which two different kinds of spores, giving rise to two different kinds of gametophytes (male and female) are produced. It is **heterosporous,** in contrast to the majority of vascular cryptogams, which are **homosporous.** The development of heterospory is an essential precursor to the evolution of seeds.

The stem of *Selaginella* is short or elongate, and dichotomously to sym-podially branched. Often there is an indefinitely elongating underground stem, the rhizome, from which more or less erect aerial branches arise. The internal anatomy of the stem is complex, with some particular specializations that need not be discussed at this level of study.

Most species of *Selaginella* have characteristic slender, leafless branches, called **rhizophores,** from which the roots originate. There has been much argument as to whether rhizophores are roots, stems, or organs *sui generis* (of their own kind). Such controversy ignores the fact

Fig. 16.18   *Lycopodium lucidulum,* a club moss of eastern North America. [Photo by Hugh Spencer, from National Audubon Society Collection/Photo Researchers, Inc.]

Fig. 16.19 *Lycopodium complanatum* var. *flabelliforme*, ground cedar, in Massachusetts. In this species the stems are flattened and photosynthetic, and the leaves are reduced to tiny scales.

that the fundamental and convenient classification of plant organs as roots, stems, and leaves cannot be expected to work perfectly among the more primitive kinds of vascular plants, in which the differences among these organs have not become wholly stabilized.

The leaves of *Selaginella* are small, generally well under 1 cm long, and have a short ligule on the upper side near the base. The more primitive species have spirally arranged, alternate leaves, as do many species of *Lycopodium*. More often the stem has basically opposite leaves, which are displaced so as to form four rows—two rows on the upper surface and one row along each margin. The leaves of the marginal rows are distinctly larger than those of the upper rows.

The sporophylls of *Selaginella* are much like the vegetative leaves and are always borne in terminal cones. Often they are arranged in four distinct vertical rows, forming a four-sided strobilus. Each sporophyll bears a single sporangium on the adaxial surface near the base, or the sporangium may be in the axil of the sporophyll.

The sporangia (Fig. 16.21) are of two types. One type, called the **megasporangium** (Greek *mega*, large), contains one to four relatively large, thick-walled spores called **megaspores.** The wall of the megasporangium often encloses the megaspores so closely that their outline governs the shape of the sporangium. The other sporangial type, called the **microsporangium,** contains many much smaller and thinner-walled spores, called **microspores.** The **megasporophylls** and **microsporophylls** are similar in appearance and usually borne in the same cone, the microsporophylls commonly above the megasporophylls, less often alongside or intermingled.

The microsporangium contains a large number of potential microspore mother cells. Some of these degenerate just before maturity, but most of them undergo reduction division, each producing a tetrad of microspores. The microspores have n chromosomes and represent the first step in the male gametophyte generation. Development and maturation of the male gametophyte results from a number of mitotic cell divisions without any increase in size, and the mature male gametophyte (Fig. 16.22) is largely or wholly contained within the old microspore wall. The male gametophyte is not photosynthetic, and the original protoplast of the microspore is its only source of nourishment.

Germination of the microspores commonly begins before they are shed from

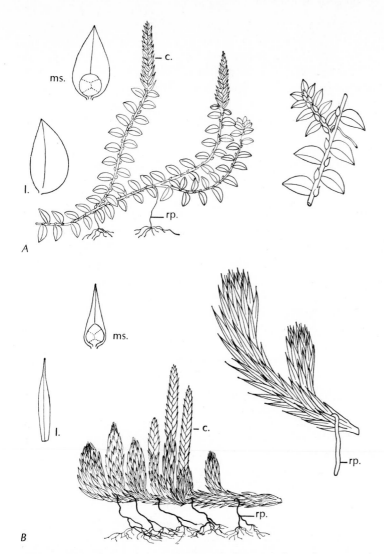

Fig. 16.20   *Selaginella,* two species; habit, × 2; branches, × 4; sporophylls and vegetative leaves, × 10 (*A*) and × 6 (*B*): c., cone; l., leaf; ms., megasporophyll; rp., rhizophore.

the microsporangium, and continues with little or no interruption after they are released. Just before maturity the male gametophyte consists of a multicellular antheridium and a single additional cell, the **prothallial cell.** The prothallial cell is considered to be an evolutionary vestige of the body of the gametophyte. The numerous sperms in the antheridium each have two whiplash flagella at one end. As the gametophyte nears maturity, the spore wall ruptures, exposing one end of the antheridium. The jacket cells of the antheridium soften and degenerate when

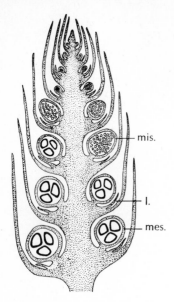

Fig. 16.21 *Selaginella*, diagrammatic representation of long section of cone (× 30): l., ligule; mes., megasporangium; mis., microsporangium.

ing megaspore mother cell undergoes reduction division, generally producing a tetrad of megaspores. Sometimes one or more of the megaspores do not develop to maturity, so that the megasporangium has less than four megaspores. The megaspores have n chromosomes and represent the first step in the female gametophyte generation.

The megasporangium opens and the megaspores usually fall out, but sometimes they remain loose in the sporangium throughout the development of the female gametophyte. Development and maturation of the female gametophyte result from a number of mitotic cell divisions without much increase in size. The protoplast becomes multinucleate, and partitions are then formed, beginning at the top. The megaspore wall eventually ruptures at one end, and the top of the mature gametophyte (Fig. 16.23) protrudes. The exposed part of the gametophyte often

moistened, or concurrently with spermatogenesis, allowing the sperms to escape.

The megasporangium of *Selaginella* contains a considerable number of potential megaspore mother cells. Usually all but one of these degenerate. The remain-

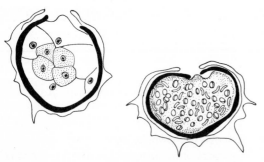

Fig. 16.22 Male gametophytes of *Selaginella kraussiana* (× 1500.) [After Stagg.]

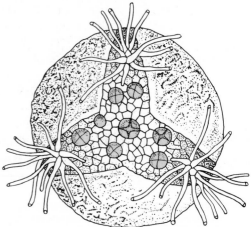

Fig. 16.23 Female gametophyte of *Selaginella martensii*, top view. (× about 125.) Note the archegonia, distinguishable by the four neck cells, and the clusters of vestigial, nonfunctional rhizoids. [After Bruchmann.]

becomes green and may develop rhizoids, but the amount of food made is relatively small. The main source of nourishment of the female gametophyte is the original protoplast of the megaspore, including its stored food. The female gametophyte generally produces several archegonia, but fertilization of the egg in one of these seems to inhibit further development of other archegonia.

Fertilization takes place in the normal pteridophytic way, with the sperms being attracted to the egg by chemical stimuli and requiring a film of water in which to swim. Obviously the male gametophyte must be very close to the female gametophyte in order for the process to work. Wind and simple gravity are the only

means of dispersal for the microspores or male gametophytes, and thus great numbers of them are necessary. Evolutionary progress requires at least occasional cross-fertilization, in order that new and potentially useful combinations of genes may be produced (see Chapters 29 and 30), and a number of microspores that would be sufficient to drop down and provide for the megaspores in the same cone would not be enough to promote the necessary outcrossing.

The fertilized egg has 2n chromosomes and represents the first step in the sporophyte generation. By a series of mitotic divisions the fertilized egg gives rise eventually to the embryo sporophyte. The early growth of the embryo sporophyte

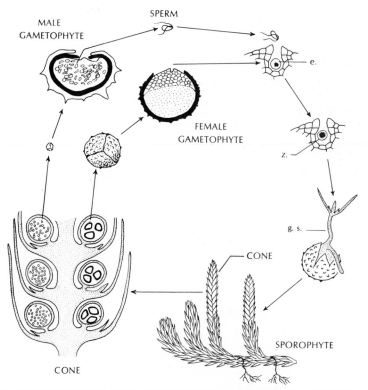

Fig. 16.24 Life cycle of *Selaginella*: e., egg; g.s., germinating sporophyte; z., zygote.

draws on the reserve food stored in the body of the female gametophyte. Continued growth leads to the establishment of the sporophyte as an independent plant with its own connection to the substrate, concurrent with the death and degeneration of the gametophyte.

In earlier chapters we noted that female gametes tend to be relatively large, non-motile, and provided with a considerable reservoir of stored food, in contrast to the smaller, motile male gametes that have little or no stored food. In *Selaginella* and other heterosporous plants we see a repetition of this same sort of sexual differentiation, at the level of the whole gametophyte instead of merely the gametes. The female gametophyte is relatively large, not readily dispersible, and provided with a reservoir of stored food. The male gametophyte is smaller, more dispersible (though not by its own efforts), and nearly or quite without stored food. The next step in the evolution of a seed is the retention of the female gametophyte on the old sporophyte, so that the parent sporophyte can serve as a continuing source of food in the early development of the new embryo sporophyte. The fossil record clearly shows that modern seed plants come from a different phyletic line than *Selaginella*, but *Selaginella* is a good modern example of a necessary way station between homosporous vascular cryptogams and seed plants.

Lepidodendrales

The Lepidodendrales are a group of fossil lycopods known from the Upper Devonian to the Permian period. They are similar to *Selaginella* in being heterosporous and in having ligulate leaves, but they differ in being much larger, commonly of tree size. The stem increased in thickness year after year by a process of **secondary growth,** comparable to the secondary growth of angiosperm stems described in Chapter 19.

The leaves of the Lepidodendrales are long and narrow, sometimes more than 15 cm long, and usually with a single unbranched midvein. The leaves were deciduous, that is, they fell away from the stem some time after reaching maturity. The older parts of stems are commonly covered with characteristic scars marking the positions of former leaves.

The rhizophore system of the Lepidodendrales is unique. The stem, which is unbranched for some distance above the ground, breaks up at the base into four large branches which undergo repeated dichotomies as they spread outward. The true roots originated endogenously (i.e., from internal tissues) from these branches, which are generally called rhizophores. Rhizophores of the Lepidodendrales have been compared to the more slender rhizophores of *Selaginella*, but the homology is doubtful.

One of the most common genera of the Lepidodendrales is *Lepidodendron* (Fig. 16.25). Large specimens reached a height of well over 30 m. The straight, tall trunk is dichotomously branched above to form a broad, rounded crown. The leaves are narrow, commonly 10–15 cm long, and arranged in a close spiral. The older parts of the trunk are covered with closely set, spirally arranged scars marking the position of leaves that have fallen.

Lepidodendron did not have seeds, and the female gametophytes of at least some species were rather similar to those of *Selaginella*. Some other members of the Lepidodendrales did have seeds, of a sort. The female gametophyte was retained within the megasporangium, and the whole megasporangium, with associated

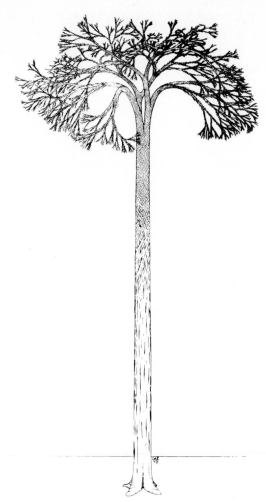

Fig. 16.25 *Lepidodendron* reconstruction by Donald A. Eggert. [This figure, reproduced through the courtesy of Dr. Eggert, appeared in *Palaeontographica* **108B:**79, 1961.]

structures, was eventually shed from the plant. Unlike modern seeds, however, these Lepidodendralean "seeds" were open at the tip, exposing the female gametophyte, and the sperm evidently reached the egg by swimming in a film of water, in the usual pteridophytic fashion. Thus they were or were not seeds, according to one's criteria.

The rhizophore system of several genera of Lepidodendrales is indistinguishable from that of *Lepidodendron*. When they are found alone, without the remainder of the sporophyte, these rhizophore systems are given the generic name *Stigmaria*. *Stigmaria* (Fig. 16.26) is a good example of what, in paleobotany, is called a **form genus.** A single name is necessarily used to cover fossil structures of a particular type, even though it is realized that these are merely detached parts belonging to several genera that differ in other respects.

EQUISETOPHYTA

The **equisetoids** (i.e., members of the division Equisetophyta) are vascular cryptogams with whorled leaves, jointed stems, and a terminal strobilus. The group deserves some attention because it was abundant and diversified in Paleozoic times, especially the Carboniferous period. Although the modern (and many fossil) species all have simple leaves with an unbranched midvein, the fossil evidence clearly shows that these leaves represent reduced branch systems.

Equisetum

All modern equisetoids belong to the single genus *Equisetum* (Fig. 16.27), with about 20 species widely distributed throughout most of the world, except Australia. Most of them occur in wet places, or in shallow water from which the stems emerge, but some grow in ordinary, well-drained soil. The epidermal cells of the stem contain appreciable amounts of silica, and the species with relatively firm stems were used in pioneer communities for scouring pots and pans, whence the common name **scouring rush.** The common name **horsetail** is also often

Fig. 16.26 *Lepidodendron* stumps on Stigmarian bases; natural sandstone casts at Victoria Park, Glasgow. An enclosure has been built to preserve these fossils from weathering and vandalism.

used especially for species with branching stems.

The aerial stems of *Equisetum* arise as branches from a rhizome that creeps along beneath the surface of the ground. Ordinarily the stem is green and photosynthetic, less than 2 cm thick, and not more than 1 or 2 m tall. The aerial stems of some species are annual, dying down each year; in other species they are perennial, lasting through several years. In the common species *E. arvense* (Fig. 16.28) the stems are of two types. The fertile stems, which arise early in the spring, are only a few inches tall, unbranched, soft, and without chlorophyll. They soon produce a terminal cone, and then wither and die. The sterile vegetative stems, which persist throughout the season, arise from the rhizome after the fertile stems are mature. They are several decimeters tall, much branched, green, and photosynthetic.

Stems of *Equisetum* are hollow and conspicuously jointed, with a whorl of small, narrow leaves at each joint. The basal part of each internode is meristematic, and increase in length is brought about by the activity of these **intercalary meristems** as well as by the apical cell at the stem tip. The meristematic tissue consists of soft, thin-walled cells, and the stems can readily be pulled apart at the nodes. The stem is longitudinally ribbed and grooved, with the ribs and grooves tending to alternate in successive internodes. Each leaf is directly aligned with a rib of the internode beneath it.

The leaves of *Equisetum* are slender and usually less than 3 cm long, each with a single, unbranched midvein. The leaves in each whorl are joined laterally to form a sheath around the stem. The tips of the leaves are usually free from each other, so that the sheath is toothed at the apex. The leaves have little or no chlorophyll and seem to be mere vestigial structures whose only function is to protect the intercalary meristems and branch primordia of the stem.

Branches of the stem of *Equisetum* arise from small branch primordia which alternate with the leaves at the surface of the stem at each node. The branches break

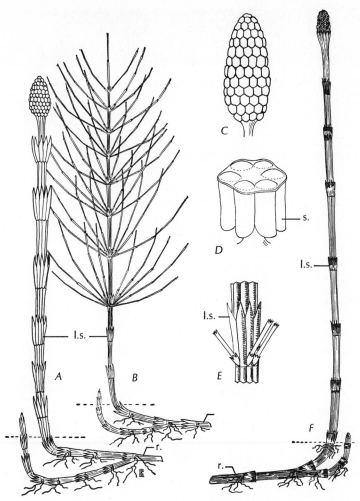

Fig. 16.27 *Equisetum.* A-E, *Equisetum arvense:* A, fertile shoot
(× 1); B, sterile shoot (× 1/4); C, strobilus (× 2); D,
sporangiophore (× 20); E, part of stem (× 3); F, *Equisetum
hyemale* (× 1/4): l.s., leaf sheath; r., rhizome; s., sporangium.

through the basal part of the leaf sheath,
so that on casual inspection they seem to
be borne just beneath the leaves.

Roots of *Equisetum* are slender and
rather short, seldom over 2 or 3 cm long,
with or without a few lateral branches.
The roots originate at the nodes of the
rhizome and subterranean parts of the
aerial branches; their precise ontogeny is
still debatable. Unlike the stems, they
have a protostelic central cylinder with a
solid core of xylem.

The sporangia of *Equisetum* are borne
on sporangiophores that are grouped into
a terminal cone commonly 1 to several
centimeters long. The sporangiophores
are arranged in successive whorls on the
cone axis in the same way as the leaves on

Fig. 16.28   *Equisetum arvense*, the common horsetail, showing mature fruiting stem and young vegetative stems. (About natural size.) [Photo by Grant Haist, from National Audubon Society Collection/Photo Researchers, Inc.] (*Inset*) Cone of the same species. (× 4.) [Photo by Herman H. Giethoorn, from National Audubon Society Collection/Photo Researchers, Inc.]

the vegetative part of the stem. The term sporangiophore is used, instead of sporophyll, because there may never have been a time in the evolutionary history of the group when these structures looked like ordinary leaves (see Fig. 16.6).

Each sporangiophore has 5 to 10 sporangia attached to the lower side of the cap in a ring around the stalk. The sporangia are more or less cylindric and commonly 2 to several millimeters long. The numerous spores are all alike in appearance, but there may be a functional difference in some species, with some spores producing male gametophytes and others producing gametophytes that are at first female but eventually produce antheridia if none of the eggs is fertilized.

Fossils essentially similar to *Equisetum* are referred to the genus *Equisetites* and are known from as far back as the Upper Carboniferous period. Some other fossils are apparently transitional between the Equisetales and Calamitales.

Calamitales

The Calamitales (Fig. 16.29) are fossils that more or less resemble *Equisetum* except for their larger size, well-developed secondary thickening, and different cone structure. The cone has whorls of sterile appendages alternating with the whorls of sporangiophores. Large plants possibly reached a height of 30 m, with a stem 6 dm thick, but a height of 6–10 m was probably more common. The leaves were small and narrow, not more than about 3 cm long, and sometimes more or less connate (joined together) as in *Equisetum*, but their structure indicates that they were photosynthetic.

The Calamitales are known from the late Devonian to early Permian period, and were abundant throughout the Car-

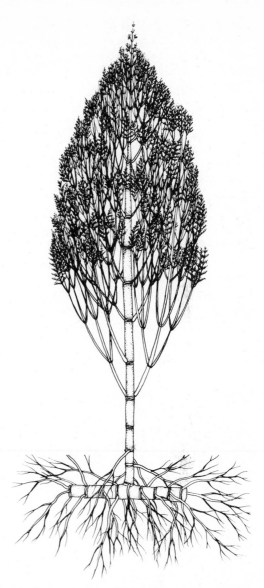

Fig. 16.29 *Calamites* reconstruction, after Hirmer. [Courtesy of Donald A. Eggert.]

boniferous. There are numerous species, with considerable differences in detailed morphology. Some species were homosporous, others heterosporous, and *Calamocarpon* produced seeds comparable to those of some Lepidodendrales. The plants were readily fragmented, and it is often difficult to match up parts that have been separately preserved as fossils. A number of different generic names have been applied to Calamitalean fossils representing various parts of the plant, but in the broad sense they are sometimes all referred to the genus *Calamites*. If the plants existed today, so that they could be more thoroughly studied and compared, they would doubtless be classified into several different genera.

The equisetoid type of stem structure, with a necessarily soft meristem at each node, is not well adapted to the tree habit, and it should not be surprising that the Calamitales were ultimately unsuccessful in competition with other kinds of trees.

ECONOMIC IMPORTANCE

Modern pteridophytes are of relatively little economic importance. Some of them are grown for ornament (especially certain ferns), and others are collected for Christmas decorations. They furnish some forage for livestock, and a few are more or less edible for humans; some others are mildly poisonous. A drug derived from the rhizomes of certain species of the fern *Dryopteris*, especially *D. filix-mas*, the "male fern," is a traditional anthelmintic, used to expel intestinal worms, especially tapeworms, but it has now largely been replaced by synthetics. Spores of *Lycopo-dium* have been used as flash powder in the past, and are still sometimes used to coat pills, giving them a smooth surface that promotes easy swallowing.

Aside from their evolutionary role as ancestors of the seed plants, the great importance of the vascular cryptogams comes from their abundance during the Carboniferous period. The Calamitales and the Lepidodendrales, in particular, were the principal contributors to the vegetable deposits of the Carboniferous period that became transformed into coal.

SUGGESTED READING

Banks, H., Early vascular land plants: Proof and conjecture, *BioScience* **25**:730–737, 1975.

Doyle, W. T., *The biology of higher cryptogams*, Macmillan, London, 1970.

Kidston, R., and W. H. Lang, On old red sandstone plants showing structure, from the Rhynie chert beds, Aberdeenshire, Parts i–v, 1917–1921, *Trans. Roy. Soc. Edin.* **51**: 761–784; **52**:603–627, 643–680, 831–854, 855–902. The classic papers on fossils from the Rhynie beds.

Mickel, J. T., *How to know the ferns and fern allies*, Brown, Dubuque, Iowa, 1979. An illustrated key for identification, covering North America north of Mexico.

Sporne, K. R., *The morphology of pteridophytes; The structure of ferns and allied plants*, 4th ed., Hutchinson, London, 1975. Textbook style.

Chapter 17

Gymnosperms (Pinophyta)

Giant Sequoia, *Sequoiadendron giganteum,* in the Mariposa grove, Yosemite National Park, California. [National Park Service photo by Ralph H. Anderson.]

INTRODUCTION TO
THE SEED PLANTS

The divisions Pinophyta (gymnosperms) and Magnoliophyta (angiosperms) differ from all other plants, except some fossils of the lycopod and calamite groups, in that they produce seeds. Even the lycopod and calamite seeds lack one of the essential features of the seeds of modern plants, as has been noted in Chapter 16.

All seed plants are heterosporous. The characters that distinguish seed plants from heterosporous cryptogams relate to the ontogeny of the male and female gametophytes and the embryo sporophyte. The megaspore of a seed plant is permanently enclosed in the megasporangium, and the female gametophyte that develops from the megaspore completes its development while still enclosed within the megasporangium wall. The developing female gametophyte is nourished by the enclosing sporophyte. The egg or eggs produced by the female gametophyte may or may not be associated with definite archegonia, but in any case they are enclosed within the megasporangium. The embryo sporophyte that develops from the fertilized egg characteristically begins its growth while still enclosed within the body of the female gametophyte.

The structure and function of the reproductive parts of seed plants were partly deciphered before the correlation of these parts with the reproductive structures of vascular cryptogams was understood. Some of the reproductive parts of seed plants, therefore, have names that do not reflect their homologies. The wall of the megasporangium of a seed plant is called the **nucellus.** The nucellus is enclosed by one or two additional layers of tissue, the **integuments.** The integuments have no obvious counterparts among heterosporous cryptogams, and their evolutionary origin is still debatable. (The most generally accepted view is that the single integument of most gymnosperms represents a set of sterile branches—essentially telomes—joined by their lateral margins to form one or two cups around the megasporangium. The situation in angiosperms is more complex.) The megasporangium, plus the enclosing integument or integuments, is called an **ovule** (Fig. 17.1). There is a small opening through the integuments at one end of the ovule. In the less modified types of ovules this opening, called the **micropyle** (from Greek *mikros*, small, and *pyle*, gate), is at the opposite end from the stalk. At the micropylar end of the ovule the megasporangial wall (the nucellus) is generally well defined; at the other end it is often not clearly differentiated from the integuments.

The microspores of seed plants generally undergo one or more mitotic divisions (within the enclosing spore wall) before they are released from the microsporangium in which they are produced. At the stage in which they are shed from the microsporangium, the resulting young male gametophytes are called **pollen grains** (Fig. 17.2). The microsporangium itself is called a **pollen sac.** The nuclear divisions that occur in the development of pollen grains from microspores do not result in any significant increase in the volume of the protoplast, and the wall of the pollen grain is merely the matured wall of the microspore.

The pollen grains must be transported in one way or another to the ovule or its vicinity if they are to function. In the gymnosperms and some angiosperms this transfer, called **pollination**, is generally accomplished haphazardly by the wind. Large numbers of pollen grains must

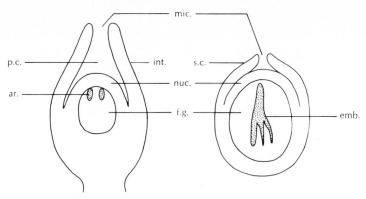

Fig. 17.1 Diagram of gymnosperm ovule and seed: ar., archegonium; emb., embryo; f.g., female gametophyte; int., integument; mic., micropyle; nuc., nucellus; p.c., pollen chamber; s.c., seed coat.

therefore be produced if any are to find their mark.

The sperms of most seed plants (ginkgo and the cycads are notable exceptions) are

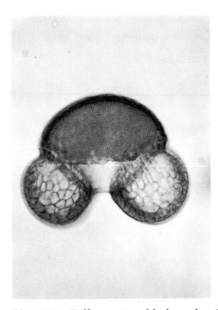

Fig. 17.2 Pollen grain of lodgepole pine, *Pinus contorta*, much enlarged. Pines are pollinated by wind, and the two air sacs contribute to the buoyancy of the pollen grain. [Photo courtesy of Alan Graham.]

delivered to the egg through an outgrowth from the pollen grain, called the **pollen tube** (see Fig. 17.15). There is no definite antheridium, and sooner or later the sperms are free in the cytoplasm of the tube. The pollen tube digests its way through the tissue enclosing the female gametophyte, and breaks open at the end, delivering the sperms.

The ripened ovule is called a **seed**, and the integuments of the ovule become the **seed coat.** The mature seed characteristically represents three successive generations: (1) the integuments and nucellus of the old sporophyte; (2) the female gametophyte; and (3) the embryo sporophyte. The nucellus and the body of the female gametophyte may or may not be evident in the mature seed, depending on the taxonomic group. The micropyle of the ovule is often visible as an imperfection in the mature seed coat.

The ovules and young seeds of Magnoliophyta (flowering plants) are characteristically enclosed within a structure called an ovary, and the flowering plants are therefore called **angiosperms** (Greek *angeion*, vessel, or enclosing vessel, and sperma, seed). The ovules and young seeds of Pinophyta are exposed to the air, so that the pollen grains can be carried directly to the micropyle. The Pinophyta are therefore commonly called **gymnosperms** (Greek *gymnos*, naked). This basic difference between gymnosperms and angiosperms was first pointed out in 1825 by the Scottish botanist Robert Brown (Fig. 17.3), but it took several decades before botanists in general were convinced that the gymnosperms and angiosperms should be treated as separate, coordinate groups of seed plants.

Gymnosperms always have a multicellular female gametophyte, consisting of several hundred to many thousand cells, which develops into the food-storage tissue of the seed. We shall see in Chapter 25 that the angiosperms, on the contrary, have a much reduced and modified female gametophyte that does not develop into a food-storage tissue. Other differences between the two groups, reflecting in general the more advanced evolutionary status of the angiosperms, are best considered in connection with the latter group.

BROAD CLASSIFICATION OF GYMNOSPERMS

The modern gymnosperms are an evolutionary remnant of a once much larger and more diversified group. There are less than 600 modern species in all, and about 300 of these belong to the single order

Fig. 17.3 (*Left*) Robert Brown (1773–1858), Scottish botanist and naturalist, who pointed out the distinction between gymnosperms and angiosperms. [Courtesy of the New York Botanical Garden.] (*Right*) Charles B. Beck (1927–   ), American paleobotanist, outstanding modern student of fossil gymnosperms.

Pinales. Some of the modern species of Pinales are widespread, abundant, and economically important, as noted on later pages, but the present-day role of gymnosperms in the vegetation of the earth cannot compare to their role in the Carboniferous, Permian, Triassic, Jurassic, and Lower Cretaceous periods. Any proper understanding of the gymnosperms therefore requires considerable attention to the fossil members.

There are three well-marked major groups of gymnosperms, here treated as the subdivisions Cycadicae, Pinicae, and Gneticae. The Cycadicae and Pinicae have been distinct at least since the Lower Carboniferous period, and they have in recent decades often been considered to constitute separate divisions, under the names Cycadophyta and Coniferophyta. Discovery and study of fossils during the 1960s, especially by the American paleobotanist Charles B. Beck (1927–   , Fig. 17.3), have led to the growing conviction that the Pinicae and Cycadicae both evolved from a group of Upper Devonian, heterosporous, fernlike (but woody) plants that were themselves not much advanced beyond the ancestral rhyniophytes. These probable ancestors of the gymnosperms have come to be called **progymnosperms** by many botanists. It is sometimes convenient to use the English words **cycadophytes** and **coniferophytes** for the Cycadicae and Pinicae, without implying that they should have divisional status.

The Gneticae consist of only three genera (*Gnetum*, *Ephedra*, Fig. 17.4, and *Welwitschia*) so different from each other, in spite of a series of technical similarities, that botanists now customarily put each genus in an order by itself. The Gneticae have a negligible fossil record and appear to be much more recent than

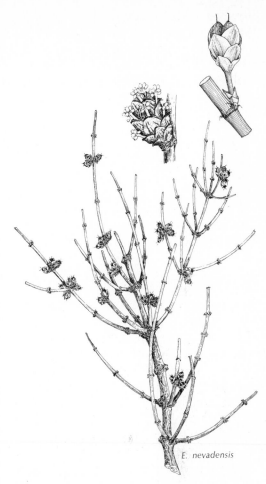

E. nevadensis

Fig. 17.4 *Ephedra nevadensis.* Habit, × 1/3; male and female cones, × 1 3/4. [Drawings by Kay H. Thorne (habit) and Alma Hochhauser (cones), from *Intermountain flora,* vol. 1, Hafner, New York, 1972, copyright by the New York Botanical Garden.] The source of the decongestant drug ephedrine, *Ephedra* has also been used locally for beverages, whence the name Mormon tea, used for it in western United States.

the other two main groups of gymnosperms. Opinion is still divided as to whether they are more nearly related to the Cycadicae or the Pinicae. Although they are of considerable interest to botanists, they have never played an important role in the vegetation of the world, and they are not further discussed here.

PINICAE

General Features

The Pinicae (coniferophytes) are gymnosperms with usually freely branched, woody stems, vigorous secondary growth, and mostly simple, usually small, leaves. In some of the fossil members of the order Ginkgoäles the leaves are dichotomously several times divided, but the pinnately compound leaves that characterize most of the Cycadicae are unknown in the Pinicae.

The male and female reproductive structures of coniferophytes are borne separately on the same or different individual plants. The organs on which the microsporangia are borne are commonly called **microsporophylls,** although it may reasonably be doubted that the modified telomes or telome systems from which they evolved were ever proper leaves. These microsporophylls are aggregated onto an axis to form a simple strobilus (cone), roughly comparable to the cones of horsetails and club mosses. The seed-bearing organs of coniferophytes are variously modified but never distinctly leaf-like. Most often they are arranged in cones of complex structure, the relatively large ovule-bearing scales being axillary to the smaller primary scales.

Classification

There are about 400 living species of Pinicae. These may be classified into three orders: the Pinales (often called Coniferales), Taxales, and Ginkgoäles. A fourth order, the Cordaitales, is known only from fossils. The Pinales, Taxales, and Cordaitales are more similar among themselves than any one of them is to the Ginkgoäles. This similarity is expressed by grouping the first three orders into a class Pinopsida, in contrast to the class Ginkgoöpsida, which contains only the order Ginkgoäles. The fossil record suggests that the Ginkgoöpsida and Pinopsida have been distinct since their origin from progymnosperms in the Upper Devonian period. *Ginkgo biloba* (see Fig. 2.2), the only living species of Ginkgoöpsida, is the last remnant of a long evolutionary line.

Within the class Pinopsida, the order **Cordaitales** is clearly the oldest group, originating from progymnosperms toward the end of the Devonian or the beginning of the Carboniferous period and continuing into the Permian period. The Cordaitales (Fig. 17.5) were trees up to about 30 m tall, with long, narrow, undivided leaves sometimes as much as 1 m long. The leaf venation was dichotomously branched and free (without cross-connections), as in some of the modern conifers such as *Agathis* and *Araucaria*. The trunk had a large pith, commonly 1 to 4 or even to 10 cm thick. The seeds were borne in cones that differ somewhat in structure from cones of other gymnosperms.

The Pinales (conifers) arose from the Cordaitales, probably during the Upper Carboniferous period. The Taxales diverged somewhat later, probably in the Triassic period, from the Cordaitales or from primitive Pinales. Both the Pinales and the Taxales have a very slender pith, in contrast to the thick pith of the Cordaitales. The Taxales, exemplified by *Taxus* (Fig. 17.6), the yew, have a well-developed fleshy covering surrounding the individual seeds, which are terminal

Fig. 17.5 Reconstruction of *Cordaites*. [Courtesy of the Field Museum of Natural History.]

or subterminal on short, axillary shoots. The Pinales, on the other hand, have dry seeds borne in cones.

Pinales

Distribution, Habit, and Age. The Pinales, or conifers (often called Coniferales) are the largest and most familiar group of modern coniferophytes. They occur from the Arctic to the Antarctic Circle, but are most common in temperate and cold-temperate regions, especially in the Northern Hemisphere, where they form extensive forests. The largest genus is *Pinus* (pine) (Fig. 17.7) with about 90 species, confined to the Northern Hemisphere. *Araucaria* is a common genus of the Southern Hemisphere; various species are called Chilean pine, Brazilian pine, Norfolk Island pine, and monkey puzzle. Some other familiar genera of the Northern Hemisphere are *Abies* (fir), *Cedrus* (Fig. 17.8, true cedar, including the cedar of Lebanon), *Juniperus* (juniper, "cedar"), *Larix* (larch), *Picea* (spruce), *Pseudotsuga* (*P. menziesii* is the Douglas fir), *Thuja* (*T. plicata* is the western red cedar), and *Tsuga* (hemlock).

Conifers are usually trees, less often shrubs, typically with **excurrent branching** (i.e., with a well-defined central axis and smaller lateral branches). Individual trees of *Sequoiadendron giganteum*, the California bigtree or Sierra redwood, are the largest living things, reaching a height of nearly 100 m and a diameter of more than 10 m.

All woody plants have an inherently limited life span. Sooner or later growth slows down and the tree becomes more susceptible to attack by natural enemies, such as fungi and insects. Ages of 100 to 200 years are common among the conifers, and some species characteristically live much longer. The Sierra redwood, which has very few natural enemies, sometimes reaches an age of 4000 years. The Great Basin bristlecone pine (*Pinus longaeva*, see Part I halftone) reaches an even greater age, approximating 5000 years.

Stem and Root. Stems and roots of conifers are basically similar to those of woody dicotyledons, as described in

Fig. 17.6 Yew, *Taxus baccata;* a branch with berrylike seeds. (About half natural size.) [New York Botanical Garden photo.]

Chapters 19 and 20, although the conifers are anatomically less complex in some respects. Most notably, the conifers do not have the vessels that commonly occur in angiosperm xylem, and they also lack the companion cells found in angiosperm phloem. On the other hand, conifer wood (xylem) commonly does contain specialized resin ducts, which are infrequent in angiosperm wood. When the tree is injured, the resin flows out over the injured surface, sealing it off from possible infection.

Leaves. The leaves of conifers are always simple and usually very narrow, although some of the genera that occur chiefly in the Southern Hemisphere, such as *Agathis* and *Araucaria*, have relatively broad leaves with numerous subparallel, dichotomously branched veins. Typically the leaves are from 1 to 10 or 15 cm long and less than 3 mm wide. Such leaves are called **needles** (Fig. 17.9). In most junipers, and in the various other native American conifers that are called cedars,

as well as in some other genera, the leaves of mature plants are reduced to small green scales appressed to the twigs and covering them.

Conifer leaves are typically alternate in a rather close spiral, less often opposite or whorled. In the larches, the true cedar, and some other genera, most of the leaves are borne on specialized short spur shoots, although some of them are on the more elongate ordinary shoots. In *Pinus* these short spur shoots have become further specialized, and all the foliage leaves (except in seedlings) are borne on them. Each spur shoot of *Pinus* bears a bundle of one to five needles (Fig. 17.10), according to the species. The bundle is surrounded at the base by a bundle sheath consisting of several small leaves that are modified into thin, membranous scales wrapped around the spur shoot. The whole spur shoot with its attached needles is eventually deciduous.

Conifers produce some new leaves every year, but the duration of the leaves varies with the genus and the species. In

Fig. 17.7 Longleaf pine, *Pinus palustris*, in North Carolina.

Fig. 17.8 Atlas cedar, *Cedrus atlantica*, in the Botanical Garden of Geneva, Switzerland.

Larix and *Pseudolarix* they die and fall off after a single season, and the tree is bare during the winter. In *Taxodium* and *Metasequoia* the leafy twigs of the season fall off in the autumn, and the tree is also bare in the winter. In other conifers the leaves of one year persist at least until the leaves of the next year are developed, and the tree is evergreen. Sometimes the leaves live as long as 10–15 years.

Cones. The reproductive structures of conifers are borne in unisexual cones (strobili). Most genera are monoecious, with both kinds of cones borne on the same tree, but *Juniperus* and some other genera are dioecious, with male and female cones borne on separate trees.

The male cones (Fig. 17.9) are axillary or terminal on short branches. They are usually about 1 cm long or less, ranging

Fig. 17.9 Twig of white pine, *Pinus strobus*, in northern New York, showing male cones and fascicles of needles. The new leaves of the year are just beginning to expand, above the male cones. (About 2/3 natural size.)

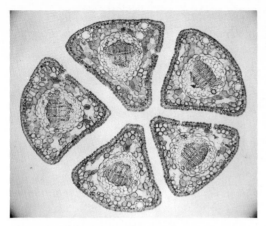

Fig. 17.10 *Pinus strobus* needles, in cross section. [Courtesy CCM: General Biological Inc., Chicago.]

phyll, and range from about 2 to 15 in number, 2 being the most common number in the familiar genera of the Northern Hemisphere. The microsporangium contains numerous microspore mother cells, which undergo reduction division to produce tetrads of microspores. The microspores have n chromosomes and are the first cells in the male gametophyte generation. The most common n number in conifers is 12.

The female cone is generally a compound structure, the ovules being borne not on the primary appendages (the **bracts**) of the cone axis, but on outgrowths (the **ovuliferous scales**) from the axils of the bracts. The ovules are mostly on the upper (adaxial) side of the ovuliferous scale, usually two per scale. In *Pinus* (Fig. 17.11) the bract is much shorter than the ovuliferous scale which it subtends, although it is readily visible when the cone has opened. In *Pseudotsuga* (Fig. 17.12) the bract is longer than the ovuliferous

up to 10 cm or more in some species of *Araucaria*. The microsporophylls are attached directly to the central axis of the cone in a close spiral or a series of whorls. The microsporangia are borne on the lower (abaxial) surface of the microsporo-

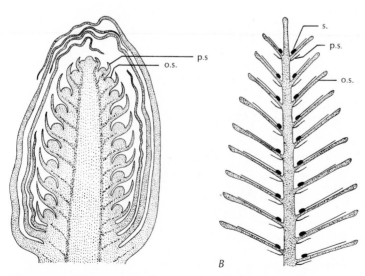

B

Fig. 17.11 Diagrammatic long section of (*A*) young pine cone (× 20); (*B*) mature pine cone (× 1): o.s., ovuliferous scale; p.s., primary scale; s., seed.

Fig. 17.12 Twig of Douglas fir, *Pseudotsuga menziesii*, showing male and female cones. (About 1/2 natural size.) [U.S. Forest Service photo by W. D. Brush.]

scale, but narrower and thinner. In *Araucaria* the bract is the principal cone scale, and the "ovuliferous scale" of other conifers is reduced to a mere "ligule" on the upper surface of the bract. (Technically, the ligule is the free tip of the ovuliferous scale, the base of this scale being fused to the bract.) In *Agathis* the ligule is obsolete, and the cone appears to be simple in structure like the male cones. The female cones of *Juniperus* and some related genera are small and superficially berrylike, with only a few small, thickened cone scales, but the basic morphology is the same as in other conifers.

The female cone of conifers is usually interpreted as comparable to a whole cone cluster of the Cordaitales, with each small cordaitalean cone having been reduced to a single ovuliferous scale in the axil of a

bract. A number of fossils provide intermediate steps that bolster this interpretation.

The conifer ovule (Fig. 17.13) consists of a megasporangium plus an enclosing tissue, the integument. The integument fits closely around the megasporangium wall (the nucellus) except at the end opposite the stalk, and often these two tissues are fused toward the stalk end. At the end opposite the stalk there is a cavity, the pollen chamber, between the nucellus and the integument. The pollen chamber opens to the surface via a passageway (the micropyle) through the integument.

The ovule contains a single megaspore mother cell, which undergoes meiosis to form a tetrad of megaspores (Fig. 17.14). Usually only one megaspore is functional, and the others soon degenerate. The megaspore has *n* chromosomes and is the first cell of the female gametophyte generation.

Pollen. The conifer microspore has a

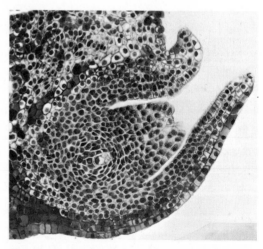

Fig. 17.13 Long section of young pine ovule, showing megaspore mother cell; micropyle and integuments ascending at the right. [Courtesy of Triarch Incorporated, Box 98, Ripon, Wis. 54971.]

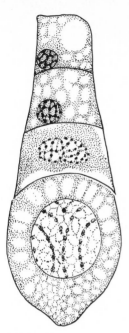

Fig. 17.14 Linear tetrad of megaspores in *Pinus laricio.* (× 1000.) Only the lowest megaspore is functional. [After Ferguson.]

two-layered wall. Usually one or more nuclear divisions occur inside the microspore wall before the young male gametophyte, now called the pollen grain, is shed from the microsporangium. In *Pinus* (Fig. 17.2) and some other genera, the outer wall of the pollen grain bulges away from the inner wall to form two large air sacs, increasing the volume of the grain without changing its weight, thus contributing to its buoyancy and facilitating its distribution by the wind. All conifers are normally wind pollinated, and some of them produce tremendous quantities of pollen.

The pollen grain of *Pinus* (Fig. 17.15) has two small cells, the prothallial cells, at one side; a somewhat larger cell, the generative cell, lies next to the prothallial cells; and a still larger tube cell occupies the remainder of the space enclosed by the wall of the pollen grain. The prothallial cells soon degenerate, and they play no part in the further development of the gametophyte; they are mere evolutionary vestiges of the vegetative body of the gametophyte.

Female Gametophyte. The megaspore in the conifer ovule gives rise by a series of mitotic divisions to the female gametophyte, which usually has several thousand cells. The archegonia, usually several in number, are borne at or near the nucellar end of the gametophyte (i.e., the end where the nucellus separates the gametophyte from the pollen chamber.) Each archegonial initial gives rise to a single large egg, a ventral canal cell (or nucleus), and several (typically eight) neck cells. The well-defined jacket of the archegonium is derived by differentiation from ordinary vegetative cells of the gametophyte.

Pollination and Fertilization. When the ovule is ready for pollination, some of the cells of the nucellus degenerate, forming a mucilaginous pollination drop that reaches the orifice of the micropyle. Pollen grains that land on this pollination drop are drawn down into the pollen chamber as the drop dries up, and the pollen grain comes to rest on the nucellus. Germination of the pollen grain to form a pollen tube typically follows. In some species, pollen grains that come to rest on the ovuliferous scale not too far from the micropyle can germinate and function as well as if they were in the pollen chamber.

The pollen tube is formed by an outgrowth from the pollen grain on the side opposite the prothallial cells. The nucleus of the tube cell moves into the pollen tube and, if it has not already done so, the generative cell divides, forming a stalk cell (next to the prothallial cells) and a

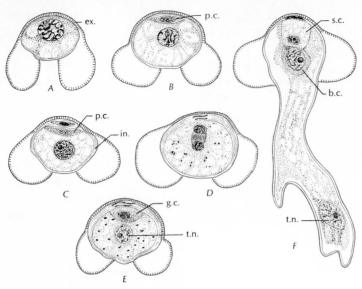

Fig. 17.15 Pollen development in *Pinus laricio*: b.c., body cell; ex., exine; g.c., generative cell; in., intine; p.c., prothallial cell; s.c., stalk cell; t.n., tube nucleus. (× 500.) [After Coulter and Chamberlain, *Morphology of gymnosperms*, 1910, courtesy of the University of Chicago Press.]

body cell. The body cell divides again, forming two sperm cells which are released into the pollen tube. The sperms do not have a well-defined cell wall, and often their cytoplasm, if any, cannot be distinguished from that of the pollen tube. In any case they have no cilia or flagella (in conifers).

The pollen tube continues to grow slowly through the nucellus, releasing digestive enzymes into the surrounding tissue and causing it to degenerate. After a period of from a few weeks to more than a year, the pollen tube reaches the female gametophyte. The tip of the tube bursts, discharging the sperms next to the egg. Fusion of a sperm and an egg constitutes fertilization. The fertilized egg has 2n chromosomes and is the first cell of the sporophyte generation. Usually the fertilization of an egg in one archegonium inhibits fertilization and further development in other archegonia of the same ovule.

Seeds and Seedlings.   The fertilized egg gives rise to the embryo of the seed by a series of mitotic divisions. The developing embryo has an elongate, usually coiled suspensor which pushes it deeper into the body of the female gametophyte. Later the suspensor tends to degenerate, forming a mere cap over the radicle of the embryo. In addition to the suspensor, the embryo in the mature seed consists of the radicle, the hypocotyl, the cotyledons, and the plumule (epicotyl), just as in angiosperms. The radicle is the embryonic root. The hypocotyl is the transition region, just beneath the cotyledons, between the stem and the root. The cotyledons are embryonic leaves, and the epicotyl is the

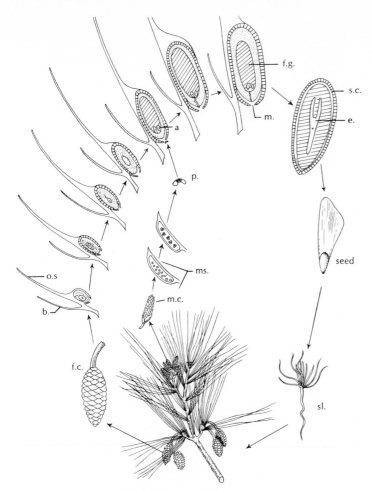

Fig. 17.16   Life cycle of *Pinus*: a., archegonium; b., bract (primary scale); e., embryo; f.c., female cone; f.g., female gametophyte; m., micropyle; m.c., male cone; ms., microsporophyll; o.s., ovuliferous scale; p., pollen grain; s.c., seed coat; sl., seedling.

embryonic shoot. Most conifers have several cotyledons (unlike angiosperms), but some have only two or three.

The mature embryo is embedded in the body of the female gametophyte, which is the principal food-storage tissue of the seed. (In this respect the seeds of conifers and other gymnosperms differ from those of angiosperms.) The nucellus tends to degenerate as the seed ripens, but a vestige of it is often visible as a cap over the micropylar end of the gametophyte. The seed coat, developed from the integument of the ovule, is usually dry and firm. Often the seed has a rather large, flattened, winglike extension at the distal (upper) end. The wing appears on first inspection to be an expanded part of the seed coat, but it actually develops from the tissue of the ovuliferous scale.

Conifer seeds are eaten by birds, squirrels, and other animals. The seeds of the piñon pine (*Pinus edulis* and related species) of the southwestern United States and northern Mexico are quite palatable, and some Old World species also have relatively large, edible seeds.

The seeds of some conifers are shed as soon as they are ripe, and are ready to germinate immediately. Others require a rest period of weeks, months, or even years. The cones of some species, such as *Pinus contorta* (lodgepole pine) of the western United States and Canada and *P. banksiana* (jack pine), its eastern counterpart, commonly remain on the tree for several or many years, opening only very tardily. After a forest fire, the cones release their seeds, and even-aged stands of these species usually indicate a previous fire.

The radicle of a germinating conifer seed elongates and pushes through the seed coat to form the primary root of the seedling, after which the hypocotyl elongates, carrying the seed coat, cotyledons, epicotyl, and the remains of the female gametophyte up into the air. The tips of the cotyledons remain enclosed within the seed coat for some time before it is pushed off. The cotyledons are generally green and much like ordinary small needles. The first true leaves (after the cotyledons) of all the needle-leaved and scale-leaved conifers have the needle form and are borne separately on the main shoot axis, regardless of their form and position in mature trees.

Fossils.   The conifers are well represented as fossils from the Upper Carboniferous to the present. The Paleozoic conifer fossils, such as *Lebachia*, are vegetatively much like the modern forms, but their cones are intermediate to the cordaitalean type. Except in the structure of the female cones, the Araucariaceae are among the most primitive modern conifers, and fossil wood (*Araucarioxylon*) resembling that of *Araucaria*, from rocks of Mesozoic and more recent age, is very similar to that of the Paleozoic Cordaitales.

Economic Importance

The Pinicae, particularly the conifers (Pinales), are highly important to modern society, although perhaps not essential. They are the principal source of lumber for a wide variety of uses, from matches to construction timbers, and from shingles and siding to boxes and cedar chests. The western American species *Pseudotsuga menziesii*, the Douglas fir (Fig. 17.17), is one of the most important timber trees in the world. Many species of *Pinus* are extensively logged, and various kinds of *Abies*, *Larix*, *Picea*, *Thuja*, and *Tsuga*, among others, are also much used for lumber in North America. In the Southern Hemisphere *Agathis* and *Araucaris* are similarly important.

Wood pulp for use in making paper is obtained from conifers as well as from some angiosperm trees, such as aspen and birch. Tremendous quantities of the rather small trees that make up the boreal spruce-fir forests are used in the production of newsprint. The smaller pines of the southeastern United States have also come into prominence for this purpose, now that a satisfactory process has been devised to remove the resin from the pulp and permit the production of a white paper. Much of the pulp used in the manufacture of rayon is obtained from spruce wood.

Fig. 17.17 Douglas fir more than 75 m tall, in Oregon. [U.S. Forest Service photo by Weldon Heald.]

Crude **turpentine** is a mixture of resins obtained from certain pines and some other conifers. The tree is tapped by slicing into the bark, and the liquid exudate is collected in a cup attached to the tree. The mixture gradually solidifies when exposed to air. On distillation it yields oil of turpentine and a hard, amber-colored resin called rosin. Oil of turpentine is the principal solvent for most paints and varnishes and has some minor medical uses related to its stimulant, astringent, antispasmodic, and diuretic properties. Rosin is an ingredient of varnishes, paints, soaps, and a variety of other products; it is used on bowstrings and baseballs to increase surface friction, and in paper to impart gloss and water resistance. In the days of sailing ships, crude turpentine was extensively used in caulking and waterproofing, and turpentine products are still known as **naval stores.** In the United States the southeastern species *Pinus palustris* (longleaf pine), *P. caribaea* (slash pine), and *P. taeda* (loblolly pine) are the principal sources of crude turpentine.

Most of the ornamental evergreen trees and shrubs planted in the United States belong to the Pinicae. Various species of *Pinus, Picea, Abies, Tsuga, Thuja, Juniperus, Chamaecyparis,* and *Taxus,* among others, are familiar sights around homes throughout most of the United States, and well-formed individuals of *Ginkgo biloba* are among the most beautiful of all specimen trees.

CYCADICAE

General Features

The Cycadicae (cycadophytes) are gymnosperms with more or less fernlike, usually large and compound leaves that are typically borne in a crown atop an unbranched stem (Fig. 17.18). The stem has a

Fig. 17.18 *Encephalartos transvenosus,* a cycad, in Transvaal, Republic of South Africa. [Photo by W. H. Hodge.]

large pith, a well-developed cortex, and rather scanty vascular tissue. There is usually some secondary thickening, but the cortex remains a prominent part of the mature stem. The seeds are borne on scarcely to strongly modified megasporophylls which, in most modern species and some of the fossils, are aggregated into a simple terminal cone.

Classification

There are four principal orders of cycadophytes, which are here considered to constitute three classes. The orders Lyginopteridales (also called Cycadofilicales) and Caytoniales belong to the class Lyginopteridopsida. The class Bennettitopsida has only the order Bennettitales (also called Cycadeoidales), and the class Cycadopsida has only the order Cycadales. The order Cycadales is represented by both fossil and living species; the others are known only from fossils of Mesozoic or late Paleozoic age.

Lyginopteridales

The Lyginopteridales, often called **seed ferns** or **pteridosperms,** are fernlike plants with seeds (Fig. 17.19). They are known from rocks of Lower Carboniferous (or possibly late Devonian) to Jurassic age, being most abundant in the Carboniferous. Pteridosperms were among the most important contributors to the masses of decaying vegetation in Carboniferous time that became transformed into coal, and most of the numerous fossil leaves that have caused the Carboniferous period to be called the Age of Ferns are actually pteridosperms. Pteridosperm leaves were evidently firmer than most fern leaves, and the cuticle is more resistant to alteration under chemical treatment than the cuticle of modern or fossil ferns. In these respects the seed ferns resemble the modern cycads.

Bennettitales

The Bennettitales are a group of fossil cycadophytes known from rocks of Triassic to Cretaceous age. They were evidently common during most of the Mesozoic era; and it is this group, rather than the true cycads (Cycadales), that has caused the

Fig. 17.19   A seed fern, reconstructed.

Mesozoic era sometimes to be called the Age of Cycads. The Bennettitales, often called **cycadeoids,** are an offshoot of the pteridosperms, characterized by their well-developed, structurally complex cones with highly modified megasporophylls. It is generally agreed that the Bennettitales have no modern descendants.

Cycadales

The Cycadales, or cycads, are cycadophytes with the microsporophylls simple and aggregated into a strobilus. There are 9 modern genera with about 65 species, all restricted to tropical and subtropical regions. *Zamia floridana* (Fig. 17.20), of Florida, is the only cycad native to the United States. Other species of *Zamia* occur from Mexico and the West Indies to northern South America. Fossil cycads are known from as far back as the early Permian, and possibly the late Upper Carboniferous. It is thought that the cycads evolved from the seed ferns.

Cycads are often much like small tree ferns in appearance, with a simple trunk and a crown of pinnately compound leaves, but the leaves are much firmer than those of most ferns. The arborescent forms are usually called palms by nonbotanists, and even in areas where tree ferns are well known they are not generally called ferns.

In contrast to the Bennettitales, the cycads are strictly dioecious (Greek, two houses) that is, the megasporophylls and microsporophylls are borne on separate individuals. The Bennettitales, with both types of sporophylls on the same plant, were monoecious.

The modern cycads are not of much economic importance. They are often grown for ornament in warm regions, and *Cycas revoluta* is regularly cultivated in greenhouses for its attractive leaves, which are used in funeral wreaths and on Palm Sunday. Many cycads are poisonous, and *Dioön edule* is reported to cause serious losses of cattle in parts of Mexico. A special kind of flour, now called Florida arrowroot starch, is prepared from the thickened base of *Zamia floridana,* after the poisonous principle has been leached out.

Fig. 17.20 *Zamia floridana,* with female cones, in Florida. (*Inset*) Female cone of another species of *Zamia.* [Photos by W. H. Hodge.]

SUGGESTED READING

Allen, G. S., and J. N. Owens, The life history of the Douglas Fir, *Environment Canada,* Forestry Service, Ottawa, 1972.

Beck, C. B., Reconstruction of Archaeopteris and further consideration of its phylogenetic position, *Amer. J. Bot.* **49**:373–382, 1962. An important paper bearing on the evolutionary ancestry of the gymnosperms.

Denison, W. C., Life in tall trees, *Sci. Amer.* **228**(6):74–80, June 1973. An ecosystem outlook, with emphasis on conifers.

Schulman, E., and W. R. Moore, Bristlecone pine, the oldest known living thing, *Nat. Geographic* **113**:355–372, 1958. Good *National Geographic* style.

Sporne, K. R., *The morphology of gymnosperms,* Hutchinson, London, 1965. A textbook.

Chapter 18

Angiosperms: Introduction

Tulip tree, *Liriodendron tulipifera*, a native of eastern United
States. (About natural size.) *Liriodendron* belongs to the
Magnoliaceae, one of the most archaic families of
angiosperms. [Photo by F. E. Westlake, from National
Audubon Society Collection/Photo Researchers, Inc.]

The Magnoliophyta, commonly called the **angiosperms** or **flowering plants,** are the most complex as well as the most familiar of all plants. The name angiosperm (from Greek *angeion*, vessel, and *sperma*, seed) refers to the fact that the young seeds are enclosed within a special structure, the ovary, instead of being exposed directly to the air, as in gymnosperms (from Greek *gymnos*, naked). The divisional name **Anthophyta** (Greek *anthos*, flower, and *phyta*, plants) has sometimes been used for the Magnoliophyta, in reference to the characteristic flowers that are associated with seed production. Many angiosperms have large and showy flowers, but in others the flowers are small and inconspicuous. All the garden flowers and crops, the familiar broad-leaved trees and shrubs, the cereal grains and other grasses, and the ordinary garden and roadside weeds are angiosperms.

The two classes of angiosperms, the Magnoliopsida and Liliopsida, have often been called Dicotyledonae and Monocotyledonae, respectively, in reference to the seed structure as noted below. The English names **dicotyledons** and **monocotyledons,** or simply **dicots** and **monocots,** are well established in informal usage. The two groups differ in several respects, but no one of the differences is absolutely constant. The most consistent single difference is the one suggested by the names: In dicotyledons the embryo of the seed has two **cotyledons** (seed leaves), whereas in monocotyledons it has only one. These two classes are contrasted in Chapter 27, following the study of the structure and life cycle of angiosperms. Most of the characteristic features of the monocots are evolutionary modifications from the condition in dicots.

Because the angiosperms are the most common and familiar plants, botanical terminology tends to center around them. Many terms and concepts developed for the angiosperms are equally applicable to at least some of the other divisions of plants, especially to other vascular plants (i.e., plants with a well-developed conducting system); but often the structures of plants in these other groups differ from those of angiosperms, and it is sometimes difficult to decide just how far the terms should be stretched.

In angiosperms and nearly all other vascular plants, the vegetative plant body, which typically has 2n chromosomes, customarily consists of three general kinds of parts, called roots, stems, and leaves (Fig. 18.1). These may be considered the primary organs of the plant; other kinds of organs are modifications or appendages of these three. **Roots** are typically the organs that anchor the plant in the soil and absorb water and minerals. The **leaves,** which are generally thin and flat, are usually the principal photosynthetic organs of the plant, although the stems are often also photosynthetic. The **stems,** which are typically cylindrical and usually branched, display the leaves and the flowers and eventually the seeds. The stem and leaves are collectively called the **shoot.** Roots, stems, and leaves often have additional or different functions from those just listed, and they are sometimes so highly modified as to be recognizable only with difficulty. Sometimes one or another of these three basic organs is entirely absent. For example, the aquatic flowering plant bladderwort (*Utricularia*) lacks roots, and *Wolffia*, a highly reduced floating aquatic, lacks both roots and leaves.

Angiosperms differ greatly in length of life. Plants that live only one year are called **annuals;** plants that live two years are called **biennials;** and plants that live

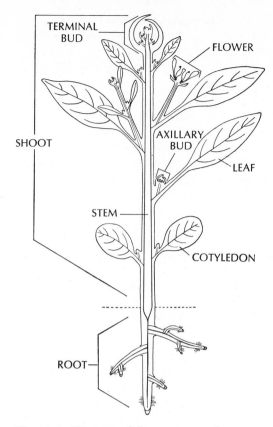

TERMINAL
BUD

FLOWER

SHOOT

AXILLARY
BUD

LEAF

STEM

COTYLEDON

ROOT

Fig. 18.1 Diagram of the structure of a typical dicot. [Adapted from Sachs.]

three years or more are called **perennials.** Typical annuals, such as corn and petunias, complete their life cycle in a single growing season; they come up from seed in the spring, mature within a few months or even weeks, and die. Winter annuals, such as winter wheat, come up in the fall and set seed the following year, but still complete their life cycle in less than 12 months.

TISSUES

Angiosperms are multicellular plants, generally with several different kinds of cells, which are organized into groups called **tissues.** The cells of any one tissue have a common origin and function. Tissues in which the cells are all essentially alike are called **simple tissues.** Tissues in which different kinds of cells contribute to a common function are called **complex tissues.** Any group of tissues which collectively form an externally differentiated body is called an **organ.** The primary organs of the vegetative body of an angiosperm are, as noted above, the roots, stems, and leaves.

Some kinds of tissues occur in all three kinds of organs; others are usually much more restricted. Such tissues as parenchyma, collenchyma, and sclerenchyma are defined largely by the nature of the cells of which they are composed; often these simple tissues are parts of complex tissues that contain several kinds of cells. Parenchyma, collenchyma, and sclerenchyma cells sometimes also occur singly in complex tissues of other types.

Tissues such as cortex, pericycle, and pith are defined by their position, regardless of the types of cells of which they are composed. These tissues are discussed in later chapters. Several tissues have both a characteristic position or positions and a characteristic structure. The most notable of these are epidermis, endodermis, meristem, xylem, phloem, and cork.

Parenchyma

Most tissues are specialized for one or sometimes two or more functions, the cells having a characteristic form and structure according to their function. Tissues composed of relatively unspecialized cells are called **parenchyma.** The wall of a parenchyma cell is usually thin, without a secondary layer, and is usually composed largely of cellulose and pectic compounds. The cell is living when functional and generally has a well-developed central vacuole. Parenchyma cells vary in shape, but they are seldom much elongate; often they are more or less isodiametric, although this tends to overstate the case. Adjacent cells are usually in contact over only parts of their surface, leaving large or small intercellular spaces.

Closely packed parenchyma cells are generally polyhedral, with a number of flat surfaces. Ideally, they have 14 surfaces, with 8 hexagonal and 6 square faces, but this form, called the *orthic tetrakaidecahedron* (Fig. 18.2) is seldom precisely realized. Experiments with bubbles and compressed shot by the American botanist Edwin B. Matzke (1902–1969) and his students indicate that, under conditions of mutual pressure, the orthic tetrakaidecahedron is the most economical shape, with the lowest surface-volume ratio, if no interspaces are to be left among bodies which in the absence of pressure would be spherical and of equal size.

Parenchymatous tissues (Fig. 18.3) generally occur in all organs of a plant, and they carry on a variety of functions. The name itself (from Greek *para*, beside, and *en* + *chein*, to pour in), reflects the ancient concept of parenchyma as something that is figuratively poured around the other, more specialized tissues in the various organs. Morphologically, physiologically, and phylogenetically, parenchyma is the basic or fundamental kind of tissue, and the other types may be considered as modifications of it. Parenchyma cells often have chloroplasts, and most photosynthetic tissues are parenchymatous. Sometimes it is useful to distinguish chlorophyllous parenchyma under the name chlorenchyma.

Collenchyma

Collenchyma (Fig. 18.3) is a living tissue composed of more or less elongate cells with thick primary walls that consist largely of cellulose, along with the usual pectic compounds. The protoplast has a well-developed central vacuole, and the

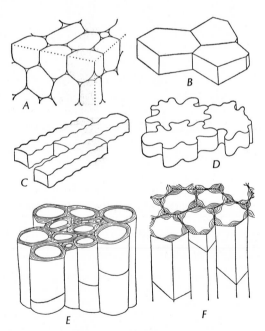

Fig. 18.3 Some types of tissues, showing only the cell walls or cell outline, without contents. *A*, Parenchyma; *B-D*, epidermis; *E*, *F*, collenchyma.

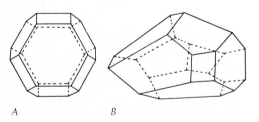

A   B

Fig. 18.2 (*A*) An orthic tetrakaidecahedron, the theoretically ideal shape of parenchyma cells. (*B*) The shape of an actual parenchyma cell from a dicot stem, showing an approach to the orthic tetrakaidecahedron. [After Matzke.]

cytoplasm often contains chloroplasts. The cells are typically prismatic, but the more elongate types, which are sometimes as much as 2 mm long, taper at the ends. The thickening of the wall is usually unequally distributed, often chiefly in the angles of the cell. The term *collenchyma* (from Greek *kolla,* glue) refers to the characteristically glistening wall.

The principal function of collenchyma is mechanical strengthening. It is usually the first supporting tissue in young stems and leaves, and it is important in many mature ones as well. It is much less often found in roots. The cells combine considerable tensile strength with flexibility, and they can continue to grow after the wall has thickened; they are, therefore, ideally adapted to the support of growing organs.

Sclerenchyma

Sclerenchyma (from Greek *skleros,* hard) is a tissue composed of thick-walled cells whose sole or principal function is mechanical strengthening. The cells have a definite secondary wall that is often lignified, and they commonly lack living contents at maturity. In both these respects sclerenchyma differs from collenchyma, another strengthening tissue. Sclerenchyma is distinguished from xylem (a complex tissue composed largely of thick-walled, lignified, dead cells) by its function and position. Xylem is important in the conduction of water and minerals as well as in mechanical support, whereas sclerenchyma has no conducting function.

Sclerenchyma cells (Fig. 18.4) vary greatly in size and shape. Long, slender ones are called **fibers,** a term also applied to long, slender xylem cells. (Xylem fibers might, in fact, be considered as scleren-chyma cells, although the term is seldom so applied.) Short or irregular sclerenchyma cells are called **sclereids,** or **stone cells,** although there is no sharp distinction between these and fibers. Sclereids of unusual shape that occur singly among cells of more ordinary type are often called **idioblasts.** Sclereids often occur as scattered cells in otherwise parenchymatous tissue, but fibers nearly always occur in organized groups. Collenchymatous or parenchymatous tissues sometimes become sclerenchymatous in age; that is, they develop secondary walls and thereafter tend to lose their protoplasts.

Epidermis

Epidermis (Fig. 18.3) is the characteristic outermost tissue of leaves and of young roots and stems. In older roots and stems, especially of woody plants, the epidermis is often replaced by other tissues. Epidermis consists of cells that fit closely together, nearly or quite without intercellular spaces except for a special type of intercellular space called the stomate. Stomates are found in the epidermis of most stems and leaves, but usually not in roots. They are discussed in Chapter 21. Epidermal cells vary in shape, but the inner surface is usually flat, and the outer surface flat or bulged. The depth of the cell, from interior to exterior, is usually less than any other dimension, and the cells are therefore said to be tabular in shape. The epidermal layer is typically only one cell thick, and (except for the cells bounding the stomates) the cells are in most cases all about alike. Sometimes, however, the epidermis is two to several cells thick, or composed of two or more kinds of cells, or both.

Epidermal cells are usually living when functional. They have a well-

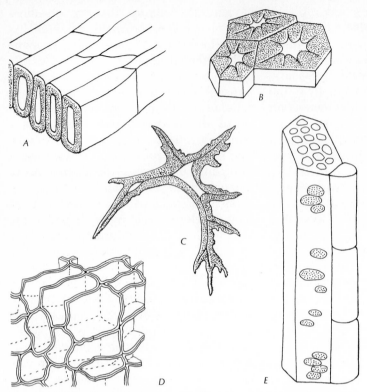

Fig. 18.4   Some types of cells. *A-C,* Sclerenchyma cells, with the walls stippled; *A,* fibers; *B,* stone cells; *C,* an idioblast; *D,* cork cells; *E,* sieve tube and companion cells of phloem, the sieve tube showing a well-developed sieve plate at the end, and some less well-developed lateral sieve areas.

developed central vacuole, and, except for the guard cells adjoining the stomates, they seldom have chloroplasts. The outer walls of epidermal cells of stems and leaves, but usually not of roots, are impregnated with a waxy material called **cutin.** The cutin also forms a continuous layer, the **cuticle,** over the outer surface of the epidermis. Cutin retards the passage of water and water vapor. Except for the cutin, the walls of epidermal cells are usually composed largely of cellulose and pectic compounds.

Epidermis of roots is chiefly an absorb-

ing tissue, absorbing water and minerals used by the plant. Epidermis of stems and leaves is largely a protective tissue, usually providing some mechanical strength as well as inhibiting the evaporation of water from internal tissues. Sometimes it has other functions as well.

Meristem

In most animals and in many of the less complex kinds of plants, all or nearly all living cells can divide, and growth is diffuse. In angiosperms and other vascular plants, however, cell division tends to be restricted to certain tissues that occupy particular places in the plant body. Any plant tissue characterized by cell division is called a **meristem.** Any stem, and each of its branches, is ordinarily tipped by a group of cells that make up the **apical meristem** (see Fig. 19.1). The main root and branch roots also have apical meristems. The apical meristem of a root might better be described as subapical, inasmuch as it is covered by a tissue called the rootcap. All tissues derived directly by differentiation from an apical meristem are called **primary tissues.** Mature primary tissues often retain some capacity for cell divisions, generally at a rate much slower than that in the apical meristem.

Increase in thickness of a stem or root, after the primary tissues have matured, is usually due to cell divisions from **lateral meristems** that lie within the stem or root, usually nearer to the surface than to the center. The most important lateral meristem is the **vascular cambium,** often called merely the cambium. The cells formed from the cambium mature into xylem and phloem, the characteristic conducting tissues of the plant. (Some primary tissues also consist of xylem and phloem.) Another kind of lateral meristem is the **phellogen,** or **cork cambium,** which produces cork and a parenchymatous tissue called phelloderm.

All tissues derived from lateral meristems are called **secondary tissues.** Many angiosperms, especially the monocots, lack a cambium and are composed wholly of primary tissues. Some tissues, such as the epidermis, are always primary tissues; others, such as phelloderm, are always secondary. Xylem and phloem may be either primary or secondary; the primary tissues of vascular plants characteristical-

ly include xylem and phloem, but these are also produced by cambial activity.

A third type of meristem, the **intercalary meristem,** is sometimes inserted between primary tissues in a position different from that of lateral meristems. The leaves of grasses, for example, commonly have an intercalary meristem at the base of the blade. Tissues derived from intercalary meristems are generally similar to those derived from apical meristems and are therefore considered to be primary tissues.

The characteristics of meristematic cells vary in different kinds of meristems and in different kinds of plants. Apical meristems typically consist of rather small, nearly isodiametric cells with thin cellulose walls, dense cytoplasm, rather large nuclei, and no central vacuole. The plastids are usually in the proplastid stage, and intercellular spaces are generally wanting. Cambium cells, on the contrary, are often much elongate, and they are more likely to be thick-walled than are the cells of an apical meristem.

Xylem

Xylem (Greek *xylon,* wood) (Fig. 18.5) is a complex tissue that functions in the conduction of water and minerals, and in mechanical support. Some food may also be stored in the xylem. The fundamental cell type of xylem is the **tracheid.** Tracheids are long, slender cells, tapered at the ends, with well-developed, lignified secondary walls, a well-developed lumen, and without living contents at maturity. The space enclosed by the wall of a cell, especially a dead cell, is called the **lumen.**

Tracheids are divided into several categories according to the pattern of the secondary wall. Tracheids in which the secondary wall consists of a series of

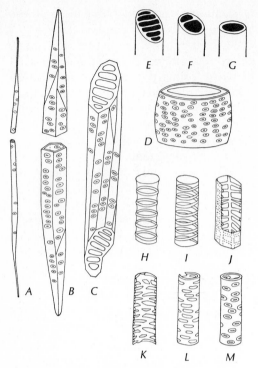

Fig. 18.5   Xylem cells. *A,* Fiber; *B,* tracheid; *C, D,* vessel elements; *E–G,* ends of vessel elements, with progressively less obstructed opening; *H–M,* types of secondary thickening of xylem cells; *H,* annular; *I,* spiral; *J,* scalariform; *K,* reticulate; *L,* reticulate-pitted; *M,* pitted.

separate rings are called annular tracheids; tracheids in which the secondary wall tends to form a continuous spiral are called spiral tracheids; tracheids in which the secondary wall tends to be arranged in the form of rungs and uprights of a ladder are called scalariform tracheids; those in which the secondary wall forms a continuous network are called reticulate tracheids; and those in which the continuous secondary wall is merely interrupted by small openings (pits) are called pitted tracheids. These several types grade into each other.

Annular and spiral tracheids can be stretched to some extent before breaking, and the first tracheids to mature near the growing points of stems and roots are usually of these types. In more mature tissues, the pitted tracheid is usually the more abundant type.

Evolutionary modification of the tracheid has proceeded in two opposite directions: (1) toward support at the expense of conduction, and (2) toward conduction at the expense of support. Cells exemplifying the former modification are called **xylem fibers;** those exemplifying the latter are called **vessel elements.** Xylem fibers are proportionately longer and more slender than tracheids, with a thick, merely pitted secondary wall and a small lumen. Fibers of more or less similar structure also often occur in other tissues but are not regarded as modified tracheids.

A series of vessel elements placed end to end constitutes a **vessel.** In its most typical form, a vessel may be compared to a series of barrels stacked end to end after the end walls have been knocked out. Vessel elements (Figs. 18.5, 18.6, 18.7) are usually broader and often shorter than tracheids, with a large lumen and often with a relatively thick secondary wall. The pattern in which the secondary wall is deposited undergoes the same variation as in tracheids, and the same types are recognized. Vessels are often as much as 1 m long, but they have complete end walls at the top and bottom.

Vessels are more effective conducting elements than tracheids, because water can move from one end of a vessel to the other without having to pass through a partition. There are all gradations between tracheids and vessel elements. Many vessels have slanting end walls with distinct perforations (Fig. 18.8). The

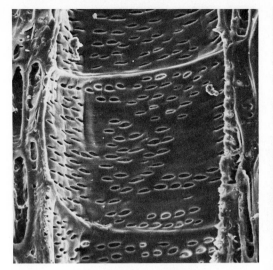

Fig. 18.6 Vessel elements of elm, *Ulmus*, with pitted walls. (× 500.) [Scanning electron micrograph courtesy of B. A. Meylan and B. G. Butterfield; from *Three-dimensional structure of wood*, copyright by Chapman & Hall Ltd., London, 1972.]

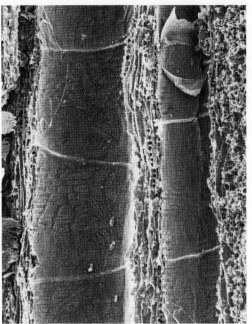

Fig. 18.7 Wood of oak, in tangential view, showing two vessels (the large tubes), some wood rays (end-on view, with isodiametric cells), and some fibers (small, vertically oriented cells, here indistinct). (× 75.) [Electron micrograph courtesy of Myron C. Ledbetter, Biology Department, Brookhaven National Laboratory.]

presence of these perforations in the end walls separating adjacent cells is the ultimate distinction between vessel elements and tracheids.

Tracheids, vessels, and fibers are all dead when functional. Xylem of stems and roots generally also contains some living cells. These are usually called parenchyma cells, although often they are not typical of parenchyma in all respects. The distribution of parenchyma cells in the xylem is discussed in later chapters, as are also the mechanics of water movement.

Phloem

Phloem (Greek *phloos*, bark) is a complex tissue that functions in the conduction of food and other dissolved substances, including minerals. Phloem is usually associated with xylem in plant organs, and together these two tissues constitute the vascular system of the plant.

The fundamental type of cell in the phloem is the **sieve element.** Sieve elements are long, slender, thin-walled cells which at maturity have cytoplasm and a central vacuole but no nucleus. The cytoplasm is continous from one sieve element to another through clusters of perforations in the wall, called sieve areas. These connecting strands of cytoplasm are hollow, lining the pores rather than filling them. The vacuolar contents are thus continuous from one sieve element to the next.

In angiosperms sieve elements are

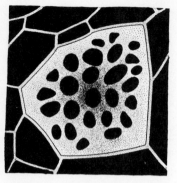

Fig. 18.9   Sieve plate of squash, *Cucurbita maxima,* with the outlines of surrounding cells. (× 100.) [From W. H. Brown, *The plant kingdom,* Ginn, Boston, 1935, 1963. Courtesy of M. A. Brown.]

Fig. 18.8   Simple perforation plate in vessel of *Knightia excelsa,* New Zealand honeysuckle. (× 600.) [Scanning electron micrograph courtesy of B. A. Meylan and B. G. Butterfield; from *Three-dimensional structure of wood,* copyright by Chapman & Hall Ltd., London, 1972.]

joined end to end to form **sieve tubes,** and the sieve areas tend to be concentrated on the slanting or transverse end walls of the sieve elements. These end walls with sieve areas are called **sieve plates** (Fig. 18.9). Any sieve areas that may be present on the lateral walls of the sieve elements are usually poorly developed as compared to the sieve plates.

The cytoplasm of sieve elements is in a degenerating condition, and it lacks a vacuolar membrane. The early metabolic processes and subsequent protoplasmic breakdown in the sieve tubes lead to the formation of a characteristic proteinaceous slime which lies in the vacuole. If the sieve tube is injured or killed, the slime quickly blocks the sieve plate, filling the pores.

In dicotyledons and many monocotyledons the degenerating protoplast includes some unusual proteinaceous fibrils, called P-protein. The function of these fibrils is still uncertain. Gymnosperms and vascular cryptogams do not have them.

The primary cell wall of sieve elements is composed largely of cellulose, and usually there is no secondary wall. **Callose** is deposited on and around the sieve areas, especially in the fall or after wounding, often eventually closing off the perforations. Callose is a high polymer of glucose, differing from cellulose in the way the glucose residues are attached to each other. Extensive accumulation of callose, forming a well-defined plug over the sieve plate, generally marks the end of the conducting activity of the sieve element. Thereafter the protoplasm tends to degenerate, and often the plug eventually separates from the sieve plate and disappears.

Sieve elements of vascular plants in general are ordinarily interspersed with more or less modified parenchyma cells that retain a functional nucleus. The cytoplasm of these parenchyma cells is continuous with that of the associated sieve tubes by numerous small plasmodesmata, and it has been suggested that the nucleus of the parenchyma cell serves for the cytoplasm of the associated sieve element as well as for its own cytoplasm. In nearly all angiosperms these parenchyma cells are rather highly specialized and have a direct ontogenetic relationship with the associated sieve elements. Modified parenchyma cells of this type are called **companion cells.** The sieve element and its associated companion cell(s) arise from the same individual meristematic cell. A single companion cell may extend the whole length of the sieve element, or there may be a vertical file of two to several companion cells next to each sieve element. The eventual death and degeneration of the protoplasm of the sieve element are usually correlated with the death and degeneration of the protoplast(s) of the associated companion cell(s).

In addition to the sieve tubes and companion cells, the phloem of angiosperms usually contains fibers, ordinary parenchyma cells, and sometimes other kinds of cells as well. The fibers are usually sclerenchymatous, with thickened, often lignified secondary walls, and without living contents at maturity; but phloem fibers with a persistent protoplast are also known.

The mechanics of movement of dissolved substances in the phloem are discussed in Chapter 24.

Cork

Cork (Fig. 18.10) is a waterproofing protective tissue, dead when functional, characterized by the **suberization** of the cell walls. The secondary wall consists largely of **suberin,** a waxy substance; the primary wall, originally composed chiefly of cellulose, tends to become impregnated with suberin as well. Cork cells are usually more or less prismatic; in cross sections of stems and roots, they often appear to be rectangular or box shaped, but careful studies show that they tend to be tetrakaidecahedral, with an average of 14 surfaces. Cork is ordinarily a secondary tissue, derived from a particular kind of meristem, called the **phellogen** or **cork cambium;** and cork cells tend to be arranged in regular rows aligned with the cells of the phellogen, nearly or quite without intercellular spaces. Cork commonly occurs in the outer part of woody stems and roots, forming or helping to form a protective outer covering after the epidermis has broken down.

SECONDARY METABOLITES

Most plants, or at least most of the vascular plants, produce several kinds of secondary metabolites. These are substances which do not appear to be necessary for the ordinary, routine metabolic functions that are common to plants in general. Secondary metabolites have little in common chemically. They include such diverse substances as alkaloids, terpenes, saponins, mustard oils, cyanogens, iridoid compounds, tannins, many flavonoids, and others. Each of these groups has many chemical variants. Terpenes, for example, are a chemical family of hydrocarbons, including such substances as rubber, resins, balsams, essential oils, and carotenoids.

Secondary metabolites were at one time thought to be mainly metabolic wastes, cast aside and often sequestered in special cells or ducts. More recently it has become clear that many, perhaps most, of them serve a protective function, discouraging potential predators and pathogens either by their unpleasant taste or by their poisonous quality or both.

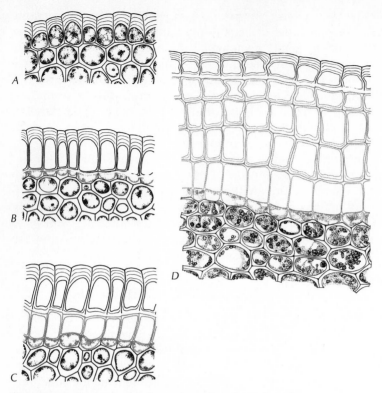

Fig. 18.10   Stages in the formation of cork in oleander, *Nerium oleander.* (*A*) Epidermis and adjacent cortex; the exposed outer surface of the epidermal wall is much thickened in this species. (*B*) The epidermal cells have elongated and divided to produce an internal layer of phellogen, next to the cortex. (*C*) The phellogen has produced a single layer of cork cells just beneath the epidermis. (*D*) The uppermost row of cells is epidermis, the next four rows are cork, and the sixth row is the phellogen. (× 300.) [From W. H. Brown, *The plant kingdom*, Ginn, Boston, 1935, 1963. Courtesy of M. A. Brown.]

They are weapons in the constant and shifting war between predator and prey.

Many plants have some specialized, elongate cells or chains of cells (**laticifers**) that contain latex, a colorless to more often white, yellow, or reddish liquid, often slightly viscous, characterized by the presence of colloidal particles of terpenes dispersed in water. The kinds and amounts of terpenes differ in different species, but rubber and resins are particularly common. The latex of commercial rubber trees (*Hevea brasiliensis*) contains up to 50 percent rubber. Latex may also contain other substances in addition to the terpenes.

Resin is frequently borne in intercellular channels (**resin ducts**) that originate either by separation of cells or by disintegration of cells. Tannins are often concentrated in tanniferous cells, and essential oils in oil cells.

The evolutionary significance of secondary metabolites is discussed in Chapter 30.

SUGGESTED READING

Ewan, J. (ed.), *A short history of botany in the United States,* Hafner, New York, 1969, Suitable for browsing.

Ray, P. M., *The living plant,* 2nd ed., Modern Biology Series, Holt, Rinehart & Winston, New York, 1972.

Chapter 19

Angiosperms: Stems

Three-dimensional structure of wood of *Laurelia novae-zelandiae*, a member of the archaic family Monimiaceae. (× 150.) Note vessels (large cells), with slanting, scalariform (ladderlike) perforation plates that have numerous cross-bars; fibers, the vertically elongate, thick-walled cells associated with the vessels; and wood rays, formed of radially elongate cells. [Scanning electron micrograph courtesy of B. A. Meylan and B. G. Butterfield; from *The structure of New Zealand woods,* Crown copyright, 1978.]

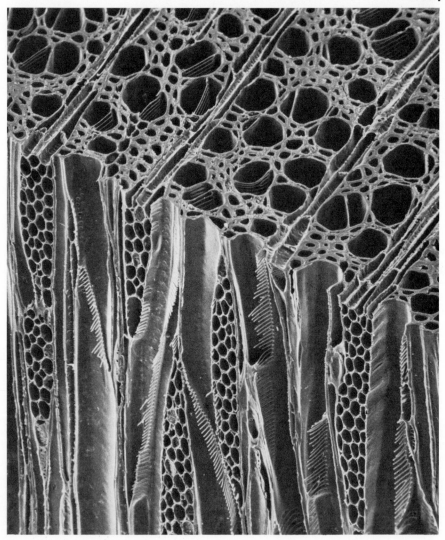

INTRODUCTION

Some External Features

Stems are the organs of a plant on which the leaves are borne. They are usually more or less round in cross section, commonly tapering toward the tip. The leaves are usually arranged in a regular pattern. A place on the stem where a leaf is (or has been) attached is called a **node,** and the part of the stem between two successive nodes is called an **internode.** The point of the angle formed by the leaf or leafstalk and the upward internode of the stem is called the **axil** of the leaf. Normally each axil bears a bud, the **axillary bud,** which is capable of developing into a branch shoot. Even when the leaves are highly modified, or much reduced and inconspicuous, the axillary bud, marking the node, is often evident. It may be said that stems characteristically have nodes and internodes, and that branch stems (of angiosperms) characteristically arise from axillary buds. Both of these features are in contrast to those of roots.

Apical Growth

The tip of the stem is ordinarily occupied by an **apical meristem.** In an actively growing stem, the apical meristem is protected by the developing young leaves which extend up and around it from below (Fig. 19.1). As these leaves expand and spread away from the stem with increasing maturity, new young leaves are progressively formed by the apical meristem, and these in turn protect the growing point for a time before they mature. The apical meristem and its protective leaves constitute the **terminal bud.**

The apical meristem of woody stems is often dormant for much of the year, and the dormant meristem is usually covered

Fig. 19.1   Long section of growing stem tip of *Elodea*. [Courtesy of E. J. Kohl, Lakeside Biological Products, Ripon, Wis.]

and protected by some modified, small, firm leaves, the **bud scales.** When the apical meristem resumes its growth after a period of dormancy, the bud scales spread away from the growing point and fall off, typically leaving a ring of bud-scale scars around the twig (Fig. 19.2). Twigs of many woody plants show successive rings of bud-scale scars marking the annual increments in length. There are all degrees of variation, from dormant buds that are merely covered by ordinary young leaves whose development is arrested, to the more common type in which the bud scales are distinctly different from ordinary leaves. Axillary buds of woody plants, as well as terminal buds, are commonly covered by bud scales.

Growth in length of the stem is due to cell divisions occurring in and near the apical meristem, followed by enlargement

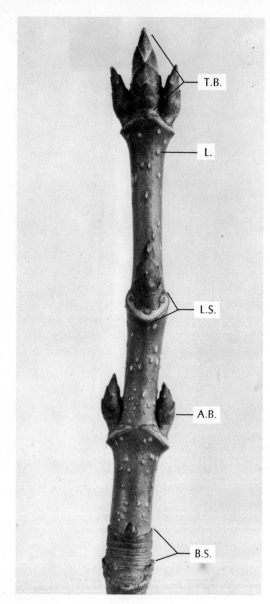

Fig. 19.2 Twig of sugar maple, *Acer saccharum*, in winter after the leaves have fallen: A.B., axillary bud; B.S., bud-scale scars; L., lenticel; L.S., leaf scar; T.B., terminal bud. [New York Botanical Garden photo.]

of the cells that do not remain meristematic. The apical meristem is thus constantly being carried upward and onward by the growth of the new cells it produces. The region just beneath the apical meristem, marked by rapid enlargement (especially elongation) of the cells, is called the **region of elongation,** or more precisely, the region of cell enlargement. The boundary between the apical meristem and the region of elongation is rather vague; some capacity for cell division is often retained even by mature primary tissues.

Leaves originate from the apical meristem of a stem, appearing first as mere bumps (the **leaf primordia**) on the flanks of the apical meristem. The axillary buds also originate from the apical meristem, usually after the formation of the subtending leaf primordia. Sometimes the formation of the axillary bud is delayed so long that it develops from essentially mature primary tissues of the stem just above the leaf. Under some circumstances, especially after injury, adventitious buds are produced away from the axils. The term **adventitious** is applied to structures that originate from mature nonmeristematic tissues, especially if such a development would not ordinarily be expected.

Each axillary bud is potentially the terminal bud of a new branch shoot. Sometimes the bud continues its growth without interruption, developing immediately into a leafy branch or one or another kind of modified shoot (such as a flower). More often the bud becomes dormant after developing to a length of usually less than 1 cm. It may then remain permanently dormant, or it may sooner or later resume active growth, depending on the species and the environmental conditions. The number of axillary branches on a plant seldom represents more than a small fraction of the total number of axillary buds.

The dormancy or activity of the axillary buds is governed largely by hormones produced by the plant. Actively growing shoots commonly produce hormones that tend to inhibit the growth of axillary buds. If the shoot tip is cut off, interrupting the supply of inhibitory hormone(s), one or more axillary buds usually begin to grow, forming a new shoot which in turn inhibits the development of other buds. The phenomenon is called **apical dominance.** Plant hormones are discussed further in Chapter 26.

Anatomy

The nature and arrangement of the primary tissues of the stems of angiosperms differ in different groups, but nearly all have several features in common. They are bounded externally by an epidermis. They contain a network of longitudinal strands of vascular tissue, the **vascular bundles,** embedded in a usually parenchymatous matrix, the **fundamental tissue.** Each vascular bundle contains both xylem and phloem, commonly with the xylem toward the center of the stem and the phloem toward the outside.

Stems in which the primary tissues are similar in composition and arrangement often differ in the amount and distribution of secondary tissues. Woody stems differ from herbaceous stems in the higher proportion of wood, that is, hard tissue composed largely of thick-walled cells. In woody dicots, as in conifers, xylem is the principal or only woody tissue, and the terms wood and xylem are often used interchangeably. Much of the wood of woody monocots, however, is not xylem (see p. 365).

HERBACEOUS DICOT STEMS

General Features

A cross section of a typical herbaceous dicot stem, before or soon after secondary growth has started (Fig. 19.3), shows a central pith successively surrounded by a ring of vascular bundles, a cortex, and an

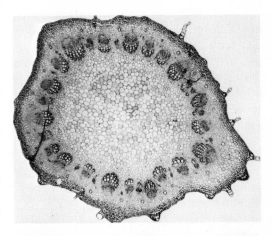

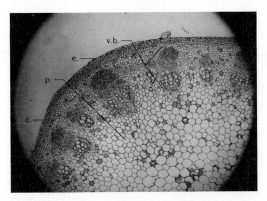

Fig. 19.3 Cross section of a sunflower stem. (*Above*) The whole stem. [Courtesy of E. J. Kohl, Lakeside Biological Products, Ripon, Wis.] (*Below*) A segment of the stem: c., cortex; e., epidermis; p., pith; v.b., vascular bundle, with the xylem toward the center and the phloem toward the outside. [Courtesy CCM: General Biological, Inc., Chicago.]

epidermis. All these are primary tissues or groups of tissues, formed by growth and maturation of cells derived from the apical meristem. The pith, the cortex, and the tissue surrounding the vascular bundles collectively constitute the fundamental tissue.

Pith

The pith is usually parenchymatous, with starch-forming leucoplasts (amyloplasts), but without chloroplasts. Sometimes, especially in age, it shows specialized cells in which tannins or crystals of one sort or another are deposited, or some of the cells may be modified into sclereids or other specialized types. Resin ducts, lacticifers, or both are also present in some species; in addition to their presence in the pith, these may occur in any internal tissue of a stem, root, or even a leaf.

The most common function of the pith is food storage. Often it loses all function and dies or degenerates as the plant matures.

Vascular Cylinder

The vascular bundles appear in cross section as separate units, but they actually fork and rejoin in a fairly regular pattern, forming a network in the shape of a hollow cylinder. This cylinder, together with the pith it encloses, is called the **central cylinder,** or **stele.** The spaces among the vascular bundles are filled with parenchyma, connecting the parenchyma of the pith with the parenchyma of the cortex. These interfascicular areas of parenchyma are called **medullary rays,** or **pith rays.** Often they are wider than the bundles themselves.

Some of the bundles diverge from the stele and pass out through the cortex into the leaves. These are the **leaf traces.** Often

there are three traces for each leaf. The place of a departing leaf trace in a stele is taken by a new bundle formed by the fusion of branches from adjacent bundles on each side.

The vascular supply of a branch shoot usually begins as two **branch traces** which arise from the vascular bundles of the stele in the same manner as the leaf traces. Typically the branch traces flank the leaf trace (or the central leaf trace, if there is more than one) and depart from the stele at nearly the same level.

Each vascular bundle contains xylem and phloem (Fig. 19.4). Typically the

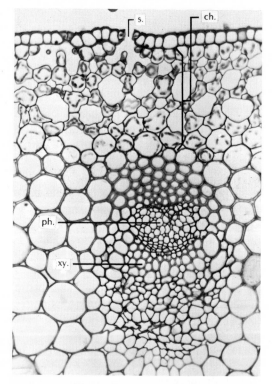

Fig. 19.4 Cross section of part of stem of buttercup, *Ranunculus:* ch., chlorenchyma; ph., phloem; s., stomate; xy., xylem. [Courtesy of E. J. Kohl, Lakeside Biological Products, Ripon, Wis.]

xylem forms a strand on the side toward the pith, and the phloem forms a strand on the side toward the cortex. These two tissues are usually separated by a single layer of **procambium,** the ontogenetic forerunner of the cambium. The basic structure of xylem and phloem was discussed in Chapter 18. It should be noted that the xylem of angiosperms characteristically contains vessels as well as tracheids and/or fibers. A prominent strand of fibers, technically part of the phloem, is often also present on the cortical side of a vascular bundle. Bundles with such fibers are often called **fibrovascular bundles.**

Endodermis

In most herbaceous dicot stems, the vascular bundles abut directly on the cortex. In a few species, however, the cortex is delimited at the inner margin by an endodermis, as in roots. Endodermis is discussed in Chapter 20.

Cortex

The cortex is usually parenchymatous, at least toward the inside, but toward the surface of the stem it often contains some collenchyma or sclerenchyma. These latter tissues, which help to support the stem, occur as individual vertical strands, or a network, or a complete sheath. Frequently some or all of the outer part of the cortical parenchyma bears chloroplasts and carries on photosynthesis. Food storage is another common function of the cortical parenchyma.

Epidermis

The epidermis is the outermost tissue of an herbaceous dicot stem. It retards evaporation of water from internal tissues and furnishes some protection from mechanical injury and infection. The epidermis of stems generally has some stomates (discussed in Chapter 21), but usually not so many as leaves.

Secondary Tissues

In some herbaceous dicot stems, the procambium remains parenchymatous or eventually matures into vascular elements, and there is no secondary growth. More often the procambium becomes a definite cambium that produces secondary tissues. The cambium may be restricted to the vascular bundles (**fascicular cambium**), or it may also extend from one bundle to the next (**interfascicular cambium**), forming a complete sheath. Typically the secondary tissues produced by the cambium are xylem and phloem, but in some species only the fascicular cambium produces vascular tissues, and the interfascicular cambium produces parenchyma.

Most cell divisions in the cambium are **periclinal** (parallel to the cambial surface). Of the two daughter cells formed in a division, one remains a cambial cell and the other usually matures directly or eventually into either a phloem cell (if external to the cambium) or a xylem cell (if internal to the cambium). Often these cells cut off from the cambium undergo one or more periclinal divisions before maturing into xylem or phloem, but only the cambium remains permanently meristematic. In a cross-sectional view of the stem, the cells of the secondary phloem, and especially of the secondary xylem, appear to be arranged in regular radial rows, in contrast to the less regular arrangement of the cells of the primary xylem and phloem.

Herbaceous dicots with an active cambium, such as the common sunflower, produce a considerable amount of secondary xylem by the end of the season, and in cross section the xylem forms a continuous broad band. If such stems lived through the winter and continued to produce secondary tissues year after year, they would be woody stems. In herbaceous stems, the cortex and epidermis usually retain enough capacity for cell division, so that the increase in circumference caused by the secondary growth does not damage the outer tissues; but in woody stems, cell division in the outer tissues does not keep pace with the increase in diameter (and circumference) caused by secondary growth, and the outer tissues are progressively ruptured and sloughed off. Herbaceous stems usually die down to the ground each year (especially in temperate and boreal regions), in contrast to woody stems, which live year after year. Woody stems either grow continuously or more often go into dormancy for part of the year (the winter, in temperate and boreal regions; the dry season, in many tropical regions) and resume growth when favorable conditions return. All these differences are subject to exceptions or difficulties in interpretation, however, and the distinction between woody and herbaceous stems is not at all absolute.

WOODY DICOT STEMS

General Features

The basic structure and organization of the primary tissues of a woody dicot stem are essentially similar to those of an herbaceous stem. On the average, however, woody stems have a more nearly continuous primary vascular cylinder, with narrower interfascicular areas. The cambium is a continuous layer, forming secondary xylem and phloem in the same way as in herbaceous stems.

Fig. 19.5 Sugar maple, *Acer saccharum*, in winter, at the New York Botanical Garden, showing the characteristic deliquescent branching of angiosperm trees. The main stem melts away into a series of branches, instead of continuing as a definite central axis. [New York Botanical Garden photo.]

Secondary Xylem

The secondary wood (Figs. 19.6, 19.7) usually consists of vessels, fibers and/or fiber-tracheids, and parenchyma cells. Fiber-tracheids are intermediate in structure between fibers and tracheids. True tracheids, which make up most of the wood of gymnosperms, are much less common in angiosperms, being found mainly in some of the more archaic families such as the Winteraceae. Many (rarely all) of the parenchyma cells are arranged in ribbon-shaped bands which, in cross-sectional view of the stem, form lines radiating toward the outside. These bands are called **wood rays** or xylem rays. A

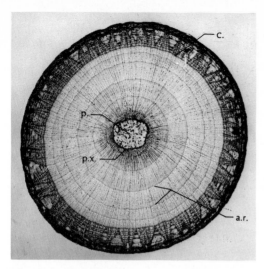

Fig. 19.6 Cross section of a 5-year-old stem of basswood, *Tilia*: a.r., annual ring; c., cambium; p., pith; p.x., primary xylem. [Courtesy CCM: General Biological, Inc., Chicago.]

wood ray is one to several cells wide and one to many cells high.

The ray cells of the secondary wood, like the other cells, are formed by division from the cambium. New rays are initiated by the cambium from time to time, so that the number of rays is greatest next to the cambium, and least toward the pith. The lateral distance between the rays is thus maintained at a fairly constant figure (usually well under 1 cm) by the progressive insertion of new rays between previously established ones at points progressively farther from the pith.

Most woody dicots have vertical parenchyma cells as well as rays. These are in contact with one another as well as with the vessels and fibers, and they form an interconnecting system with the rays. Every vessel is in contact with one or more parenchyma cells somewhere along its length.

The cambium continues to produce

secondary xylem and phloem year after year, so that the stem increases progressively in thickness. In tropical regions without a pronounced dry season, secondary growth is often more or less continuous, but in regions with more marked seasonal differences, the cambium undergoes alternate periods of activity and dormancy.

Xylem cells produced in the spring, when moisture is abundant and growth is rapid, tend to be relatively large and thin walled, with a large lumen. As the season progresses and the water supply diminishes, growth slows down, and the mature cells are smaller and thicker walled. Usually there is an abrupt contrast between the last-formed xylem cells of one year and the first-formed ones of the next year. These annual growth layers, which appear as a series of concentric rings when the stem is viewed in cross section, are called **annual rings.** Annual rings vary from hardly visible to the naked eye to sometimes as much as 2 or 3 cm thick, but thicknesses of 1 mm to 1 cm are much more common. The age of a tree may be determined by counting the annual rings, although there is a possibility of error because of occasional false rings formed during unusual years when growth is interrupted by a dry spell, and then resumed later in the season.

The parenchyma cells of the xylem remain alive and continue to function normally for some years after they are produced. These functions include food storage, lateral conduction (in the rays), and possibly some service in keeping the vessels functional in water conduction.

Sooner or later the metabolic activity of the parenchyma cells is modified, with the result that the other xylem cells are infiltrated with various organic substances such as oils, gums, resins, tannins, and aromatic and coloring materials. Due

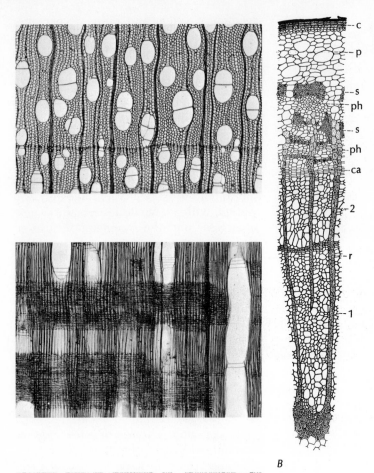

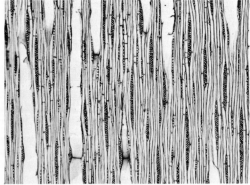

A

B

Fig. 19.7   (*A*) Wood of river birch, *Betula nigra:* (*top*) Transverse section; (*middle*) radial section; (*bottom*) tangential section. [Photos courtesy of the U.S. Forest Products Laboratory.] (*B*) Cross section of part of a 2-year-old stem of tulip tree, *Liriodendron tulipifera.* Near each side is a wood ray extending to the pith; in the center is a wood ray that does not reach the pith: 1, wood of first annual ring; 2, wood of second annual ring; c., cork; ca., cambium; p., parenchyma; ph., phloem; s., sclerenchyma fibers of the phloem. (× 60.) [From W. H. Brown, *The plant kingdom,* Ginn, Boston, 1935, 1963. Courtesy of M. A. Brown.]

to these changes, the vessels (and tracheids, when these are present) lose their conducting function, retaining only the mechanical function of support. The parenchyma cells then die and their protoplasm degenerates.

The central core of xylem that has lost its water-conducting function is called **heartwood,** and the outer sheath which continues to conduct water and dissolved substances is called **sapwood.** The heartwood is generally darker in color than the sapwood. The boundary between the two is commonly irregular, extending several annual rings closer to the cambium at some points than at others. The inner part of the sapwood is continuously changed into heartwood as more sapwood is formed by divisions of the cambial cells. The heartwood thus constitutes a progressively expanding core, whereas the sapwood tends to remain about the same thickness throughout the life of the tree. A large tree may have a ring of sapwood only a few centimeters thick surrounding a core of heartwood a meter or more thick.

The eventual death of the xylem parenchyma cells and the associated change of sapwood into heartwood furnish another example of the general principle that the nonmeristematic cells of a multicellular plant body do not have an unlimited life. The specialization of structure and function of the cells is always metabolically imperfect, and sooner or later metabolic imbalances and the accumulation of waste materials lead to death. Cells of the meristems largely avoid these difficulties by the constant production of new cells, so that the metabolic wastes are continuously distributed away from the meristem. Unicellular organisms commonly pass some of their wastes into the surrounding medium, often by means of contractile vacuoles, but perhaps even unicells

would not be potentially immortal were it not for the distribution of wastes resulting from cell division.

Bark

The part of a woody stem external to the cambium is called the **bark.** In a fairly young stem, this includes the epidermis, cortex, and primary phloem as well as the secondary phloem. In older stems, these tissues have generally been lost, and the bark consists wholly of secondary phloem plus the other tissues (mainly cork) that have been formed in it.

The secondary phloem is often a very complex tissue, with more or less regularly arranged zones of sieve tubes, parenchyma cells, and fibers. As in the xylem, the parenchyma cells form horizontal rays and a more diffuse vertical system intermingled with the conducting elements. The phloem rays are generally continuous with the xylem rays, interrupted only by the cambium, and often some of them are notably wider than others. The phloem fibers often form interconnected vertical strands, providing strength combined with flexibility.

The first phellogen (see Fig. 18.10) usually differentiates during the first year, not long (if at all) after the vascular cambium becomes active. Typically some cortical parenchyma cells just beneath the epidermis become meristematic and begin to function as phellogen (cork cambium), but in different species the phellogen may originate in any of the tissues external to the cambium. In birches, beeches, and some other trees, the phellogen forms a continuous sheath, but more often it is a series of overlapping scales. The phellogen and the tissues it produces are collectively called the **periderm.**

The phellogen functions in much the

same way as the vascular cambium, except that the two tissues it produces are cork (externally) and phelloderm (internally). In some species the phellogen cuts off cells only toward the outside, and no phelloderm is formed. Cork is also called phellem. Cork cells, as we have noted, are suberized, and a layer of cork cuts off the water supply of all tissues external to it. These outer tissues die and finally slough off. The first layer of phellogen may persist and remain active for many years, as in beeches, but more often additional layers develop from parenchyma cells farther in, eventually in the secondary phloem, and the outer cork layers then also tend to slough off.

The part of the bark between the cambium and the innermost cork layer(s) contains both living and dead cells and constitutes the live bark, in which conduction of foods and other solutes occurs. The part of the bark from the inner edge of the innermost cork layer(s) to the surface contains only dead cells and constitutes the dead bark, which protects the inner tissues from mechanical injury and loss of water. The live bark is usually less than 1 cm thick, but the dead bark of mature trees is commonly several centimeters or even several decimeters thick.

The continued increase in thickness of the tree, due to cambial activity, causes the dead bark to rupture and fall off. Eventually an equilibrium is reached, with the loss of dead bark from the outside about equaling the formation of new bark at the inside. The appearance of the bark of a mature tree depends in part on the duration of the initial phellogen, the size and shape of the initial and later phellogen layers, and the degree of cohesiveness of the cells making up the cork layers. The pattern is often distinctive, and many species can be recognized from the bark alone (leaving aside the botanical in-joke about the dogwood).

The dead bark of stems and roots is generally marked by scattered **lenticels** (Fig. 19.8). These are slightly raised areas of loosely arranged, nearly or quite unsuberized cells, formed largely by particular parts of the phellogen. Very often the lenticels arise just beneath the stomates. The lenticel phellogen, which is continuous with the ordinary phellogen, produces cells at a relatively rapid rate, and the lenticel becomes slightly raised above the general surface of the periderm. Lenticels are typically lenticular (like a double-convex lens) in external outline, whence the name. They are usually oriented vertically or horizontally on the stem, according to the species, and they vary from barely visible to as much as 1 cm or even 2.5 cm long. In trees with deeply fissured bark, the lenticels are usually at the bottom of the fissures. It is thought that the function of lenticels is to allow a slow

Fig. 19.8 Cross section of part of the stem of elderberry, *Sambucus canadensis*, showing lenticel. [Photo courtesy of E. J. Kohl, Lakeside Biological Products, Ripon, Wis.]

exchange of gases between the internal tissues and the atmosphere.

HERBACEOUS MONOCOT STEMS

The herbaceous monocot stem typically has numerous small vascular bundles arranged in two or more rings, or scattered throughout the fundamental tissue (Fig. 19.9). In the latter case there is neither a well-defined cortex nor a well-defined pith. The fundamental tissue is usually parenchymatous, but often there is a collenchymatous or sclerenchymatous layer not far from the epidermis. The tissue just internal to the epidermis is often photosynthetic.

Usually only a small part of the tissue of the monocot stem is formed directly from the apical meristem. A **primary thickening meristem** (Fig. 19.10), shaped like an inverted thimble, is continuous at the top with the small apical meristem. The leaf primordia and axillary buds originate from the apical meristem, but the thickening of the apical region of the stem is due chiefly to divisions in the cap of the primary thickening meristem. Activity of this meristem decreases progressively in the region of elongation (corre-

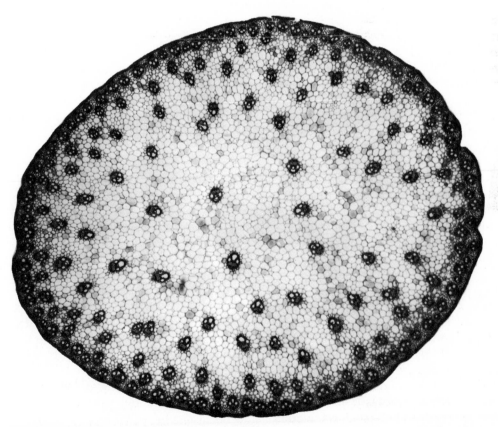

Fig. 19.9 Cross section of stem of corn, *Zea mays*, showing scattered vascular bundles. [Courtesy of E. J. Kohl, Lakeside Biological Products, Ripon, Wis.]

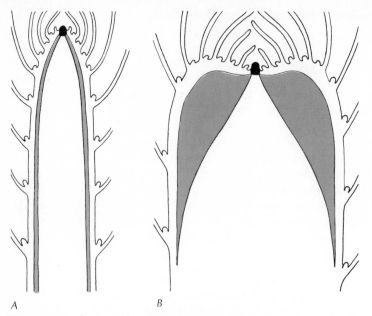

Fig. 19.10 (*A*) Diagram of characteristic dicot stem, in long section, with the procambium gray. (*B*) Diagram of characteristic monocot stem, in long section.

sponding to the sides of the thimble), and usually ceases in the region of maturation or in the mature region.

In a few herbaceous monocots, such as *Aloe* or *Sanseveria*, the thickening meristem continues to function more or less indefinitely. In the mature region of the stem, this meristem, which is continuous upward with the primary thickening meristem, is called the cambium. Unlike the cambium of dicots, it does not pass through the individual vascular bundles. The cambium of monocots produces parenchyma and vascular bundles toward the inside, and usually only a small amount of parenchyma toward the outside. The vascular bundles of monocots are called **closed bundles** (i.e., closed to further growth), as contrasted to the **open bundles** that characterize most dicots.

EVOLUTION OF STEM TYPES

The woody dicot stem, long lived and with active cambium, is regarded as the primitive type in flowering plants. The herbaceous dicot stem, short lived and with less active or no cambium, is considered to be an evolutionary modification of the woody type. This change has occurred repeatedly in different groups of dicots. A few groups have reverted to the woody habit after having become herbaceous. The common sagebrush (*Artemisia tridentata*) of the western United States is a familiar woody plant that very probably had herbaceous ancestors.

The herbaceous monocot stem, with polycyclic or scattered vascular bundles and no fascicular cambium, is evidently derived from the herbaceous dicot type.

The woody monocot stem is derived from the herbaceous one by a change of the fundamental tissue from parenchyma to collenchyma or sclerenchyma, associated with the persistence of the stem over a period of years. The wood of monocots is chiefly a primary tissue, consisting of the vascular bundles together with the tissues in which they are embedded. The wood of woody dicots, on the other hand, consists of secondary xylem plus a little primary xylem. The vast majority of monocots are herbaceous, but the woody habit has been achieved in more than one group. The palms are the most familiar woody monocots, but some of the grasses (e.g., the bamboos) are rather woody, and the Joshua tree (*Yucca brevifolia*) of the southwestern United States is a relative of the lilies.

MODIFIED STEMS

The stem is the oldest organ in the evolutionary history of vascular plants. The most primitive vascular plants, such as *Cooksonia* and *Rhynia* of Silurian and Devonian age, respectively, consisted essentially of a dichotomously branching stem with terminal sporangia. The evolutionary differentiation of this primitive stem into a more complex plant body with roots, stems, and leaves evidently occurred largely in the Devonian period. The distinction among these three basic organs has been maintained through evolutionary history since that time. It is equally true, however, that special evolutionary modifications of these organs have repeatedly arisen in various groups of plants. The whole stem or stem system of the plant may be modified, as in twining stems, succulent stems, and essentially leafless, strongly photosynthetic stems. Or only some of the branch stems may be modified, for example, to form thorns or tendrils.

Stems as Perennating and Reproductive Structures

Reproduction that does not involve sexual processes is called **vegetative reproduction.** Many angiosperms reproduce vegetatively as well as by forming seeds (which normally result from a sexual process). Leaves, stems, and roots may all carry on vegetative reproduction in various kinds of plants, but stems are much more often involved than roots or leaves. All the individuals derived by vegetative reproduction from a single individual constitute a **clone.** Except for the possibility of mutations, all members of a clone are genetically identical. A few kinds of angiosperms have lost the power of sexual reproduction entirely and exist in nature only as clones.

A modified stem that creeps along below the surface of the ground, sending down adventitious roots and producing new erect stems at intervals, is called a **rhizome.** The leaves of rhizomes are generally much reduced and, of course, without chlorophyll, commonly appearing as mere small scales at regular intervals.

Rhizomes are commonly perennating structures, as well as a means of reproduction. The aerial stems of many herbs die down to the ground each fall and develop anew each spring from the rhizome. The rhizome branches at intervals, and eventually dies off from behind, thus producing a clone of separated but genetically identical individuals. Plants that perennate by means of rhizomes are potentially immortal.

Rhizomes may be long and slender, as

in Kentucky bluegrass and other sod-forming grasses, or thickened and fleshy, as in *Iris*. Some slender rhizomes have strongly thickened parts, especially at the end, called **tubers.** The edible organ of the "Irish" potato (*Solanum tuberosum,* a native of northern South America) is a large tuber. Tubers have relatively little vascular tissue and a high proportion of parenchyma. The food stored in the parenchyma is used the following year to give the plant a quick start.

Some kinds of plants, especially among the monocots, perennate by corms or by bulbs. A **corm** is a very short, stout, erect underground stem, as in the crocus or the gladiolus (Fig. 19.11), in which food is stored. At the end of the growing season, the corm goes into dormancy, and the other parts of the plant usually die. The corm may be covered by one or more layers of leaf bases (the remains of the foliage leaves), but these are usually dead and dry, not serving as storage organs. At the beginning of the new growing season, the corm sends down adventitious roots, and the terminal bud develops into an aerial flowering stem. As the season progresses, one or more new corms usually form from axillary buds of the old one, and it is these new corms that survive the winter.

A **bulb** differs from a corm in that the principal storage organs are the thickened leaves (bulb scales) that surround the short, erect stem. The bulb scales may be very short and differentiated from the foliage leaves, as in tulips and onions, or they may be thickened bases of ordinary leaves, as in daffodils. A new bulb may be produced each year, or the same bulb may persist year after year, with the aerial flowering stems being produced from axillary buds. In nonbotanical usage, the term bulb is commonly extended to cover corms as well as true bulbs.

Fig. 19.11 *Gladiolus* corms. (*Left*) Side view of mature corm, with leaf bases, attached to the old corm of the previous year. (*Right*) Top view, after the leaves have been stripped off. [New York Botanical Garden photos.]

Artificially Induced Reproduction by Stems

Many kinds of plants, both woody and herbaceous, can be propagated by **cuttings** (Fig. 19.12). A small stem is merely cut off and the cut end placed in water or moist soil. Adventitious roots develop from the covered part, and a new plant is established. Willows are very easily propagated by cuttings, as is also the cultivated *Coleus*. The formation of adventitious roots on a cutting can be encouraged by the judicious application of indole-acetic acid, or any of several other chemicals, to the cut end (see Chapter 26).

Most kinds of angiosperms can be reproduced by **grafting** (Fig. 19.13), but only the woody species are usually so treated. The cultivars of most commercial fruit and nut trees are individual clones that are perpetuated by grafting. A twig (the **scion**) of one plant is transferred to the

Fig. 19.12 Vegetative propagation. A cutting of dumb cane, *Dieffenbachia seguine,* has been rooted in sand. [Photo courtesy of T. H. Everett.]

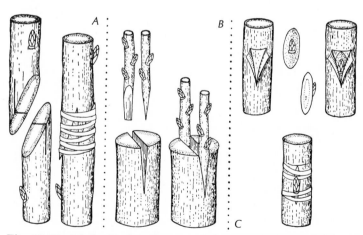

Fig. 19.13 Methods of grafting (*A, B*) and budding (*C*). Budding is a special kind of grafting.

stem of another plant (the **stock**). Equivalent tissues of the scion and stock are juxtaposed, particular attention being paid to the cambium. The scion and stock are then bound firmly together and protected from evaporation while the living tissues have a chance to grow together. The cambium of the scion and stock must unite if the graft is to be successful. Water conduction through the graft is likely to be impeded to some extent until the healed cambium produces a continuous sheath of new xylem.

SUGGESTED READING

Fritts, H. C., Tree rings and climate, *Sci. Amer.* **226**(5):92–101, May 1972.

Hartman, D. T., and D. E. Kester, *Plant propagation: Principles and practices*, 3rd ed., Prentice-Hall, Englewood Cliffs, N.J., 1975.

Plant life. A Scientific American book, Simon & Schuster, New York, 1957. A collection of popular articles by outstanding botanists, originally appearing separately in the *Scientific American* from 1949 to 1957. Informative and easy reading, covering a wide variety of topics from photosynthesis to strangler trees to hybrid corn.

Chapter 20

Angiosperms: Roots

Germinating seed of *Phaseolus coccineus*, the scarlet runner bean. Note the root hair zone, well back from the tip. [Photo by Jeanne White, from National Audubon Society Collection/Photo Researchers, Inc.]

GENERAL CHARACTERISTICS

Position and External Morphology

Roots are the characteristic underground organs of angiosperms and most other vascular plants. In contrast to stems, roots typically grow in the direction of gravity. Roots are therefore said to be **positively geotropic,** whereas typical stems are **negatively geotropic.** Tropisms (Greek *tropos,* turn) are discussed in Chapter 26.

Like stems, roots are usually subcylindric, tapering toward the tip. Because their direction of growth is influenced by pebbles and other obstructions in the soil, and often by differences of moisture and other conditions as well, roots are usually much more crooked than stems. This is particularly obvious in young, slender roots; small kinks or bends may eventually be obscured as roots increase in thickness.

Unlike stems, roots do not bear leaves; therefore they lack nodes and internodes. Branches of angiosperm roots originate from internal tissues (as explained on a subsequent page) rather than from axillary buds as do branches of stems. There is some correlation between the position of the branch roots and the arrangement of the internal tissues, but there is nothing in the external structure to indicate the points at which the branches will appear.

The apical meristem of a root is covered by a special protective tissue, the **rootcap,** which is typically less than 1 mm long.

Functions

The two principal functions of typical roots are **anchorage** and the **absorption** of water and minerals from the soil. Conduction of water and minerals upward to the stem, being an implicit corollary of absorption, is not always listed as a separate function. A third important function of many, but not all, roots is the **storage** of food. Various types of specialized roots may have other functions.

A function of roots that is often overlooked or insufficiently appreciated is the synthesis of various organic compounds from raw materials. In many species nitrogen ions are converted into organic nitrogen compounds before being translocated to the shoot. In different species, diverse other substances produced in the root are passed on to the shoot. Among these substances are the nicotine of tobacco and various other alkaloids of other plants, as well as certain hormones. Conversely, the roots depend on the shoot for photosynthate and miscellaneous necessary metabolites such as auxin, thiamin, and niacin.

Types of Root Systems

The nature of the root system (Fig. 20.1) varies in different kinds of plants. Plants in which the primary root of the seedling eventually gives rise to most or all of the whole root system are said to have a **primary root system.** Plants in which most or all of the roots develop adventitiously from underground stems, or from the base of the aerial stems, are said to have an **adventitious root system.** Among plants with primary root systems, there are all variations from a **taproot system,** with a central taproot that is larger than any of its branches, to a **fibrous root system,** in which the primary root is quickly deliquescent into several or many roots all about the same size. Adventitious root systems are usually also fibrous, but both adventitious roots and taproots are sometimes thickened and fleshy, serving

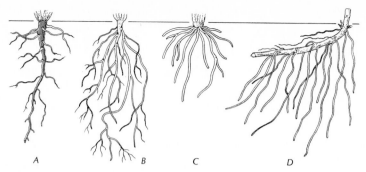

Fig. 20.1 Types of root systems. (A, B) Primary; (C, D) adventitious; (A) tap; (B, C, D) fibrous.

as storage organs. Most vascular crypto-gams and all monocots, as well as a considerable number of dicots (especially those with creeping rhizomes) have an adventitious root system. Many dicots and most gymnosperms, on the other hand, have a primary root system, of either the taproot or the fibrous type.

Different kinds of plants exploit different levels in the soil. The roots of grasses are usually rather shallow; most of the roots of the common lawn grasses are in the upper few inches of soil, and the dense sod that they form (together with the rhizomes) can be peeled in a layer only about 1 dm thick. In a number of common crop plants, the principal development of the root system is at a depth of 1–1.5 m, soil permitting. Root systems more than about 3 m deep are uncommon, even among trees.

The depth of penetration of roots is influenced by the nature of the soil and by other environmental conditions, as well as by their inherent tendencies. A hard-pan or a layer of rock may block further growth. Soils saturated with water do not have enough oxygen to support the root growth of most plants, and roots do not generally descend below the lowest level

to which the water table sinks during the growing season. Roots of hydrophytes (plants especially adapted to wet places or to open water) are not checked by the water table, however. Conversely, roots of most plants will not grow through very dry soil, but roots of some xerophytes (plants adapted to very dry conditions) will penetrate several meters of dry soil before reaching a moister zone.

The development of the root system is also influenced by differences in the amount of water at different levels or in different areas of soil. The best development and most extensive branching of the roots generally occurs in soil that is moist but not saturated. Cultivated plants may be encouraged to develop relatively shallow or relatively deep root systems, according to whether they are watered frequently and lightly or less often and more heavily. Sewer pipes, furnishing a combination of abundant moisture and adequate aeration. provide a favorable habitat for root growth, especially of species adapted to moist or wet soils. Roots of trees such as willows and sycamores often penetrate joints in the pipes and proliferate so extensively as to interfere with the water flow.

PRIMARY TISSUES

Primary Growth

Growth in length of roots is governed by an apical meristem, much as in stems, although the detailed anatomy of the meristem is different. The apical meristem of roots is really subapical rather than truly apical, being covered by a rootcap.

The **rootcap** is composed of living parenchyma cells. The cells near its outer surface generally have mucilaginous walls. The mucilage apparently acts as a lubricant, easing the progress of the root tip through the soil. The outer cells of the rootcap are continually being sloughed off and lost, and the walls of the next layers of cells in turn become mucilaginous. New cells originate on the inner side of the rootcap from the apical meristem.

The primary tissues of the root are derived from the apical meristem by cell division and subsequent differentiation. The successive zones (Fig. 20.2) behind the apical meristem are:

1. **Region of elongation**
2. **Region of maturation**
3. **Region of mature primary tissues**
4. **Region of secondary tissues** (in many roots)

The *region of elongation* is usually only a few (1–10) millimeters long, in contrast to the same region in stems, which is commonly several centimeters long. It is characterized by enlargement—especially elongation—of the cells. A relatively small number of cell divisions and the beginnings of differentiation of the various cell types also occur in this region. The region of elongation merges at one end with the apical meristem and at the other with the region of maturation.

The *region of maturation* is usually one to several centimeters long. Here the cells,

having reached essentially full size, gradually assume their mature form and function. The region of maturation is often known as the root-hair zone, because of the characteristic **root hairs** (Fig. 20.3) borne by many of the epidermal cells. Each root hair originates as a bump on an epidermal cell and elongates to form a projection often several millimeters long. It is cylindrical in form, but commonly bent this way and that in its passage among the soil particles. Root hairs may be regarded as a means of increasing the absorbing surface of the root, a view formulated by the German botanist Franz Meyen (1804–1840) in 1838. The nucleus usually moves from the body of the cell out into the root hair, and this change of position may be correlated with the absorbing function of the hair.

It is convenient to regard the beginning of root-hair formation as the dividing line between the region of elongation and the region of maturation. Certainly no significant elongation occurs after the root hairs have begun to form; if it did, they would be sheared off by the movement of the root through the soil. Root hairs may occur on all the epidermal cells in the root-hair zone, or only some of them, according to the species, and the cells on which they occur may or may not be otherwise obviously different from the others. Often there are several hundred root hairs for every square millimeter of the epidermal surface from which they originate.

Anatomy of Primary Tissues

Roots have a number of anatomical features in common with stems, but also show some differences. In the zone of primary permanent tissues, before any secondary tissues have been formed, a cross section of a typical root (Fig. 20.4)

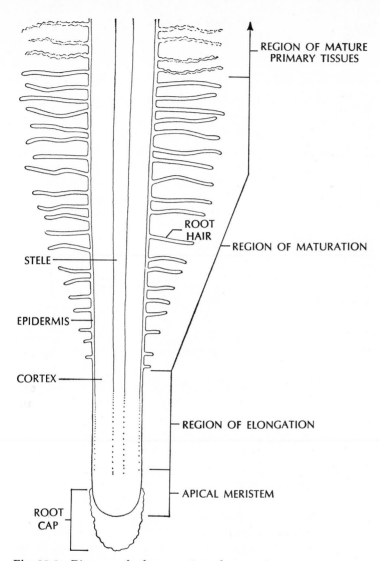

REGION OF MATURE
PRIMARY TISSUES

ROOT
HAIR

STELE

REGION OF MATURATION

EPIDERMIS

CORTEX

REGION OF ELONGATION

APICAL MERISTEM

ROOT
CAP

Fig. 20.2   Diagram of a long section of a root tip.

shows a solid *stele* (without a central pith), surrounded successively by the *pericycle*, the *endodermis*, the *cortex*, and the *epidermis*. The xylem of the stele commonly forms a deeply ridged and grooved central strand. In cross section it tends to be roughly star shaped, with several arms radiating from a common center. The xylem at the tips of the arms matures first, and maturation progresses toward the center. Stems of some of the more primitive kinds of vascular plants resemble roots in having the primary xylem mature from the outside toward the center. Stems of nearly all angiosperms and most gymnosperms, on the contrary,

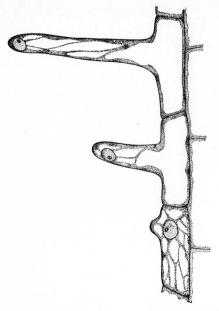

Fig. 20.3 Development of root hairs in millet, *Panicum miliaceum.* (× 400.) [From W. H. Brown, *The plant kingdom,* Ginn, Boston, 1935, 1963. Courtesy of M. A. Brown.]

have the primary xylem next to the pith maturing first, with maturation progressing toward the outside.

The primary phloem of the root forms a series of separate strands alternating with the ridges or arms of primary xylem. The protoxylem (the first primary xylem to mature) and the primary phloem are thus on *alternate radii,* instead of on the same radii as they are in stems. Discrete vascular bundles containing both xylem and phloem, such as are regularly seen in stems, do not occur in roots.

Usually there are one or more layers of parenchyma between the primary xylem and primary phloem of a root. Some of these parenchyma cells may eventually mature into a cambium, much as in stems, or they may remain permanently parenchymatous, depending on the species.

The **stele** of a root is usually delimited by a definite endodermis. Usually there is a thin, parenchymatous layer, one or a few cells thick, between the endodermis and the vascular tissue; this is the **pericycle.** The pericycle is important chiefly as the tissue in which branch roots commonly originate. Because of the ontogeny of the tissues, botanists find it convenient to consider the pericycle as the outermost layer of the stele, and the endodermis as the innermost layer of the cortex, but it is often also convenient to consider these tissues separately.

The **endodermis** (Fig. 20.5) forms a continuous sheath, one cell thick, without intercellular spaces. The cells are typically six sided and brick shaped, with the long axis vertical (parallel to the axis of the stele) and the intermediate axis tangential to the stelar surface. The tangential walls are usually bulged, so that the cell may look more nearly elliptic than rectangular in cross section.

The two tangential walls of an endodermal cell (those that abut on the pericycle and the cortex) usually have no special features, but the four radial walls have a modified area, the **casparian strip,** which completely encircles the cell. Attention was first called to these strips in 1865–1866 by the German botanist J. X. Robert Caspary (1818–1887). The casparian strips are usually thickened, and whether thickened or not they are generally impregnated with a waxy substance similar (but not quite identical) to suberin.

The casparian strips tend to inhibit the passage of water and solutes through the radial walls of the endodermal cells. In order to move between the stele and the cortex, substances must therefore pass through the tangential wall, into the cytoplasm, and out through the other tangential wall. The movement of materials

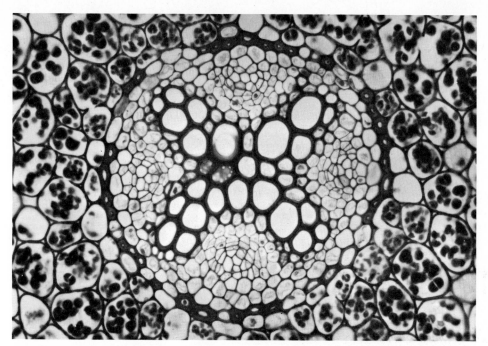

Fig. 20.4 Cross section of root of buttercup, *Ranunculus*, showing central cylinder and part of the surrounding cortex. The strands of phloem lie between the radiating arms of xylem. Most of the endodermal cells are deeply stained, and the cortex bears numerous dark-stained starch grains. [Courtesy CCM: General Biological, Inc., Chicago.]

between the stele and the cortex is thus subjected to protoplasmic control. This may be of some significance to root pressure, as noted in Chapter 22.

The **cortex** of a root usually consists largely of parenchyma, often with some sclerenchyma as well, seldom with any collenchyma. The sclerenchyma frequently forms a complete cylinder near the epidermis, or just outside the endodermis. Laticifers or resin ducts are present in many species.

The **epidermis** of roots generally consists of a single layer of thin-walled, elongate, living cells without intercellular spaces. The formation of root hairs by epidermal cells has already been noted. Young epidermal cells with root hairs lack

a cuticle. Older epidermal cells often develop a cuticle and become thick walled and even lignified. In roots of many angiosperms and gymnosperms the outermost part of the cortex, just beneath the epidermis, eventually matures into a lignified or even suberized layer, one to several cells thick, which contributes to or takes over the protective function of the epidermis after the root hairs have degenerated.

SECONDARY TISSUES

Secondary Growth

Roots of most monocots lack secondary growth. Among dicots, secondary growth in roots varies from vigorous to scanty or

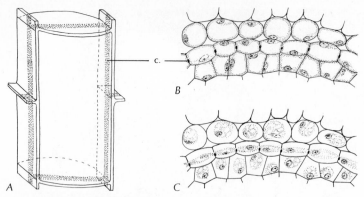

Fig. 20.5   Structure of the endodermis. (*A*) A single cell, showing only the wall. (*B*) Cross section of part of endodermis, with cortex above (outside) and pericycle below (inside). (*C*) Same as above, but with the cells plasmolyzed, showing the attachment of the cytoplasm to the casparian strips; c., casparian strip. [Adapted from K. Esau, *Plant anatomy*, 1953; courtesy of John Wiley and Sons, New York.]

none. Dicot trees and shrubs regularly have woody roots with well-developed secondary growth. Perennial herbs with little or no secondary growth in the stem sometimes have woody roots; conversely, herbs, such as goldenrod (*Solidago*) with vigorous secondary growth in the stem may have an adventitious root system in which there is little if any secondary growth.

Secondary growth in roots and stems is essentially similar.

Anatomy of Secondary Tissues

As compared to stems, the xylem of roots tends to have larger and more numerous vessels with thinner walls, as well as fewer fibers, more parenchyma, and more abundant rays. The phloem has less sclerenchyma and more storage parenchyma. The cambium opposite the primary xylem ridges commonly produces wide vascular rays. These differences between stems and

roots are correlated with differences in their functions. More food is usually stored in roots than in stems, and the roots, being surrounded by soil, have less need for rigid supporting tissues.

Annual rings are usually less pronounced in roots than in stems, partly because the seasonal changes in environment to which the stem is subjected are buffered, for roots, by the soil. Long-lived woody roots develop a distinct core of heartwood surrounded by sapwood, just as do stems.

Most perennial roots, especially those with an active cambium, sooner or later develop a periderm. The cork cambium typically originates in the pericycle as a continuous sheath, forming a continuous layer of cork, along with some internal phelloderm. All tissues external to the cork then die and decay. In some roots new phellogens are formed progressively farther into the secondary phloem, as in stems. The periderm of roots, like that of stems, has scattered lenticels. Environmental conditions of the soil are conducive to rapid decay of dead tissues, and even long-lived woody roots seldom have a thick bark.

BRANCH ROOTS

Origin

Roots that originate as branches of other roots are called **secondary roots.** The primordia that give rise to secondary roots, unlike those that give rise to branches of stems, generally originate some distance behind the apical meristem, typically in the region of maturation.

Branch-root primordia nearly always originate internally, typically in the pericycle. Some cells of the pericycle become meristematic and by a series of divisions produce the apical meristem of a branch root. The new apical meristem soon develops a rootcap, and the branch root (Fig. 20.6) pushes its way through the outer tissues of the root into the soil. In mature roots, from which the pericycle has been sloughed off, branches commonly originate in the cambium.

Adventitious roots borne on stems usually originate internally in much the same way as secondary roots. In young stems the root primordium commonly develops in the interfascicular parenchyma, that is, the medullary rays. In older stems it may develop near or in the cambium of a ray. Stem-borne adventitious roots are restricted to the nodes in many species, but in others they occur along the internodes as well.

Arrangement

Branch roots are not so regularly arranged as leaves and axillary buds, and what arrangement there is tends to be obscured by the twisting and turning of the root as it

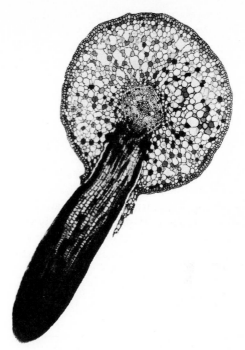

Fig. 20.6 Origin of a branch root in willow, *Salix*. [Courtesy of E. J. Kohl, Lakeside Biological Products, Ripon, Wis.]

grows through the soil. When plants are grown in water or in loose soil, it can generally be seen that the secondary roots are in vertical rows; these typically correspond in number and position to the primary xylem ridges of the stele.

Number

The root system of an angiosperm or other vascular plant is ordinarily much branched, with numerous small ultimate branches. A single plant of rye (*Secale cereale*), after 4 months of growth in a box containing 2 cubic feet of soil, had nearly 14 million roots and over 14 billion root hairs. The combined length of the roots was 387 miles and of the root hairs about 6,000 miles. The total surface of the root hairs was over 4,000 square feet, and the total surface of the remainder of the root system more than 2,500 square feet. A single cubic centimeter of soil in which bluegrass (*Poa pratensis*) was growing has been estimated to have about 150,000 root hairs and 400 square centimeters of root surface, with the root system making up 2.8 percent of the volume of the soil.

The intricate branching and large surface area of the root system are significant both in anchoring the plant and in exploiting the supply of water and minerals in the soil. A concomitant result is that the soil itself is held in place. The value of a plant cover in minimizing soil erosion is well known.

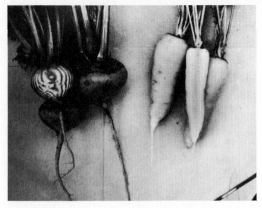

Fig. 20.7   Fleshy taproots of beet (*left*) and carrot (*right*). [New York Botanical Garden photos.]

ROOT-STEM TRANSITION

Inasmuch as the primary vascular tissues of roots are organized differently from those of stems, there must be a transition region in which these tissues are rearranged. Typically the transition occurs in the **hypocotyl** (the part of the axis between the top of the root and the cotyledon(s) of the seedling). In plants in which the cotyledons are never brought above the ground, the hypocotyl generally remains very short, and the transition region then often extends through one or more internodes of the stem.

MODIFIED ROOTS

Storage Roots

Most roots store some food as well as serving in anchorage and absorption. Carbohydrates, especially starch and sucrose, are the foods usually stored. Sugar beets commonly contain 15–20 percent sucrose.

One of the most common types of modified root is the fleshy storage root (Fig. 20.7). Many biennials produce a fleshy storage taproot during the first year

and use up the stored food the following year when flowers and seeds are formed. Carrots, beets, parsnips, turnips, and radishes are familiar examples, although some of these may be induced to mature the first year. In many plants with fleshy taproots, including carrots and beets, the fleshy structure includes the hypocotyl as well as the root proper, and the two parts can be distinguished only by careful anatomical study.

Fleshy thickened roots, or parts of roots, in an otherwise fibrous root system are called tuberous roots, in reference to their superficial similarity to tubers. In sweet potatoes (Fig. 20.8) and some other plants, tuberous roots are also reproductive organs, producing adventitious stems from internal tissues the following year.

All storage roots have in common an abundance of parenchyma that is thoroughly permeated by vascular elements. In other respects their anatomy varies.

Mycorhizae

The roots of many vascular plants harbor a filamentous fungus, forming a fungus-root combination called a **mycorhiza.** My-

Fig. 20.8 Tuberous root of sweet potato, *Ipomoea batatas*, producing roots below and adventitious stems above. [Photo courtesy of T. H. Everett.]

corhizae commonly have shorter, thicker, and frequently more crowded branches than ordinary roots, and they have few or no root hairs. Some, all, or none of the smaller roots of a plant may be mycorrhizal, depending on the species and the environmental conditions. Often the roots in the upper part of the soil, where organic matter is abundant, are mycorrhizal, whereas those in the underlying mineral soil are not. The roots of some kinds of plants are necessarily and always mycorrhizal; some are or are not with varying degrees of benefit and harm according to

the circumstances. Probably more than half the kinds of vascular plants usually have mycorrhizae under natural conditions. The conifers (especially pines), orchids, and heaths (Ericaceae) are usually strongly mycorrhizal, and successful cultivation of species of these groups generally requires the maintenance of conditions favorable for the growth of the fungus. Plants that are heavily dependent on their mycorrhizal fungus for survival are said to be **mycotrophic.**

The fungi involved in mycorrhizae include various kinds of basidiomycetes, ascomycetes, and zygomycetes, especially the basidiomycetes. The fungus may also occur independently of the mycorrhizae, or it may be largely or wholly restricted to mycorrhizal associations. The host may bear only one or several different kinds of fungi, often more than one in the same root. Often a particular mycorrhizal fungus is limited to only one kind or a few closely related kinds of hosts. Truffles and many kinds of mushrooms are especially noteworthy in this regard, although the common field mushroom, *Agaricus campestris*, is not mycorrhizal. The association of certain fungi with the roots of oaks was known to Theophrastus (372?–287? B.C.), the Greek "Father of Botany."

The mycorrhizal fungus is partly in the soil and partly in the root, often forming a tangle of filaments (the mantle) over the surface so that the root is not in direct contact with the soil (Fig. 20.9). The mantle formed on tree roots by truffle fungi was noted by the French mycologist Louis René Tulasne (1815–1885) in 1841. The fungus mantle often has numerous short, radiating branches that resemble root hairs.

The fungus is typically limited to the outer part of the root, not extending into the endodermis. The filaments are tolerated by the outer tissues but are digested

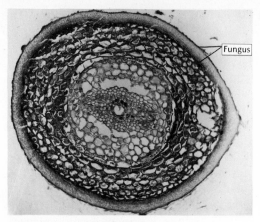

Fungus

Fig. 20.9 Cross section of root of Indian pipe, *Monotropa uniflora*, with its fungus mantle. [Photo courtesy of J. Arthur Herrick.]

and absorbed by the host as they extend inward or into areas of greater metabolic activity (such as the apical meristem). In some plants, however, the fungus permeates the host, and indeed the presence of the fungus in the seed coat is necessary for germination of some kinds of orchids under natural conditions.

The ecological relationships between the fungus and the root vary according to the species and the environmental conditions. Typically the fungus digests starch grains stored in the host cells, absorbing the digestion products and using them in its own metabolism. Water and minerals are absorbed by the fungus and passed on to the host. Mycorhizal absorption is often more efficient than that which would be carried on by the unaided root. The increased absorbing surface provided by the extensive development of the mycorhizal fungus in the soil is believed to be necessary under natural conditions for many forest trees, to supply enough water to balance the loss by evaporation.

Mycorhizal fungi can also tap a mineral supply that would not be available to the unaided host. Nitrogen, phosphorus, and calcium, among other necessary elements found in humus, are not directly available to plant roots until the complex organic compounds in which they occur have been broken down to yield soluble inorganic compounds. Mycorhizal fungi attack the humus, however, digesting and absorbing the organic compounds. Digestion of some of the internal filaments of the fungus releases compounds that are then used by the host. Furthermore, it seems probable that in tropical forests much of the recycling of nitrogen and other minerals involves mycorhizal fungi which feed on the leaves that fall to the forest floor. Under such conditions, the bacterial stages in the nitrogen cycle, leading to the production of nitrates, are largely bypassed. Under experimental conditions, the absence of mycorhizal fungi in pines and some other plants can be compensated for by fertilization.

Mycorhizae evidently originated, in an evolutionary sense, by change from a simple host-parasite relationship to a modified sort of parasitism in which the fungus more or less compensates for the food it uses by providing an increased supply of water, or necessary nutrients, or both. Doubtless the relationship is often even more complex, with each symbiont having become dependent on the other for certain substances that it has lost the ability to produce for itself. In some orchids, such as coralroot (*Corallorhiza*), and certain heaths, such as Indian pipe (*Monotropa*) (Fig. 20.10) and pinedrops (*Pterospora*), the flowering plant lacks chlorophyll and is wholly dependent on the fungus for its food supply as well as for water and minerals. The role of the fungus in this type of mycorhiza was first suggested by the German botanist Wilhelm Pfeffer (1845–1920) in 1877. The

Fig. 20.10 Indian pipe, *Monotropa uniflora*, a plant that has become dependent on its mycorhizal fungus for food.

benefit, if any, derived by the fungus in such cases is obscure; possibly it depends on its "host" for certain vitamins or enzymes.

It has recently been shown that the mycorhizal fungus of at least some nongreen plants also extends into the roots of nearby trees. The nongreen mycotroph is thus a sort of indirect parasite on the tree, with the fungus acting as a conduit for the transfer of nutrients. The term mycoparasite has been devised for such plants.

Other Modified Roots

The function of some fleshy roots as reproductive structures has already been mentioned. Slender roots may also serve a reproductive function. The Canada thistle (*Cirsium arvense*, a native of Eurasia) and many other plants, including a number of noxious weeds, have long, slender, deep-seated horizontal roots that function in the same way as creeping rhizomes, sending up stems at intervals (Fig. 20.11). The stems arise adventitiously from internal tissues, in much the same way as branch roots. Black locust (*Robinia pseudoacacia*) and silver poplar (*Populus alba*) are familiar trees that produce adventitious stems from some of the roots. Roses, blackberries, and a number of other plants are commonly propagated by root cuttings.

Many plants have adventitious **prop roots.** In maize these develop in whorls at the lower nodes of the stem, spreading out and entering the soil at an angle. Mangrove (Fig. 20.12) and some other tropical trees have large prop roots as much as several meters long, which help to anchor the trees firmly in the marshy soil in which they grow. Prop roots of the banyan tree (*Ficus benghalensis*) descend vertically from the widely spreading branches and become columnar supports resembling ordinary trunks.

Poison ivy and many other climbing plants have numerous short adventitious roots on the aerial stems. These roots tend to penetrate the crevices and cracks in the bark of trees, or in brick or stone walls, and may be provided with small adhesive terminal disks.

Tropical epiphytes often have slender, hanging aerial roots with a more or less modified epidermis. In many orchids and aroids (members of the family Araceae) the epidermis of these roots is developed into a multiseriate layer called the velamen, consisting of compactly arranged dead cells with thickened walls. The velamen absorbs rain or dew deposited on

1 ft.

2 ft.

3 ft.

4 ft.

5 ft.

6 ft.

Fig. 20.11   The root system of part of a single clone of Canada thistle, *Cirsium arvense*. [After Hayden.]

Fig. 20.12   Thicket of mangrove, *Rhizophora mangle*, showing numerous prop roots. [Photo courtesy of José Cuatrecasas.]

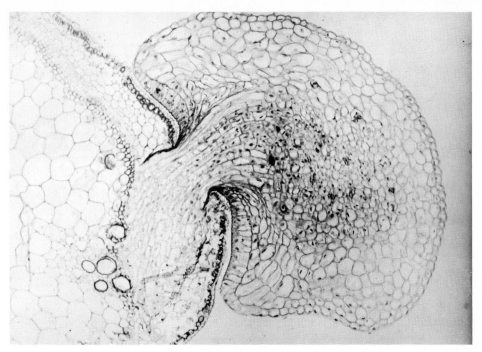

Fig. 20.13 Section showing attachment of the haustorium of dodder, *Cuscuta*, to its host. [Courtesy CCM: General Biological, Inc., Chicago.]

it, and the internal tissues absorb water from the velamen. During dry weather the cells of the velamen dry out and are filled with air.

Parasitic plants such as dodder (*Cuscuta*) and mistletoe have modified roots, called **haustoria** (Fig. 20.13), which enter the tissue of the host, establishing connection with its vascular system and absorbing nutrients. Dodder is wholly dependent on the host for food as well as for water and minerals. Mistletoe, on the other hand, is green and makes at least some of its own food, depending on the host chiefly for water and minerals. The term *haustorium* is also applied to root-like hyphae of parasitic fungi that penetrate the cells of the host and absorb food.

SUGGESTED READING

Epstein, E., The role of roots in the chemical economy of life on earth, *BioScience* **27**: 783–787, 1977.

Hacskaylo, E., Mycorrhiza: the ultimate in reciprocal parasitism? *BioScience* **22**:577–583, 1972.

Harley, J. L., *The biology of mycorrhiza*, 2nd ed., Leonard Hill, London, 1969.

Torrey, J. G., and L. J. Feldman, The organization and function of the root apex, *Amer. Sci.* **65**:334–344, 1977.

Went, F. W., and N. Stark, Mycorrhiza, *BioScience* **18**:1035–1039, 1968.

Chapter 21

Angiosperms: Leaves

Leaves persisting on tulip tree, *Liriodendron tulipifera*, around a street light on a cold and windy November day at the New York Botanical Garden. A few days later these leaves had all fallen.

EXTERNAL STRUCTURE AND ARRANGEMENT

General Features

Leaves (Fig. 21.1) are the characteristic photosynthetic organs of most vascular plants. They are appendages of the stem, usually having a distinct stalk, the **petiole,** and a thin, flat, expanded **blade** with an upper (adaxial) and a lower (abaxial) surface. Leaves in which the blade is attached directly to the stem, without a petiole, are **sessile.** Often there is a pair of small appendages at the base of the petiole, one on each side; these are the **stipules.**

A large proportion of monocots have sessile leaves with a relatively broad, more or less sheathing base that extends most or all of the way around the stem. In the grasses and sedges and some other monocots, this sheathing base is developed into a well-defined **leaf sheath** that encloses the stem for some distance above the node.

Arrangement

The leaves on a stem are usually arranged in a definite pattern and are typically oriented so that each leaf is exposed to the light with a minimum of interference from its neighbors. The exact pattern or arrangement varies from species to species, and sometimes even on different parts of the same individual. Leaves are said to be **alternate, opposite,** or **whorled,** according to whether one, two, or more than two are borne at each node.

Duration

Leaves have an inherently limited life span, usually only a single growing season (or part of one), seldom more than a few years. Woody plants, in which the aerial stems persist year after year, may be classed as **evergreen** or **deciduous,** according to the duration of the leaves. In evergreen plants, the leaves formed during one season persist at least until the leaves of the following season are pro-

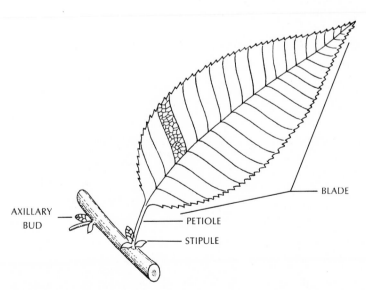

Fig. 21.1 A typical leaf.

AXILLARY BUD

PETIOLE

STIPULE

BLADE

duced; in some evergreens the leaves last several years. Nearly all conifers (gymnosperms) are evergreen, as are most angiosperm trees of moist tropical regions. Only a few angiosperms of temperate regions, such as holly (*Ilex*), live oak (*Quercus* spp.), and sagebrush (*Artemisia tridentata*) are evergreen. There are also some evergreen herbs with basal leaves that persist throughout the winter.

The leaves of deciduous trees and shrubs fall off at the end of the growing season, leaving the stems temporarily bare. Like other physiological responses of plants, leaf fall is governed by a set of interacting environmental factors, including the photoperiod (day length), temperature, and moisture supply. In temperate climates, the leaves commonly develop in the spring and are shed in the autumn. One might assume that temperature plays the dominant role here, and it is certainly the ultimate cause, but for many species, at least, the photoperiod is the actual trigger. The leaves do not begin to develop in the spring until the days reach a certain length, and in the autumn leaf fall is initiated when the day length falls to a crucial point. In such plants, leaf fall can be delayed for some weeks by a night light, as illustrated in the photograph at the beginning of the chapter.

Abscission

Leaves of most vascular cryptogams and herbaceous angiosperms usually wither on the stem and are eventually shed through decay and external causes. Leaves of most gymnosperms (especially conifers) and woody angiosperms (with the notable exception of some woody monocots), on the other hand, are generally shed as a result of changes in a specialized transverse **abscission zone** (Fig.

21.2) at the base, often before the leaf has died. Abscission is the natural separation from plants of leaves, flowers, or fruits.

Abscission of the leaves can be brought about, at least in many plants, by any of several environmental factors, including water deficit, low temperature, reduced light intensity, and decreased day length. Abscission can also be induced by destroying a large part of the blade, or by treating it with various chemicals, such as ethylene and carbon tetrachloride (see also Chapter 26).

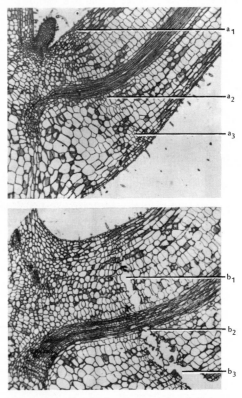

Fig. 21.2 Abscission in petiole of *Coleus*. Developing abscission layer (small cells) at a_1, a_2, a_3. Abscission in progress at b_1, b_2, b_3. Note beginning of destruction of vascular tissue at b_2. [Photomicrographs courtesy of Tillman Johnson.]

Simple and Compound Leaves

The form and structure of the leaf blade vary in different kinds of plants (Fig. 21.3), and even to some extent on different individuals of the same kind. An extensive terminology has been developed to describe the kinds of leaves, due to the

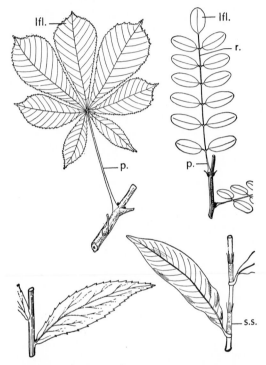

Fig. 21.3 Types of dicot leaves. (*Upper left*) Opposite, palmately compound, exstipulate leaf of horse chestnut, *Aesculus hippocastanum* (× 1/12). (*Upper right*) Alternate, pinnately compound leaf of black locust, *Robinia pseudoacacia,* with opposite leaflets, and with the stipules modified into short spines (× 1/4). (*Lower left*) Alternate, simple, sessile, exstipulate leaf of goldenrod, *Solidago rugosa* (× 1/3). (*Lower right*) Alternate, simple, subsessile leaf of smartweed, *Polygonum*, with sheathing stipules (× 1/3); lfl., leaflet; p., petiole; r., rachis; s.s., sheathing stipules.

usefulness of these characters in recognizing and distinguishing different species. If the blade is all in one piece, the leaf is said to be **simple;** the margin may be **entire** (smooth), or variously **toothed** or **lobed.** If the blade is composed of several separate leaflets, the leaf is **compound.** Compound leaves in which the leaflets are attached nearly at the same point, as in the horse chestnut (*Aesculus hippocastanum*), marijuana (*Cannabis sativa*), and lupines (*Lupinus*), are **palmately compound.** Compound leaves in which the leaflets are attached on opposite sides of an elongate rachis, as in walnuts (*Juglans*), hickories (*Carya*), ashes (*Fraxinus*), roses (*Rosa*), and black locust (*Robinia pseudoacacia*), are **pinnately compound.** Trifoliolate leaves, with three leaflets, may be either pinnately compound, as in poison ivy (*Toxicodendron radicans* and related species, Fig. 21.4), or palmately compound, as in white clover (*Trifolium repens*), according to whether or not there is a definite rachis extending beyond the two lateral leaflets.

Distinction Between Compound Leaves and Leafy Stems

Compound leaves may sometimes be confused with leafy stems, but the distinction is not usually difficult. Compound leaves bear axillary buds just as do simple leaves, but the leaflets do not. The position of the axillary buds, therefore, can be used to mark the base of the leaf. A stem typically ends in a terminal bud with an apical meristem, although in some woody plants the tip is abscised so that the stem ends abruptly just above one of the axillary buds. Leaves, on the other hand, never have a terminal bud, and in seed plants they almost never have an apical meristem. Very often a compound leaf has a terminal leaflet like the other leaflets.

Fig. 21.4   Poison ivy, *Toxicodendron radicans*, climbing on a tree trunk. [Photo by Hal H. Harrison, from National Audubon Society Collection/Photo Researchers, Inc.]

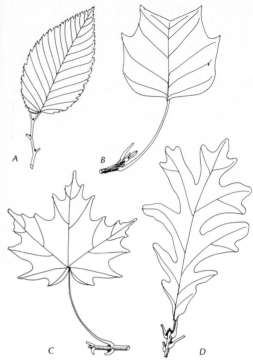

Fig. 21.5   Form and venation of dicot leaves. (*A*) American elm, *Ulmus americana*, with ovate, acute, pinnately veined, doubly serrate leaves. (*B*) Tulip tree, *Liriodendron tulipifera*, with pinnately veined, broadly retuse, four-lobed leaves. (*C*) Sugar maple, *Acer saccharum*, with palmately veined, palmately lobed, coarsely toothed leaves. (*D*) White oak, *Quercus alba*, with pinnately veined, pinnately lobed leaves.

Venation

Leaves of angiosperms characteristically have a branching vascular system derived from the leaf traces of the stem. The principal veins (vascular bundles) are usually readily visible to the naked eye and very often stand out on the lower surface of the leaf. The midrib, in particular, is usually prominent. The smaller veins are evident or obscure according to the kind of plant.

The arrangement of veins in leaf blades varies in detail from species to species, with a strong tendency toward a general difference between monocots and dicots. In most dicots the main veins are either pinnately or palmately arranged (Fig. 21.5), with several orders of smaller branches, and some of the smaller veins form a more or less obvious network (Fig. 21.6). Dicots are therefore usually said to have **net venation. Pinnately veined** leaves, such as those of elms and beeches, have an evident central midvein, with the primary branches departing at intervals along its length. **Palmately veined** leaves, such as those of maples, sycamores, and begonias, have several main veins diverging from the base, with the central one

Fig. 21.6 Veins in a small part of the leaf of lime, *Citrus aurantifolia*, showing the fine network and blind ends of the smallest veins. [From W. H. Brown, *The plant kingdom*, Ginn, Boston, 1935, 1963. Courtesy of M. A. Brown.]

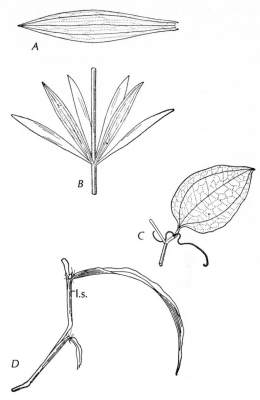

Fig. 21.7 Characteristic monocot leaves. (*A*, *B*) Whorled leaves of *Lilium philadelphicum*; (*C*) crabgrass, *Digitaria*, with a basal sheath (l.s.) and a flat, terminal blade; (*D*) catbrier, *Smilax*, with a petiole, stipular tendrils, and a broad blade.

usually not much, if at all, larger than the others.

In most monocots the leaf has several or many nearly parallel veins (Fig. 21.7) which run the length of the leaf, and the branches that interconnect these longitudinal veins are very small and inconspicuous. Monocots are therefore usually said to have **parallel venation.** Most monocots have relatively long, narrow leaves, with closely spaced longitudinal veins, but some broader-leaved forms, such as *Smilax, Trillium*, and many aroids, do have an evident network of smaller veins between the more widely spaced main ones. Thus although the different types of venation can usually be recognized, no hard and fast line can be drawn between the parallel and the net type, and even less between pinnate and palmate.

There is some reason to suppose that the typical parallel-veined leaf of monocots is, in evolutionary origin, a flattened petiole with the blade suppressed. In this view, monocots with broad, net-veined leaves have undergone a secondary expansion of the petiole tip. Some modern aquatic plants, such as species of arrowhead (*Sagittaria*), show diverse leaf forms, according to the depth of the water, that bolster this interpretation.

Stipules

The basal pair of appendages called stipules are a characteristic part of the leaves of many angiosperms, but they are wanting from gymnosperms and most vascular cryptogams. Their form and function differ from species to species. Most often they are small, green, and rather weakly photosynthetic. In bedstraw (*Galium*), which has sessile leaves, the stipules are enlarged and resemble leaf blades. In black locust (*Robinia pseudoacacia*) they are modified into short, stout spines. In *Smilax* (greenbrier, catbrier) they are modified to form tendrils. In the tulip tree (*Liriodendron*), they form the protective scales of the terminal bud; each leaf, as it expands and diverges from the growing stem, reveals the stipules of the next leaf, which completely cover the bud. The stipules of many plants fall off as the leaf matures, and in many others they are very much reduced. Often there are no stipules at all. Leaves without stipules are said to be exstipulate.

The evolutionary origin of stipules is debatable. Current thinking, based largely on the comparative morphology of living species, is that they probably originated early in the evolutionary history of angiosperms as petiolar flanges serving to protect the terminal bud. In many of the more advanced angiosperms the stipules have been converted to other functions, or reduced to insignificance, or entirely lost.

ANATOMY

A leaf blade ordinarily consists of an upper and lower epidermis, enclosing a chlorophyllous mesophyll that is traversed by a vascular system (Fig. 21.8). These three tissue systems will be discussed in sequence.

Epidermis

The nature and general structure of epidermal tissues have been discussed in Chapter 18. Epidermis of leaves generally serves a protective function, regulating the evaporation of water from the internal tissues and providing a partial barrier to infection and mechanical injury. It also contributes to the even distribution of water and solutes in the leaf. The extensive lateral contact area among the epidermal cells, as contrasted to that of the mesophyll cells, is doubtless responsible for the greater conducting function of epidermis. Sometimes the epidermal cells are thick walled and add significantly to the strength and rigidity of the leaf.

Leaf epidermis is ordinarily covered by a definite cuticle (see Chapter 18), which retards the evaporation of water. The cuticle is usually thicker on the upper epidermis than on the lower, and thicker on plants grown in the sun than in shade.

Often the epidermis is provided with **trichomes** of one sort of another, and is then said to be **pubescent.** A trichome is a small, unicellular or multicellular, often hairlike appendage of the epidermis. Differences in pubescence are very useful to taxonomists seeking precise, describable differences among related species and genera, but their functional significance to the plants is often more debatable. A dense coating of trichomes may retard evaporation under some conditions, and may also protect very young, developing leaves from insect damage. Stems as well as leaves often bear trichomes.

The only intercellular spaces in the epidermis of most leaves are the **stomates** (from Greek *stoma*, a mouth). Each stomate (Fig. 21.9) is bounded by a pair of specialized cells, the **guard cells,** which regulate the size of the opening. The cells adjoining the guard cells may be ordinary epidermal cells, or they may be more or less differentiated as **supporting cells.** In the strictest sense, the term *stomate* applies only to the space between the guard cells, and the stomate plus the guard cells and any supporting cells collectively con-

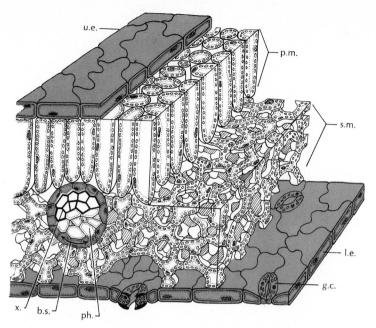

Fig. 21.8 Three-dimensional diagram of a typical leaf, with successive layers stripped away: b.s., bundle sheath; g.c., guard cell; l.e., lower epidermis; ph., phloem; p.m., palisade mesophyll; s.m., spongy mesophyll; u.e., upper epidermis; x., xylem.

stitute the stomatal apparatus. The term *stomate* is often more loosely and conveniently used, however, to apply to the whole stomatal apparatus.

The guard cells, which are usually rather narrowly kidney shaped or broadly sausage shaped, are placed side by side. When the guard cells are turgid, they are bowed apart in the middle, leaving a definite opening between them. When they lose their turgidity, they become appressed to each other throughout their length, closing the opening. The walls of the guard cells are often strongly and unevenly thickened, often with ledges along the upper and lower margins next to the aperture. Unlike ordinary epidermal cells, guard cells contain chloroplasts, and these are indirectly involved in the

mechanism of opening and closing the stomates under some conditions (see Chapter 22). Grasses and sedges have unusual, dumbbell-shaped guard cells, as shown in Fig. 21.9.

The stomates may be equally abundant on both sides of the leaf, but usually they are more numerous on the lower surface. In dicots the stomates are generally scattered at more or less regular intervals over the leaf surface (Fig. 21.10), but in monocots they commonly occur in longitudinal lines between the veins. Stomates of xerophytes (plants adapted to dry conditions) are often individually sunken or grouped in sunken stomatal pockets.

The number of stomates per unit area varies rather widely on different leaves and different individuals of the same

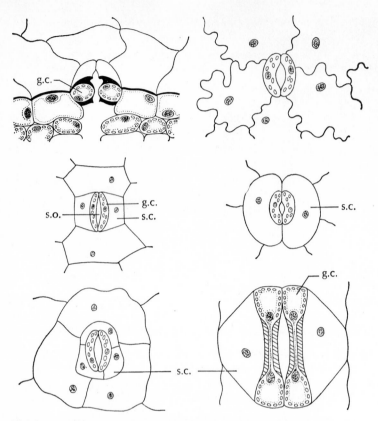

Fig. 21.9 Characteristic types of stomates: g.c., guard cell; s.c., supporting cell; s.o., stomatal opening. The unusual shape of stomates of grasses and sedge is illustrated at the lower right.

species, according to the environment, as well as from species to species. Common figures for the lower epidermis are about 15–1000 per square millimeter, the lower numbers typically occurring in plants with larger stomates. The stomatal opening is very small, commonly 7–40 microns long and 3–12 microns wide when fully open. The fully open stomates usually cover about 0.5–2 percent of the lower epidermis. The role of stomates in controlling transpiration is further discussed in Chapter 22.

The epidermal cells, other than those of the stomatal apparatus, may be all alike, or there may be one or more specialized types in addition to the ground mass. Scattered cork cells, fibers, secretory cells, and crystal-containing cells are not uncommon. Most monocot leaves have some **bulliform** (bubble-shaped) **cells** scattered over the surface or arranged in longitudinal strips. Bulliform cells are relatively large, with thin walls, large vacuole, and little solid content. In some species, changes in the turgor of the bulliform cells govern the lateral rolling and unrolling of the leaf under varying moisture conditions, but in others they have no apparent function.

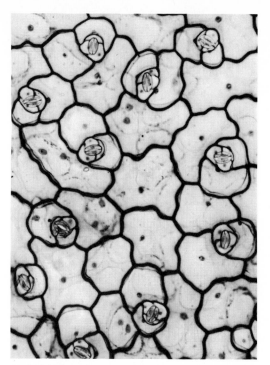

Fig. 21.10 Epidermis of *Sedum* leaf, showing stomates. [Courtesy CCM: General Biological, Inc., Chicago.]

Mesophyll

The mesophyll of a leaf usually consists largely or wholly of chlorenchyma, that is, chlorophyll-bearing parenchyma. Typically it is rather sharply differentiated into an upper layer, the **palisade parenchyma** (palisade chlorenchyma, palisade mesophyll) and a lower layer, the **spongy parenchyma** (spongy chlorenchyma, spongy mesophyll).

The palisade parenchyma typically consists of one or two tiers of elongate, more or less cylindric or prismatic cells just beneath the upper epidermis and perpendicular to it. The cells are fairly closely set, but usually more of the wall adjoins intercellular space than adjoins other cells. Often the cells taper toward

the base, so that the intercellular spaces are wider near the spongy parenchyma than near the upper epidermis.

The spongy parenchyma typically consists of irregularly shaped cells with a high proportion of intercellular space. Many of the cells have knobs or arms that extend out and join other cells. Commonly there is an extra large space, the stomatal chamber, next to each stomate. The spongy layer is usually more than one cell thick, but the cells are not arranged in definite tiers. Cells of the spongy parenchyma generally contain fewer chloroplasts in proportion to their size than do those of the palisade layer.

In addition to being traversed by a vascular system, the mesophyll may also contain laticifers, resin ducts, or oil cavities. It may also contain scattered idioblasts of one sort or another. An idioblast is any single cell that is very different from the cells surrounding it. The functions of idioblasts are varied and often obscure.

Mesophyll is structurally well adapted to the primary function of leaves, which is photosynthesis. The densest tissue and the greatest concentration of chloroplasts are toward the upper surface, which receives the most light. The intercellular space is greatest near the stomates, which are the gateways to the outer air, and diminishes progressively toward the upper surface. Since essentially all the carbon dioxide used in photosynthesis enters through the stomates and is distributed through the intercellular space, the need for intercellular space as a passageway is greatest near the stomates and diminishes progressively toward the upper surface. Evaporation from the palisade cells, which receive more light than the spongy tissue, is partly checked by the relatively small amount of wall surface that is in contact with intercellular space; it is from these exposed wall surfaces that evaporation takes place.

The characteristic modifications in the structure of mesophyll induced by the environment may also be interpreted as adaptive, although there is more room for argument here. The double palisade layer of many sun leaves (in contrast to shade leaves) absorbs light efficiently, and the smaller proportion of intercellular space reduces the evaporating power of the sun, but distribution of the necessary carbon dioxide to the upper palisade cells is also restricted. In any case the leaf does not directly attempt to adapt its structure to the environment. Differences in the environment influence the chemical and physical processes occurring in the cells of the growing leaf, and changes in these processes affect the structure of the mature leaf. Insofar as these responses are beneficial, they have survival value and tend to be preserved (by natural selection) during the course of evolution.

Vascular System

The veins of angiosperm leaves generally form an interconnected branching system, as noted on an earlier page. A vein typically consists of a strand of xylem and a strand of phloem, collectively surrounded by a **bundle sheath.** The phloem is usually on the lower (abaxial) side and the xylem on the upper. Most veins consist only of primary tissue, but the larger veins of many dicots show some secondary growth.

The midrib and often also some of the larger lateral veins of the leaf are usually embedded in a mass of compact parenchyma, rather than being surrounded by a mere bundle sheath. Often there is some strengthening tissue, usually collenchy-ma, sometimes sclerenchyma, associated with the parenchyma. It is generally the parenchyma and the supporting tissues that stand out on the lower surface of the leaf and mark the position of the vein. The remainder of the veins usually have a distinct bundle sheath and are embedded in the upper part of the spongy mesophyll, just beneath the palisade layer.

The bundle sheath is ontogenetically part of the mesophyll, but it is usually discussed along with the vascular system because of its close morphologic and physiologic connection to that system. Usually it consists of a single layer of elongate, thin-walled, living cells, completely surrounding the vein. The long axis of the cells parallels that of the vein, and there are no evident intercellular spaces. The bundle sheath may be thought of as a means of increasing the area of contact between the vascular elements and the more loosely arranged mesophyll cells. Its principal function is to facilitate the conduction of water and solutes from the veins to the mesophyll, and also of solutes from the mesophyll to the veins. In many dicots plates of cells similar to those in the bundle sheath extend part way or completely to one or both epidermises. In some species the bundle sheaths, or their extensions, or both, are thick walled (collenchymatous or sclerenchymatous) and contribute to the mechanical strength of the leaf. In some kinds of plants the bundle sheath plays a special role in photosynthesis, as discussed under C_4 photosynthesis in Chapter 24.

Hydathodes

Leaves of many kinds of plants have special structures called **hydathodes** which permit the extrusion of liquid water under conditions of high water intake and low evaporation. The **guttation drops** that form at the hydathodes are usually seen in the early morning after a cool night and are often mistaken for dew. Hydathodes commonly occur along the margins and tips of leaves; sometimes they form definite teeth (projections from the margin), but most leaf teeth are not hydathodes. Each hydathode has one or more pores which more or less resemble large stomates except that they are permanently open. The nature and arrangement of the tissue beneath the pore vary in different species. Typically a small vein containing only tracheids ends blindly in the loose parenchyma of the hydathode, and there is a large intercellular space, resembling a stomatal chamber, just beneath the pore. Guttation water is further discussed in Chapter 22.

SPECIALIZED LEAVES

We have noted that the principal function of most leaves is photosynthesis. Photosynthetic leaves sometimes also have subsidiary functions, such as storage of food or water. Sometimes the additional functions become more important than photosynthesis, and some highly specialized leaves are not photosynthetic at all. Some specialized types of leaves are discussed in the following paragraphs.

Storage Leaves

Members of the family Crassulaceae and some other families commonly have strongly thickened, succulent leaves with water-storage regions consisting of large parenchyma cells with big central vacuoles that contain hydrophilic colloids. Such storage leaves resist desiccation, and many plants with succulent leaves occur in dry habitats. Some succulents, however, are just as susceptible to drought as other plants.

The ordinary photosynthetic leaves of daffodils and many other plants, particularly of the lily family, are enlarged at the base to form food-storage organs. Bulbs, consisting of a short, erect stem with thickened scale leaves or leaf bases, have been discussed in Chapter 19. Cotyledons, which often serve as storage organs, are discussed in Chapter 25.

Reproductive and Floral Leaves

A few kinds of plants, such as *Kalanchoe*, produce adventitious buds along the margins of the leaves. These buds drop off, develop roots, and grow into new plants. A somewhat similar phenomenon is the development of **bulbils** or bulblets in the leaf axils, or in place of the flowers. These bulbils, consisting of a short bit of stem with one or more small, thickened, modified leaves, drop off and develop into new plants. Many plants can be propagated artificially by leaf cuttings (Fig. 21.11).

Some plants, such as *Poinsettia*, in which the sepals and petals are very small or absent, have brightly colored, petal-like leaves surrounding the flower clusters.

The sepals, petals, stamens, and carpels of flowers, further discussed in Chapter 25, have traditionally been considered to be modified leaves, and are often called floral leaves. This interpretation may not be technically quite correct. Sporangium-bearing leaves (sporophylls) of seed plants may well have originated from telome systems in parallel with the evolution of true, vegetative leaves from telome systems. Stamens and carpels of angiosperms (and the petals of some kinds of flowers) must take their origin directly or indirectly in the microsporophylls and

Fig. 21.11   Leaf cuttings of African violet, *Saintpaulia*, which have taken root in moist sand. [Photo courtesy of T. H. Everett.]

megasporophylls of gymnosperms. Thus there may never have been a time in the evolutionary history of these organs when they were functionally and structurally comparable to vegetative leaves. The common evolutionary antecedent of vegetative leaves, sporophylls of gymnosperms, and floral leaves of angiosperms may well be a telome complex. (See p. 294 for an explanation of the telome theory.) On the other hand, there is no doubt that floral leaves are in an ontogenetic sense serial homologues of vegetative leaves. In the growth of the stem apex they originate in the same way as leaves, as lateral primordia on the apical meristem at sites where leaf primordia might otherwise be expected. The term *phyllome* has been coined to cover true leaves and their serial homologues such as floral leaves, regardless of possible differences in present form and evolutionary origin. This matter is here explored at some length, not to inflict a needlessly complex terminology on the student, but to illustrate the fact that

conceptually necessary generalizations in biology often mask an unsuspected complexity. The careful phraseology of generalizations throughout this textbook reflects an effort to avoid what some biologists have called the necessity to lie a little to the freshmen.

Climbing Leaves

Peas (*Pisum sativum*), sweet peas (*Lathyrus odoratus*), and some other legumes have some of the leaflets of a compound leaf modified into slender, twining tendrils. The stipular tendrils of *Smilax* have already been noted. Similar climbing leaves occur in various other less familiar plants.

Insect-Catching Leaves

Plants of several families which are only distantly related have modified leaves that entrap insects, which are then digested and partly absorbed. These insectivorous plants commonly grow in marshy or aquatic habitats, in which the supply of available nitrogen is severely limited. Nothing larger than an insect is in any danger from them, and stories of man-eating trees are pure fabrication.

Sarracenia, the pitcher plant (Fig. 21.12), has pitcher-shaped leaves lined with stiff, downward-pointing hairs that lead unwary insects into a basal pool of water that contains digestive enzymes. The leaves of sundew (*Drosera*, see Fig. 26.21) are covered with long, stout, spreading hairs, each hair being tipped by a shining, sticky globule containing digestive enzymes. An insect alighting on a leaf is held fast by the sticky secretion of the hairs it touches, and the nearby hairs bend over and often completely cover it. After the insect has been digested, the hairs return to their normal position. *Dionaea*,

Fig. 21.12   Pitcher plant, *Sarrácenia purpurea*. [New York Botanical Garden photo.]

the Venus's-flytrap (see Fig. 1.3) has the leaves "hinged" along the midrib. When two or more of the several trigger hairs on the surface of the blade are touched, the leaf snaps shut and the fringed margins interlock. A few other kinds of plants have still other means of catching insects.

Protective Leaves

Winter buds of most deciduous trees are covered by small, modified leaves called **bud scales.** Bud scales are usually firmer than ordinary leaves, with more sclerenchyma and without chlorophyll. Often they are resinous. When the meristem resumes activity in the spring, the bud scales generally full off, leaving a ring of closely set bud-scale scars around the twig.

The leaves of some kinds of plants, such as cacti, are modified into slender, sharp spines that protect the plant from grazing animals. The spines of many cacti are so numerous and so oily that they can readily be burned off, leaving an unprotected plant body that is palatable to livestock.

AUTUMNAL COLORATION

Leaves of deciduous trees in the temperate zone gradually cease to function as autumn advances, and eventually they are abscised, as noted on an earlier page. A severe frost will kill the leaves of most angiosperms at any time, but the autumnal senescence often occurs well in advance of killing frosts. Leaves of many woody plants take on yellow to red colors during this time, especially if there is a prolonged period of warm, sunny days and cool nights.

Senescence of the leaf commonly does not affect all parts at an equal rate. In many species the chlorophyll is depleted while the carotenoid pigments are still abundant, so that the leaf turns yellow. If tannins are also present, as in some beeches and aspens, the leaf may be bright golden-yellow. In some other species, especially some of the maples and oaks, the phloem tends to become inactive while photosynthesis is still going on, so that sugars accumulate in the leaf. An excess of sugars is commonly associated with the formation of red or purplish anthocyanins, regardless of the season. Leaves with anthocyanins in addition to the photosynthetic pigments are usually dull reddish, but if the anthocyanins persist after the deterioration of the other pigments, as often happens in the fall, very showy colors may result. Autumnal brilliance thus results from the maladjustments associated with senescence and death of leaves.

Brilliant autumnal coloration is not an automatic feature of forested regions. It depends on both the climate and the particular kinds of trees. Some years and some regions produce much better displays than others. Autumnal coloration in Europe is usually much less spectacular than in the northeastern United States.

SUGGESTED READING

Ryder, V. L., On the morphology of leaves, *Bot. Rev.* **20**:263–270, 1954.

Schnell, D. E., *Carnivorous plants of the United States and Canada*, Blair, Winston-Salem, N.C., 1976. Beautiful color photographs; botanically accurate.

Chapter 22

Angiosperms: Water Relations

Saguaro cactus, *Cereus giganteus,* in the desert of southern Arizona.

Water is the chief component of all physiologically active protoplasm. The volume of water in the vacuole of a normal cell usually much exceeds the volume of the protoplast itself, and any great loss of water from the vacuole generally sends the protoplast into a state of relative inactivity, if it is not actually fatal. Under normal atmospheric conditions, water evaporates from any exposed moist surface, living or dead. The exchange of gases necessary for photosynthesis requires (in land plants) that moist cell surfaces be exposed to the air, thus inevitably resulting in some loss of water. The means by which water is absorbed and translocated, and by which its evaporation is prevented or retarded, are therefore highly important in the physiology of the plant.

ABSORPTION BY ROOTS

Roots are the organs of vascular plants through which most or all of the water is generally absorbed. The expanded absorbing surface provided by root hairs and by the fungal component of mycorhizae is mentioned in Chapter 20.

The root-hair zone is the most important water-absorbing part of the root, at least in plants that do not have extensive mycorhizae. Below the root-hair zone, absorption is limited by the smaller surface area and the lack of an effective conducting system to carry the water away. Above the root-hair zone, the surface of the root is usually suberized, and the rate of water intake is much reduced.

Absorption of water by roots is due chiefly to the difference in water potential between the cell sap and the soil water, according to the principles elucidated in Chapter 6. The concentration of solutes in the cell sap is ordinarily higher than that in the soil water, and other factors are seldom great enough to upset the resulting osmotic relationship. The osmotic potential of root-hair cells and other epidermal cells of roots is commonly in the range of -3 to -5 atm, and the turgor pressure is commonly in the range of 1 to 4 atm. The net water potential of the cells, after allowing for the modifying effect of turgor pressure, typically remains in the range of -1 to -4 atm. The osmotic potential of the soil solution, on the other hand, is seldom more than a fraction of an atmosphere, and under normal moisture conditions the matric forces (attraction between water molecules and soil particles) add only another small fraction of an atmosphere to the water potential. Thus there remains a difference of from a fraction of an atmosphere to several atmospheres between the water potential of the cell sap and that of the soil solution, and water moves into the cell.

When the soil is moist, roots readily absorb water from it if there is the usual difference in concentration of solutes in the soil water and in the cell sap. As the plant continues to draw water from the soil, and as natural evaporation continues, the film of water around each soil particle becomes progressively thinner and more tightly held. Eventually the remaining water is so strongly held that it cannot be absorbed by the plant, regardless of the relative concentrations of solutes. A convenient and important measure that approximates or just precedes this stage is the **permanent wilting percentage,** or **wilting coefficient,** defined as the percentage of water (in relation to dry weight) to which a soil is reduced when the plants growing in it have just reached a condition of "permanent" wilting. A "permanently" wilted plant, in this context, is one that will not recover its turgidity unless water is supplied to the soil. It should be understood that leaves with a high proportion of mechanical tissue, such as those of many broad-leaved evergreens, can be "wilted" in a physiological sense without showing the characteristic flaccidity of other wilted leaves. Some desert plants can maintain life and slowly withdraw water from the soil for some time after "permanent" wilting has occurred, but not much water is removed from the soil under these conditions.

Although the amount of water in the soil at the permanent wilting point varies greatly in different soils, it does not vary much for different kinds of plants growing in the same soil (except in highly alkaline soils, discussed below). Individual species seldom diverge as much as 10 percent from the general average in the amount of water they leave in the soil at the permanent wilting point. The reason for the similarity of different kinds of plants in their ability to absorb water from the soil is clear. The tenacity with which water is held by the soil, as measured by the water potential, begins to increase rapidly as the wilting percentage is approached, and beyond the wilting percentage it increases very rapidly indeed. A large increase in the water potential of the cell is therefore necessary to effect only a slight reduction in the permanent wilting percentage. Any very significant reduction of the wilting percentage would require an impossibly high water potential in the cell.

Halophytism

In most soils under ordinary conditions there are not enough solutes in the water to interfere much with the absorption of water by the roots. The concentration of solutes in the soil moisture of alkaline or saline soils, on the other hand, is often greater than that in the cell sap of most plants. Plants that have a relatively high water potential (osmotic potential, less the turgor pressure), enabling them to grow in such soils, are called **halophytes.** The high osmotic potential of the cell sap of halophytes is caused partly by the high concentration of solutes, and partly by the presence of hydrophilic colloids. Osmotic potentials as high as −100 atm are frequently recorded for halophytes. These figures should be regarded as very rough approximations, however, because of difficulties in the technique of measurement.

As might be expected, there are varying degrees of halophytism in different kinds of plants. Species of shadscale (*Atriplex*), greasewood (*Sarcobatus vermiculatus*), and winter fat (*Eurotia lanata*) are important moderate halophytes in desert areas of the interior western United States. In more salty areas these give way to extreme halophytes, such as samphire (*Salicornia*) and pickleweed (*Allenrolfea*) (Fig. 22.1); and in much of the extensive salt flats west of Great Salt Lake in Utah, no plants at all are found.

Many halophytes, especially the more extreme ones, are succulent, with soft tissues that have an obviously high water content. The succulent habit is discussed later in this chapter.

Continued irrigation of fields frequently leads to increased salinity of the soil and decreased growth of crops, ultimately making the soil too salty for agriculture. The irrigation water does not sink deep enough into the soil to reach any permanent water table that may exist, and eventually most or all of it evaporates from the soil or from plants that have absorbed it. The solutes carried in by the irrigation

Fig. 22.1 Hummocks of pickleweed, *Allenrolfea*, on salt flats near Great Salt Lake in Utah. Photo taken in late May, when the new growth for the year was just well started.

water remain in the soil, and additional amounts are carried upwards from the subsurface soil, as water that has percolated downward for some distance rises again and evaporates. The build-up of salts can be postponed or countered by occasional flushing with larger amounts of water, but this practice increases the salinity of the rivers and exacerbates the problem when the water is used downstream for irrigation. Agriculture along the Colorado River near its mouth in northwestern Mexico has suffered severely from the increasing salinity of the river water, resulting in large part from such flushing in the United States. Irrigation agriculture in past societies has always been temporary. It remains to be seen

whether our society will be able to do better.

Metabolic Factors In Water Absorption

Absorption of water by roots is discussed in the foregoing paragraphs as a purely passive process resulting from differences in water potential, but that is not the whole story. There is in general a positive correlation between the rate of respiration and the rate of water absorption by roots. It was at one time thought that respiratory energy might be used directly in water absorption, but after decades of experimentation and controversy this view is now generally rejected. The relationship, instead, is indirect. The intake of minerals (discussed in Chapter 23) is accomplished against the concentration gradient by the expenditure of respiratory energy. The water rising in the xylem carries some of the minerals with it, and more minerals must then be taken in to maintain the requisite concentration for water absorption. Otherwise the water potential of the cells would drop, and absorption would be impeded or stopped.

ABSORPTION BY PLANT PARTS OTHER THAN ROOTS

Although roots are the organs through which water is usually absorbed, any plant organ that is not thoroughly waterproof may absorb water when wet. It has been experimentally demonstrated that some species can absorb water vapor from saturated or nearly saturated air, and the leaves of redwood (*Sequoia sempervirens*) evidently absorb a considerable amount of water from the fog in which they are frequently bathed along the California coast. The leaves of a few kinds of desert plants can even absorb water from air that is merely moist but not saturated. Epiphytes very often absorb most of their water through the leaves and stems during and after rainstorms.

MOVEMENT OF WATER INTO THE XYLEM

Most of the water absorbed by the epidermal cells of the root is passed from cell to cell across the immature cortex and into the conducting elements of the xylem. Much of the movement takes place through the cell walls, which are wholly permeable to water, rather than through the protoplasts of the intervening cells. The detailed mechanism of the transfer is

not yet fully understood, but the principle is simple enough. There must be a progressive gradient of water potentials (see Chapter 6), with the epidermal cells having the lowest water potential and the xylem cells the highest (as measured in negative numbers). We shall see on a subsequent page that water in the xylem is often under tension instead of pressure; this tension contributes to the water potential of the xylem cells.

ASCENT OF THE SAP

The mechanisms by which the xylem sap is raised from the roots to the leaves has been intensively studied. Most botanists are agreed that a theory developed largely by the British botanist H. H. Dixon (1869–1953, Fig. 22.2) and expounded in a series

Fig. 22.2 H. H. Dixon (1869–1953), British botanist, whose theory of cohesion is generally accepted as the principal explanation for the rise of water in plants. [Courtesy of the University of Dublin.]

of papers dating from 1894, is essentially correct. Root pressure also has some significance, especially in herbs and shrubs. Other mechanisms that have been suggested are probably negligible, although the possibility cannot yet be excluded that the living parenchyma cells of the xylem are an essential accessory in the process.

Dixon's Theory of Cohesion

The basic principle behind Dixon's theory of cohesion is the attraction of water molecules for each other. In large masses of water this attraction is not very obvious, but in long, slender tubes it is readily demonstrable, becoming greater as the tube grows more slender. A column of water in a tube the thickness of a xylem vessel has a strong resistance to being broken, and a pull at the top of the column is transmitted throughout its length. The water can thus be pulled upward, much as a rope can be pulled up through a pipe. The loss of any water from the cells of the leaf, either by evaporation, or by use in photosynthesis or other metabolic processes, transmits a strain to the adjacent cells and thus eventually to the water columns in the xylem, so that the water tends to move up through the xylem and from cell to cell of the leaf, replacing the loss.

Experimental measurements and calculations indicate that the tensile strength of water in the xylem elements is more than enough to withstand the stress of being pulled to the top of the tallest trees, even when the resistance from friction against the cell walls is figured in. The characteristic vessels of angiosperm xylem are of course more efficient water conductors than the tracheids of angiosperms and gymnosperms, but the cell wall of both vessels and tracheids is fully permeable to water, and thus the water columns remain functionally intact.

There is no doubt that the water in the xylem is often in a state of tension. This can be demonstrated in living plants by jabbing an individual vessel of an appropriate herb with a fine needle, after careful dissection to expose the xylem. The water column can be observed to jerk apart at the point of rupture. Tension in the xylem sap can also be demonstrated by other means, but it is hard to measure.

Significance of Xylem Parenchyma in the Ascent of the Sap

The water columns in the xylem of woody plants do finally break after a period of years, and the older sapwood is progressively converted into nonconducting heartwood, as described in Chapter 19. The precise causes of these changes are not yet fully understood. A point of view once held by many botanists is that the living parenchyma cells of the xylem have some special function in keeping the conducting cells in operating condition, or even in the actual conduction, and that the eventual senescence and death of these parenchyma cells necessarily results in cessation of conduction. The characteristic structure of xylem, with every conducting cell in contact with at least one parenchyma cell, certainly lends support to this view; but attempts at a more direct demonstration of this function have given negative or inconclusive results, and botanists now tend to be skeptical about the whole idea.

Possibly the relationship is the reverse of what has been suggested, and it is the breakage of the water columns in the vessels and tracheids that leads to the death of the associated parenchyma cells.

The dying parenchyma cells then produce the resins that turn the sapwood into heartwood and help ward off decay.

Root Pressure

Under conditions of ample moisture, adequate soil aeration, and low transpiration, water may be absorbed so vigorously by the roots that a positive pressure, rather than a tension, develops in the xylem sap. This pressure, called **root pressure,** causes the sap to rise in the xylem. When the sap is under pressure, it will exude from any cut surface, and often from hydathodes and stomates as well. The formation of guttation drops (Fig. 22.3) is commonly due to root pressure. Measurements of the pressure of xylem sap seldom exceed 2 atm, and even this amount can seldom be maintained when there is much transpiration. Root pressure, therefore, must be regarded as of relatively minor importance in the rise of sap, superseding the Dixon mechanism only under favorable conditions, and even then seldom causing the sap to rise more than about one or two meters above the ground.

Conditions favorable for the development of root pressure in woody plants (and consequent exudation from cuts) often occur early in the spring, before the leaves have expanded, thus fostering the popular idea of a rise of sap in the spring. Maple syrup is made from the sap collected from the sugar maple (*Acer saccharum*) under such conditions. In this species, as in some other maples and other trees, a considerable amount of sugar is stored over the winter in the sapwood.

TRANSPIRATION

Only a small fraction of the water absorbed from the soil is permanently retained by the plant. The rest is lost, chiefly by evaporation into the air. Evaporation and consequent loss of water from a living plant is called **transpiration.**

Physical Basis of Transpiration

Evaporation is a physical phenomenon that results from the escape of more molecules of water (or any other liquid) into the air than are returned to the evaporating surface from the air. When a balance is reached, so that as many molecules return from the air to the liquid as escape from the liquid into the air, the air is said to be **saturated.** Under ordinary conditions, the walls of living plant cells, unless cutinized or suberized, are permeated with water and present a moist surface to any intercellular spaces. The air in the intercellular spaces, in turn, is nearly or quite saturated with water vapor.

The part of the air pressure that is due to water vapor is called the **actual vapor pressure;** the vapor pressure at the saturation point is called the **saturation vapor pressure.** The ratio between the actual

Fig. 22.3 Guttation drops on a strawberry leaf. [Photo courtesy of J. Arthur Herrick.]

vapor pressure and the saturation pressure under the same conditions is called the **relative humidity** and is usually expressed as a percentage figure. The saturation vapor pressure of water increases directly with the temperature, although the relationship is not strictly linear; an increase in temperature from 20° to 30°C nearly doubles the saturation vapor pressure.

Rate of Transpiration

Most of the evaporation of water from living plant tissues occurs into the air of the intercellular spaces. The water vapor of the intercellular spaces then diffuses out into the air mainly through the stomates, lenticels, and any other openings to the surface that may exist. The speed of diffusion through a given opening is governed largely by the steepness of the gradient between the vapor pressure of the air in the intercellular spaces and that of the outside air. It also increases directly with the temperature, rising about 20–30 percent with each increase in temperature of 10°C. Thus the potential rate of transpiration, under a given condition of relative humidity, increases very markedly with the temperature, because of the greater difference between the actual vapor pressure and the saturation pressure at higher temperatures, and also because of the greater speed of the molecules at higher temperatures. The relationship between temperature and saturation vapor pressure is itself a reflection of the differing speeds of molecules at different temperatures.

Any part of the plant, even the roots, may transpire under appropriate conditions, but by far the greatest amount of transpiration usually occurs through the leaves, due to the presence of stomates and to the large surface area exposed to the sun and air. The epidermis of many stems also has stomates, but the surface area of stems is generally less than that of leaves, and older stems commonly develop a more or less waterproof periderm marked by scattered lenticels that permit only minimal transpiration. The older parts of roots are also protected by a waterproofing layer, and soil conditions seldom favor much evaporation from the absorbing parts.

The rate of transpiration from leaves is governed by the environmental conditions and by the structure and physiological condition of the leaf. Under conditions favorable for transpiration, the rate in most angiosperms is 0.5–2.5 (rarely 5) gm per square decimeter of leaf area per hour. The leaf area here is calculated by simple length-width measurements, disregarding the additional surface presented by the opposite side of the leaf. Many herbaceous plants will transpire several times their own volume of water in a single day. It has been estimated that a field of corn may transpire enough water during the growing season to cover the field to a depth of about 4 dm, and the rate of transpiration from a deciduous forest is probably even higher.

Importance of Transpiration

The universal occurrence of transpiration in land plants has led to the supposition that it must have some value, and several values have been suggested. None of the suggested uses withstands examination, however, and it is now believed that transpiration is merely a necessary evil. The exchange of gases necessary for photosynthesis requires the exposure of moist cell surfaces to the air, and this exposure inevitably results in transpiration. Instead of being useful, transpiration is one of the most serious hazards to the growth of land plants.

Control of Transpiration

The features of the plant that influence the rate of transpiration per unit area are chiefly the size, number, position, and degree of opening of the stomates (Fig. 22.4), the thickness of the cuticle, the amount and distribution of intercellular spaces in the mesophyll, and the proportion of hydrophilic colloids in the cell. The stomates are the principal avenues through which water vapor escapes from,

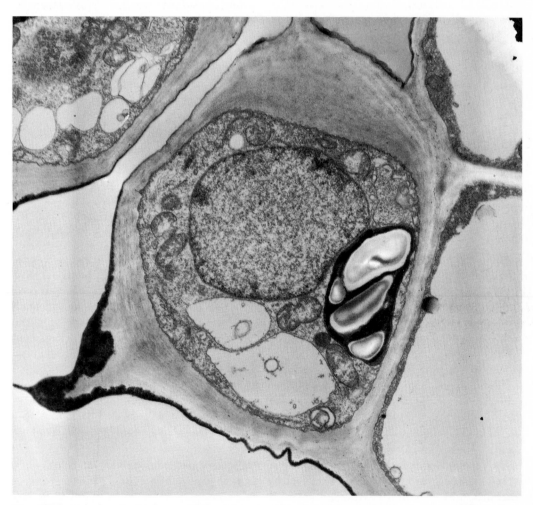

Fig. 22.4 Guard cell of tobacco-leaf epidermis. The stomatal opening here comes in from the lower left. The dark body with white enclosures (at right) is a chloroplast with included starch grains. The large nucleus lies just to the left of the chloroplast. Two vacuoles (bottom) and several mitochondria can also be seen. (× about 10,000.) [Electron micrograph courtesy of T. E. Jensen.]

the plant. Although the cuticle of the epidermis and the suberized layers of stem and roots are not absolutely moistureproof, the amount of transpiration that they permit is small compared to that permitted by the open stomates. It has been experimentally demonstrated that numerous small openings permit much more rapid diffusion than a single large opening of the same total area, the diffusive capacity of pores in certain size ranges being proportional to the perimeter rather than to the area. Thus although the fully open stomates commonly cover only about 0.5–2 percent of the lower epidermis of a leaf, it appears that their combined diffusive capacity is often more than half that of an opening the size of the leaf itself.

Transpiration through the stomates is so great that most plants could not long survive under natural conditions if the stomates were permanently open; the stomates are, in fact, closed much of the time. The precise responses to environmental conditions vary in different kinds of plants, but in general the stomates close or remain closed whenever (1) there is any serious water deficiency in the leaves; (2) there is little or no light; or (3) the temperature is low. Any one of these three conditions usually induces and maintains closure of the stomates, regardless of the other two, although, as we shall see, the stomates of succulents are typically open at night. The net result of these responses to the environment is that stomates are usually open only when photosynthesis creates a need for exchange of gases with the air, and then only if the water supply is reasonably adequate.

The means by which the environmental factors govern the changes in the guard cells have been intensively investigated, with only partial success. It is clear that in general they induce changes in the osmotic pressure of the guard cells, resulting in changes in turgor, which in turn cause changes in shape. An increase in turgor of the guard cells, either absolutely or with respect to that of the other epidermal cells, causes the stomatal opening to widen, and vice versa. Oftentimes there is a threshold effect, so that the stomates open or close rapidly, in a minute or less.

Changes in the osmotic pressure of the guard cells, when the supply of water is adequate, are due principally to reversible changes in the concentration of solutes, especially in the proportion of soluble to insoluble carbohydrates. The guard cells, unlike ordinary epidermal cells, regularly have chloroplasts. They generally contain a considerable amount of carbohydrate, and changes in the proportion of soluble to insoluble carbohydrates influence the concentration of solutes and thus the osmotic pressure.

These changes in carbohydrates depend on complex balances in other factors, including the pH (acid-alkaline balance; see Chapter 23) of the cell sap, which in turn is controlled by factors such as the rate of photosynthesis and the degree of turgor. A relatively high concentration of carbon dioxide favors closing of the stomates, possibly through its effect on the acid-alkaline balance of the cell sap. (Carbon dioxide reacts with water to produce the weak acid H_2CO_3.) A low concentration of carbon dioxide, such as would follow photosynthesis in the guard cells, favors opening of the stomates.

The supply of solutes in the guard cells is also regulated to some extent by influx and outflow of potassium ions, with resultant changes in osmotic potential and turgor. That of course brings us to the still unanswered question of what controls the movement of the potassium. As so often in science, the answer to one question poses another.

The effects of the structural and physiological features of the plant on the rate of transpiration are for the most part fairly obvious. A thick cuticle, a high proportion of hydrophilic colloids, and a low proportion of cell-wall surface exposed to intercellular space all tend to reduce the rate. Restriction of the stomates to the lower surface of the leaf, as in many plants, minimizes the direct effect of the sun in promoting transpiration. Features that tend to limit transpiration are further discussed on subsequent pages in connection with xerophytes.

WATER STRESS

The water supply is less than optimal for most plants much of the time, and the cells are less than fully turgid. Under conditions of mild water stress, growth slows down. Synthesis of proteins and cell wall components in rapidly growing tissues is retarded. A slightly greater stress also impedes cell division. A logical interpretation of these facts is that plants have evolved controls that slow down increase in cell size when low turgor prevents cell expansion. The smaller size of cells in summer wood, as compared to spring wood, reflects the fact that cell enlargement is more sensitive to water stress than is cell division.

A further increase in water stress causes the stomates to close. The closure reduces the loss of water, but it also interferes with uptake of carbon dioxide and puts a brake on photosynthesis. Progressive increase in water stress causes a substantial decline in the rate of respiration and of translocation of photosynthate, followed by a slowing of ion transport and an increase in the level of

hydrolytic enzymes. Still greater stress causes imbalances among amino acids, with accumulation of proline. Water columns in the xylem may be broken, and senescence of older leaves is accelerated. Finally, the cell may become plasmolyzed, with the plasma membrane shrunken away from the wall over most of its surface.

These responses to water stress are to a large degree reversible as the supply of water is increased. Some changes, however, cannot be reversed. Broken water columns in the xylem cannot ordinarily be rejoined if they are beyond the reach of root pressure. Senescent leaves cannot be rejuvenated, and in most kinds of plants continued plasmolysis of cells soon kills them.

HYDROPHYTES, MESOPHYTES, AND XEROPHYTES

Plants may be classified ecologically as hydrophytes, mesophytes, or xerophytes, according to the abundance of water in their habitat. **Hydrophytes** grow in water; **xerophytes** grow in very dry places; **mesophytes** grow on land but not in extremely dry sites. The vast majority of angiosperms are mesophytes. There is every degree of transition among these three types, but it is possible to note some features characteristic of each.

Hydrophytes (Fig. 22.5) have an abundant supply of water, but a scanty supply (at least for the submerged parts) of oxygen. Their structure generally reflects an adaptation to these conditions. The submerged parts usually lack stomates and have little if any cutin, but they commonly have an extensive system of air cavities, often reaching even into the roots. Much of the oxygen released in photosynthesis

Fig. 22.5 Hydrophytes at the New York Botanical Garden. Parrot's feather, *Myriophyllum brasiliense,* in the immediate foreground; water hyacinth, *Eichhornia crassipes,* behind the *Myriophyllum;* papyrus, *Cyperus papyrus,* the tall plants toward the rear; water lily, *Nymphaea,* left foreground; Oriental lotus lily, *Nelumbo nucifera,* emergent leaves toward the rear at the left. [New York Botanical Garden photo.]

is retained in the air cavities and is available for use in respiration.

The roots of hydrophytes grow well in saturated soil and absorb water from it, but the roots of most mesophytes and xerophytes do not penetrate very far into saturated soil and absorb little or no water when the soil is saturated. The inability of mesophytes and xerophytes to grow and absorb in saturated soils is apparently due to the low oxygen supply and consequent interference with respiration and other metabolic processes that maintain conditions favorable for absorption. The ability of hydrophytes to tolerate saturated soils

is apparently due, in some cases, to an aerating system, and in others to a greater ability to carry on anaerobic respiration, or to a combination of these two factors.

The aerial parts of hydrophytes are subject to environmental conditions approaching those of mesophytes, and the similarity of environment is reflected in the structure. Cuticle and stomates are usually well developed, and aerating tissue is generally much less prominent.

Xerophytes face, in an exaggerated form, the same type of conditions that confront mesophytes: the supply of soil moisture is likely to be inadequate, and the aerial parts need some protection from evaporation that will not also prevent exchanges of gases with the atmosphere. The standard adaptation to these conditions, in both mesophytes and xerophytes, is the protection of the photosynthetic tissue by an epidermis that is covered by a cuticle and interrupted by stomates. Older or nonphotosynthetic tissues are often covered by a periderm with scattered lenticels. Most xerophytes also have some additional means of combating drought, as indicated below.

Xerophytes may mostly be divided into three ecological (not taxonomic) groups: (1) ephemerals, (2) succulents, and (3) drought-enduring species.

Ephemerals (Fig. 22.6) are short-lived annuals that complete their life cycle during a relatively short period (often only a few weeks) when the soil is moist and survive the dry periods as seeds. They are a conspicuous feature of the landscape in the spring during a good year in the southwestern United States and in similar areas elsewhere in the world. During the period of their growth, ephemerals are no more drought resistant than typical mesophytes, and they are often classified as drought-escaping mesophytes. The ephemerals belong to many different taxonomic families, all of which also contain many ordinary mesic species.

Succulents (Fig. 22.7) are characterized by the accumulation of reserves of water in the fleshy stems or leaves, due largely to the high proportion of hydrophilic colloids in the protoplasm and cell sap. The photosynthetic parts generally have a very thick cuticle, and sometimes the leaves are very much reduced, as in cacti.

Succulents commonly have the stomates open at night, and during dry weather they may be open *only* at night. This pattern does not restrict photosyn-

Fig. 22.6 *Monoptilon bellidiforme,* an ephemeral annual, in the Mojave desert of California. Annuals in the Mojave desert are typically winter annuals, which germinate in the fall, grow slowly and inconspicuously during the winter, and burst into bloom in the spring, setting seed and dying soon thereafter. In some deserts, such as the Chihuahuan desert farther east, which have a different seasonal pattern of precipitation, ephemeral annuals commonly germinate, flower, set seed, and die within a period of a few weeks. [Photo by Rupert C. Barneby.]

thesis as much as might be expected. Carbon dioxide is taken up and chemically bound into some organic acids in the

Fig. 22.7 Miscellaneous succulents. (*From left to right*) *Opuntia, Aloe obscura, Echeveria corderoyi, Crassula argentea, Agave horrida.* [New York Botanical Garden photo.]

cells during the night. In the daytime these acids break down, liberating carbon dioxide, which is used in the "dark reactions" of photosynthesis to make carbohydrate. Thus the energy-trapping early steps of photosynthesis are carried on behind tightly closed stomates during the day, and the carbon dioxide that is to be incorporated into photosynthate is taken on at night, when the stomates can be open without an undue amount of transpiration. The photosynthetic pattern in which carbon dioxide is taken on at night is called **crassulacean acid metabolism** (or **CAM**), from the family Crassulaceae, in which it is well developed. CAM is further discussed in Chapter 24.

Several quite different families of angiosperms, such as the Cactaceae, Crassulaceae, and Aizoaceae, are characteristically succulent, and several others, such as the Euphorbiaceae and Liliaceae,

also contain many succulent species. The whole order Caryophyllales, to which the Aizoaceae and Cactaceae belong, shows a tendency toward producing succulent species and other species with a high water potential. The family Chenopodiaceae, of this order, has many halophytic species, and a good number of these halophytes are also succulent.

Drought-enduring species are characterized by the ability to endure desiccation without irreparable injury. Creosote bush (*Larrea tridentata*, Fig. 22.8), a characteristic species of the deserts of southwestern United States and northern Mexico, is highly drought enduring. It is evergreen, retaining many of its small, firm leaves throughout the dry season. During dry periods, the water content of the leaves may sink to less than 50 percent of their dry weight, in contrast to the leaves of most woody mesophytes, in

Fig. 22.8   Creosote bush, *Larrea tridentata,* in the desert of southern Arizona.

which the water content ranges from 100 to 300 percent of the dry weight. Under such conditions the leaves are of course largely dormant; only when the water supply again becomes adequate do they return to physiological activity.

Many drought-enduring species have firm leaves with a high proportion of sclerenchyma or other strengthening tissue. Such plants are called **sclerophylls.** When moisture is plentiful, there is no great difference in transpiration rates between sclerophylls and typical mesophytes. The strengthening tissue apparently serves mainly to prevent mechanical injury while the leaf is physiologically wilted during periods of drought.

A variety of other adaptations aid drought resistance of some species. Frequently the aerial part of the plant is very small in proportion to the root system. The cuticle is commonly rather thick, and the stomates may be sunken in pockets or protected by pubescence. (Pubescence is not restricted to xerophytes, however, and

its significance in restricting transpiration has been overestimated.) The leaves sometimes fall off during the dry season, minimizing the loss of water during the period of shortest supply. The deciduous habit of most mesophytic temperate-zone angiospermous trees is of course merely another expression of this same sort of adaptation to drought. In tropical regions with pronounced wet and dry seasons, leaf fall of deciduous trees is related to the dry season rather than the temperature.

PHYSIOLOGICAL DROUGHT

When a considerable amount of water is present in the soil but not readily available for absorption, there is said to be a **physiological drought.** A high concentration of solutes in the soil water, as discussed on an earlier page, is a typical cause of physiological drought. Another is the freezing of the soil water in the winter. Some transpiration continues to occur even from the bare branches of deciduous trees in the winter, and winter killing of trees in the northern United States and in Canada is often due to physiological drought rather than to the direct effect of low temperature on the protoplasm.

SUGGESTED READING

Briggs, G. E., *Movement of water in plants,* Blackwell Scientific Publications, Oxford, England, 1967.

Gates, D. M., Heat transfer in plants, *Sci. Amer.* **213**(6):76–84, December 1965.

Hadley, N. F., Desert species and adaptation, *Amer. Sci.* **60**:338–347, 1972.

Kramer, P. J., and T. T. Kozlowski, *Physiology of woody plants,* Academic Press, New York, 1979.

Levitt, J., The mechanism of stomatal action, *Planta* **74**:101–118, 1967.

Penman, H. L., The water cycle, *Sci. Amer.* **223**(3):98–108, September 1970.

Scholander, P. F., Tensile water, *Amer. Sci.* **60**:584–590, 1972.

Solbrig, O., and G. H. Orians, The adaptive characteristics of desert plants, *Amer. Sci.* **65**:412–421, 1977.

Tibbitts, T. W., Humidity and plants, *Bio-Science* **29**:358–363, 1979.

Zelitch, I., Control of leaf stomata—their role in transpiration and photosynthesis, *Amer. Sci.* **55**:472–486, 1967.

Chapter 23

Angiosperms: Mineral Nutrition

California pitcher plant, *Darlingtonia californica,* in a bog near the coast in Oregon. Pitcher plants characteristically grow in boggy places. In such habitats the supply of available nitrogen is limited, and pitcher plants augment their supply by digesting insects that are caught in their pitcherlike leaves.

415

The elements essential for plant growth and their roles in cellular metabolism have been discussed in Chapter 4. Among land plants, the soil is the usual source for all these elements except carbon and some of the oxygen. The carbon is obtained directly from the air, as carbon dioxide. Some of the necessary oxygen is obtained from carbon dioxide and some from the water and other substances in the soil. The ability of many legumes and some other higher plants to augment their supply of nitrogen through symbiosis with nitrogen-fixing bacteria is well known, as is also the insectivorous habit of some bog plants, but the majority of land plants are directly dependent on the soil as the sole source of nitrogen.

In plant physiology it is customary to apply the term *minerals* to the inorganic substances, except water, that are ordinarily obtained from the soil. In this chapter we shall consider the occurrence, absorption, and transportation of minerals, and something of their roles in the metabolism of the plant.

Most minerals are absorbed in the form of ions, which are electrically charged particles that result from dissociation of molecular components in solution. Some of the molecules of the compound calcium nitrate $Ca(NO_3)_2$, for example, dissociate in solution to form calcium (Ca^{2+}) and nitrate (NO_3^-) ions. Ions with a positive electric charge, as indicated by the plus (+) sign, are called **cations,** and those with a negative charge, as indicated by the minus (−) sign, are called **anions.**

THE MEANING OF pH

One of the important factors influencing the availability of minerals is the degree of acidity or alkalinity of the soil solution. A very small fraction of the molecules of water dissociate to form H^+ and OH^- ions. The H^+ ions are the essential particles of an acid, and OH^- ions are the essential particles of a base. When there is an equal number of H^+ and OH^- ions in solution, as in pure water, the solution is neutral; if there are more H^+ than OH^- ions, it is acid; if there are more OH^- than H^+ ions, it is basic. In pure water the fraction of molecules that dissociate is about 1/10,000,000, a figure which may be more conveniently expressed as $1/10^7$. If more H^+ ions are added from any source, the number of OH^- ions diminishes, because some of the OH^- ions become bound to the new H^+ ions. If more OH^- ions are added, the number of H^+ ions decreases. The ratio of H^+ ions to intact molecules of water can thus be used as an indicator of the degree of acidity or alkalinity (basicity) of a solution. The power of 10 represented by the denominator of the fraction is used as a convenient measure of the proportion of H^+ ions, and is called the pH (an abbreviation for "*Potenz* (= power) Hydrogen"). Thus a pH of 7 indicates a neutral solution, with a ratio of $1/10^7$ (= 1/10,000,000) between H^+ ions and molecules of intact water. A pH of less than 7 indicates an acid solution, and a pH of more than 7 indicates a basic solution. The higher the pH, the fewer H^+ ions in solution, and vice versa.

It should be noted that pH is a logarithmic rather than a simple arithmetical measure. At a pH of 6, there are 10 times as many H^+ ions as at a pH of 7, and at a pH of 5, there are 100 times as many H^+ ions as at a pH of 7; likewise, pH's of 8 and 9 indicate concentrations only 1/10 and 1/100 as great, respectively, as a pH of 7.

Different kinds of acids and bases differ in both their degree of solubility in water and in the extent to which they dissociate in solution. A strong acid or base is one in which a high proportion of the molecules dissociate, thus releasing a high proportion of H^+ or OH^- ions.

OCCURRENCE OF MINERALS

Soil ordinarily consists of a mixture of organic and inorganic soil particles, together with some water, organic and inorganic solutes, and air. In a broad sense, all inorganic material in the soil is mineral matter; thus a soil with little or no organic matter is called a mineral soil. Only particles of molecular and submolecular size can be absorbed by plants, however, and it is with these very small particles that we are here concerned.

Mineral ions occur in the soil both free in solution and adsorbed to soil particles of colloidal size. The adsorption of ions is due to the electrical imbalance (charge) of the clay particles and organic compounds that make up the colloidal fraction of the soil. The imbalance of the clay particles is usually due to an oversupply of electrons, and these particles, therefore, have a particular affinity for the cations of the soil solution.

The principal cations found in most soils are H^+, Ca^{2+}, Mg^{2+}, K^+, and Na^+, here given in the usual order of decreasing tenacity with which they are held by the soil particles. Many other cations, including those of copper, iron, manganese, molybdenum, and zinc are generally present in small amounts. It will be noted that Ca^{2+} is one of the most tenaciously held ions. One effect of the addition of lime to the soil is the replacement of some of the other adsorbed ions by calcium ions; the displaced ions are released into the soil solution, from which they are more readily taken into the roots.

The principal anions found in most soils are Cl^-, SO_4^{2-}, HCO_3^-, $H_2PO_4^-$, NO_3^-, and OH^-. Except for phosphates, most anions readily leach out of the soil. In neutral and alkaline soils, the phosphate ions generally combine with calcium or magnesium to form compounds that are not highly soluble. In acid soils, the $H_2PO_4^-$ ions tend to replace OH^- ions in the mineral kaolinite and in hydrated oxides of iron and aluminum.

The supply of a particular element available to a plant varies in different kinds of soils and under different conditions. Boron, iron, manganese, and zinc tend to be converted to insoluble or otherwise unavailable forms in soils with a pH of 7 or above. Waterlogged soils are usually deficient in available nitrogen, although the total nitrogen content is often much greater than in ordinary soils. Phosphorus is most readily available in nearly neutral or slightly acid soils and becomes progressively less available in more basic or more strongly acid soils. Addition of phosphate to the soil may bring on an iron shortage by the formation of insoluble iron phosphate. Under continuous farming without fertilization, the supply of all mineral nutrients is depleted, but deficiencies in nitrogen, phosphorus, and potassium are likely to be the most serious, partly because these elements are used in such large quantities.

ABSORPTION

Differential Absorption

Ions may be absorbed by the roots either from the soil solution, or (with greater difficulty) directly from the soil particles to which they had been adsorbed. They are not generally absorbed in the same proportions as they are present in the soil solution or on the colloidal surfaces. Some ions, on the contrary, are much more vigorously absorbed than others,

Fig. 23.1 Effects of mineral deficiencies in tobacco: 1, nitrogen; 2, phosphorus; 3, potassium; 4, calcium; 5, magnesium; 7, boron; 8, sulfur; 9, manganese; 10, iron. Number 6 was supplied with a full nutrient solution. [Photo courtesy of J. E. McMurtrey, U.S. Agricultural Research Service.]

and different kinds of plants sometimes differ greatly in their affinity for various kinds of ions. The rate of absorption of individual kinds of ions is also influenced by the kind and amount of other ions present in the soil solution and in the cell.

Differential absorption is only partly effective in governing the amounts and proportions of minerals absorbed. The higher the concentration of a particular kind of ion, the more of it is likely to be absorbed. Toxic concentrations of the trace elements are easily obtained experimentally and sometimes occur in nature as well.

Differential absorption of minerals would modify the pH of the protoplasm and the cell sap through the same sort of

processes described below for soils, except that the plant has a delicate and effective **buffer system** that tends to counteract any such changes. Any change in the pH shifts the equilibrium in some reversible reactions among the cell constituents, resulting in the restoration of the pH nearly or quite to the original point. The buffer system of soils is less sensitive and less effective.

Mechanics of Absorption

The mechanisms by which minerals are absorbed from the soil are only partly understood. Simple diffusion is not an adequate explanation, inasmuch as the concentration of mineral solutes in the cell sap is generally greater than that of the soil solution.

After a great deal of study, it has become clear that the bulk of mineral absorption occurs against the concentration gradient and that respiratory energy is used in the process. Either aerobic or anaerobic respiration is effective, as long as energy is released beyond that required for other metabolic processes. In most land plants, however, anaerobic respiration does not provide enough energy for rapid absorption, and plants that normally grow in well-aerated soils do not usually absorb either water or minerals freely when the soil is saturated with water.

Experiments under conditions of moderate or slow respiration have usually indicated that the cytoplasmic membrane is not very readily permeable to most mineral solutes. Under conditions of more rapid respiration, however, it evidently permits the entry of many solutes, while restricting their exit. It is thus clear that a directional difference in the permeability of the cytoplasmic membrane is maintained by the expenditure of respiratory energy. All this is in accord with current concepts of the structure and function of the plasma membrane, as discussed in Chapter 4.

The general term *cytoplasmic membrane* is used advisedly in the preceding paragraph, because the roles of the plasma membrane, vacuolar membrane, and the intervening cytoplasm are not yet clear. One hypothesis is that energy is expended at the surface of the plasma membrane to incorporate the ions into compounds that migrate through the cytoplasm (by diffusion and cytoplasmic cyclosis) to the vacuolar membrane, where they break down, releasing energy and liberating the ions into a vacuole. This hypothesis is in accord with some evidence that, in the absence of active absorption, minerals in the cytoplasm are easily lost to the soil solution, whereas those in the vacuole are more firmly retained.

The supply of cations for absorption into the cytoplasm is governed to some extent by an ionic exchange mechanism. Actively absorbing roots are actively respiring and consequently liberating carbon dioxide. Some of this carbon dioxide reacts with the water of the soil solution to form carbonic acid, H_2CO_3, which in turn dissociates to some extent, liberating H^+ ions. The root surface and the soil water immediately surrounding it therefore have a relatively high supply of H^+ ions. Some of these replace other cations that had been adsorbed to colloidal soil particles, transferring these other ions to the

Fig. 23.2 Effect of copper deficiency in tobacco. [Photo courtesy of R. A. Steinberg, U.S. Department of Agriculture.]

soil solution or directly to the root surface, as the case may be.

Mineral cations adsorbed to the outer surface of the cell wall of the absorbing part of the root are believed to be transferred through the wall by further exchange with H^+ ions, in a process somewhat similar to diffusion, except that the movement of the ions is restricted to a very small orbit adjacent to the absorbing surface. It will be recalled that the cell wall is not a homogeneous layer but rather is composed of overlapping micelles with a good deal of interspace.

The supply of anions for absorption by the root may also be influenced by an ionic exchange mechanism, but the situation here is less clear.

Control of Soil pH

The pH of the soil solution is governed by a complex balance of factors relating especially to the kinds, amounts, and proportions of anions and cations with which the H^+ and OH^- ions may associate, the extent to which the molecules dissociate in solution, and the degree of solubility of the intact molecules. Differential absorptions of ions by the roots modifies this balance and consequently affects the pH of the soil solution.

Under ordinary farm and garden conditions, with some part of the plant being harvested and removed from the soil, the net effect of the mineral absorption is almost always to decrease the pH of the soil. The various mineral cations adsorbed to the clay particles are progressively replaced by H^+ ions, making the soil more acid. The clay micelle with its adsorbed ions may be thought of as a sort of chemical compound, an acid that dissociates to some extent to provide the H^+ ions which are the essential constituents of an acid. As the pH decreases, the phosphate and some other anions become progressively less available, and the concentration of aluminum ions often reaches toxic proportions.

In order to maintain soil fertility under conditions of continuous cropping, it is necessary periodically to replenish the supply of minerals and to raise the pH to about 6 or 6.5. Usually the pH is increased by adding calcium carbonate ($CaCO_3$) to the soil. Lime ($Ca(OH)_2$) is equally effective but more expensive. In farming practice, it is customary to expand the term *lime* to include calcium carbonate.

When a soil is limed, the calcium carbonate dissociates to some extent in the soil solution, forming Ca^{2+} ions and CO_3^{2-} ions. Some of the H^+ ions of the clay micelles are then exchanged for Ca^{2+} ions. The H^+ ions released from the clay combine with the carbonate ions of the soil to form carbonic acid (H_2CO_3), a weak acid that tends to break up into water and carbon dioxide. The H^+ ions are thus progressively removed from the soil solution, and the pH increases. As the exchange of H^+ and Ca^{2+} ions continues, the supply of H^+ ions for replacement decreases, and the supply of Ca^{2+} ions to replace them also decreases. Eventually a balance is reached, with some H^+ as well as Ca^{2+} and other ions adsorbed to the clay. The more heavily the soil is limed, the more H^+ ions of the clay are replaced and bound up in water, and the higher the pH becomes.

Too high a pH of the soil solution is as much to be avoided as too low. At a pH of more than 7, the iron, manganese, boron, and zinc tend to be converted to unavailable forms. Deficiency of available iron is one of the most common causes of chlorosis (yellowing of the leaves) in basic soils.

Fig. 23.3 Effect of iron deficiency in tobacco. Plants of this same age, supplied with a full nutrient solution, were in bloom. [Photo courtesy of R. A. Steinberg, U.S. Department of Agriculture.]

Mineral-Absorbing Capacity of Different Kinds of Cells

The ability to absorb minerals against the concentration gradient by the use of respiratory energy seems to be confined largely to meristematic or immature cells. As the cells mature and lose their capacity for growth and division, they usually also lose their capacity to absorb minerals against the gradient. The only rapidly growing cells of the root that are exposed to the soil are the outer (pre-epidermal) cells of the region of elongation, and the youngest cells of the root-hair zone, with growing root hairs. It is in this region that most of the absorption of minerals occurs.

Role of Mycorhizae in Mineral Absorption

Experimental work on mineral absorption by vascular plants has been done almost entirely with plants that do not have mycorhizae under the conditions of the experiment. In mycorhizal plants much of the mineral absorption is presumably carried on by the fungus, especially if the fungus forms a mantle over the root surface. The elements usually absorbed by higher plants in mineral form are also absorbed by the fungus as components of more complex organic compounds; doubtless some of these compounds, or their metabolic descendents, are often transferred to the higher plant symbiont. There is some reason to believe that in tropical forests most of the minerals are recycled through mycorhizae, rather than being converted into free ions that are absorbed by roots.

Mineral Absorption by Aerial Organs

Absorption of minerals by aerial organs of most plants is ordinarily insignificant under natural conditions. Epiphytes such as Spanish moss (*Tillandsia usneoides*, Fig. 23.4), which absorb water from the exposed aerial surface when wet, are, of course, dependent on the solutes of that water for their minerals.

All of the necessary minerals, including nitrogen, appear to be readily absorbed when sprayed on the leaves in dilute solution. Too strong a solution will cause reverse osmosis and eventual protoplasmic damage, just as in the soil water. Chlorosis caused by iron deficiency can be overcome by spraying with ferrous sulfate. Pineapples in Hawaii and citrus fruits in California are among the crops frequently so treated. Copper, manganese, boron, and zinc are also supplied to some crops in some areas by spraying. The blooming season of annual garden flowers can often be extended by judicious application of spray fertilizers.

TRANSPORTATION

Both experimental and circumstantial evidence indicate that most of the upward

Fig. 23.4   Spanish moss, *Tillandsia usneoides*, hanging from a tree near Tampa, Florida. [New York Botanical Garden photo.]

movement of minerals ordinarily occurs in the xylem sap. Once they enter the xylem sap, the mineral ions are carried upward in the transpiration stream. The endodermis, with its suberized casparian strips, then tends to keep the minerals from leaking back into the cortex.

Further experiments have clearly shown that when movement through the xylem is artificially interfered with, the minerals move upward freely in the phloem. Phloem is primarily a food-conducting tissue, and the mechanics of movement of solutes through it are discussed in Chapter 24.

Most minerals move upward and are distributed in the same form in which they enter the plant. Nitrogen and phosphorus and probably also sulfur, however, are carried both as inorganic ions and as organic compounds that have been made in the roots. The organic compounds as well as the mineral ions move in both the phloem and the xylem.

The means by which minerals pass from cell to cell on their way from the absorbing cells of the root surface to the xylem are still controversial. Several explanations, mostly relating to differing rates of metabolic activity, have been proposed, but the matter is not yet satisfactorily resolved.

FUNCTIONS OF THE NECESSARY MINERAL ELEMENTS

The functions of the elements necessary for plant growth have been discussed at the cellular level in Chapter 4. The division of labor in a multicellular plant causes a greater requirement for certain elements in some tissues than in others, and in these plants an insufficiency of any particular element causes characteristic visible symptoms.

The roles of the various necessary elements are interrelated in a very complex way, so that increase or decrease in the supply of any one element influences directly or indirectly the need for one or several others. Sodium, although not itself considered necessary, can substitute for part of the necessary potassium. An inadequate supply of phosphorus can be partly compensated for by an extra supply of magnesium. An increase in the supply of calcium increases the need for boron, copper, magnesium, manganese, and potassium; and an inadequate supply of calcium interferes with the absorption of nitrates. Many other such correlations could be listed. The exact concentrations and proportions of the various minerals necessary for optimum growth of any particular species are probably seldom, if ever, attained, but there is usually a considerable range of variation within which reasonably normal growth occurs.

Species and individuals differ in the amount of a particular element necessary for best growth, and in the vigor with which they absorb it from the soil. An amount of available iron that permits normal growth of alfalfa (*Medicago sativa*) will result in chlorosis in garden beans (*Phaseolus vulgaris*), and a supply of potassium just adequate for an apple orchard is not enough for peaches. Aluminum and selenium are among the elements that are vigorously and necessarily accumulated by certain species, but are poisonous to others. Different strains of maize differ considerably in the concentration of soil calcium necessary for proper growth, and similar variations occur in other species. Some species are known as *calciphiles,* requiring a relatively high concentration of calcium in the soil; some others are *calciphobes.* Species with a wide latitude of tolerance for calcium concentrations under favorable conditions may be distinctly calciphilic or calciphobic under more marginal conditions.

Some elements are more readily removed from the tissues in which they are being used than others. Deficiencies of relatively immobile elements such as cal-

Fig. 23.5 *Stanleya elata*, in Death Valley, California. This plant requires selenium. [Photo courtesy of Rupert C. Barneby.]

cium, iron, and manganese usually show up first in the younger, more rapidly growing parts, such as the apical meristems and young leaves. The more mobile elements, such as potassium and magnesium, tend to be carried away to the growing tissues when in short supply, and the symptoms of deficiency show up first or most prominently in mature leaves.

Nitrogen is one of the elements often in short supply. Plants deficient in nitrogen are unthrifty, with yellowish-green rather than rich green leaves, and the plants tend to mature early, while the aerial part is still poorly developed. A common symptom of less severe shortage is the firing (development of a reddish-brown color, followed by death) of the lower leaves, although species differ in this regard, and loss of the lower leaves is normal for many species.

Nitrogen is necessary for the synthesis of proteins, and a shortage therefore interferes directly with the formation of new protoplasm that is necessary for growth.

Nitrates are more rapidly converted into organic compounds than most other minerals that enter the root, and, in case of shortage, they are mostly used in the root, so that the symptoms of shortage appear first in the shoot. Plants deficient in nitrogen therefore tend to have the root system excessively developed as compared to the shoot (Fig. 23.6).

Plants deficient in *phosphorus* usually have deep green leaves, but they grow slowly and maturity is delayed. Flowers and fruits develop poorly or not at all. The roles of nitrogen and phosphorus in metabolism are somewhat antagonistic, so that an oversupply of one causes a relative deficiency of the other. An excess of nitrogen over phosphorus favors vigorous vegetative growth and delay or suppression of flowering and fruiting, whereas an excess of phosphorus over nitrogen favors early maturity and vigorous flowering and fruiting. A fertilizer for flower gardens should have a higher proportion of phosphorus, in relation to nitrogen, than one

Fig. 23.6   Relative growth of tops and roots of squash and wheat seedlings that grew in cultures with insufficient nitrogen and with adequate nitrogen (N). [Photos courtesy of Mary E. Reid, Boyce Thompson Institute for Plant Research.]

Fig. 23.7 Effects of calcium deficiency on young plants of romaine lettuce. Normal plant at left, calcium-deficient plant at right. [Photo courtesy of James Vlamis.]

for lawns. Much of the phosphorus of a mature plant is concentrated in the seeds and fruits, where it accumulates during their development.

Potassium, another element often in short supply, plays such varied roles in metabolism that the symptoms differ considerably under varying degrees of shortage. A mild shortage tends to interfere with protein synthesis, without affecting photosynthesis, so that carbohydrates accumulate in the leaves. The excess of carbohydrates causes the development of anthocyanins, so that the leaves often turn dark reddish. More severe potassium shortage also interferes with photosynthesis.

Calcium and *boron* play mutually antagonistic roles, especially in meristematic and immature tissues. Any shortage of one of these elements, or oversupply of the other, upsets the delicate balance necessary for proper growth, often killing the terminal buds and root tips. Any lateral buds that may be stimulated by the death of the terminal buds also fail to develop properly and eventually die, so that the plant often becomes much branched and bushy. Less severe imbalance causes contortions and malformations of the leaves without killing the meristems.

Deficiency of *iron, magnesium,* or *manganese* usually causes a mottled chlorosis, with green strips along the veins and yellowish or whitish interveinal areas. When the shortage is severe, the chlorotic areas often die and disintegrate.

Two principles should be evident from the foregoing discussion. For each of the necessary minerals there is a basic minimum, below which metabolism cannot proceed normally, or cannot proceed at all. Above that minimum, the balance among the mineral nutrients is often as important as the individual totals. Excess supplies even of the necessary minerals can be toxic.

SUGGESTED READING

Deevey, E. S., Jr., Mineral cycles, *Sci. Amer.* **223**(3):148–158, September 1970.

Epstein, E., *Mineral nutrition of plants: Principles and perspectives*, Wiley, New York, 1972. High-level presentation.

Kellogg, W. W., et al., The sulfur cycle, *Science* **175**:587–596, 1972.

Chapter 24

Angiosperms:
Manufacture, Transportation,
and Storage of Food

Sugar maple, *Acer saccharum*, in Massachusetts. In this, as
in other plants, the optimum light for photosynthesis is
determined by the balance of optima for the individual
leaves. Maximum photosynthesis occurs when the outermost
leaves have more than optimum light and the innermost
ones less than optimum.

PHOTOSYNTHESIS

The chemistry of photosynthesis is discussed in Chapter 7. In this chapter we shall consider some of the environmental factors affecting photosynthesis.

Water

Most of the water absorbed by land plants is lost through transpiration. Some of the remainder is used in various metabolic processes and as a principal constituent of any new protoplasm or cell sap that is formed. Only a relatively small amount, usually less than 1 percent of the water absorbed, is used directly in photosynthesis. Under conditions that permit reasonably normal physiological activity, there is probably never a significant direct shortage of water as a raw material for photosynthesis. Shortage of water does indirectly retard the rate of photosynthesis, however, even if all other conditions are favorable. Furthermore, water shortage usually causes the stomates to close, thus reducing the rate of entry of carbon dioxide, the other essential raw material for photosynthesis.

Carbon Dioxide

Carbon dioxide makes up only a small proportion of the air, a little over 0.03 percent on the average (see p. 583). The actual concentration in plant habitats varies from about half to several times this value, due to local influences such as removal of carbon dioxide by photosynthesis and addition of carbon dioxide by respiration—especially of soil microorganisms.

Carbon dioxide is absorbed chiefly through the cell walls of the mesophyll. It dissolves in the water at the surface of the cell walls, entering the protoplasm both directly and as H_2CO_3, the unstable acid formed by reaction with water. Much of the CO_2 and H_2CO_3 is transformed immediately into other unstable compounds on entry into the protoplast, so that there is probably no need to absorb against the concentration gradient. These compounds in turn readily give up carbon dioxide to the chloroplast, where it is used in photosynthesis.

Physiological processes of any sort are generally governed by several or many interacting environmental factors. Under a given set of conditions for all factors but one, the remaining factor has a minimum below which the process does not occur, an optimum at which it occurs fastest or best, and a maximum above which it does not occur. The minimum effective value of any particular factor or set of factors is often called the **threshold.** The optimum value for a factor differs under different conditions, and even the minimum and maximum are subject to some change. Thus the optimum light intensity for photosynthesis is higher for plants well supplied with water and carbon dioxide than for plants inadequately supplied.

If none of the factors governing the rate of a process is very far from the optimum, improvement of any one or several factors increases the rate of the process. Any factor that is too far from the optimum, however, may limit the rate, regardless of how much the other factors are improved. The factor which, through being too far from the optimum, exerts the most drag on the rate of a process under a given set of conditions is often called the **limiting factor.**

The amount of carbon dioxide in the air is often the limiting factor for photosynthesis. Under conditions of rapid evaporation, water may be the limiting factor (through its effect on other metabolic processes). On a summer day, decreasing light becomes the limiting factor as the sun goes down and darkness follows; the rate of photosynthesis falls progressively with increasing darkness until the amount of light falls below the threshold, and no further photosynthesis occurs until the next day. On a winter day in an evergreen forest, temperature is commonly the limiting factor for photosynthesis.

A moderate increase in the carbon dioxide content of the air, up to several times as great as normal, usually results in an increased rate of photosynthesis, if other conditions are reasonably favorable (Fig. 24.1) Much higher concentrations tend to retard photosynthesis, the optimum varying according to the kind of plant, the length of the experiment, and the environmental conditions. Some plants that show an initial increase in the rate of photosynthesis with increased carbon dioxide drop back again within a few hours, because of closure of the stomates

triggered by the carbon dioxide. Concentrations of more than 0.5 percent are harmful to most species, and a concentration of 0.3 percent (10 times normal) was found to be harmful to tomato plants within two weeks, after an initially favorable effect. Some kinds of plants in greenhouses and even in open fields can be effectively "fertilized" by addition of carbon dioxide to the air, often increasing the yield by as much as 100–200 percent, but such treatment is uneconomical except possibly for a few greenhouse crops in some areas.

Light

Sunlight is ordinarily the source of energy for photosynthesis, although artificial light is also effective. The nature of sunlight is discussed in Chapter 3.

Light from practically all the visible spectrum is effective in photosynthesis, although there is usually a marked drop in the effectiveness of the green and yellow

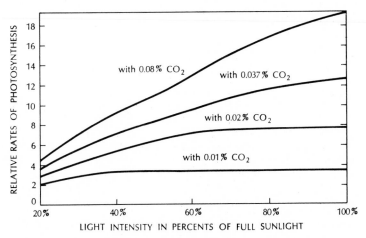

Fig. 24.1 Rates of photosynthesis in relation to light intensity and the concentration of carbon dioxide. [Data from W. H. Hoover, E. S. Johnson, and F. S. Brackett. From Transeau, Sampson, and Tiffany, *Textbook of botany*, rev. ed., Harper & Row, New York, 1953.]

portions, between 500 and 600 nm, because much of the light in this part of the spectrum is reflected or transmitted rather than absorbed (Fig. 24.2). Radiation in the ultraviolet or infrared range is of relatively little value in photosynthesis.

Leaves commonly absorb about half the total radiant energy to which they are exposed, including about 50 percent of the visible light. The exact proportion of the visible light absorbed varies according to the thickness of the leaves and other factors but is seldom less than 25 percent or more than 75 percent. The remainder is partly reflected and partly transmitted through the leaf.

The intensity of sunlight varies with the latitude, the season, the time of day, the altitude, and the atmospheric conditions. Clouds, dust, and water vapor in the air all cut down the light intensity. The exposure and pitch of a slope have obvious effects on the amount of light hitting a given area of ground surface. The quality of light (i.e., the distribution of wavelengths over the spectrum) is influenced by most of these same factors, but under natural conditions the differences in intensity are far more important.

The intensity of sunlight at noon on a clear summer day in the temperate zone commonly ranges from about 80,000 to 100,000 or even 120,000 lux.* In midmorning or midafternoon it is often only half as great. On a cloudy winter day it is often less than 10,000 lux, even at noon. Under a tree with a rather open crown, the light intensity is usually from about 1/10 to 1/20 that of full sunlight.

The optimum light intensity for photosynthesis in individual leaves of ordinary land plants commonly ranges from about 1/10 of full sunlight in shade species to 1/3 or 1/4 of full sunlight in maize and in trees such as apples and pines, if other conditions are reasonably favorable. The optimum light intensity for the whole plant, however, is often markedly higher than that for ordinary leaves. The interior leaves of a densely leafy tree may receive

*A lux is a meter-candle, the intensity of light received at a distance of 1 meter from a candle of specified size.

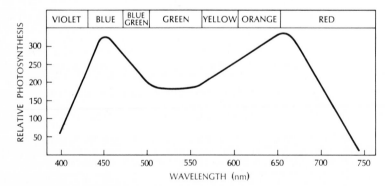

Fig. 24.2 Action spectrum of light in photosynthesis (i.e., the relative rates of photosynthesis in different wavelengths of light of equal intensity). [After B. S. Meyer and D. B. Anderson. From Transeau, Sampson, and Tiffany, *Textbook of botany*, rev. ed., Harper & Row, New York, 1953.]

less than 1 percent as much illumination as the upper ones. The rate of photosynthesis for the whole plant increases with the light intensity until the reduction in rate by overilluminated leaves counterbalances the increase in rate by the underilluminated ones. The optimum light intensity for many trees approaches or equals that of full summer sunlight if other factors do not intervene. Some social parallels might be seen in the difference between the optimum light intensity for any individual leaf and that for the tree as a whole.

Too high an intensity of light may depress the rate of photosynthesis in several ways. One way is through increasing transpiration, causing the stomates to close and cut off most of the carbon dioxide supply. High light intensity is also associated with an increased rate of photorespiration, which uses up photosynthate without a commensurate return of respiratory energy. Photorespiration is discussed in Chapter 8.

Only a small proportion of light absorbed by leaves is changed into potential chemical energy by photosynthesis. The proportion varies according to the quality and intensity of the light, the other environmental factors, and the kind of plant. At light intensities somewhat less than the optimum, as much as 10–20 percent of the absorbed light energy may be fixed in photosynthesis. In full sunlight the efficiency of individual leaves may be only 1/5 or 1/10 as great. These figures reflect the net energy gain of the leaves, as determined experimentally by the heat of combustion, and thus really measure the excess of photosynthesis over respiration and photorespiration (if any), rather than the total amount of photosynthesis.

We have noted that leaves commonly absorb only about 50 percent of the visible light that falls on them, that only about 40 percent of the radiant energy from sunlight is in the visible spectrum, and that only the energy of the visible spectrum is available for photosynthesis (see Chapter 3). Thus the efficiency of the leaf in using the radiant energy to which it is exposed is roughly only 1/5 as great as its efficiency in using the light it has absorbed. Actual figures of the photosynthetic efficiency of individual leaves in terms of the total radiant energy to which they are exposed ranged from 0.6 to 7.7 percent in one set of experiments using several different kinds of common plants under varying light intensity.

The efficiency of natural vegetation or a cultivated crop in fixing solar energy depends on the proportion of light that falls on the plant rather than on the ground surface, as well as on the efficiency with which the light that falls on the plant is used. The effect of respiration must also be considered for the entire plant, rather than just for the chlorophyllous part. The efficiency of plants is low indeed when measured in these terms.

Maize is one of the most efficient crops. In the mid-1950s it was calculated that an acre of maize in north-central Illinois fixed about 1.2 percent of the solar energy that fell on it during a growing season of 100 days, after allowing for loss by respiration. About 1/3 of the net photosynthetic product was grain, so that the efficiency of maize in converting solar energy into chemical energy of the grain was only about 0.4 percent. Measured on an annual basis, the efficiency was even less, since the crop occupied the field for less than 1/3 of the year. Present-day methods of culture, with the plants more closely set and more heavily fertilized, double or triple these figures. It has recently been estimated that on an annual basis a typical field of maize now converts about 1 percent of the solar radiation into biomass, sometimes as much as 2 percent. A deciduous forest in the temperate zone may convert about 0.5 percent, a tall-grass prairie about 0.1 percent, and a desert 0.05 percent or less. The ocean is less productive than the land. Occupying a little more than 70 percent of the surface of the globe, it probably contributes only about 1/3 of the net photosynthesis. Probably only about 0.1 percent of the energy received from the sun by the earth is fixed in photosynthesis. All of these figures are only rough estimates, some rougher than others, but they give some idea of the efficiency of conversion of solar energy into food.

Temperature

The low-temperature threshold for photosynthesis varies from as low as $-35°C$ in some northern conifers to about $5°C$ in many tropical angiosperms. In the winter, temperature is doubtless often the limiting factor for photosynthesis in evergreen angiosperms and conifers, but during the ordinary growing season some other factor is almost always limiting, and variations in temperature have less effect. Too great an intensity of light promotes destructive breakdown of various protoplasmic constituents, in a process called photo-oxidation.

Under experimental conditions with a plentiful supply of carbon dioxide and with all other factors maintained in a favorable balance, the effect of variation in temperature is very evident (Fig. 24.3). The optimum then varies not only with the species but also with the duration of the experiment. The highest initial rate is often attained at temperatures as high as $35°C$, but in most species the rate falls off if such temperatures are maintained for even a few hours. The cause of the decline is not fully understood. One suggestion is that the initially favorable effect of high temperature, due to the general principle that the speed of a reaction tends to increase with temperature, is eventually counterbalanced by the progressive destruction or inactivation of some of the necessary enzymes. An alternative or complementary explanation is that eventually some side reactions associated with photosynthesis are favored over photosynthesis itself.

Some desert plants (Fig. 24.4) have a notably high optimum temperature for photosynthesis, as much as $47°C$. This is associated with the use of the C_4 pathway in photosynthesis, as discussed in the following section.

CAM and C_4 Photosynthesis

In typical photosynthesis, as discussed in Chapter 7, carbon dioxide is absorbed from the atmosphere, taken into the cytoplasm in solution, and fed into the Calvin cycle in the production of carbohydrate. Phosphoglyceric acid (PGA), a 3-carbon compound, is a key member of the Calvin cycle. It has become convenient to refer to photosynthesis in which carbon follows this standard pathway as C_3 photosynthesis.

A pair of chemically allied variants from the standard C_3 pattern send the carbon dioxide through a series of 4-carbon compounds before releasing it into the Calvin cycle. These variant patterns are called **crassulacean acid metabolism (CAM)** and C_4 **photosynthesis.** In both of them the absorbed carbon dioxide reacts initially with a 3-carbon compound (phospho-enol-pyruvate, abbreviated PEP) to produce oxalacetate (oxalacetic acid),

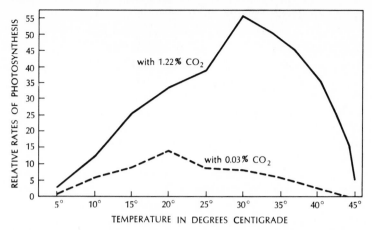

Fig. 24.3 Rates of photosynthesis in potato leaves in relation to temperature and the concentration of carbon dioxide. [Calculated from data by H. G. Lundegardh. From Transeau, Sampson, and Tiffany, *Textbook of botany*, rev. ed., Harper & Row, New York, 1953.]

which is then transformed into malate (malic acid) or aspartate (aspartic acid). Oxalacetate, malate, and aspartate are all 4-carbon acids. At some subsequent time the malate (or aspartate) gives up a mole-

Fig. 24.4 *Tidestromia oblongifolia,* a member of the amaranth family (Amaranthaceae), in Death Valley, California. This species, which follows the C_4 pathway in photosynthesis, is reported to have an optimum temperature for photosynthesis of 47°C. [Photo courtesy of Robert W. Pearcy and Harold Mooney.]

cule of carbon dioxide and reverts to pyruvate, a 3-carbon acid that can be transformed into phospho-enol-pyruvate and renew the cycle. The released carbon dioxide is fed into the Calvin cycle. The distinctive feature of both CAM and C_4 photosynthesis is that 4-carbon compounds are used as carriers of the carbon dioxide on its way into the Calvin cycle. On a purely logical basis both CAM and C_4 photosynthesis (as the term is customarily used) have an equal claim to being called C_4 photosynthesis, but they were studied and named separately before their common features were realized.

The basic chemical difference between CAM and C_4 photosynthesis is that in CAM the carbon dioxide is released into ordinary mesophyll cells, whereas in C_4 the 4-carbon carrier is transferred into the bundle sheath before giving up carbon dioxide. There are also some other differences that do not seem to be inherent in the primary difference here described.

In CAM the carbon dioxide is usually

taken in through the stomates mainly at night, and the stomates are typically closed during the day. Thus, as pointed out in Chapter 22, CAM may be regarded as a means of minimizing the loss of water inherent in absorbing carbon dioxide through open stomates. CAM can also operate with the stomates open during the day, and it can coexist with C_3 photosynthesis. CAM is widespread in succulent groups. It takes its name from the family Crassulaceae, in which it was first intensively studied.

In C_4 photosynthesis the stomates are ordinarily open during the day and closed at night, as in C_3 plants, and ordinary C_3 photosynthesis characteristically occurs in the same leaf along with C_4. The bundle sheath of the veins in the leaves of C_4 plants characteristically has chloroplasts (unlike CAM and most strictly C_3 plants), and some of the mesophyll cells are often closely packed around the bundle sheath in a radial arrangement. This **Kranz anatomy** (from German *Kranz*, a ring) (Fig. 24.5) was noted and named long before C_4 photosynthesis was discovered.

For reasons as yet not understood, C_4 plants have little or no photorespiration. Photorespiration is an apparently wasteful dissipation of energy (see Chapter 8), the function of which is still obscure. It is doubtless significant that C_4 plants are especially common in hot climates with intense sunlight, conditions that favor photorespiration in other plants.

C_4 photosynthesis is widely (though not randomly) scattered taxonomically

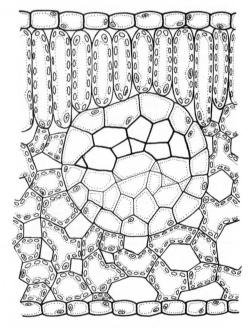

Fig. 24.5 Diagrammatic cross section of portion of a leaf showing Kranz anatomy (*left*), associated with C_4 photosynthesis, and more ordinary anatomy (*right*) associated with C_3 photosynthesis. In the diagram of Kranz anatomy, note the thick-walled cells of the bundle sheath, with chloroplasts present especially on the side toward the vein, and note the radially arranged, palisadelike cells around the bundle sheath.

among the angiosperms, and occurs in both monocots and dicots. A considerable number of grasses (including both maize and sugar cane) are C_4 plants.

From a theoretical standpoint, C_4 photosynthesis appears to be a modification of CAM. The succulent habit favors the evolution of CAM, and CAM can be further modified into C_4. Thus it should not be surprising that the order Caryophyllales (to which cacti, pinks, and four-o' clocks belong) in which the succulent habit is common, has many C_4 plants as well as CAM plants. The evolutionary route to C_4 photosynthesis in various groups of grasses is less obvious, inasmuch as CAM is as yet unknown (and surely not common) in that family.

It must be supposed that each of these three pathways of carbon in photosynthesis (C_3, C_4, and CAM) is selectively advantageous under some conditions, but not others. CAM and C_4 use metabolic energy in taking carbon dioxide through the 4-carbon route. As a rough approximation, it may be suggested that this loss of energy is outweighed in CAM plants by the ability to restrict transpiration, and in C_4 plants by the suppression of photorespiration. Doubtless there are other complicating factors. Possibly CAM and C_4 are more effective than C_3 in extracting carbon dioxide from the atmosphere, even at the cost of some expenditure of energy.

TRANSLOCATION OF FOODS

Under reasonably favorable conditions, the products of photosynthesis accumulate more rapidly during the day than they can be used or removed to other parts of the plant. Glucose, the usual primary product of photosynthesis, is soluble in water and thus affects the osmotic balance of the cell. In most angiosperms, much of the glucose is converted into starch as fast as it is produced. Starch is insoluble and does not affect the osmotic balance of the cell. During the night much of the starch is converted into sucrose or other soluble sugars and distributed to the nonphotosynthetic parts of the plant.

Phloem

Sucrose and other organic solutes are usually translocated through the phloem. The solution in the vacuoles of the sieve tubes commonly contains 10–25 percent of dry matter (in solution), of which 90 percent (or more) is sucrose.

The vacuolar contents flow, under pressure, from one sieve element to the next. Often the pressure is as much as 20 atm. The means of establishing such a pressure are not fully understood, but clearly a vital process is involved; translocation stops if the cells are killed.

Given the existence of such a pressure system, it is useful to think of movement through the phloem in terms of source and sink. Materials are put in wherever they are made, and withdrawn wherever they are used. If the solvent (water) is withdrawn along with the solute, the contents of the sieve tube will flow in the direction of use. Such a concept of the system helps one to understand the repeated observations that the same solute can move in opposite directions in the phloem at different times, and that different solutes can move in opposite directions at the same time in different sieve tubes in more or less adjacent parts of the phloem. Each vascular bundle, and to some extent even each individual sieve tube, constitutes its own distribution system.

In our present era of complicated and precise instrumentation in scientific experiments, it is interesting to note the simplicity of the studies that supplied the clinching evidence, in the early 1960s, that the phloem solution is indeed under pressure. Aphids insert their long, slender beaks into the stem or leaf. When its probing beak penetrates a sieve tube, the aphid settles down and allows the food to pump in. If the body of the aphid is cut off from the beak, the phloem solution continues to exude from the severed end of the beak for some time, commonly several days. If the phloem were not under pressure, no such exudation would occur.

As the student should by now have learned to expect, the simple story of mass flow of the contents of the sieve tubes under pressure is not the *whole* story. Other factors certainly operate, but efforts to identify and understand these other factors are still at an early stage. Any more sophisticated future explanation of the function of sieve tubes will have to take mass flow under pressure as an integral part of the story.

Mass flow in actively functioning sieve tubes commonly occurs at a rate of 50–60 cm per hour, sometimes as much as 150 cm. Inasmuch as the openings in the sieve plate typically cover only about 10 percent of the plate, a rate of 150 cm per hour through the body of the sieve tube would require some 1500 cm (15 m) per hour through the openings in the sieve plate. A rate of 1500 cm per hour is more than 4 mm per second.

Sooner or later the openings in the sieve plate are narrowed and finally completely blocked by deposits of callose, and the sieve tube ceases to function. In dicotyledons sieve tubes commonly function for only a year or less, although the phloem may live several years longer as a storage tissue, with the sieve tubes more or less collapsed. In some of the woody monocots, such as palms, the sieve tubes last much longer, even 50 years or more.

Other Tissues

Not all the living cells of a plant are in direct contact with the phloem. Therefore some movement of foods must occur through cells or tissues of other kinds. The familiar principles of diffusion and cyclosis seem adequate to explain such movement, especially since the distance to be traveled is usually only a few cells wide and since the cytoplasm of adjacent cells is connected by plasmodesmata.

The greatest distance that food must travel outside the phloem is usually from the cambium to the innermost part of the sapwood of a tree, often 1 dm or more. This lateral conduction occurs largely in the wood rays, which are composed of parenchyma cells. An inadequate food supply at the end of the line may be one of the factors contributing to the progressive change of sapwood to heartwood.

Storage of sugar in the sapwood of trees during the winter and early spring easily gives rise to the misimpression that the xylem is an important food-conducting tissue. Early in the spring, when the water supply is abundant and the leaves have not yet expanded, a considerable amount of sugary sap will often ooze out if the wood is cut into. Doubtless much of the stored sugar is later carried upward with the transpiration stream, but the total amount of food moved in the xylem is insignificant as compared with the amount in the phloem.

STORAGE OF FOOD

Food is stored in many forms, including carbohydrates, proteins, and fats, in various organs of the plant. Many plants store some carbohydrates as soluble sugars that are freely transported; otherwise the food stored in the cell is made there from simple sugars or disaccharides, or from these plus other components. We noted in Chapter 4 that simple sugars are not the only immediate products of photosynthesis, but they are still the primary building blocks that are variously modified and recombined to form the major share of the organic components of the cell.

Forms of Stored Food

Carbohydrates are by far the most abundant form of stored food in angiosperms and other embryophytes, and starch is by far the most common storage product. Starch grains are formed singly (or in small numbers) within the leucoplasts, often becoming so large that the body of the leucoplast forms only a thin layer around the grain. A leucoplast with its included starch grain(s) is often called an **amyloplast** (Figs. 4.18, 24.6). Usually the grain appears to consist of a series of concentric layers, with the inner part of each layer denser than the outer part. The shape of the grain varies with the species, and the source of the starch in flour can often be identified by microscopic examination, especially with polarized light.

Some plants store polysaccharides derived from fructose. The most common fructosan, as such compounds are called, is inulin, a white, powderlike substance that forms a colloidal sol in water and is dispersed in the cell sap of the cells in which it accumulates. Inulin is especially common in the family Asteraceae, being found, for example, in dahlia, dandelion, goldenrod, Jerusalem artichoke, and salsify. Inulin is more often found in roots and tubers than in aerial stems, and some plants that store inulin in the roots also have some starch in the stems.

The hemicelluloses (polysaccharides that are common constituents of cell walls) are often used as reserve food, especially by the germinating embryo of the seed. The hemicellulose of cell walls

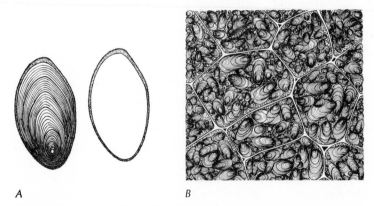

A   *B*

Fig. 24.6   (*A*) Single grain of potato starch within an amyloplast, and the amyloplast with the starch grain removed. (× about 750.) (*B*) Section showing a few cells of a potato tuber; the cells contain large and conspicuous starch grains and a few small granules of protein. (× about 150.) [From W. H. Brown, *The plant kingdom*, Ginn, Boston, 1935, 1963. Courtesy of M. A. Brown.]

in the xylem of trees is often used as food when growth is resumed in the spring.

Monosaccharides, disaccharides, and trisaccharides are also often stored and used as food. Sweet corn is rich in glucose, a monosaccharide. Fructose, another monosaccharide, is stored in the flesh of many fruits, as well as elsewhere. As much as 20 percent of the fresh weight of the roots of sugar beet and the stems of sugar cane may consist of sucrose, a disaccharide that yields one molecule of glucose and one of fructose when digested. Raffinose, a trisaccharide that yields one molecule each of glucose, fructose, and galactose, is also stored in many plants.

The very numerous proteins and fats that are stored as food by plants are not readily classifiable into groups comparable to the groups of carbohydrates.

Organs in Which Food Is Stored

Food may be stored in any organ of the plant, although there is not ordinarily much storage in the flower parts. The storage organ may or may not be obviously modified. Temporary storage of starch in the leaves is a normal corollary of photosynthesis. The stems and roots of trees, and the roots or underground stems of perennial herbs necessarily store some food, which is called on when the plant resumes growth after a period of dormancy.

Many plants have special modified roots, stems, or leaves, usually underground, which serve as storage organs. The scales of an onion bulb are modified leaves; the Irish potato is a familiar underground stem, and the sweet potato is a root. Many familiar biennials have storage roots (sometimes joined to a massive hypocotyl) (see Fig. 20.7). Underground storage organs are usually high in carbohydrate, which is readily digestible and available, and low in protein and fat, which are more economical of space; space is not generally at a premium in such organs.

Fleshy fruits commonly have a supply of stored carbohydrates. Often, especially in plants of temperate regions, the amount of food stored in the fruit is not very great, and the sweetness is provided chiefly by fructose, a monosaccharide that is much sweeter than glucose or sucrose. Food stored in the pericarp of a fruit is not generally used directly by the plant, but serves instead as part of the attraction to animals that distribute the seeds.

Seeds, in which space is usually at a premium, commonly have a higher proportion of proteins (Fig. 24.7) and fats than do vegetative storage organs. Proteins and fats have more potential energy, per unit of volume or weight, than carbohydrates. Protoplasm is essentially proteinaceous, with a considerable fat content as well. If the seed had little or no stored protein or fat, it would be necessary to use carbohydrates in their manufacture, and also as a source of energy for the process. If the manufacture is carried out before the seed is mature, the need for stored carbohydrate as an energy source is reduced. Even with some of the necessary proteins and fats already provided, the germinating seed requires a supply of stored energy. In terms of space and weight, this is more economically provided by fats than by carbohydrates. Nevertheless, carbohydrates as well as proteins and fats are major components of the stored food of most seeds, and often there is more carbohydrate than the total of protein and fat.

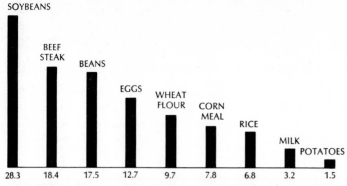

Fig. 24.7   Percentage of protein in various foods. [From Transeau, Sampson, and Tiffany, *Textbook of botany*, rev. ed., Harper & Row, New York, 1953.]

SUGGESTED READING

Botkin, D. B., Forests, lakes, and the anthropogenic production of carbon dioxide, *BioScience* 27:325–331, 1977.

Brown, L. R., Human food production as a process in the biosphere, *Sci. Amer.* **223**(3): 160–170, September 1970.

Esau, K., Explorations of the food conducting system in plants, *Amer. Sci.* **54**:141–157, 1966.

Gates, D. M., The flow of energy in the biosphere, *Sci. Amer.* **225**(3):88–100, September 1971.

Radmer, R., and B. Kok, Photosynthesis: Limited yields, unlimited dreams, *BioScience* **27**:599–605, 1977.

Sharon, N., Carbohydrates, *Sci. Amer.* **243**(5): 90–116, Nov. 1980.

Spanner, D. C., and F. Moattari, The significance of P-protein and endoplasmic reticulum in sieve elements in the light of evolutionary origins, *Ann. Bot.* **42**:1469–1472, 1978.

Woodwell, G. M., The energy cycle of the biosphere, *Sci. Amer.* **223**(3):64–74, September 1970.

Chapter 25

Angiosperms:
Reproduction

Germinating seed of *Phaseolus vulgaris,* the common bean.
(About 5 times natural size.) [Photo by Jerome Wexler, from
National Audubon Society Collection/Photo Researchers,
Inc.]

A **flower** is a specialized short shoot with modified leaves. Some of these leaves characteristically produce sexual reproductive structures, leading ultimately to the formation of seeds. A flower bud differs from an ordinary vegetative bud in the kind of leaves it produces, and in that its terminal growth is limited because the meristem either ceases to function or matures into some other kind of tissue. The floral leaves lack axillary buds, and the internodes of the flower always remain very short. (Note the caveat on terminology of floral leaves on p. 395.)

Flowers play an essential role in the production of seeds. Many of them are large and showy, although others are so small and inconspicuous as to escape notice by one not searching for them. Experience of taxonomists during the past several centuries has shown that differences in the structure of the flower furnish some of the most reliable guides to identification and relationships in the angiosperms. Some of these differences used in classification are obviously important to the plant itself; others, although very useful taxonomically, are of little or no apparent significance to the plant.

INITIATION OF FLOWERS

Many vegetative buds sooner or later change into flower buds. Even the terminal bud of a rhizome is not always permanently vegetative; in *Iris* and many other rhizomatous plants, the terminal bud turns up into the air each year and becomes a flower bud, the rhizome continuing to elongate by the development of an axillary bud. In many plants, such as tulips, all active buds sooner or later become flower buds, so that each stem is terminated by a flower or an inflorescence (cluster of flowers). In many other plants,

some buds are permanently vegetative, and only certain buds, often in particular positions, become flower buds.

From various experiments in different kinds of plants it appears that the change from vegetative to floral growth is governed by hormones. It was at one time thought that there might be a specific flowering hormone. Prolonged and unsuccessful search for this *florigen* (as it was called, in absentia) has led to the present belief that flowering is regulated by a balance among several hormones that also have other functions. Gibberellin (see Chapter 26) promotes flowering, but auxin, abscisic acid, and possibly other growth-regulating substances also have a role.

In some kinds of plants, the internal factors governing the transformation of vegetative buds into flower buds are not much influenced by the environmental conditions. The bud produces a more or less limited number of leaves and then changes into a floral bud that produces a flower or an inflorescence. In many other kinds of plants, the duration of the vegetative period and the change from vegetative to floral growth can be markedly shortened or prolonged by changing the environment.

Effect of the Photoperiod

The environmental factor that most often governs flowering, at least in temperate regions, is the length of day, or **photoperiod.** The effect of the photoperiod was first demonstrated by the American botanists (Fig. 25.1) W. W. Garner (1875–1956) and H. A. Allard (1880–1963) in 1920, and later studies by these and other botanists have shown the general importance of photoperiodism. In such studies, the natural day length is extended by artificial

Fig. 25.1 H. A. Allard (1880–1963) (*left*) and W. W. Garner (1875–1956) (*right*), American botanists, who discovered photoperiodism. [Courtesy of the U.S. Agricultural Research Service.]

light or shortened by covering the plants or placing them in a dark room. The supplemental light need not be very strong; in most cases 1/500 to 1/2000 of the light intensity of a bright summer day is enough, and sometimes an intensity only a few times higher than that of bright moonlight is effective.

Some plants are short-day plants, in which flowers are initiated as soon as the day length falls below a critical photoperiod of usually about 12–14 hours. Short-day plants usually bloom in the fall or the spring. Chrysanthemums, poinsettias, and violets are familiar short-day plants. Commercial growers of chrysanthemums carefully control the effective day length for the plants, using black covering cloths or electric lights as necessary to bring them

into flower on any chosen day of the year. Easter lilies can be managed in the same way, but more often the temperature is controlled, under natural daylight, to prepare them for sale at the chosen time.

Long-day plants, such as lettuce (Fig. 25.2), clover, potatoes, and most grains, initiate and produce blossoms only when the day length is more than a critical period of usually about 12–14 hours. Under artificial conditions, long-day plants flower even when continuously illuminated with no night. Long-day plants usually bloom in the summer.

Intermediate-day plants, such as Indian grass (*Sorghastrum nutans*) and some other grasses, bloom only when the day length is within a certain range, remaining vegetative if the day is too long or too

Fig. 25.2 Effect of the photoperiod. (*Above*) Lettuce, a long-day
Salvia, a short-day plant. [Photos courtesy of the Boyce Thom
Research.]

short. Such plants have two critical photoperiods, an upper one and a lower one, rather than only one.

Day-neutral plants, such as dandelions, tomatoes, and cotton, bloom during a wide range of day lengths. Like long-day plants, they actually have a critical photoperiod below which they will not flower, but this period is usually shorter than the day lengths that occur during the growing season.

Short-day plants might perhaps better be called long-night plants, and long-day plants called short-night plants, because it is the length of the night, rather than the day, that is generally controlling. Interruption of the night by only a few minutes of light commonly has the same effect on flowering as long days.

Different stages in reproductive growth often have different photoperiodic requirements. In some short-day plants, the critical photoperiod is progressively shorter for the initiation of flower buds, the actual development of the flowers, and the setting of seed; these are necessarily fall-blooming plants. Such differences emphasize the complex and constantly changing nature of the chemical balances that govern growth and reproduction.

Species with a wide latitudinal distribution commonly show some variation in photoperiodic response in different parts of their range, and often there is some variation even within a local area, so that some individuals will bloom before others. The short-day species may even include races. Cultivated crops often include a number of races. day, and intermediate-day odic responses. different photoperiodic responses single cultivar. made at intervals. of a short-day plant are made at intervals. will all bloom at about the season, they will all bloom at about the same time, but

different cultivars may bloom at different times, even if sown at the same time and treated in the same way.

Effect of Temperature

The second most important environmental factor influencing flowering is temperature. Aside from the photoperiodic response, some plants flower only (or best) at relatively high temperatures, some at relatively low temperatures, and some at intermediate temperatures, whereas others are indifferent to temperature. A regular cyclical variation, corresponding to the usual variation between day and night, is necessary for some plants. Blooming of lettuce, some species of phlox, and peppers is favored by high temperatures, whereas blooming of celery, beets, onions, and carrots is favored by low temperatures.

The photoperiodic response of plants may be changed or even reversed by changes in temperature. In one experiment, Heavenly Blue morning glory behaved as a long-day plant at a night temperature of 13°C, as a short-day plant at a night temperature of 20°C, and as a day-neutral plant at a night temperature of 18°C.

It will be seen that temperature and photoperiod combine to form a seasonal control over flowering for most species under natural conditions. The control is often less than perfect, however. *Forsythia* is a short-day plant that produces a profusion of flowers very early in the spring, but our own bushes in New York regularly have a few flowers in the autumn as well. *Erigeron divergens*, a wild daisy, customarily blooms in the spring like other species of its genus, but I once saw it blooming vigorously in the fall in Idaho.

Fig. 25.2 Effect of the photoperiod. (*Above*) Lettuce, a long-day plant. (*Below*) Scarlet sage, *Salvia*, a short-day plant. [Photos courtesy of the Boyce Thompson Institute for Plant Research.]

short. Such plants have two critical photoperiods, an upper one and a lower one, rather than only one.

Day-neutral plants, such as dandelions, tomatoes, and cotton, bloom during a wide range of day lengths. Like long-day plants, they actually have a critical photoperiod below which they will not flower, but this period is usually shorter than the day lengths that occur during the growing season.

Short-day plants might perhaps better be called long-night plants, and long-day plants called short-night plants, because it is the length of the night, rather than the day, that is generally controlling. Interruption of the night by only a few minutes of light commonly has the same effect on flowering as long days.

Different stages in reproductive growth often have different photoperiodic requirements. In some short-day plants, the critical photoperiod is progressively shorter for the initiation of flower buds, the actual development of the flowers, and the setting of seed; these are necessarily fall-blooming plants. Such differences emphasize the complex and constantly changing nature of the chemical balances that govern growth and reproduction.

Species with a wide latitudinal distribution commonly show some variation in photoperiodic response in different parts of their range, and often there is some variation even within a local area, so that some individuals will bloom before others. The same species may even include short-day, long-day, and intermediate-day races. Cultivated crops often include a number of races with different photoperiodic responses. If successive sowings of a single cultivar of a short-day plant are made at intervals during the season, they will all bloom at about the same time, but different cultivars may bloom at different times, even if sown at the same time and treated in the same way.

Effect of Temperature

The second most important environmental factor influencing flowering is temperature. Aside from the photoperiodic response, some plants flower only (or best) at relatively high temperatures, some at relatively low temperatures, and some at intermediate temperatures, whereas others are indifferent to temperature. A regular cyclical variation, corresponding to the usual variation between day and night, is necessary for some plants. Blooming of lettuce, some species of phlox, and peppers is favored by high temperatures, whereas blooming of celery, beets, onions, and carrots is favored by low temperatures.

The photoperiodic response of plants may be changed or even reversed by changes in temperature. In one experiment, Heavenly Blue morning glory behaved as a long-day plant at a night temperature of 13°C, as a short-day plant at a night temperature of 20°C, and as a day-neutral plant at a night temperature of 18°C.

It will be seen that temperature and photoperiod combine to form a seasonal control over flowering for most species under natural conditions. The control is often less than perfect, however. *Forsythia* is a short-day plant that produces a profusion of flowers very early in the spring, but our own bushes in New York regularly have a few flowers in the autumn as well. *Erigeron divergens*, a wild daisy, customarily blooms in the spring like other species of its genus, but I once saw it blooming vigorously in the fall in Idaho.

Effect of Other Factors

The effect on flowering of the ratio between available nitrogen and available phosphorus has been noted in Chapter 23. A relatively high proportion of phosphorus favors early change from vegetative to reproductive growth, and a high proportion of nitrogen tends to delay the change.

A number of chemical compounds induce flowering when applied to the plant externally. Acetylene and ethylene promote flowering in pineapples and some other plants. Small amounts of indoleacetic acid promote flowering, at least in some plants, whereas larger amounts inhibit it.

Many kinds of plants, such as pansies, produce more flowers over a longer period of time if the old flowers are plucked off without being allowed to produce seed. Our understanding of the mechanism of this response is still very rudimentary.

STRUCTURE OF FLOWERS

The Floral Organs

A typical flower (Fig. 25.3) has four kinds of modified leaves—the sepals, petals, stamens, and carpels—attached to a stem tip called the **receptacle.** The sepals are on the outside, the petals next, the stamens next, and the carpels are in the center. One or more of these kinds may be missing, but the order (in normal specimens) is invariable.

The **sepals** are the outermost set of floral leaves. Typically they are green or greenish, and more or less leafy in texture. They cover and enclose the other flower parts before the flower has opened, protecting them from injury. In some plants, especially members of the lily and iris families, such as tulips, iris, and gladioli, the sepals are brightly colored and much like the petals. All the sepals are collectively called the **calyx.** Not infrequently they are joined toward the base by their

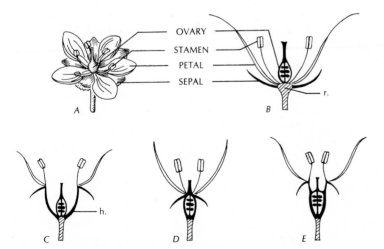

Fig. 25.3 Characteristic types of flowers: *A, B,* hypogynous; *C,* perigynous; *D,* epigynous; *E,* epigynous with prolonged hypanthium: h., hypanthium; r., receptacle.

margins, forming a calyx tube with terminal calyx lobes or teeth. The most common number of sepals is five in the dicotyledons, and three in the monocotyledons, but other numbers occur.

The **petals** characteristically form the second set of floral leaves, just internal to the sepals. Typically they are brightly colored (or white) and attract insects or birds to the flower. The function of insects and birds in transferring pollen is described on a subsequent page. All the petals are collectively called the **corolla.** Sometimes the petals are joined together toward the base to form a corolla tube (the corolla then said to be **sympetalous**), or the lower parts of the sepals, petals, and stamens may be joined to form a **hypanthium** which may itself be joined to the ovary or form a cup around it. Typically the petals are of the same number as the sepals and set in a whorl on alternate radii with them. When the petals are all alike, as in buttercups and petunias, the corolla is said to be **regular;** when they differ among themselves, as in pansies and sweet peas, the corolla is **irregular.**

Sometimes it is useful to refer to the calyx and the corolla collectively as the *perianth* and to its members as *tepals,* especially when the sepals and petals are not clearly differentiated from each other.

The **stamens** are in effect the male organs of the flower. Each stamen usually consists of a slender **filament** and a terminal **anther.** The anther usually consists of two **pollen sacs,** joined together by a slender **connective,** which is merely a prolongation of the filament. The development and function of the pollen, which is borne in the pollen sacs, are discussed on subsequent pages. The stamens are often of the same number as the petals, but larger or smaller numbers are not infrequent.

In many plants one or more of the stamens are modified in one way or another and lack functional anthers. Such a modified stamen is called a **staminode.** In the most common type of staminode, the filament is expanded to form a petal-like blade; the extra petals of many "double" flowers, such as cultivated roses, are actually staminodes. The ordinary petals of many flowers probably originated by evolutionary modification of some of the outer stamens; that is, they are also essentially staminodes.

The **carpels** are in effect the female organs of the flower. In its most typical form the carpel is a modified leaf, bearing a row of **ovules** along each edge, and folded lengthwise so that the ovules are enclosed rather than being exposed to the air. The development of ovules into seeds is discussed on subsequent pages.

The flower may have one to several or many carpels, and these may be separate from each other or joined together. Each carpel (if the carpels are separate) or the group of carpels (if the carpels are united) is called a **pistil.**

The simplest situation is illustrated by peas and other legumes, in which the flower has a single carpel folded together to form a simple pistil. If a pea pod is carefully opened along the seed-bearing suture and laid out flat, it can be seen that half the ovules (peas) are attached along one side, and half along the other. The suture that does not bear seed is the midrib of the carpel.

In some plants, such as members of the buttercup family and the rose family, the flower has several or many separate carpels, each forming a simple pistil containing one or more ovules.

In well over half the kinds of angiosperms, the flower has a single compound pistil, composed of two or three to less often four or five (seldom more) carpels. Tulips and other lilies have three-carpellate pistils. If three pea pods were joined together with the seed-bearing sutures of all three juxtaposed, the resulting compound structure would be comparable to the three-chambered pistil of lilies.

A pistil of any kind (either simple or compound) is ordinarily differentiated into three externally visible parts, the ovary, style, and stigma. The **ovary,** at the base, contains the ovules. The **stigma,** which receives the pollen, is usually elevated above the ovary on a slender **style.** A compound pistil very often has the style cleft at the summit to form as many stigmas as there are carpels.

Most flowers are bisexual (**perfect**); that is, they have both stamens and pistil(s). Unisexual (**imperfect**) flowers, with sta-

mens but not pistils, or vice versa, are not uncommon, however. Maize, for example, has highly reduced and specialized unisexual flowers, the male ones borne in the tassel at the top of the stem, the female ones in the ears farther down. An ear of corn is a group of female flowers which lack both calyx and corolla. The individual corn grains are the ovaries, and the silk consists of the styles.

Hypogynous, perigynous, and epigynous flower types are diagrammed in Fig. 25.3. These features are very useful to plant taxonomists, but of more doubtful significance to the plants themselves.

Most flowers produce nectar, a sweet liquid gathered and eaten by insects or other animals that act as unknowing pollinators. The structures or tissues that produce nectar are called **nectaries.** The nature of nectaries differs in different kinds of flowers. Most often they are reduced and modified stamens, but parts of the sepals, petals, ovary, hypanthium, or receptacle may also be differentiated as nectaries.

REPRODUCTIVE PROCESSES

Development of the Ovule

The life cycle of angiosperms was worked out, and the terminology developed, before the basic homologies with other embryophytes were understood. The carpels of angiosperms are essentially megasporophylls, and the ovules that they bear are essentially megasporangia provided with an additional outer covering (the integuments).

An ovule that is ready for fertilization ordinarily consists of a female gametophyte, a **nucellus** (the wall of the megasporangium, enclosing the female gametophyte), one or two **integuments** partly enclosing the nucellus, and a basal stalk.

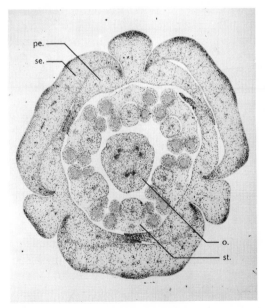

Fig. 25.4 Cross section of flower bud of lily: o., ovary; pe., petal; se., sepal; st., stamen. [Courtesy CCM: General Biological, Inc., Chicago.]

The integuments arise as rings of tissue near the base of the young nucellus. Growing faster than the nucellus, they soon completely enclose it, except for a small opening (the **micropyle**) at the tip. Ovules of angiosperms are very often bent back on themselves, so that the micropyle is alongside the stalk, as shown in Fig. 25.5.

Each ovule ordinarily produces only a single megaspore mother cell. The megaspore mother cell, which has 2n chromosomes and represents the last stage in the sporophyte generation, undergoes reduction division to produce four megaspores that are usually arranged in a row to form a linear tetrad. The megaspores each have n chromosomes and represent the first stage in the female gametophyte generation. Three of the four megaspores ordinarily degenerate and disappear, leaving a single functional megaspore which gives rise to the female gametophyte by a series of mitotic divisions.

The female gametophyte of angiosperms is called the **embryo sac** (Figs. 25.6, 25.7). Typically it consists of eight nuclei embedded in a mass of cytoplasm that is only partly differentiated into cells, with some tenuous plasma membranes but without cell walls. Species with as few as 4 and as many as 128 nuclei in the embryo sac are known, and there are some species in which more than one megaspore contributes to the development of the embryo sac.

Typically the two nuclei formed in the first division of the megaspore migrate to opposite ends of the developing embryo sac. Each nucleus then divides into two, and each of these into two more, so that the embryo sac has four nuclei at each end. One nucleus from each end then migrates to the center of the embryo sac, so that three nuclei remain at each end and two lie near the center. The three nuclei at the end farthest from the micropyle are called the **antipodal nuclei.** They have no apparent function and are usually interpreted as evolutionary vestiges of the vegetative body of the female gametophyte. The two nuclei in the center are often called the **polar nuclei,** in reference to their origin at the two ends or poles of the embryo sac. Their function is discussed on a later page. One of the three cells at the micropylar end of the embryo sac is the **egg.** The other two are called **synergid cells,** in reference to their apparent function in assisting fertilization. Presumably the synergid cells represent vestiges of the archegonium or of the vegetative body of the gametophyte.

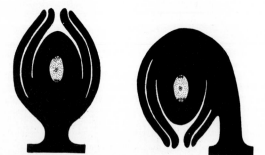

Fig. 25.5 Orthotropous (*left*) and anatropous (*right*) ovules; the latter type is the more common.

Development of the Pollen

The stamen is essentially the microsporophyll of an angiosperm, and the pollen sacs (Fig. 25.8) are the microsporangia. Usually there are two pollen sacs, each of which is divided by a lengthwise partition so that the anther contains four locules (chambers).

The microspore mother cells, often called **pollen mother cells,** have 2n chro-

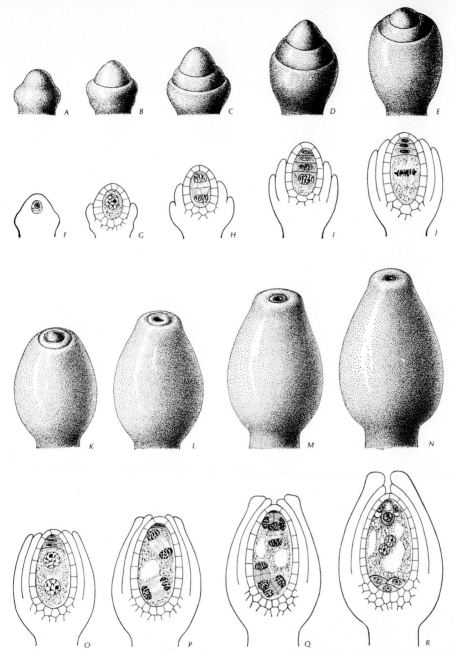

Fig. 25.6 Diagrammatic representation of development of an ovule: *A–E, K–N,* growth of integuments; *F, G,* megaspore mother cell; *H, I,* reduction division, forming a linear tetrad of megaspores; *J,* three megaspores compressed, one enlarged, its nucleus dividing; *O,* binucleate stage of embryo sac, upper megaspores degenerating; *P,* four-nucleate stage; *Q,* eight-nucleate stage; *R,* mature female gametophyte, ready for fertilization, with egg and two synergid nuclei near the micropyle, two polar nuclei in the center, and three antipodal cells at the end opposite the micropyle. [From W. H. Brown, *The plant kingdom,* Ginn, Boston, 1935, 1963. Courtesy of M. A. Brown.]

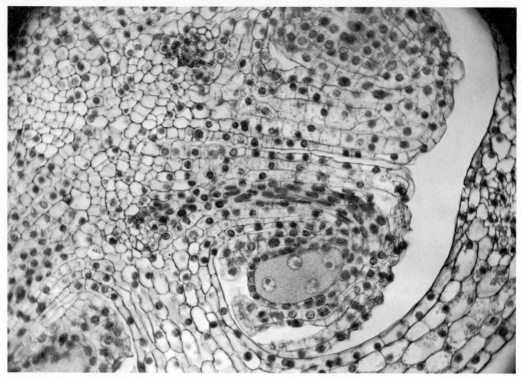

Fig. 25.7 Young ovule of *Fritillaria* (a lily) in the ovary, showing the four-nucleate stage of embryo sac development. [Courtesy CCM: General Biological, Inc., Chicago.]

mosomes and represent the last stage in the sporophyte generation, comparable to the megaspore mother cell of the ovule. Each microspore mother cell undergoes reduction division to produce a usually spherical tetrad of microspores. The microspores have n chromosomes and represent the first stage in the male gametophyte generation.

The microspores of a tetrad usually separate from each other before developing, with little or no increase in size, into **pollen grains** (Fig. 25.9). Each pollen grain is a young male gametophyte, with usually two nuclei (these being produced mitotically from the original microspore nucleus). The **generative nucleus,** with a small amount of differentiated cytoplasm

around it, floats freely in the cytoplasm of the **tube cell,** which fills the grain. The generative cell, which later gives rise to the sperms, is thus enclosed within the tube cell, which later produces the pollen tube that carries the sperms to the egg. The tube cell has its own nucleus; the tube nucleus and the generative nucleus are the two nuclei of the pollen grain.

The wall of the pollen grain consists of an **exine,** or outer layer, and an **intine,** or inner layer; each of these, in turn, consists of two or more sublayers. Typically the exine has a roof layer (tectum) held above the foot layer by small columns (columellae), and is said to have a tectate-columellate structure. The space amongst the columellae is filled by substances that

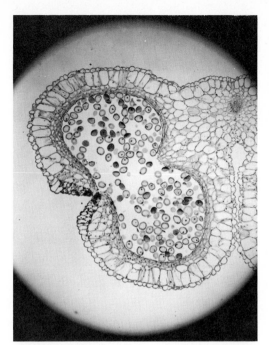

Fig. 25.8 Cross section of part of lily anther, showing one pollen sac, with pollen ready to be discharged; the partition between the two chambers of the pollen sac has broken down, and the pollen sac is ready to dehisce. [Courtesy CCM: General Biological, Inc., Chicago.]

When the pollen grains are mature, the anther opens and releases them. Usually the partition between the two pollen chambers of each sac breaks down to some extent, and the sac dehisces (opens) along the furrow between the two pollen chambers.

Pollination

The transfer of pollen from the anther to the stigma is called **pollination.** Most angiosperms are normally or frequently cross-pollinated, but self-pollination is not uncommon. Cross-pollination has the advantage of producing new and different combinations of existing genes (see Chapters 29 and 30), but self-pollination is a more certain method, not subject to the vagaries of wandering breezes or flighty insects. Many species have special mechanisms that prevent or restrict effective self-pollination, most notably self-sterility.

Insects are the most common agents of cross-pollination (Fig. 25.10), but many flowers are pollinated by wind, and a considerable number are pollinated by birds. A few kinds of flowers are pollinated by bats (or other small mammals), and a few others by water currents. All these pollinators sometimes also transfer the pollen merely from one flower to another on the same plant, so that if the plants are self-fertile they are still, in effect, self-pollinated.

Bees, wasps, butterflies, and moths usually visit flowers in search of nectar and/or pollen. In collecting the nectar at the base of the flower, the insect almost unavoidably brushes against the anthers and the stigma. Some of the pollen remains on the insect and may be brushed off on the stigma of some other flower visited later. The stigma is generally a bit

influence the germination of the pollen grain after it has been transferred to a stigma. Usually there are one or more slits or circular openings, the germ pores, in the exine, through which the intine bulges out when the grain germinates. The pollen of most dicots has three such apertures and is thus said to be triaperturate, but that of most monocots and some of the more archaic dicots has only one (uniaperturate). The exine is often provided with minute spines or low ridges. These and other variations in the structure of the wall, as well as the shape of the pollen grain, often provide useful taxonomic characters.

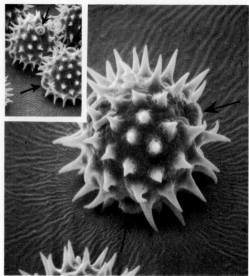

Fig. 25.9 Pollen grains. (*Top left*) Easter lily, *Lilium longiflorum*, showing the elongate single furrow commonly found in monocots. (× about 600.) (*Right*) Common sunflower, *Helianthus annuus*, with three germ pores, as in most dicots. (× about 15,000 and about 6,000.) [Scanning electron micrographs courtesy of JEOL U.S.A., Inc.] (*Bottom right*) *Cuphea koekeana*, a Mexican dicot. (× about 3,500.) [Scanning electron micrograph courtesy of Alan Graham.]

sticky, and a pollen grain that touches it is likely to remain there. The insect does not consciously or intentionally pollinate the flower, but pollination is effected just the same.

On any individual trip from the hive or nest, a bee usually visits flowers of only one kind, ignoring others, unless the supply is very restricted. The flower constan-cy of bees is one of the factors restricting interspecific hybridization. Many species that will produce hybrids when artificially pollinated only rarely hybridize in nature, because the pollen does not usually get transferred from one species to the other.

Some kinds of flowers are open to all comers and attract a wide range of insect

Fig. 25.10 Flower of *Barbacenia involucrata*, a relative of the lily family, pollinated by carpenter ants (*Campanotus*). [Photo courtesy of Edward S. Ayensu, Smithsonian Institution.]

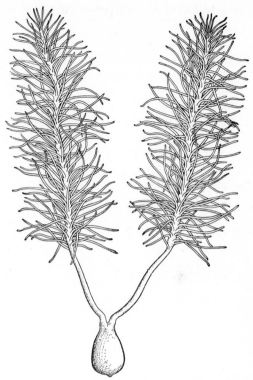

Fig. 25.11 Pistil of a grass, showing feathery stigmas adapted to wind pollination. ($\times$ 100.) [From W. H. Brown, *The plant kingdom*, Ginn, Boston, 1935, 1963. Courtesy of M. A. Brown.]

visitors; others are usually visited by only one or a few species, to which they are especially adapted. In many regions the bumblebee is the only effective pollinator of red clover; ordinary honeybees cannot reach the nectar. Charles Darwin accurately predicted the discovery of a moth with a tongue 11 inches long in Madagascar, because this length would be required to reach the nectar of a certain orchid (*Angraecum sesquipedale*) in a moth-pollinated genus. Some orchids have flowers that simulate the females of certain species of wasps, thus attracting the males, which inadvertently transfer the pollen in attempting to perpetuate their own kind. Many other adaptations to specific pollinators exist.

Wind-pollinated plants generally produce large amounts of pollen, enough to overcome the tremendous waste inherent in the method. Most wind-pollinated plants, such as ragweed, have small and inconspicuous flowers, often without petals, and many of them have relatively very large, brushlike stigmas (Fig. 25.11). Many or most grasses are wind-pollinated, although many others are self-pollinated instead.

Many people show an allergic reaction, called hay fever, to the pollen of one or more kinds of plants. Probably almost any species could so affect somebody under the proper circumstances, but only wind-pollinated plants generally shed enough pollen into the air to affect large numbers of people. The several species of ragweed (*Ambrosia*) are the most serious cause of hay fever in the United States.

Events Leading to Fertilization

The sticky fluid covering the surface of the stigma contains water, sugars, and other substances. In this environment the pollen grain germinates almost immediately. Many kinds of pollen grains can be induced to germinate by putting them in a dilute sugar solution, but others require more specialized conditions that are generally provided only by the stigma of their own species.

The intine (inner wall) of the pollen grain bulges out, usually through one of the openings in the exine, producing a **pollen tube** that penetrates the stigma and grows down through the style. The protoplasm of the pollen grain flows into the tube, leaving the empty grain still perched on the stigma. The protoplast of the pollen tube produces digestive enzymes that decompose the surrounding tissue of the style. Some of the digestion products are absorbed and used as food by the protoplast of the pollen tube.

The tube nucleus (the nucleus of the tube cell) usually lies near the growing tip of the pollen tube, with the generative nucleus a little farther behind. During the growth of the pollen tube, the generative nucleus divides mitotically, giving rise to two sperms, each with a small amount of differentiated cytoplasm, but without definite walls or locomotor structures. (In some species the generative nucleus divides earlier, so that the mature pollen grain has three nuclei instead of two.)

Eventually the pollen tube reaches and penetrates the ovule. In most species it enters through the micropyle, but in some it enters through the side or base. The micropyle in the angiosperms is merely an evolutionary vestige from their gymnospermous ancestry.

On reaching the embryo sac the pollen tube enters one of the synergid cells and

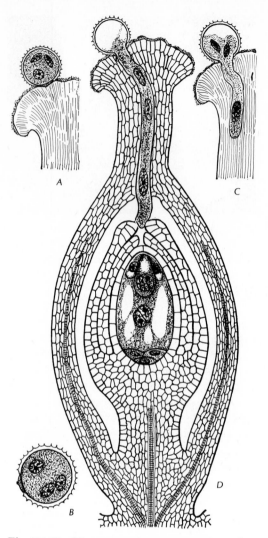

Fig. 25.12 Diagrammatic representation of an ovary with a single, erect ovule, showing development of the pollen tube: *A*, pollen grain on stigma; *B*, a pollen grain with a tube nucleus and two sperm nuclei; *C*, germinating pollen grain; *D*, pollen tube entering the micropyle. [From W. H. Brown, *The plant kingdom*, Ginn, Boston, 1935, 1963. Courtesy of M. A. Brown.]

opens at or near the end, discharging the sperms. The tube nucleus and the cytoplasm of the pollen tube then die and degenerate, leaving the sperms free in the synergid cell. The synergid cell likewise dies and degenerates (sometimes even before the pollen tube penetrates it). It appears that some of the products of degeneration are important to the subsequent behavior of the sperms, but the details are still obscure.

One of the two sperms fuses with the egg. The fertilized egg has 2n chromosomes and represents the first stage in the new sporophyte generation. By a series of mitotic divisions, the fertilized egg later gives rise to the embryo of the seed.

The other of the two sperms fuses with the two polar nuclei near the middle of the embryo sac, forming a **triple-fusion nucleus** with, ordinarily, 3n chromosomes. Often the two polar nuclei fuse before the sperm nucleus reaches them, but the end result is the same. The triple fusion nucleus, or **endosperm nucleus,** typically gives rise, by a series of mitotic divisions, to the **endosperm** of the seed, a food-storage tissue. In many plants, however, the endosperm degenerates before the seed is mature, and the food is stored in the embryo, the nucellus, or even in the seed coat.

The events following discharge of the sperms into the embryo sac are often referred to as **double fertilization,** because the true fertilization (fusion of the sperm and egg) is accompanied by the other fusion process (fusion of the other sperm with the polar nuclei), which also resembles fertilization. Double fertilization is a characteristic feature of angiosperms and is not known to occur in any other group of plants. It was independently reported in 1898 by the Russian botanist Sergei Gavrilovich Navashin (1857–1930, now better known for Navashin's solution, a mixture used for certain cytological preparations) and in 1899 by the French botanist Jean Louis Leon Guignard (1852–

1928). Here again we see that when the time is ripe for a discovery, someone—and often more than one—will make it.

The use of a triploid rather than a haploid or diploid tissue for food storage has no apparent special value and is usually though not unanimously thought to be a mere evolutionary happenstance. (Haploid tissue adequately performs the same function in gymnosperm seeds.) Most botanists doubt that such a process would be likely to originate and become fixed more than once (i.e., in separate evolutionary lines), and the regular occurrence of double fertilization is one of the features that contribute to the belief that the angiosperms are a truly natural group.

A number of species of angiosperms, in a wide range of families, can by-pass reduction division and fertilization in the ovule, and set seed asexually. The embryo then usually originates from the nucellus. The setting of seed without fertilization is called **apomixis.** The taxonomic and evolutionary consequences of apomixis are more significant than might at first be assumed, and are of great interest to botanists and plant breeders, but they are beyond the scope of this text.

FRUITS

After the ovules are fertilized, they develop into **seeds,** and the ovary in which they are borne ripens into a **fruit.** More formally, the fruit may be defined as the ripened ovary, together with any other structures that ripen with it and form a unit with it. It will be noted that the botanical definition of a fruit is broader than the popular definition. Botanically, a string bean or a corn grain is as much a fruit as is a peach or a strawberry. The old chestnut, "Is a tomato a fruit or a vegetable?" raises no

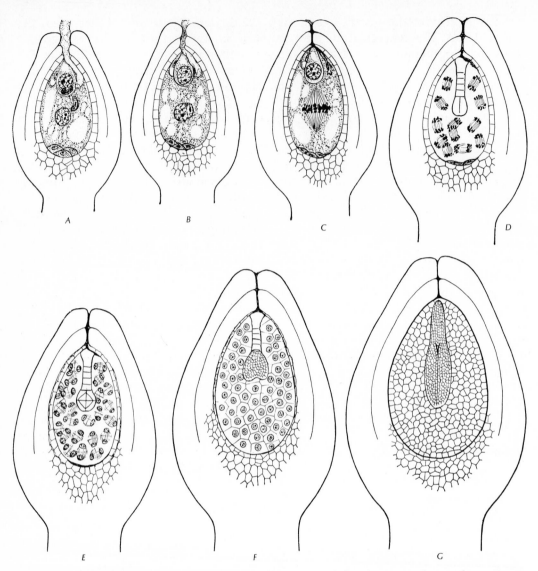

Fig. 25.13   Diagrammatic representation of fertilization and development of an embryo. *A*, The pollen tube has entered through the micropyle, and a curved sperm nucleus lies in the egg just below the egg nucleus; the two polar nuclei have fused to form a single nucleus and the second sperm is near the fusion nucleus. *B*, The sperm and egg nuclei are nearly fused, and the second sperm is fusing with the nucleus formed by fusion of the polar nuclei, to form the endosperm nucleus. *C*, The endosperm nucleus is dividing. *D*, A young embryo has developed from the fertilized egg, and numerous endosperm nuclei are dividing. *E*, A more advanced stage. *F*, The cotyledons of the embryo are beginning to differentiate, and the endosperm has become cellular. *G*, A seed in which the embryo is surrounded by endosperm and this by two integuments. [From W. H. Brown, *The plant kingdom*, Ginn, Boston, 1935, 1963. Courtesy of M. A. Brown.]

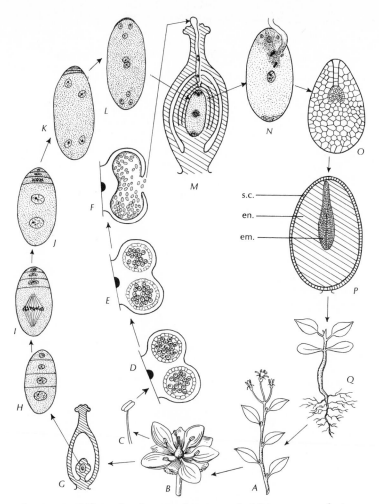

Fig. 25.14 Life cycle of an angiosperm. *A*, Mature sporophyte;
B, flower; *C*, stamen; *D–F*, development of pollen; *G*, pistil; *H–L*,
development of embryo sac; *M*, growth of the pollen tube toward
the embryo sac; *N*, release of the sperms into the embryo sac; *O*,
developing embryo and endosperm; *P*, mature seed; em., embryo;
en., endosperm; s.c., seed coat; *Q*, seedling.

botanical problem. Botanically, the tomato is a fruit; so is a chestnut. Vegetable, on
the other hand, is a nontechnical term, not
much used botanically except in such
expressions as "the vegetable kingdom."

The form, texture, and structure of
fruits are almost infinitely variable. Some

are fleshy, some dry; some have many
seeds, some have few or only one; some
open when ripe, releasing the seeds (**dehiscent fruits**), some remain closed (**indehiscent fruits**). These differences, and
others, are useful to taxonomists, and they
have some importance to the plant as

well, being correlated to some extent with the means of seed dispersal.

SEED DISPERSAL

Wind, animals (including birds), water, and explosive dehiscence of fruits are the principal means of seed dispersal. The whole plant, the fruit, or only the seeds may be transported.

Tumbleweeds, such as Russian thistle (*Salsola kali*) and Jim Hill mustard (*Sisymbrium altissimum*) are carried about by the wind during the fall and winter, dropping seeds as they go. These plants have much-branched stems, commonly forming a rounded mass that has considerable wind resistance but little weight.

Fleshy fruits are commonly eaten by birds and mammals. The seeds may be discarded without being eaten, but more often they are eaten along with the fruit and are later passed out of the alimentary tract. The seed coat or the inner layer of the fruit wall in many fleshy fruits is so hard and impervious to water that the seeds do not germinate readily unless they are subjected to the softening action of digestive juices.

Indehiscent dry fruits usually have only one or a few seeds, and seed dispersal involves the whole fruit. Winged fruits, such as those of maple, ash, and elm (Fig. 25.15), are adapted to distribution by wind. Many of the familiar plants with winged fruits are trees, and the fruit

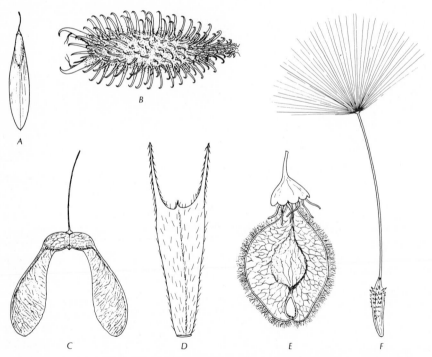

Fig. 25.15   Some dry, indehiscent fruits. *A*, Ash, *Fraxinus* (× 1); *B*, cocklebur, *Xanthium* (× 2); *C*, maple, *Acer* (× 1); *D*, beggar ticks, *Bidens* (× 6); *E*, elm, *Ulmus* (× 4); *F*, dandelion, *Taraxacum* (× 6). Cocklebur and beggar ticks are adapted to distribution by animals, the others to distribution by wind.

starts its journey some distance above the ground. The one-seeded fruits of dandelion (Fig. 25.16) and many other members of its family (Asteraceae) have a terminal tuft of long hairs formed by the modified calyx.

Many indehiscent dry fruits are adapted to distribution by animals. The fruit may have hooked or barbed projections that catch in the fur of passing animals or the clothing of people. Acorns and some other fruits are commonly buried by squirrels as a reserve food supply for use during the winter. Many of these escape rediscovery, and the seeds germinate in place.

Some dry indehiscent fruits are lighter than water and are carried about by streams. The wide distribution of coco-

nuts on remote tropical islets depends partly on the fact that the fruits float in the ocean (Fig. 25.17).

The seeds of dehiscent fruits are individually distributed by the same means as are indehiscent fruits, plus the additional means of explosive dehiscence. Tensions develop in the drying wall as the fruit ripens, causing it to burst open along lines of weakness. The fruits of Scotch broom (*Cytisus scoparius*), touch-me-not (*Impatiens*), and many other plants dehisce so violently that the seeds are shot several feet through the air.

Many seeds, such as those of milkweed (*Asclepias,* Fig. 25.18) and cottonwood (*Populus*), are provided with a tuft or covering of long, loose hairs that help keep them afloat in the breeze. In *Catalpa* and some other plants the seed coat is produced into a broad, flat wing. Many conifers also have winged seeds. The seeds of many weeds are light enough to float or be carried along in water, and the distribution of such seeds by irrigation water is a serious problem in some parts of the western United States.

It should be noted that plants with no

Fig. 25.16 Fruiting head of dandelion, *Taraxacum officinale*. (Nearly twice natural size.) The one-seeded fruits are dispersed by the wind. [Photo by Lynwood M. Chace, from National Audubon Society Collection/Photo Researchers, Inc.]

Fig. 25.17 Coconut, a one-seeded fruit adapted to distribution by water. The "meat" of the coconut is endosperm. [Photo by W. H. Hodge.]

Fig. 25.18 Fruits of milkweed, *Asclepias syriaca*. Each fruit consists of a single carpel, which opens along the seed-bearing suture at maturity. The wind-borne seeds have a tuft of hairs at one end. [Photo by John H. Gerard, from National Audubon Society Collection/Photo Researchers, Inc.]

obvious specializations for dispersal often manage to get around anyway. Seeds of marsh plants are often carried long distances in the mud on the feet of migrating waterfowl. Violent windstorms may carry seeds, fruits, or even whole plants that are not affected by ordinary breezes. Many plants have been introduced to America in the ballast dumped from ships.

Other things being equal, plants with the most efficient means of seed dispersal will most quickly occupy any new site that becomes available. Some of our commonest weeds do not do well in competition with undisturbed native vegetation but survive because their efficient mechanisms of seed dispersal combined with a large production enable them to quickly occupy any disturbed habitat from which the original vegetation has been removed.

Seed dispersal is, of course, only one of the factors involved in the perpetuation of a plant population. Other factors, such as longevity of the plant, number of seeds produced each year, the amount of food stored in the seed, and the ability to compete with other plants are also important. The distribution and abundance of a species are determined by a balance of all factors, and inefficiency in one respect can often be compensated for by efficiency in others.

SEED STRUCTURE

The seed is the ripened ovule. Seeds vary in size, from barely visible, as in many orchids, to about 1 dm thick, as in coconuts, or sometimes even larger. The two essential parts are the **seed coat,** developed from the integuments of the ovule, and the **embryo,** ordinarily developed from the fertilized egg. Often the embryo is embedded in or accompanied by the **endosperm,** a food-storage tissue derived from the triple-fusion nucleus of the ovule. Usually the nucellus is no longer recognizable in the mature seed, but in some plants it develops into a special food-storage tissue. Generally there is not even a recognizable vestige of the embryo sac. A little reflection will show that the seed spans three generations—the old sporophyte, the female gametophyte, and the young sporophyte.

The embryo is a miniature plant, consisting of a short axis with one or two (rarely more) attached leaves, the **cotyledons.** The part of the axis above the cotyledon(s) is the **epicotyl,** or **plumule,** which becomes the terminal bud of the seedling. The part of the axis immediately below the cotyledon(s) is the **hypocotyl,** a root-stem transition region. The hypocotyl is prolonged at the base into the **radicle,** which becomes the primary root of the seedling. The radicle may or may not be externally differentiated from the hypocotyl; sometimes it consists merely of a group of meristematic cells at the tip.

Dicotyledons typically have two cotyledons (whence the name) that are essentially opposite in position (see chapter-opening photograph). Monocotyledons have only one cotyledon, and this one is often highly modified. Botanists are now agreed that the monocotyledons had ancestors with two cotyledons.

During its early stages (Fig. 25.19), the embryo of a developing ovule becomes differentiated into a basal stalk, or **suspensor,** attached at the micropylar end of the embryo sac, and a terminal body which becomes the proper embryo of the mature seed. The suspensor tends to degenerate as the seed matures, and it is represented in the ripe seed, if at all, by a mere remnant at the end of the radicle. The suspensor has no evident function in modern angiosperms; it is regarded as an evolutionary vestige from cryptogamous and gymnospermous ancestors in which it may have served to force the embryo sporophyte deeper into the body of the female gametophyte.

The triple-fusion nucleus formed in the embryo sac typically develops by a series of mitotic divisions into the endosperm, a triploid food-storage tissue in the seed. Often, however, the endosperm degenerates before the seed is mature, and the

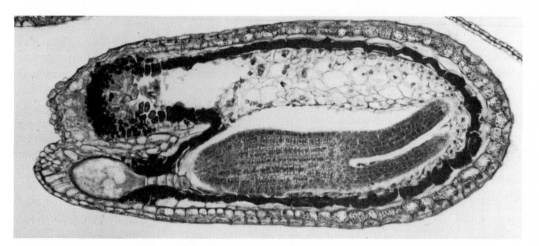

Fig. 25.19 Long section through a young seed of shepherd's-purse, *Capsella bursa-pastoris.* The embryo, lengthwise near the bottom, has a well-developed suspensor at the left; the endosperm is beginning to degenerate, and will be wanting from the mature seed. [Courtesy CCM: General Biological, Inc., Chicago.]

food is stored in some other part of the seed. In peas, garden beans, peanuts, and other legumes, for instance, the food is stored chiefly in the thickened cotyledons. It should also be noted that the so-called endosperm which is the food-storage tissue of gymnosperm seeds is actually the body of the female gametophyte.

The food stored in the seed includes proteins, carbohydrates, and fats, the proportions varying with the species. The seeds of legumes often have a relatively high protein content and are among the best vegetable sources of protein. In sweet corn much of the carbohydrate of the endosperm remains as sugar instead of being converted into starch. The desirable table quality of the corn results from a genetic deficiency that interferes with the formation of starch.

GERMINATION

Mature seeds of most kinds of plants normally undergo a rest period before developing into new plants. During its early stages of growth, before it has become wholly independent of the food stored in the seed, the new plant is called a **seedling.** The processes occurring from the time the embryo resumes its growth until the seedling is established are collectively called **germination.**

Conditions Required for Germination

All seeds require a supply of moisture and oxygen for germination, as well as a suitable temperature. Some seeds also require light, although light inhibits the germination of some other kinds of seeds.

Seeds ordinarily have a relatively low water content, and the physiological processes necessary for germination occur only when the proportion of water is increased. Most seeds germinate best when the moisture content of the soil is near the field capacity.

Germinating seeds respire rapidly, and a supply of oxygen is necessary. In garden peas and some other kinds of seeds, the seed coat is relatively impervious to oxygen even when wet, and growth during the early stages of germination is supported largely by anerobic respiration. Even seeds of this sort eventually need oxygen, however, and aerobic respiration becomes dominant as soon as the seed coat is ruptured by the growth of the embryo.

The optimum temperature for germination varies according to the species and the environmental conditions. For any species there is a maximum and minimum above or below which germination will not occur. The minimum for wheat is only a little above the freezing point, and the maximum is about 35°C. The maximum and minimum for corn, a species of subtropical origin, are about 10 degrees above those for wheat.

The effect of light on germination differs in different species. Many kinds of seeds, especially those of various epiphytes, require light. Some others, including those of numerous grasses, germinate better when exposed to light than when kept in total darkness. Germination of many other kinds of seeds, such as those of onion and some other members of the lily family, is retarded or prevented by light. The effect of light may also vary according to the other environmental conditions and the past history of the seed. Seeds of some species of bluegrass (*Poa*) which ordinarily respond favorably to light are indifferent to it after a period of dry storage.

The influence of light on seed germination is exerted through the phytochrome pigments, which are discussed in Chapter 26.

Dormancy

Freshly ripe seeds of many plants fail to germinate even under favorable environmental conditions. Such seeds are said to be **dormant.** Dormancy may be due to any of several factors, alone or in combination. The most common of these are:

1. Seed coat impervious to water, or oxygen, or both
2. Seed coat mechanically resistant to expansion of the embryo
3. Rudimentary or immature embryo
4. Necessity for further physiological changes (afterripening) in a fully developed embryo
5. The presence of chemicals that inhibit germination

Under natural conditions seeds with a very impervious or firm coat cannot germinate until the coat has softened. A bony inner layer of the fruit wall, such as the pit in cherries and other stone fruits, has the same effect. The eventual softening may result from decay, from partial digestion in the alimentary tract of an animal, from changes in the colloidal structure of cell walls caused by repeated wetting and drying, from mechanical rupture of the cells by freezing and thawing, or from any combination of these factors. Dormancy of seeds with hard or impervious coats can also be broken by cracking or deeply scratching the seed coat, a process known as **scarification.** Sulfuric acid is often used for scarification, but the treatment must be carefully controlled to avoid damaging the embryo.

In many kinds of plants the embryo does not develop as rapidly as the rest of the seed, so that the embryo in the "ma-ture" seed is still immature. Many members of the buttercup family, for example, have only a rudimentary embryo in the newly ripe seed. Such seeds will not germinate until the embryo has had time to develop more fully.

Many kinds of seeds with fully mature embryos require a period of **afterripening** and will not germinate when newly ripe, even if the seed coat is removed and all external conditions are favorable. Apple, peach, hawthorn, iris, basswood, ash, dogwood, hemlock, and pine are familiar examples. Most seeds that require a significant period of afterripening normally germinate in the spring, after having lain on the ground all winter. Under controlled conditions, afterripening can often be hastened by exposure to low temperatures, or by repeated alternation of high and low temperatures, thus compressing into a short time some of the environmental changes to which the seed is normally exposed during the dormant period.

The kinds of physiological changes that occur in the seed during afterripening evidently differ in different species. Often a particular chemical that inhibits germination must degenerate before germination is possible. Germination inhibitors occur in the fruits of some plants, as well as in the seeds. Tomato juice inhibits the germination of tomato seeds and many other seeds, even when considerably diluted with water.

Different seeds of the same species often differ in the length of dormancy. A single crop of seeds from a desert annual will commonly produce some seedlings each year for several years. The survival value of this variation among seeds is evident, since otherwise a single bad year with insufficient spring rains might wipe out all the annual plants in a large region. Many common weeds show a similar

Fig. 25.20 Results of an experiment on seeds of *Rosa rubiginosa*. The seeds in the row at the left had been stored dry. Those in the other rows had been stored in moist sand for 6 months, at the temperatures indicated, then planted in this flat in a greenhouse. [Photo courtesy of the Boyce Thompson Institute for Plant Research.]

variability in length of dormancy. The gardeners' aphorism, "One year's seeding is seven years' weeding," has a solid foundation in fact.

Longevity of Seeds

The life span of seeds varies from a few weeks to a thousand years or more, depending on the species and the environmental conditions, but is seldom more than a few decades. Seeds of silver maple (*Acer saccharinum*) ordinarily germinate soon after they are shed (in June), and the seeds that do not find suitable conditions for germination die as soon as the water content drops from an original figure of nearly 60 percent to about 30–34 percent. In nature this often happens within a few weeks.

Seeds of most of our common crop plants live only one or a few years under ordinary conditions of storage. When the seeds are stored at low temperatures and low concentrations of oxygen, respiration and other physiological processes leading to deterioration are slowed down, and the viability may be prolonged severalfold.

In general, seeds with firm, hard coats live longer in ordinary storage than other seeds. The longevity of such seeds is doubtless due in part to the low concentration of oxygen available to the embryo, but the embryo may also be inherently longer-lived as a reciprocal adaptation to the long period of dormancy enforced by the seed coat. The record for longevity is held by the oriental lotus, *Nelumbo nucifera*, a member of the water lily family with large seeds enclosed in a very hard, dry fruit-wall. Viable seeds found deep in a peat deposit in Manchuria appear to be about 1000 years old by radiocarbon dating. Fragments of wood associated with viable *Nelumbo* seeds found 6 m underground in Japan have been radiocarbon dated as about 3000 years old.

Legumes, which often have a very hard seed coat, are notoriously long-lived, although they are not known to approach the record of *Nelumbo*. The true age of viable *Lupinus* seeds from Canada that were reported to be 10,000 years old is yet to be adequately established.

Reports of germination of seeds recovered after thousands of years of storage in Egyptian pyramids have not been verified and are more nearly in the category of exploitation of gullibility than scientific record.

In 1879 the late Professor W. J. Beal of Michigan Agricultural College (now Michigan State University) began a continuing experiment on longevity of seeds. Seeds of 20 species of common plants were mixed with sand and buried in the soil in inverted open bottles. At intervals of 5, and later 10 years, a bottle was dug up and the seeds tested for germination. After 90 years, in 1970, only moth mullein (*Verbascum blattaria*) still showed some germination, and it had dropped to a level of about 20 percent. In 1980, under somewhat different conditions of culture, moth mullein showed about 40 percent germination, and one seedling of common mullein (*V. thapsus*) showed up in the lot. Presumably this represents a contaminant in the original set of seeds, since the seeds of these two species look much alike, and common mullein was not included in the experimental design. More surprisingly, in 1980 there was one seedling of common mallow (*Malva rotundifolia*), out of 50 seeds planted. Mallow had otherwise shown no germination since 1900. The

principal results of the experiment are shown in Fig. 25.21. The 1980 mallow seedling is not provided for in the figure.

The causes of death of the embryo in stored seeds are only partly understood. Death commonly occurs long before the food supply is exhausted and while the digestive enzymes are still fully potent. Loss of water evidently kills some short-lived seeds, but does not otherwise seem to have much significance, and long-lived seeds commonly have an initially low water content. Seedlings from old seeds resemble those from heat-treated or irradiated seeds, which show similar aberrations in nuclear structure, but this is an expression of deterioration rather than its cause. The one clear lead is that conditions that increase the rate of respiration shorten the life of seeds, whereas conditions that decrease the rate of respiration lengthen the life. Therefore it may reasonably be surmised that senescence and death of the embryo in seeds are due to the same factors that cause senescence and death of cells in general: physiological imbalances associated with the accumulation of waste materials.

The Germination Process

Under appropriate environmental conditions, absorption of water by the seed triggers the series of changes resulting in the development of the seedling. The embryo respires rapidly and begins to grow, drawing on the food that has been stored in the seed. In endospermous seeds, the embryo produces digestive enzymes that migrate out into the endosperm and decompose it. Cell walls, protoplasm, and granules, crystals, and globules of stored food are all attacked. The digested materials are absorbed by the cotyledons and passed on to the rest of the embryo as needed. In seeds without endo-

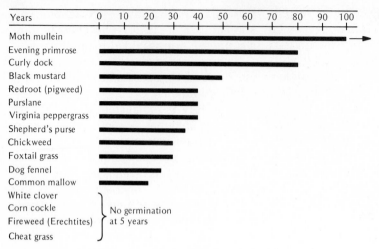

Fig. 25.21   Longevity of seeds in Beal's experiment.

sperm, the digestion of the endosperm and absorption of food by the cotyledons have often occurred before the seed was ripe, and a similar breakdown now goes on in the cotyledons; eventually the cotyledons shrivel and degenerate or fall off. In either case, the cells of the endosperm degenerate concomitantly with the absorption of food by the cotyledons.

The first part of the embryo to emerge from the seed coat is generally the radicle. The radicle is positively geotropic (see Chapter 26), and soon turns down in its growth, regardless of its initial orientation.

The radicle regularly becomes the primary root, and the epicotyl becomes the primary shoot, but the functions of the hypocotyl and cotyledons differ in different kinds of seeds. In seeds with **hypogaeous** germination, such as peas and corn, the hypocotyl remains short, the cotyledons do not emerge from the seed, and if the seed is underground, the epicotyl reaches the surface only by its own growth. In seeds with **epigaeous** germination, such as beans (Fig. 25.22) and castor

beans, the hypocotyl elongates, bringing the cotyledons and epicotyl (and commonly the remains of the degenerating seed coat) above the ground. Cotyledons that are brought above the ground in germination may thereafter function for some time as ordinary leaves, as in castor beans, or they may show little or no photosynthetic activity and soon fall off, as in garden beans.

The one function common to cotyledons of nearly all kinds of seeds is the absorption of food. Functions performed by some but not all cotyledons are food storage, photosynthesis, and protection of the epicotyl during germination. In general, cotyledons that serve in food storage do not have a prominent photosynthetic role.

The hypocotyl, if it elongates at all, commonly first appears above the ground as an upside-down U, attached at one end to the developing radicle, and at the other to the cotyledons and epicotyl. The hypocotyl then gradually straightens out and pulls the cotyledons and epicotyl above the ground. In epigaeous seeds with endo-

Fig. 25.22 Bean seedling. (Nearly twice natural size.) [Photo by Hugh Spencer, from National Audubon Society Collection/Photo Researchers, Inc.]

sperm, the cotyledons are commonly still encased in the endosperm and seed coat when they first come up.

SUGGESTED READING

Cleland, C. F., The flowering enigma, *BioScience* **28**:265–269, 1978. There may be a balance among inducers and inhibitors.

Clevenger, S., Flower pigments, *Sci. Amer.* **210**(6):84–92, June 1964.

Echlin, P., Pollen, *Sci. Amer.* **218**(4):80–90, April 1968.

Evans, L. T., *Daylength and the flowering of plants,* W. A. Benjamin, Menlo Park, Calif., 1975. Authoritative but not highly technical, not written for specialists.

Faegri, K., and L. van der Pijl, *The principles of pollination ecology,* 2nd ed., Pergamon, Elmsford, N.Y., 1972.

Grant, V., The flower constancy of bees, *Bot. Rev.* **16**:379–398, 1950.

Grant, V., and K. Grant, *Hummingbirds and their flowers,* Columbia University Press, New York, 1968.

Jensen, W. A., Fertilization in flowering plants, *BioScience* **23**:21–27, 1973.

Khudairi, A. K., The ripening of tomatoes, *Amer. Sci.* **60**:696–707, 1972. A molecular ecological approach to the physiology of fruit ripening, directed to the nonspecialist.

Percival, M. S., *Floral biology,* Pergamon Press, New York, 1966.

Salisbury, F. B., *The biology of flowering,* Natural History Press, Garden City, N.Y., 1971. Contains instructions for do-it-yourself experiments.

Chapter 26

Angiosperms: Growth and Movement

Dwarf forest of pine and oak on sterile, dry, sandy soil in the inner pine barrens of New Jersey in April, 1963, before the oaks had leafed out for the year. This area has been subject to repeated fire, and had been burned in 1947. It was burned again several years after the photograph was taken. The trees in this photo appear to be mature at a height of 1–1.5 m and are producing seeds. They are all stump sprouts from a base at or below ground level that is often several decimeters across and may show several old, burned-off stems up to about 1 dm thick. The ultimate height of these trees, had they been protected from subsequent fire, is uncertain; in more favorable and protected sites nearby the same species grow to about 20 m.

THE NATURE OF GROWTH

Characteristics

Increase in size or weight due to the formation of new protoplasm is called **growth.** Growth is reflected, and can in part be measured, by increase in the amount of protoplasm in the cell, increase in the volume of the cell, extension of the cell wall, increase in the mass of the cell wall (e.g., by the formation of a secondary wall), and increase in the number of cells. All these increases are involved in the growth of any tissue or organism, but in a given cell over a short period of time any one may be much more prominent than the others.

Growth is ordinarily permanent and irreversible, as long as reasonably normal conditions are maintained. Under starvation conditions some of the protoplasm itself may be respired, however, so that there is a net loss of weight. Volume of the cell, as an indicator of growth, must of course be measured under constant environmental conditions to avoid differences in turgor.

Some Chemical Aspects of Growth

Growth, like life itself, depends on a delicately balanced set of simultaneous and serial chemical processes. The chemical processes, in turn, depend on substances produced in the cell, on substances that migrate into the cell, and on environmental conditions. Changes in any one of these governing factors affect the rate or nature of growth.

Most or all of the constituents of a protoplast are simultaneously involved in several or many chemical reactions, and two adjacent molecules of the same substance may have quite different histories and fates. As an aid to understanding, we try to distinguish individual processes, and we are delighted when a balanced equation can be devised to express a reaction; but even these oversimplifications are often difficult to achieve.

The patterns and changes in growth result from changes in the dynamic balance of interacting substances that make up the protoplasm. Any change in the proportion of a particular protoplasmic constituent has a chain of consequences, shifting the point of equilibrium for some other chemical reaction, which in turn causes other changes, and so on. Any environmental change affects the chemical processes of the protoplasm unequally, thus shifting the balances and causing further changes. Some parts of the system are relatively stable; others are easily modified. Changes in some protoplasmic constituents have only minor effects on the equilibrium, whereas minute changes in the amount or distribution of others have profound effects. A by-product of some chemical reaction involved in growth may gradually accumulate, without significant effects, until it reaches a threshold and shifts some important balance.

In addition to making their own food, most green plants can make all the other organic substances they need. Only the inorganic raw materials—water, carbon dioxide, and various minerals—must be supplied. Individual parts of the plant, on the other hand, are generally not self-sufficient. Even when provided with food and raw materials, they usually need minute amounts of complex substances produced elsewhere in the plant.

Among the natural plant substances that occur in small amounts but have pronounced effects on growth, it is customary to apply the term *hormone* to distinctly migratory compounds, and the term *vitamin* to less migratory ones, especially if these latter have similar effects in all parts of the organism. The distinction between these two classes is not nearly so sharp in plants as the similar distinction in animals, and the noncommital term, **growth-regulating substance,** is often used to avoid the problem.

Some Developmental Aspects of Growth

One of the greatest wonders of growth is that it is so precisely regulated. The organism starts as a single cell, which grows and divides. Before very many divisions have occurred, the cells begin to differentiate from each other, both in their immediate qualities and in their potentialities for further growth. They all have the same basic heredity (all coming from the same zygote), and the environmental differences between two adjacent cells would not seem to be very great, yet the cells do become different, and different in a consistent pattern that is repeated in individual after individual.

The factors controlling the initial differentiation of the cells of the embryo are still obscure. Presumably, growth-regulating substances produced by the surrounding tissues of the ovule reach different parts of the embryo in differing amounts, and these slight differences shunt the cells into different paths of development. By the time the cotyledons, epicotyl, and hypocotyl can be recognized, the embryo has its own internal system of growth regulation, although growth regulators from the surrounding tissues of the ovule may still be required for some time.

Unlike mammals and many other animals, angiosperms and most other multicellular plants continue to grow nearly throughout the life of the organism. Some parts of the plant may die and decay while other parts are still growing, so that the total size or weight of perennial herbs may not continue to increase indefinitely; and periods of growth may be interrupted by periods of dormancy. But growth is a normal and regular part of plant metabolism, and a plant that stops growing sooner or later either goes into dormancy or dies. Vegetative activity is so closely linked with growth that the part of the year when plants are vegetatively active is called the **growing season.** The length of the growing season differs with the kind of plant as well as with the locality and the time of year.

The different parts of a plant do not always grow at the same time and relative rate. Water absorption is largely limited to the younger parts of the root, and root growth continues throughout most or all of the growing season. Most woody plants of temperate regions have a short period in the spring and early summer when the leaves are produced and the stems elongate. Thereafter the stem increases rapidly in thickness for a time, and then more slowly for the rest of the growing season. In some plants, especially annuals, the fruits and young seeds grow rapidly after growth in the rest of the plant has virtually ceased.

The differentiation of cells and tissues in higher plants is not so absolute as in higher animals, and parenchyma cells as well as meristematic cells are often said to be **totipotent,** that is, to have the capacity to change into any other kind of cell, or to divide to produce any other kind of cell. Other kinds of plant cells, once they are mature, do not so readily revert to an undifferentiated type, nor do they ordinarily produce different kinds of cells by division. Under unusual conditions, however, mature plant cells of other types do sometimes revert to an unspecialized condition and become totipotent.

The size of higher plants is much more subject to environmental control than is that of higher animals such as mammals. Even under reasonably normal conditions, the larger individuals of angiosperms are commonly three or four times as tall as the smaller ones. Some of our common weeds, such as tumbling mustard (*Sisymbrium altissimum*), may mature and set seed when only a few centimeters tall under unfavorable conditions, but reach a height of 1 m or more under favorable ones. The famous Ming trees of China, which matured when only a foot or so tall, were merely carefully tended starvelings of ordinary species, as are the more recent Bonsai of Japan. There is an inherent limit to the potential size of each plant of any kind, but it may fall far short of this limit and still appear reasonably normal.

It is a common observation that the age and degree of maturity of a plant influence the nature of its growth. We saw in Chapter 25 that the production of flowers is strongly influenced by the environmental conditions, but the age of the plant is also an important factor; woody plants seldom flower until they are several (or even many) years old. In many kinds of plants juvenile foliage is markedly different from mature foliage. Angiosperms with compound leaves at maturity often produce simple leaves at first; the garden bean is a familiar example. The common harebell (*Campanula rotundifolia*) produces short, rounded leaves at the base of the stem, and long, narrow leaves elsewhere (Fig. 26.1). Most junipers have very small, appressed, scalelike leaves at maturity, but the leaves produced during the first several years are more spreading and a

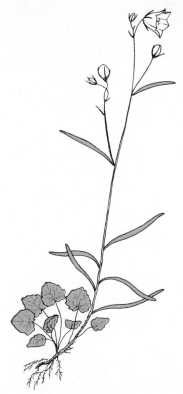

Fig. 26.1   Harebell, *Campanula rotundifolia*, showing very different basal and cauline leaves.

little longer, suggesting the leaves of other conifers. Young pine seedlings have the leaves spirally arranged on the stem, instead of clustered on short spur shoots as they are later. The latter two examples, at least, also illustrate the recapitulation principle discussed in Chapter 30.

The transition from one type of growth to another may occur at any age, depending on the kind of plant, and can often be hastened or delayed by environmental factors. Under conditions of deep shade and high moisture, harebell continues to produce juvenile leaves for some time, and flowering and fruiting are delayed. We noted in Chapter 23 that a high proportion of nitrogen causes prolongation of vegetative growth and delay in flowering and fruiting, whereas a high proportion of phosphorus has the opposite effect.

GROWTH-REGULATING SUBSTANCES

During the past several decades a great deal of attention has been devoted to plant growth-regulating substances. Most of those now known can be classified as abscisic acid, auxins, ethylene, gibberellins, kinins, and lactones. Each of these substances has complex and varying effects on growth, depending on the kind of plant, the part of the plant to which it is supplied, other internal conditions in the plant, and the environmental conditions.

The interactions among growth-regulating substances may be compared to the way grades are determined by college professors. Tests, reports, classroom discussions, and laboratory work are important, interacting factors in the determination of grades. One professor may place more emphasis on tests, another on reports, but still take some account of the other factors, or he may emphasize two of the factors and disregard or downgrade the others. The professor in a large lecture course may dispense with reports and classroom discussion entirely, and base his grades solely on test results. Still another professor must, to judge by the results, emphasize heavily some other factor whose nature is unknown to the student and is subject to much speculation.

Probably each of the several plant growth-regulating substances sets off (or blocks) a chain of reactions, each depending on the previous one. Such a series of reactions may be interrupted at any point, or shunted in a new direction, depending on the environmental conditions and the other substances in the protoplasm. The cells of the root do not necessarily respond to a particular growth-regulating substance in the same way as those of the stem, and the effects of a growth regulator may be different in the light and in the dark, or even at different temperatures.

We do not know the precise chemical action of growth-regulating substances. Our information is largely limited to their effects on particular organs of particular kinds of plants under particular environmental conditions. Some kinds of effects occur often enough to permit generalizations, but these generalizations are subject to exceptions or difficulties in interpretation, and even horticultural varieties of one species may respond differently. These difficulties are entirely expectable. All differences among cells, tissues, organs, individuals, and taxonomic groups are, in the last analysis, expressions of chemical differences.

Growth-regulating substances have both synergistic and sequential effects. The combined effect of two of them may be greater than the sum of the two supplied separately, and the effect of any of them may become dramatically greater when a certain stage of growth and differentiation is reached. By comparison, a man and a woman working together can accomplish things that would be beyond the reach of either one alone or both separately; and information about the nature and chemical structure of RNA seemed of little importance until the role of DNA as the bearer of hereditary information was discovered.

The **auxins** are the best-known plant growth-regulating substances. Natural auxin is indole-3-acetic acid (often abbreviated IAA) (Fig. 26.2), and this is usually the substance now intended when auxin

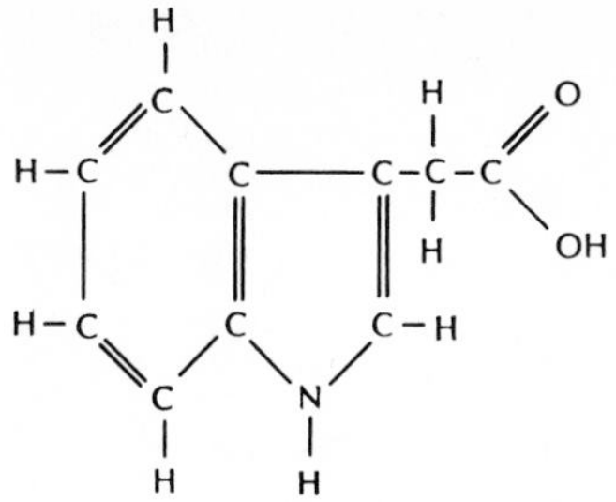

Fig. 26.2 Structural formula of indole-3-acetic acid (natural auxin).

is referred to without further qualification. IAA occurs in bryophytes, algae, and fungi, as well as in higher plants. A number of synthetic compounds related to IAA have more or less similar effects, varying in detail.

Auxin is most abundantly produced in the meristematic region of the shoot. It is also formed in root tips, leaves, cambial tissues, and sometimes other parts of the plant, migrating through the cells so that they are present in all living tissues. Lateral movement of auxins is much slower than longitudinal movement, and any destruction or interruption of the flow along one side of a stem, for example, is reflected in a lower concentration of auxin along that side for some distance below. The actual quantities involved are minute, even in the shoot apex, where the concentration is generally only a few parts per million.

An optimum concentration of auxin generally stimulates growth of cells, but the optimum varies widely in different tissues and under different conditions. Experimental evidence suggests that the concentrations normally present in growing shoot tips and root tips are near the optimum for cell division in these tissues, but are too high for much cell elongation. The lesser concentration in the region of elongation favors elongation but is not

enough to stimulate division. Growth of lateral buds is strongly inhibited by concentrations that are normal for the growing apical meristem, and apical dominance (see Chapter 19) appears to be an auxin-related phenomenon.

The existence of auxins was discovered through a long series of studies by different botanists on the response of plants to light. The knowledge that plants turn toward the light in their growth probably antedates recorded history, but the first significant experimental studies were reported in 1880 by Charles Darwin in his book, *The Power of Movement in Plants.* He found that the coleoptile of canary grass (*Phalaris canariensis*) turns strongly toward the light, with the curvature occurring well below the meristematic tip (The **coleoptile** is a modified leaf, characteristic of grass embryos, which forms a closed tube covering the epicotyl; it is the first part of the grass seedling to emerge from the ground.) If the coleoptile tip is covered with a small, light-proof cap, no bending occurs, even though the part that would otherwise curve is fully exposed. Darwin correctly concluded that the stimulus perceived by the tip is somehow transmitted to the lower part of the coleoptile. Experiments by the Danish botanist Peter Boysen Jensen (1883–1959) in 1910 and 1913 led him to suggest that the stimulus is transmitted by a chemical that is formed in the tip and migrates downward. The Dutch (now American) botanist Frits W. Went (1903–   , Fig. 26.3) spectacularly completed the proof in 1926 by briefly culturing decapitated coleoptile tips of oats on blocks of agar, which he then applied to the coleoptile stumps in various positions. Growth of the stump was always greatest on the side to which the agar block was applied, and when the block was symmetrically applied to the

Fig. 26.3 Frits W. Went (1903–), who completed the proof of the existence of plant hormones during his first year as a graduate student at the University of Utrecht.

cut top, the coleoptile grew straight upward without bending.

Light is apparently destructive to auxin in meristems. (An alternative explanation is that it causes auxin to migrate laterally to the darker side of the meristem.) The oat coleoptile is similar in this respect to shoot tips in general, with a light-sensitive meristem above a light-insensitive region of elongation. The greatest destruction (or removal) of auxin occurs on the side that receives the most light, and the auxin deficiency on that side extends well down into the region of elongation, because the auxin moves much more rapidly downward than laterally. The darker side of the stem, having a more favorable concentration of auxin, elongates faster than the illuminated side,

Fig. 26.4 Effect of α-naphthaleneacetic acid (an auxinlike synthetic compound) on a bean seedling. (*Above*) Chemical being applied to the seedling; (*below*) same plant 24 hours later. [Photos courtesy of John W. Mitchell, Agricultural Research Service.]

and the stem curves toward the light. Directional growth-response to light, as exemplified in the preceding discussion, is called phototropism. Phototropism is further discussed on p. 488.

In addition to the effects already noted, auxin influences the development of abscission layers (leading to the fall of leaves, fruits, or other structures), adventitious roots, flowers, and fruits. The effects vary in detail according to the species, the stage of growth, and the environmental conditions.

The effect of auxin in stimulating cuttings to produce adventitious roots has led to an important practical use of related synthetic compounds in commercial and home-gardening practice. Cuttings of some kinds of plants (such as willows) root freely under natural conditions, some sparingly, and some not at all. Many species that do not ordinarily root well can be induced to do so by applying IAA or any of several synthetic chemicals (Fig. 26.5). In practice some of the synthetics, such as indole-butyric acid, are more satisfactory than IAA, and such products are now widely sold in the United States. The old folk-belief that rooting is promoted by the insertion of a germinating wheat grain in a slit in the cutting turns out to be well founded, the effect being due to the auxin that spreads from the grain into the cutting.

Natural and various synthetic auxins, or auxinlike compounds, are used to promote flowering and fruiting, to retard fruit drop, to induce parthenogenetic fruiting (the fruits thus being seedless), to inhibit flowering, and for other purposes.

Several synthetic compounds related to IAA are used as weed killers. The most important of these is 2,4-D, (2,4-dichlorophenoxyacetic acid), which is readily absorbed from sprays or dusts applied to the leaves (Fig. 26.6). Relatively minute amounts of 2,4-D derange the physiology of most plants, especially the meristems, causing metabolic abnormalities, asymmetric growth, and death. Different kinds of plants differ in their susceptibility to 2,4-D. Most monocots, especially grasses, are more resistant than most dicots, and most trees are more resistant than most broad-leaved herbs,

Not treated   Treated

Fig. 26.5 Effect of indole-butyric acid (an auxinlike synthetic compound) on cuttings of holly. [Photos courtesy of the Boyce Thompson Institute for Plant Research.]

Fig. 26.6   Effect of 2,4-D. The broad-leaved plants (poison ivy and thistle) have been killed by spraying, but the grasses are not much affected. [Photo courtesy of the American Chemical Paint Company.]

but resistance also varies with the environmental conditions and the stage of growth, and even forms of the same species may differ markedly.

The **gibberellins** are another important group of plant growth-regulating substances. They occur in many slightly varying forms, of which gibberellic acid (Fig. 26.7) is one of the most common.

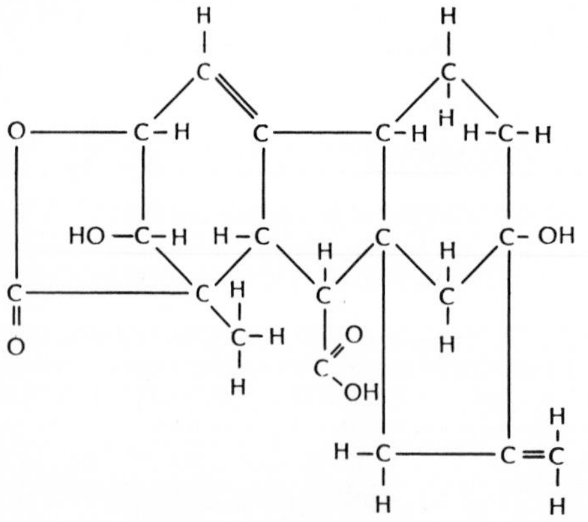

Fig. 26.7   Structural formula of gibberellic acid, the form known as GA_3.

Gibberellin was discovered in 1926 by two Japanese scientists studying the foolish seedling disease of rice. Rice seedlings, when attacked by the fungus *Gibberella fujikuroi*, grow to two or three times their normal height and then die. The fungus produces substances, now called gibberellins, which promote growth of the shoot without corresponding growth of the root, and the plant grows itself to death as the root system fails to meet the demands of the shoot.

It is now known that gibberellins occur naturally in most kinds of plants, and the foolish seedling disease results merely from an excess. Like auxin, gibberellin influences a wide variety of growth processes, with varying effects according to the kind of plant, its stage of maturity, and the environmental conditions. Some other common effects of gibberellin are to lower the low-temperature threshold for growth, to break dormancy, and to promote early flowering and fruiting and the formation of parthenocarpic fruits. Gibberellin is widely sold in America.

Abscisic acid (Fig. 26.9), another widespread and important plant growth-regulating substance, appears in general to counter the influence of gibberellin. Among other effects, it contributes to dormancy and to the senescence and abscission of leaves. The decreased photoperiod of late summer and fall may promote the formation of abscisic acid.

Lactones, another group of growth regulators, are widely distributed but not yet well understood. In general, they seem to be growth inhibitors; some of them, such as **coumarin** (Fig. 26.10), found in the seeds of some kinds of plants, inhibit germination.

Different cultivars of white sweet clover (*Melilotus alba*) differ greatly in the amount of coumarin in the seeds. The

Fig. 26.8 Effect of gibberellin on *Pelargonium*. The plants were the same size at the start of the experiment. The one at the right received four applications of gibberellic acid at two-week intervals; the one at the left received similar applications of plain water. [Photo courtesy of P. P. Pirone, New York Botanical Garden.]

Fig. 26.9 Structural formula of abscisic acid.   **Fig. 26.10** Structural formula of coumarin.

amount of coumarin that leaches into the soil from the seeds of some cultivars under natural conditions is enough to inhibit the germination and growth of other cultivars. Growth of excised root tips even of the high-coumarin cultivars is also inhibited by the amount of coumarin that leaches from the seeds; obviously something that counters the effect of coumarin is produced elsewhere in the seedlings, and the root tip depends on the other parts of the plant for its supply. The high coumarin content of these seeds seems to be another weapon in the endless competition for survival, but we do not yet know what disadvantages this weapon carries with it. The modifications necessary to withstand the coumarin may well have harmful side effects. Readers so-minded may see certain social parallels.

The production of coumarin, mentioned in the preceding paragraph, is an example of **allelopathy**—the chemical warfare between different kinds of plants. The most famous allelopathic agents are the antibiotics produced by certain bacteria and fungi, as noted in Chapter 14, but higher plants use similar competitive tactics. One of the commonest plants that chemically inhibits the growth of competitors is the bracken fern (*Pteridium aquilinum*). Individual examples have been known for many years, but a chemical approach to allelopathy is only now getting underway among botanists. It appears that at least two groups, the terpenes and the phenolic compounds, in addition to the lactones, are important allelopathic agents.

Some kinds of terpenes, phenolic compounds, and lactones are common secon-

Fig. 26.11   *Salvia leucophylla* inhibiting growth of other plants in a California grassland. The barren zone about 2 m wide around the *Salvia* is surrounded by a zone several meters wide in which growth is stunted. [Photo courtesy of Cornelius H. Muller.]

dary metabolites in plants. As noted on p. 349, many secondary metabolites are now thought to be chemical defenses against predators. Thus it might be reasonable to think of allelopathy as an extension of the chemical warfare between plants and predators to include competing plants as well. An alternative, or more probably complementary explanation is presented below.

Some substances are harmful to the organisms that produce them as well as to other organisms. We noted in Chapter 14 that the concentration of alcohol that can be obtained by fermentation is limited by the degree of tolerance of the yeast to the alcohol it produces. The decay products of peach roots have been shown to inhibit the growth of peach seedlings, and the tendency of some bunchgrasses to die out in the center may reflect a similar phenomenon. There are all gradations and combinations in nature, from conditions in which a particular substance produced by an organism has a definite survival value in restricting competition, to conditions in which the waste product or decay product is distinctly harmful to the organism that gives rise to it.

All living systems must have ways to deal with harmful by-products of their own metabolism. They must be excreted, or compartmentalized in a place where their damage is minimized, or they must be converted into harmless substances. Some allelopathic agents may have originated as disposal systems for harmful metabolic products, which later acquired an allelopathic function. If the plant could kill two birds with one stone, harming its competitors as well as disposing of poisons, so much the better. Once acquired, the allelopathic function is of course subject to improvement by natural selection.

Some students may be aware that the disposal of waste products is a problem for communities as well as for individual organisms. The human population has reached such a density, and has such a high production of poisonous wastes that our spaceship, the Earth, can no longer absorb and neutralize them effectively. If our complex social organization is to survive, we must find ways to dispose of our own wastes, and to recycle the essential elements contained in them.

Cytokinins (Fig. 26.12), ethylene (C_2H_4), and thiamin (vitamin B_1) are some of the other plant growth-regulating substances. The **cytokinins** influence many of the same processes as auxin, but not always in the same way.

Ethylene is commonly produced by ripening fruits. It hastens ripening, promotes the formation of abscission layers, and has a number of other effects, often including an increased rate of respiration. Some of the effects of auxin on growth may reflect the fact that under some

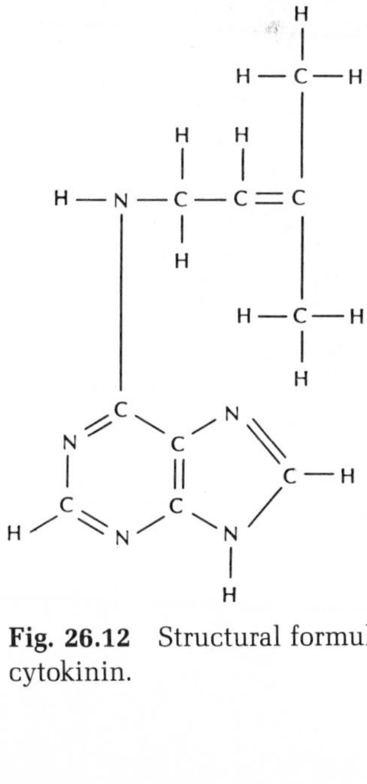

Fig. 26.12 Structural formula of zeatin, a cytokinin.

conditions it stimulates the production of ethylene. Ethylene is used commercially to ripen oranges and bananas; and ethylene chlorhydrin, a highly poisonous laboratory derivative, is used commercially to break dormancy.

Our knowledge of the role of growth-regulating substances is still very incomplete, and unsolved problems abound. For example, some leuco-anthocyanins (colorless substances chemically allied to the anthocyanins) have significant growth-regulatory effects. Many kinds of plants tend to produce anthocyanins in the leaves or stems when the cells have a high concentration of soluble sugars. Aside from their role as insect-attractants in petals, no clear function has been ascribed to anthocyanins. But might they not be a chemically inactive reserve, made under conditions of plenty, and ready to be transformed into leuco-anthocyanins as needed? Or are they merely inert disposal products, a way of getting rid of surplus leuco-anthocyanins? Or do they have some other, as yet wholly unknown function? We simply do not know.

ENVIRONMENTAL CONTROL OF GROWTH

The environmental factors affecting plant growth are so intricately related that any classification is necessarily arbitrary and imperfect. Radiant energy, temperature, water, minerals, atmospheric conditions, and gravity are some of the more obvious physical factors. Water relations, mineral nutrition, photoperiodism, and the effect of carbon dioxide in the air have been discussed in earlier chapters, and gravity is discussed in this chapter in the section on plant movements. Certain aspects of the effects of radiant energy, temperature, and atmospheric conditions remain to be discussed.

Light and Other Radiant Energy

Different wavelengths of light have different effects on plant growth, and one of the problems in growing plants under artificial light is to obtain the proper balance of various wavelengths.

For each response controlled by light there is an action spectrum (i.e., if the quantitative effects of successive wavelengths are plotted on a graph, a definite line with one or more peaks, comparable to the absorption spectrum of a pigment, is obtained). The only obvious explanation for this is that the initial effect is on some pigment(s) in the plant. Absorption of light converts the pigment into a metabolically active form, but in the absence of light of the proper wavelength it reverts to the inactive form. The precise mechanisms by which these activated pigments exert their effect are still poorly understood, but in general it may be said that they activate or deactivate enzyme systems, and/or affect membrane activities within the cells, in ways that have a chain of consequences. To at least some extent they may be thought of as growth regulators one step removed, exerting their influence through their effect on auxin, gibberellin, etc.

One of these pigments that has commanded widespread attention in recent years is **phytochrome.** After a series of studies by United States Department of Agriculture botanists, beginning about 1934, it became evident during the 1950s that red light (about 660 nm) and far-red light (about 730 nm) have opposite effects on a wide range of plant growth phenomena. The effect can be reversed back and forth as many times as one wishes by changing from one wavelength to the other. It was concluded that the effects must be due to an enzymatic pigment existing in two interconvertible forms.

This pigment was isolated by a USDA group in 1959 and named phytochrome. It is now known to be of widespread occurrence, from the algae to the flowering plants. In 1966 it was determined that phytochrome is a conjugated protein with a phycobilinlike prosthetic group consisting of an open-chain tetrapyrrol. The prosthetic group exists in two alternative forms according to the wavelength of light to which it is exposed, just as had been predicted. Red light converts most of the phytochrome to a form sensitive to far-red light (called **Pfr**), and far-red light causes the reverse change (forming **Pr**). The absorption spectra of the two forms are shown in Fig. 26.13, and the structure of the prosthetic group of the Pfr form is shown in Fig. 26.14. The Pr form differs only in the position of two hydrogen atoms.

The metabolically active form of phytochrome is Pfr, that is, phytochrome which has been changed by red light to the form more sensitive to far-red. The red-absorbing form absorbs light more effectively than the far-red absorbing form (Pfr); therefore a mixture of wavelengths such as is provided by sunlight converts most of the phytochrome to the Pfr form. In darkness Pfr tends to revert to Pr.

Pfr tends to promote certain growth responses and to inhibit others. Indeed it has opposite effects in different species on flowering and seed germination. Flowering of long-day plants is promoted by Pfr, whereas that of short-day plants is inhibited (Fig. 26.15). In about half of the several hundred kinds of seeds that have been tested, Pfr promotes germination, but in many others it has no obvious effect, and in some it is distinctly inhibitory. In general it promotes leaf expansion and the unfolding of the plumular hook during germination and it tends to inhibit excessive elongation of the stem. Etiolation (discussed on a subsequent page) is in part a reflection of the lack of adequate Pfr.

The influence of phytochrome on growth is of course modified by other factors, both hormonal and environmental. In Fig. 26.16 we see that germination of tobacco seeds is promoted by Pfr (i.e., by exposure to red light) but that germination is also influenced by temperature.

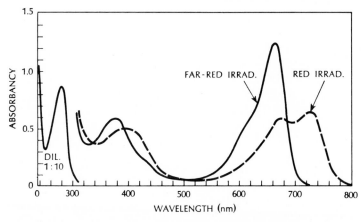

Fig. 26.13 Absorption spectra of the two forms of phytochrome. [Courtesy of H. A. Borthwick, Agricultural Research Service.]

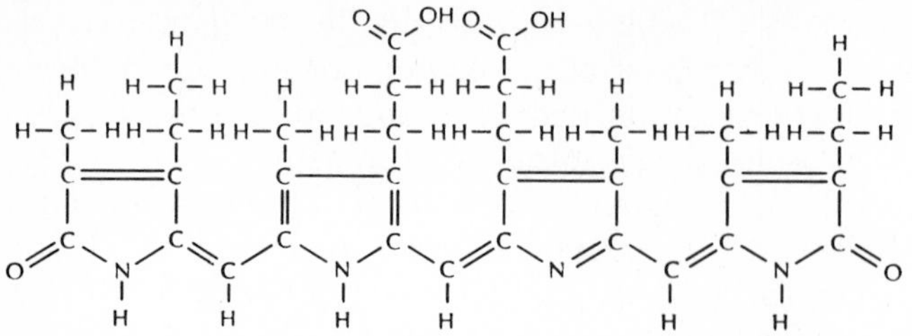

Fig. 26.14 Structure of the prosthetic group of the Pfr form of phytochrome.

Riboflavin (vitamin B_2) is a yellow prosthetic group on a protein, forming a flavoprotein. Its absorption spectrum has three peaks, all in the blue and near-ultraviolet. In its photoactivated form it can transfer energy to some other kinds of molecules in the cell. Among the compounds that can be influenced eventually are auxin and some other growth regulators. Riboflavin (in its conjugated form as a flavoprotein) is probably the light sensor in most phototropic reactions. At least in bacteria flavoproteins can also transfer energy into DNA, acting as a mutagen.

Plants grown in the dark, but with an adequate food supply, develop long,

Fig. 26.15 Effect of red and far-red light on flowering of *Kalanchoe*. (*Left*) Short-day control; (*center*) exposed at night to 1 minute of red light; (*right*) 1 minute of red light followed by 1 minute of far-red. [Courtesy of H. A. Borthwick, Agricultural Research Service.]

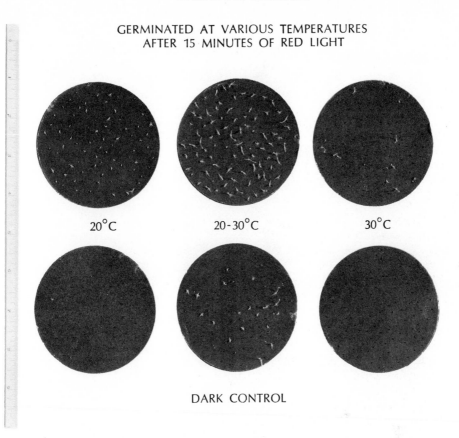

NICOTIANA TABACUM

GERMINATED AT VARIOUS TEMPERATURES
AFTER 15 MINUTES OF RED LIGHT

20°C   20-30°C   30°C

DARK CONTROL

Fig. 26.16 Effect of red light on germination of tobacco seed at various temperatures. [Courtesy of Vivian K. Toole, Agricultural Research Service.]

weak, spindly stems; the leaves of dicots generally fail to expand, and those of monocots are commonly narrower and more attenuate than normal. The shoots remain white or pale yellowish, inasmuch as chlorophyll is not formed without light. Such plants are said to be **etiolated** (from French *etioler,* to blanch). Plants with inadequate light show similar but less pronounced effects (Fig. 26.17). Red light is highly effective in preventing excessive elongation, but other wavelengths are more effective in promoting leaf expan-

sion and the formation of chlorophyll. Thus it appears that more than one photo-active system is involved.

Under natural conditions, etiolation occurs only in seedlings from deeply covered seeds and in shoots arising from underground perennating organs such as rhizomes, tubers, bulbs, and fleshy or running roots. Etiolation may thus have a survival value in enabling such shoots to reach the ground surface quickly.

Except for the direct effects of heat, the infrared and ultraviolet rays of the sun

Fig. 26.17   Etiolated bean seedlings (C), grown in total darkness, alongside plants receiving 50 percent natural light intensity (B) and 100 percent natural light intensity (A). [Photo by F. H. Norris, from Transeau, Sampson, and Tiffany, *Textbook of botany*, rev. ed., Harper & Row, New York, 1953.]

that reach the earth's surface have relatively little apparent influence on plant growth. Plants grow nearly as well under glass (which filters out most of the ultraviolet) as in the open, if other conditions are suitable. The shorter ultraviolet rays, which are mostly screened out of natural radiation by the upper atmosphere, are destructive to plants as well as to animals. X-rays are also harmful to both plants and animals.

Temperature

Any change in temperature affects the different physiological processes in the plant unequally, with consequent effects on growth. The balance between photosynthesis and respiration, for example, is highly responsive to temperature changes. In many species of plants in temperate regions, the rate of photosynthesis under field conditions is not much higher at 30°C than at 15°C because the carbon dioxide content of the air is a limiting factor. Respiration, on the other hand, is about doubled by such a change, and the higher temperature favors rapid utilization of photosynthate, rather than accumulation. We can thus envision for each species a cyclic effect of increasing temperature, from carbohydrate depletion and eventual starvation at temperatures too low for photosynthesis, to carbohydrate accumulation but slow growth, to balanced production and use of photosynthate with rapid growth, to carbohydrate depletion and eventual starvation at temperatures so high that respiration outstrips photosynthesis. The night temperature, since it affects respiration but not photosynthesis, may be even more significant than the day temperature under some circumstances.

Extremes of high and low temperature may limit growth or be fatal for various reasons. Frost injury may result either from mechanical disruption of the protoplast by internal ice crystals, or from metabolic disturbances associated with the loss of water or too rapid release of water in thawing. Many plants become more resistant to frost injury after repeated exposure to near-freezing temperatures. Many tropical species are completely unable to withstand below-freezing temperatures, but many more northern ones can, after proper cold hardening, withstand any natural extreme of low temperature.

Atmosphere

The major constituent of the atmosphere, nitrogen, is insignificant as a direct factor in the growth of most plants, being absorbed only in combined forms, such as nitrates. The second most abundant constituent of the air, oxygen, is of course necessary for respiration, but variation in the oxygen content of the air is so slight as to be unimportant, except as regards the air in the soil. Poorly aerated soils may not have enough oxygen for normal root growth. Carbon dioxide and water vapor are the only other natural components of the air that have any importance in plant growth; the effects of these have been discussed in previous chapters, as has the effect of wind in increasing transpiration.

Various harmful chemicals are released into the air as a result of man's industrial activities. The hydrocarbons from automobile exhausts sometimes reach a concentration high enough to damage plants, as well as being a prime cause of the notorious smog of many large cities. Smelters release sulfur dioxide (SO_2), and aluminum smelters release hydrogen fluoride (HF) as well. Both of these are highly toxic to plants, and in some cases all the vegetation for several miles around a large smelter is destroyed (Fig. 26.18). Justified concern about the effects of lead on both plants and animals sparks the ongoing efforts to discourage the use of lead in gasoline.

ABNORMAL GROWTH

We have seen that growth results from a complexly balanced set of chemical reactions in the protoplasm, and that the rate and nature of growth vary with the environmental conditions and the kind of plant. Under extreme or otherwise unusual environmental conditions, growth may be abnormal. Such synthetic chemicals as 2,4-D, and such natural chemicals as gibberellin, may so upset the balance that abnormal growth is quickly followed by death.

Abnormal growth may also be internally directed. A number of interspecific hybrids in *Nicotiana* (the tobacco genus) are apparently normal except for numerous small tumors along the stem (Fig. 26.19). Here the balance has obviously gone askew; some cells that should stop growing and mature into ordinary stem cells continue to grow and divide indefinitely. Doubtless other imbalances in many hybrids escape our attention.

Many insects and some bacteria that parasitize plants cause abnormal enlargements called **galls.** Often the shape and structure of the gall induced by a particular species of insect on a particular spe-

Fig. 26.18 Copper Basin, Tennessee, showing destruction of vegetation by smelter smoke. [U.S. Forest Service photo by C. R. Hursh.]

cies of plant are so constant and so complex that the insect species can be identified by the plant gall. The same host plant can produce many different kinds of galls according to the kind of parasite (Fig. 26.20), and the same insect sometimes induces different sorts of galls on different hosts. The immediate cause of the overgrowth varies according to the parasite. In general the insect probably secretes growth-stimulating substances which, in combination with the wound damage caused by feeding, subvert the host into providing a home with a built-in food supply.

MOVEMENTS

Introduction

Plants, which seem so changeless and still, are actually in continuous motion. As seen by the time-lapse camera, the stem actively twists and turns, and the flowers and leaves often repeat a series of movements like a stately dance, in a daily rhythm. Other motions may be performed but once by each of many organs, like identical routines by a series of dancers entering the stage in rapid succession.

Movements of plant parts can mostly be classified, according to the mechanism, as **growth movements** and **turgor movements.** Growth movements result from differential growth; cells in one part of the affected organ grow faster than those in another part, thus changing its position. Turgor movements are due to differential changes in turgor (and consequently size) of the cells of the affected organ. Turgor movements are readily reversible under changed conditions. Growth movements are less easily reversed; once the cells have enlarged or divided they do not return to their former size, and reversal of the movement can only be accomplished by reversal in the rates of growth.

Some growth movements are self-controlled (autonomic); others are in-

Fig. 26.19 Spontaneous tumors on the stem of a hybrid between *Nicotiana langsdorfii* and *N. suaveolens*. [Photo courtesy of Harold H. Smith, Brookhaven National Laboratory, Upton, New York.]

duced by external stimuli (paratonic). **Nutation,** the spiral twisting of the stem as it grows, is a wholly autonomic growth movement. Paratonic movements are further classified as **tropisms** and **nasties** (or **nastic movements**). Tropisms are responses to stimuli that come chiefly or wholly from one direction; the direction from which the stimulus comes affects the direction of the growth response. Nasties are responses to stimuli that more uniformly envelop the affected organ, or if unidirectional, evoke the same response regardless of the direction they come from. This classification of movements is an aid to understanding, but it is not to be expected that all plant movements will fit neatly into given categories.

Tropisms

Geotropism (response to gravity) and **phototropism** (response to light) are the most familiar tropisms, but **thigmotropism** (response to touch), **thermotropism** (response to temperature), **chemotropism** (response to chemicals), and **hydrotro-**

Fig. 26.20 Fifteen different galls on hickory leaves, caused by as many different species of insects. [Drawings by B. W. Wells, from Transeau, Sampson, and Tiffany, *Textbook of botany*, rev. ed., Harper & Row, New York, 1953.]

pism (response to water) are also well known. Tropisms may be positive, negative, or lateral, and they may differ in different organs of the plant.

Perception of gravity in the roots is a function of the root cap, at least in some kinds of plants. The primary root is usually positively geotropic, growing in the direction of gravity, and the stem is negatively geotropic, growing away from gravity. If a seed is planted upside down, so that the emerging root points upward, the root turns as it grows, and the tip soon comes to point downward. The stem, likewise, will right itself regardless of its initial position.

The means by which plants perceive gravity has been the subject of much study and speculation, without conclusive results. It appears that amyloplasts are too heavy to be carried easily by cyclosis, and tend to settle to the lowest part of the cell. In this role as falling bodies they are said to be *statoliths*. We still do not know whether or how the disturbance produced by their movement or their position at the bottom of the cell is translated into a geotropic response. Furthermore some kinds of plants show a geotropic response in the absence of amyloplasts. Perhaps in these plants some other intracellular bodies function as statoliths. The mechanism of geotropism is still open.

Phototropic responses occur in most green plants. The effect is due almost entirely to blue light, probably absorbed by a flavoprotein. Typically the stem bends toward the light as it grows; people with green thumbs know that potted plants on a windowsill need to be turned every few days if they are to grow symmetrically. The stems of some climbing plants, such as English ivy (*Hedera helix*), are negatively phototropic; placed in a pot in a window, they turn and grow away from the light, directly into the room. A little thought should suggest a relationship between this response to light (along with a normal negative geotropism) and the success of the plant as a climber on walls.

Different intensities of light often evoke opposite phototropic responses. The stems of Bermudagrass (*Cynodon dactylon*) and some other plants are positively phototropic at low light intensities, and negatively so at higher intensities. Such plants are prostrate when growing in open places where they are fully exposed to the sun, but they are more ascending or erect when shaded by other vegetation. Even some of the prostrate liverworts will grow upward at an angle in a terrarium with insufficient light.

Leaves as well as stems are often phototropic. The leaf blades of such vines as English ivy occupy practically the entire exposed surface of the wall, with a minimum of overlap, fitting together in patterns called **leaf mosaics.** Most plants with large numbers of leaves show somewhat similar, though usually less accurate, patterns, making maximum use of the available light.

The role of auxin in phototropism is discussed on earlier pages.

The most familiar examples of **thigmotropism** are shown by the tendrils or stems of some climbing vines, which curl around solid objects they touch. A slightly rough or uneven surface, providing two or more adjacent but separated stimuli, is ordinarily required. The immediate reaction, which sometimes occurs in less than a minute (and commonly within a few minutes), is probably due to differential changes in turgor, but the changes are irreversible and are followed by differential growth, so that the tendril or stem wraps around the supporting object. We have scarcely an inkling of the mechanism of thigmotropism.

The most familiar **chemotropism** in angiosperms is that shown by the pollen tube, which grows unerringly down through the style and into the embryo sac of the ovule in response to chemical stimuli. The nature of the chemical stimulus in this instance is not well understood.

Hydrotropism may be regarded as a special case of chemotropism, since water is of course a chemical. Hydrotropic curvature has been demonstrated in the roots of several kinds of plants, but apparently this response is not so common as was once believed. The seeming hydrotropism of roots of many plants results from the more rapid growth and more extensive branching of roots in moist soil than in dry soil, rather than from curvature of existing roots in the direction of moisture.

Nastic Movements

Nastic movements, in the strict sense, are growth movements, although the term is sometimes applied also to some otherwise similar turgor movements. Some nastic movements are induced by external stimuli, like tropisms, but others are internally directed. The growth movements of young leaves, bud scales, and the petals of many flowers are internally controlled. Tulips and crocus flowers are thermonastic, opening as the temperature rises, and closing again when it falls. Flowers of some other kinds of plants open or close in response to darkness. The "sleep" movements of the leaves and flowers of some plants are caused by turgor changes, but others are nastic growth movements. Nastic sleep movements may be photonastic, or thermonastic, or both, according to the species.

Both thigmonasty and chemonasty are illustrated by the tentacles of sundew (*Drosera*), an insectivorous plant of boggy places (Fig. 26.21). The tentacles are stout, spreading hairs with proteolytic enzymes in the expanded, sticky tip. The marginal tentacles turn toward the center of the leaf if they or the central tentacles are stimulated by the touch of a solid object or by any of several nitrogenous chemicals in solution. If the stimulus is not a digestible object, the tentacles generally recover and return to their normal position within a day, but under the continuous chemical stimulation of a trapped insect, they remain curved for a week or more until the prey is digested.

Turgor Movements

Turgor movements of plant parts result from differential changes in turgor (and consequently size) of some cells. The

Fig. 26.21   Sundew, *Drosera rotundifolia.* [Photo by Alvin E. Staffan, from National Audubon Society Collection/Photo Researchers, Inc.]

effective cells are often different from ordinary cells and may be concentrated in special organs, such as the pulvinus (Latin, a cushion) at the base of the leaves and leaflets of legumes and some other plants. Turgor movements are often classed with nastic movements, because the direction is not influenced by the source of the stimulus. The rolling of the leaves of many grasses in dry weather is caused by loss of turgor and collapse of some large, thin-walled, bubble-shaped cells that form longitudinal rows on one or the other surface.

The sensitive plant (*Mimosa pudica*) is famous because the leaves or leaflets quickly droop when touched (Fig. 26.22).

The speed and intensity of the movement and the amount of the plant affected vary with the intensity of the shock; the movement occurs first in the part touched and spreads progressively with diminishing force to other parts. Recovery usually takes 10–20 minutes. If a leaflet is burned in a flame, the plant reacts even more vigorously, and often all the leaves quickly droop. The movements are directly caused by differential changes in the turgor of the cells in the well-developed pulvini at the base of the leaflets and leaves, but numerous experiments have not yet clarified the mechanism of transfer of the stimulus to the pulvini, nor the chemical processes that cause the turgor

Fig. 26.22   Sensitive plant, *Mimosa pudica*. [Photo courtesy of T. H. Everett.]

changes. A very rapid movement of potassium ions into and out of some of the cells probably contributes to the changes in turgor. Current hypotheses do not adequately explain the speed of the transfer of the stimulus. A number of other legumes are sensitive to the touch, but the reaction is less spectacular.

A possible survival value of the response to touch in *Mimosa pudica* and some related species has been suggested but remains to be confirmed. These species often grow on exposed, dry ridgetops, where they may be subjected to drying winds. A strong wind may joggle the leaves enough to make them droop and fold up, reducing the rate of transpiration.

SUGGESTED READING

Braun, A. C., The reversal of tumor growth, *Sci. Amer.* **213**(5):75–83, November 1965.

Erickson, R. E., and W. K. Silk, The kinematics of plant growth, *Sci. Amer.* **242**(5):134–151, May 1980.

Evans, M. L., Rapid stimulation of plant cell elongation by hormonal and non-hormonal factors, *BioScience* **23**:71–718, 1973.

Galston, A. W., and C. L. Slayman, The not-so-secret life of plants, *Amer. Sci.* **67**:337–344, May–June 1979.

Muller, C. H., The role of chemical inhibition (allelopathy) in vegetational composition, *Bull. Torrey Bot. Club* **93**:332–351, 1966.

Rice, E. R., Allelopathy—An update, *Bot. Rev.* **45**:15–109, 1979.

Satter, R. L., and A. W. Galston, Leaf movements: Rosetta stone of plant behavior? *BioScience* **23**:407–416, 1973.

Shen-Miller, J., and R. R. Hinchman, Gravity sensing in plants: A critique of the statolith theory, *BioScience* **24**:643–651, 1974.

Steward, F. C., The control of growth in plant cells, *Sci. Amer.* **209**(4):104–113, October 1963.

Taylorson, R. B., and S. B. Hendricks, Aspects of dormancy in vascular plants, *BioScience* **26**:95–101, 1976.

Thimann, K. V., *Hormone action in the whole life of plants,* University of Massachusetts Press, Amherst, 1977. A readable textbook-level presentation reflecting an outstanding career.

van Overbeck, J., The control of plant growth, *Sci. Amer.* **219**(1):75–81, July 1968.

Wodehouse, H. W., Senescence processes in the life cycle of flowering plants, *BioScience* **28**:25–31, 1978.

Chapter 27

Angiosperms: Classification

Flowering branch of *Magnolia grandiflora,* an evergreen species of the southeastern United States. *Magnolia* is the most familiar and widespread genus among the more archaic genera of angiosperms. [U.S. Forest Service photo by W. D. Brush.]

The angiosperms, or flowering plants, with about 220,000 known species, make up more than half of the plant kingdom and form the dominant land vegetation of most of the earth. Most botanists are agreed that the angiosperms form a wholly natural group, but there is still some difference of opinion as to the taxonomic rank at which the group should be received. Some botanists consider them to constitute a distinct division Magnoliophyta, as is done in this text; others treat them as a subdivision or class of a more inclusive division such as the Tracheophyta.

DICOTS AND MONOCOTS

The angiosperms are fairly readily divisible into two groups, the dicotyledons and monocotyledons, here treated as classes under the names Magnoliopsida and Liliopsida. The difference between these groups was first pointed out by John Ray (1628–1705), although he considered it less important than the difference between woody and herbaceous plants, which is considerably downgraded in modern systems.

On the basis of both comparative morphology and the fossil record it appears that the earliest angiosperms were woody plants with an active cambium, and that the monocots originated as a very early herbaceous (probably aquatic) offshoot from primitive dicotyledons. A comprehensive speculative exposition of this concept of the ancestry of monocots has recently been advanced by the American botanist Arthur Cronquist (1919–   , Fig. 2.5). In this view, the monocots lost both their cambium and their leaf blade in association with the aquatic habitat, and either lost their vessels or diverged from the dicotyledons before the evolution of

vessels in that group. They then returned to the land as herbs, filling niches that had previously been occupied mainly by ferns and other vascular cryptogams. The evolution of vessels, a proper leaf blade, and means of secondary thickening has not been easy for monocots. Consequently they have remained a much smaller group than the dicots, showing a more obvious correlation of adaptive features with major evolutionary lines and broad-scale taxonomy.

The several differences between the dicots and monocots have been noted and discussed in the preceding chapters. No one of these differences is absolutely constant, although the number of cotyledons is nearly so. The most important contrasting characters of the two groups are summarized in the following table.

AN EVOLUTIONARY INTERPRETATION OF ANGIOSPERM CLASSIFICATION

The genera and species of angiosperms often have a characteristic aspect, and the differences among them often reflect adaptations to particular ecological niches such as habitat, method of pollination, method of seed dispersal, and the like. Many of these taxa are readily recognizable without formal botanical training. Oak, maple, hickory, goldenrod, ragweed, primrose, nettle, elm, tulip, Norway maple, sugar maple, silver maple, and black maple are among the many common names that closely coincide with botanical genera or species.

Most of the families of angiosperms, on the contrary, are not restricted to well-defined or even approximately mutually exclusive niches, and only a few of them, such as the Poaceae (grasses), Arecaceae (palms), and Cactaceae (cacti) stand out

Dicots	Monocots
Cotyledons 2 (seldom 1, 3, or 4)	Cotyledon 1 (or the embryo sometimes undifferentiated)
Leaves mostly net veined	Leaves mostly parallel veined
Vascular bundles usually borne in a ring that encloses a pith	Vascular bundles usually scattered, or in 2 or more rings
Fascicular cambium usually present	Fascicular cambium lacking; usually no cambium of any sort
Floral parts, when of definite number, typically borne in sets of 5, less often 4, seldom 3 (carpels often fewer)	Floral parts, when of definite number, typically borne in sets of 3, seldom 4, never 5 (carpels often fewer)
Pollen typically triaperturate, or of triaperturate-derived type, except in a few of the more archaic families	Pollen of uniaperturate or uniaperturate-derived type
Mature root system either primary, or adventitious, or both	Mature root system wholly adventitious

well enough by their gross appearance to have true common names. The sort of correlation between general appearance, major adaptive zone, and broad taxonomic group that students of vertebrate taxonomy are accustomed to is notable by its rarity among angiosperm families and orders, especially in the dicotyledons. Yet there are ample reasons, including the absence of interfamilial hybrids, to suppose that the basic organization of angiosperm genera into families is essentially sound and needs only to be tidied up in detail rather than to be completely recast. It appears that the principle of competitive exclusion (see Chapter 31) is much less effective among plants, which make their own food, than among animals, which are ultimately dependent on the plants. Once the angiospermous condition has been attained, the further obvious changes which help fit the plants to particular ways of making a living mostly occur so easily that they tend to mark species and

genera rather than larger groups. We have noted, however, that this may be somewhat less true of monocots than of dicots.

In their efforts to group together those genera that are most alike in all respects, and to draw the lines between groups through the gaps in variability, plant taxonomists have been forced willy-nilly to make extensive use of characters which, however stable, are not always obvious, and whose importance to the plants themselves is often dubious. Details of floral structure have proven to be particularly useful in this regard. These, like other characters, acquire their taxonomic importance a posteriori, through their usefulness in marking groups that have been perceived on the basis of all the available information, rather than from an a priori assignment of value. Characters of obvious adaptive significance, such as growth habit and adaptations to particular modes of pollination and seed dispersal, are of course not excluded from consideration, but they have proven to be insufficient to provide the major framework for the classification of angiosperms into families and orders.

Another way in which the taxa of angiosperms differ from taxa of vertebrates is the much greater prevalence of exceptions to the formal characters of angiosperm taxa. We have noted that although the monocots and dicots are obvious and obviously natural groups, none of the differences between them is fully constant. The problem is even more severe at the level of families and orders. Most members of the Scrophulariceae (the snapdragon family), for example, have a sympetalous, irregular corolla, but some species of the genus *Besseya* (Fig. 27.1), which unmistakably belongs to this group, lack a corolla entirely.

If many of the characters marking the major angiosperm taxa are of limited

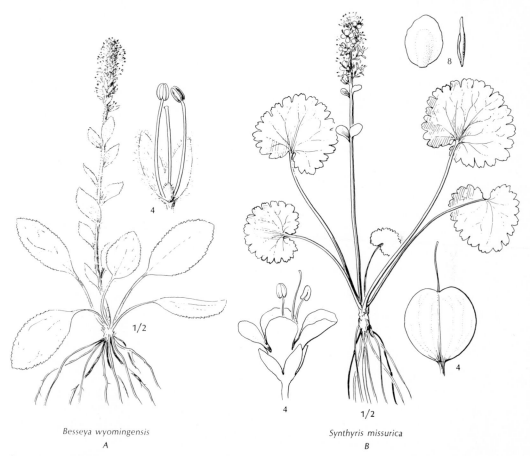

Besseya wyomingensis

A

Synthyris missurica

B

Fig. 27.1 *Besseya wyomingensis* (A) and *Synthyris missurica* (B), rather closely related members of the family Scrophulariaceae that grow in the northwestern United States. The former has no corolla, whereas the latter has a sympetalous, irregular corolla. [Drawings by Jeanne R. Janish, from *Vascular plants of the Pacific Northwest*, Part 4, by C. L. Hitchcock et al., copyright by University of Washington Press, 1959.]

adaptive significance, if adaptive changes occur relatively easily, and if competitive exclusion does not operate very effectively (all of which postulates seem to be true), then it should not be surprising that these taxa are difficult to characterize, and the characters beset with exception.

In spite of these difficulties, it has been possible to establish some 375 natural families that include practically all the known genera and species of angiosperms, and the number of genera of doubtful or controversial position is very limited. There still remains some difference of opinion as to whether certain groups should be called families, or subfamilies. Some botanists, for example, recognize a family Fabaceae (also called Leguminosae), with three subfamilies, the Faboideae, Mimosoideae, and Caesalpinioideae, whereas others recognize these same three groups as closely related families. There is general agreement, however, as to which genera should be referred to each of the three groups that collectively make up the Fabacae. No one wishes to reshuffle the genera or to distribute them among families outside the Fabaceae.

The classification of the families of angiosperms into orders is not yet as stable as the assignment of genera to families, although all systems of classification for the past century have much in common. Concepts of relationships among the orders still excite considerable disagreement among specialists, and several different modern systems of classification are competing for general acceptance. There is a general agreement on most of the evolutionary trends among the angiosperms, but the problem is to distinguish similarities due to common ancestry from those due to evolutionary parallelism, inasmuch as similar changes have evidently occurred repeatedly in diverse evolutionary lines. The more similarities one finds between two groups, the greater the likelihood that these are due to common ancestry and are the markers for a syndrome of less obvious similarities. Individual similarities, on the other hand, often run counter to the pattern and are of little taxonomic significance.

The most widely accepted overall classification of the angiosperms for many years was that of the German botanist Adolf Engler (1844–1930, Fig. 27.2) and his associates, as published in a multivolume work, *Die natürlichen Pflanzenfamilien*, from 1887 to 1899. The Germanic thoroughness of the Engler system, combined with the simplicity and apparent logic of its principles, made Engler the most influential botanist of his time. Engler tended to equate simplicity of floral structure with primitiveness. Thus he considered flowers such as those of poplars (*Populus*, Fig. 27.3) and cattails (*Typha*), which are unisexual and lack a perianth, to be primitive.

Unfortunately, Engler's evolutionary concepts have not stood the test of time. Modern phylogenists generally believe that the primitive angiosperm flower had an indefinite but fairly large number of separate parts of each kind, with the perianth not clearly differentiated into sepals and petals. Accordingly, large blocks of the Engler system are reordered in such modern systems as those of the Soviet Armenian botanist Armen Takhtajan (1910–   , Fig. 27.2, most recent comprehensive system in 1980) and the American botanist Arthur Cronquist (most recent comprehensive system in 1981).

The view that the primitive angiosperm floral type was one with numerous separate parts of each kind actually antedates Engler, going back to the system of the

Fig. 27.2 (*Left*) Adolf Engler (1844–1930), German botanist, author of a well-known system of plant classification. (*Right*) Armen Tahktajan (1910–   ), Soviet Armenian botanist, author of an influential recent classification of flowering plants.

Swiss-French botanist Augustin Pyramus de Candolle (1778–1840, Fig. 27.4). As a pre-Darwinian, de Candolle wrote (and may have thought) in terms of a series of special creations representing logically successive modifications of an original basic plan, but his ideas are readily convertible into an evolutionary interpretation. de Candolle's ideas were revived, expanded, modified, and systematically presented by the American botanist Charles E. Bessey (1845–1915, Fig. 27.4) in 1898 and 1915. Bessey's views excited much theoretical interest, but his own system based on them was never widely adopted. His concepts, nonetheless, form the foundation for the most widely accepted modern systems.

In the list on p. 500, some of the characters now believed to be primitive within the angiosperms are given in the left-hand column, and the contrasting advanced characters in the right-hand column. When it is desired to indicate successive stages of advancement, these are given in order in the right-hand column, separated by a semicolon.

FOSSILS

Aside from pollen grains, the most abundant angiosperm fossils are leaf impressions (Fig. 27.5). Seeds, dry fruits, and bits of wood are often also fossilized. Fossil flowers (Fig. 27.6) are much more rare and exist mainly as mere impressions that show the external form but not the structure. Relatively recent fossil leaves can often be identified with modern species or genera by careful comparison, but this is a

Fig. 27.3 Male (*left*) and female (*right*) catkins (flower clusters) of poplar, *Populus*. This sort of flower, consisting of only a single pistil or a few stamens, and without perianth, was believed by Engler to be a primitive type; it is now regarded as reduced. [*Left*, U.S. Forest Service photo by W. D. Brush. *Right*, photo by Clifford E. Matteson, from National Audubon Society Collection/Photo Researchers, Inc.]

risky procedure at best, and progressively more so in dealing with older fossils that are less likely to have modern counterparts. Because of these problems, concepts of relationships among the families and orders of angiosperms have in the past necessarily been based largely on comparison of living members of the groups, with only limited attention to the fossil record.

The oldest fossils that are unmistakably angiosperms are of early Lower Cretaceous age. Pre-Cretaceous angiosperms have repeatedly been reported, and as often shown to be interpretable in other ways. The Potomac beds of Maryland and vicinity contain the most extensive known Lower Cretaceous angiosperm flora. In these the proportion of angiosperms increases from small at the base to dominant at the top of the deposits, which cover a span of some 20–30 million years. Many Lower Cretaceous fossils from these and other deposits have been given names suggesting a resemblance to modern genera, such as *Nelumbites*, *Menispermites*, *Sapindopsis*, and *Celastrophyllum*, and some have even been referred to modern genera, such as *Populus* and *Sassafras*, but all such identifications must be reconsidered.

Pollen (Fig. 27.7) fossilizes better than most other plant parts, and with the recent wide availability of scanning elec-

Fig. 27.4 (*Left*) A. P. de Candolle (1778–1841), Swiss botanist, whose concepts of relationships among angiosperm families foreshadowed modern opinion. (*Right*) Charles E. Bessey (1845–1915), American botanist, noted for his dicta regarding evolutionary trends among the angiosperms.

tron microscopes it has begun to be useful in unraveling the evolutionary history of the angiosperms. An important paper published in 1969 by the American botanist James Doyle (1944–   ) indicates that earliest Cretaceous pollen is all of the primitive, uniaperturate type, and that the progressive development of more advanced types can be traced within the Cretaceous period. Subsequent studies by Doyle and other paleobotanists in various parts of the world have abundantly confirmed this interpretation. The fossil record does not require a pre-Cretaceous origin for the angiosperms, although it does not preclude the possibility that primitive angiosperms existed long before that time, especially because the most

primitive kind of angiosperm pollen is difficult to distinguish from gymnosperm pollen. A comprehensive reinterpretation of the Potomac fossils, integrating the information from pollen with that from leaves and fruits, is now well under way. It is now believed that no Lower Cretaceous angiosperms should be referred to modern genera.

By Upper Cretaceous time the angiosperms had become the dominant land vegetation, and they are abundantly preserved in the fossil record from that time to the present. Some Upper Cretaceous fossils and many early Tertiary ones appear to represent modern genera (or close allies of modern genera) in many different families.

Primitive	Advanced
1. Tropical	Temperate
2. Woody	Herbaceous
3. Cambium active	Cambium none
4. Long-lived	Short-lived
5. Vascular bundles in a cylinder, enclosing a pith	Vascular bundles scattered
6. Vessels none	Vessels present, with scalariform perforations; vessels with simple perforations
7. Sieve cells with lateral sieve areas that do not form a terminal plate	Sieve tubes with a terminal sieve plate
8. Chlorophyll present	Chlorophyll none
9. Evergreen	Deciduous
10. Leaves simple	Leaves compound
11. Leaves alternate	Leaves opposite
12. Leaves net veined	Leaves parallel veined
13. Flowers regular	Flowers irregular
14. Flowers perfect	Flowers unisexual
15. Flowers hypogynous	Flowers perigynous; flowers epigynous
16. Floral parts of each kind many, spirally arranged	Floral parts of each kind few, cyclic
17. Perianth undifferentiated	Perianth differentiated into calyx and corolla; petals and/or even sepals may thereafter be lost
18. Petals separate	Petals united
19. Stamens not differentiated into filament and anther, the pollen sacs embedded in the broad lamina	Stamens differentiated into filament and anther
20. Carpels merely folded, stigmatic along the unsealed margin	Carpel fully closed, the stigma terminal
21. Carpels separate	Carpels united, the placentation axile; placentation may thereafter become parietal, or may become free-central and then basal
22. Pollen uniaperturate	Pollen triaperturate
23. Pollen 2-nucleate at time of pollination	Pollen 3-nucleate at time of pollination
24. Pollination by beetles	Pollination by other agents
25. Fruit dehiscent	Fruit indehiscent
26. Endosperm well developed	Endosperm none
27. Cotyledons several	Cotyledons 2; cotyledon 1
28. Germination epigaeous	Germination hypogaeous

The fossil record does not provide a clear connection of the angiosperms to any other group of plants. Perhaps their closest allies are some Jurassic descendants of the seed ferns, forming an order Caytoniales, but there is still a considerable gap.

Fig. 27.5 Fossil angiosperm leaves. (About natural size.) (*Left*) *Rosa,* from Oligocene deposits in Montana. [Photo courtesy of Herman F. Becker.] (*Right*) *Ulmus,* from Miocene deposits in Oregon. [Photo courtesy of Alan Graham.]

SOME REPRESENTATIVE FAMILIES

Dicots

Magnoliaceae. The Magnoliaceae (see chapter-opening photograph) consist of 12 genera and about 220 species, widely distributed in tropical and warm-temperate regions. They are woody plants with stipulate leaves; large, mostly perfect flowers; a well-developed perianth that is not always differentiated into sepals and petals; numerous, spirally arranged stamens with uniaperturate pollen grains; and numerous, spirally arranged, separate carpels on an elongate receptacle. They are considered to be one of the most archaic families of flowering plants, although some other families in the same

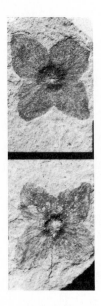

Fig. 27.6 Angiosperm fossils. (*Left*) Flowers of Oligocene age from Montana. [Courtesy of Herman F. Becker.] (*Right*) *Fagopsis*, a beechlike fossil of Tertiary age from Colorado. [New York Botanical Garden photo.] (*Inset above*) Fruit of *Thlaspi*, from Oligocene deposits in Montana, with a modern *Thlaspi* fruit laid alongside at the right. [Courtesy of Herman F. Becker.] (*Bottom*) Half fruit of maple, *Acer*, from Miocene deposits in eastern Oregon. [Courtesy of Alan Graham.]

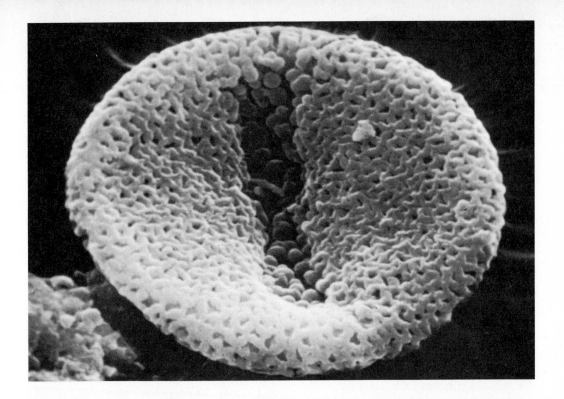

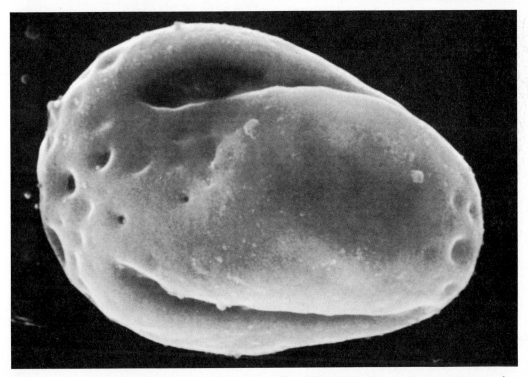

Fig. 27.7 Lower Cretaceous fossil angiosperm pollen. (*Top*) Primitive, uniaperturate type. (× 2500.) (*Bottom*) More advanced, triaperturate type. (× 3500.) [Electron micrographs by JEOL U.S.A., Inc., courtesy of James Doyle.]

order (Magnoliales) are more primitive in particular respects, such as the merely folded, unsealed carpels (Degeneriaceae) and the absence of vessels (Winteraceae and others). Many species of *Magnolia* are cultivated for ornament.

Brassicaceae. The Brassicaceae (mustard family), consisting of about 350 genera and 3000 species, grow chiefly in extratropical regions of the Northern Hemisphere (Fig. 27.8). Members of the mustard family are mostly herbs with four sepals, four petals that spread out to form a cross, six stamens (the two outer shorter than the four inner), and a characteristic type of dry, dehiscent fruit with the ovules attached to the margins of a partition that persists on the flower stalk after the walls of the fruit have fallen off. The cross-shaped flowers have given rise to the name Cruciferae that has often been used for the group. Many crucifers have a mildly peppery odor or taste, caused by a characteristic metabolic product called mustard oil. Mustard, broccoli, cabbage, cauliflower, brussels sprouts, rutabagas, and turnips are familiar cultivated forms of *Brassica*. Other cultivated members of the family include radish (*Raphanus*), watercress (*Rorippa*), horseradish (*Armoracia*), candytuft (*Iberis*), and sweet alyssum (*Lobularia*).

Fabaceae. The Fabaceae (often known by the irregular name Leguminosae), with more than 500 genera and perhaps 13,000 species widely distributed throughout the world, are one of the largest families of flowering plants, being surpassed in size only by the Asteraceae and Orchidaceae. Legumes (members of the family Fabaceae) characteristically have compound leaves, more or less perigynous flowers, five separate petals, and a solitary carpel that ripens into a dry fruit dehiscing along

Fig. 27.8 Toothwort, *Dentaria laciniata*, a common member of the family Brassicaceae in the forests of eastern United States. [Photo by Alvin E. Staffan, from National Audubon Society Collection/Photo Researchers, Inc.]

two sutures. Most species of temperate regions are herbs with 10 stamens and a characteristic type of irregular corolla (Fig. 27.9) but many of the tropical members, such as Mimosa, are trees with

Fig. 27.9 Flower structure of bristly locust, *Robinia hispida*, a legume native to eastern United States. [New York Botanical Garden photo.]

numerous stamens and a regular corolla. All degrees of transition exist between these two seemingly very distinct types. Legumes characteristically harbor nitrogen-fixing bacteria in root nodules, unlike most other kinds of angiosperms. Many legumes are cultivated: peas, beans, soybeans, and peanuts for their edible seeds; alfalfa, clover, and vetch for fodder; sweet pea, lupine, wisteria, and redbud for ornament.

Asteraceae. The Asteraceae (also known by the irregular name Compositae), with nearly 1,000 genera and perhaps as many as 20,000 species, are usually considered to be the largest family of angiosperms. They occur on all continents, most abun-

dantly in temperate and warm-temperate regions that are not densely forested. In most parts of the temperate zone from 10 to 15 percent of the species of angiosperms are composites (members of the Asteraceae). Most of the species of ordinary temperate regions are herbs, but many of those in warm-temperate and tropical regions are shrubs, and a few are real trees. The flowers of composites are grouped into compact heads that simulate individual flowers; what appears to be a single flower is actually a "composite flower." Composites have a two-carpellate ovary with a single seed, and with a sympetalous corolla seated on top of the ovary. The stamens are attached to the corolla tube and are joined together by their anthers. The calyx, when present, is highly modified, consisting of scales, hairs, or stiff bristles. Well over half the members of the family have two kinds of flowers in each head (Fig. 27.10), the central flowers being relatively small and generally perfect, with regular corolla, whereas the marginal flowers are female (without stamens) or neutral (without reproductive parts) and have an enlarged, petal-like, strap-shaped corolla. Composites are especially important as ornamentals, because of their relatively large, showy flower-heads. Asters (*Aster* and *Callistephus*), goldenrod (*Solidago*), chrysanthemum, dahlia, cosmos, marigold (*Tagetes*), lettuce (*Lactuca*), dandelion (*Taraxacum*), sunflower (*Helianthus*), thistle (*Cirsium* and related genera), and zinnia are familiar composites. Ragweed (*Ambrosia*) and cocklebur (*Xanthium*), with very much reduced, inconspicuous, wind-pollinated flowers, belong to a special evolutionary sideline within the family.

Monocots

Alismataceae. The Alismataceae are a widely distributed group of about 10 genera and 70 species. They are mostly aquatic herbs with typically three sepals, three petals, and more or less numerous (sometimes only six) spirally (sometimes cyclically) arranged stamens and pistils. The Alismataceae are of little economic importance, but they are interesting as one of the most archaic existing families of monocots. *Sagittaria* (arrowhead) is of considerable interest because some of its species show all transitions, even on the same plant, between two quite different types of leaves. The one type has a normal petiole and a well-defined, floating blade with essentially palmate venation, whereas the

other type consists of a flattened, bladeless petiole with parallel veins, looking much like the more ordinary kind of monocot leaf. Many botanists believe that the characteristic narrow, parallel-veined leaf of most monocots is in origin a flattened petiole. Because their seeds lack endosperm, and for other reasons, the Alismataceae cannot be considered ancestral to the other monocots, but *Sagittaria* does show how the typical monocot leaf may have originated.

Poaceae. The Poaceae (grasses, also known by the irregular name Gramineae) are a cosmopolitan family with about 500 genera and 8000 species. They are probably the most important single family of plants, from a human standpoint. The importance of wheat as a staple food in much of the Western world is surpassed only by that of rice, another grass, in the Orient. Corn, oats, rye, barley, sugar cane, and bamboo are some other important grasses. Grass is a staple food for grazing animals, which in turn are eaten by predators and by man. The intercalary meristem at the base of the leaf blade of grasses enables them to withstand grazing better than most other plants, because the leaf continues to grow from the base after it has been cropped off at the top. The grasses (Fig. 27.11) have very much reduced flowers, essentially without perianth, arranged in a characteristic fashion. Most grasses are wind-pollinated, or self-pollinated, and many are apomictic.

Arecaceae. The Arecaceae (palms, also known by the irregular name Palmae) consist of about 200 genera and nearly 3000 species, widespread in tropical regions of both the Old and the New World. The most familiar palms are trees with an unbranched trunk and a terminal crown of large leaves that have an anatomically

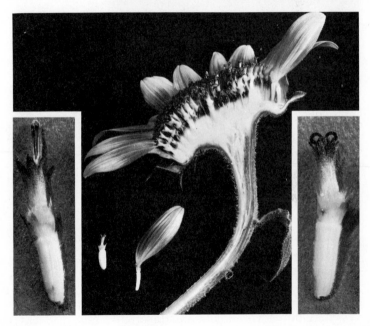

Fig. 27.10 Head and individual flowers of the common sunflower, *Helianthus annuus*, a characteristic member of the Asteraceae. [New York Botanical Garden photos.]

unique ontogeny. Some palms are climbers, with scattered leaves, or have such a short stem that the leaves appear to arise from the ground. Palms differ from the vast majority of other monocots in being woody, but as in other woody monocots the woodiness derives from the lignification of other tissues rather than from the presence of large amounts of xylem. Woodiness in palms, as in other woody monocots, is probably phyletically secondary. Palm flowers are individually most-

Fig. 27.11   The common reed, *Phragmites australis,* a grass found in moist, saline places, especially near the seacoast, over much of the world.

ly small but have an evident, biseriate perianth of six members. The flower cluster is often subtended by a large, modified leaf called a spathe. The coconut palm (*Cocos nucifera,* Chapter 28 opener) and the date palm (*Phoenix dactylifera*) are among the economically more important palms. A number of others, such as *Wash-*ingtonia, are familiar in cultivation in Florida and southern California.

Liliaceae.   The Liliaceae (lily family), with about 300 genera and nearly 4000 species, are the largest family of the order Liliales. They are widely distributed, but especially common in tropical and warm-temperate regions that are not densely forested. Most of them are perennial herbs, often from bulbs or corms, with narrow, parallel-veined leaves, a three-carpellate ovary, and three petaloid sepals that are much like the three petals. The Liliaceae (Fig. 27.12) are the most "typical" monocots, providing a mental image of what a monocot "ought" to look like. Tulips, true lilies (*Lilium*), amaryllis, day-lilies (*Hemerocallis*), asparagus, and onion (*Allium*) are familiar members of the lily family.

Orchidaceae. The Orchidaceae, with some 600 or more genera and perhaps as many as 20,000 species, are the largest family of monocots, comparable in size to the Asteraceae among the dicots. Orchids (see Part III opening photograph) are widely distributed, but are most common in the moister parts of the tropics, especially as epiphytes. They have highly modified, irregular, often very showy flowers, intricately adapted to various specialized pollinators, and very numerous, very tiny seeds with a minute and undifferentiated embryo. Most of them are strongly mycorhizal, and a few lack chlorophyll, being dependent on the fungal associate for food. Vanilla flavoring is derived from fruits of the tropical genus *Vanilla*. Otherwise orchids are important chiefly as ornamentals; *Cattleya*, in particular, is much used for corsages.

SUGGESTED READING

Anderson, E., *Plants, man, and life*, Little, Brown, Boston, 1952. A provocative, informative, outspoken book, emphasizing the origins and classification of weeds and cultivated plants, guaranteed to hold your attention.

Beck, C. B. (ed.), *Origin and early evolution of angiosperms*, Columbia University Press, New York, 1976. A symposium.

Cronquist, A., *Evolution and classification of flowering plants*, Houghton Mifflin, Boston, 1968.

Cronquist, A., *How to know the seed plants*, Brown, Dubuque, Iowa, 1979. An illustrated key for identification of families.

Rodgers, A. D., *Liberty Hyde Bailey*, Princeton University Press, Princeton, J.J., 1949. The life and times of the greatest American student of cultivated plants.

Takhtajan, A., *Flowering plants, origin and dispersal*, transl. by Charles Jeffrey, Oliver & Boyd, Edinburgh, 1969. An evolutionary approach to plant geography.

Takhtajan, A., Outline of the classification of flowering plants (Magnoliophyta), *Bot. Rev.* **46**: 225–359, 1980.

Fig. 27.12 Easter lily, *Lilium longiflorum*, showing three petals, three petal-like sepals, six stamens, and a three-lobed stigma. [Photo by Henry M. Mayer, from National Audubon Society Collection/Photo Researchers, Inc.]

Chapter 28

Angiosperms: Economic Importance

Plantation of coconut palms and bananas in southern Mexico.

Angiosperms are in a sense the foundation of our civilization. They provide man with food, clothing, and shelter. They brighten leisure, ease pain, and cure illness. They furnish the raw materials for countless industrial products. Unfortunately, they also evoke allergies such as hay fever, produce poisons that kill livestock and unwary nature-enthusiasts, and bring misery and degradation to millions trapped by drug addiction.

FOOD

The greatest importance of angiosperms is as food. Excepting fish, all the important staples of a diet are either angiosperms or are derived from animals that in turn feed largely on angiosperms.

Origin of Food Crops

The cultivation of angiosperms as food crops began long before recorded history. The wild ancestors of some of these plants, such as carrots, lettuce, and sunflowers, are readily identifiable, but most of the important food crops have been so modified under cultivation that their ancestry cannot easily be deciphered. All of them have wild relatives, but such plants as wheat, rice, and many others do not seem to be direct descendants of any existing wild species. Interspecific hybridization as well as selection has undoubtedly played a part in the evolution of crops, and the hybridization has often been accompanied by polyploidy (increase in the number of complete sets of chromosomes).

In each hemisphere there were several centers in which domestication of crops principally took place, after which the crops became widely cultivated elsewhere. The important centers in America were the highlands of Mexico and the highlands of Central America and northwestern South America—the regions where the Aztec, Maya, and Inca civilizations arose. The important centers in the Old World were southeastern Asia, the eastern Mediterranean region, the mountainous region from northern India to Transcaucasia, and possibly Ethiopia—regions noteworthy also as early centers of civilization. The development of agriculture, with its associated sedentary instead of nomadic habits, is a necessary precursor and stimulus to the rise of civilization. By about 7000 B.C. several genera of plants were being cultivated in each of at least three different regions: *Areca* (betel-nut palm) *Cucumis* (gourd, melon), *Piper* (pepper), *Pisum* (pea), *Prunus* (stone fruits such as plums), and *Vicia* (vetch, broad bean) or *Phaseolus* (bean) in southeastern Asia; *Hordeum* (barley), *Lens* (lentil), *Pisum*, *Triticum* (wheat), and *Vicia* in the Middle East; and *Capsicum* (pepper—the food, not the spice) and *Cucurbita* (pumpkin, squash) in Mexico. In the New World, *Zea* (corn) and *Phaseolus* followed within one or two thousand years thereafter. The common names here given indicate the nature of the group but not necessarily the precise species in early cultivation.

Cereal Crops

The grass family (Poaceae, also called Gramineae) outranks all other plants combined as a source of food. Edible grains obtained from cultivated grasses are called *cereals*. Wheat, rice, corn (maize), oats, barley, and rye are well-known cereal crops.

The most important of all crops in the temperate regions is *wheat*. From long before Jesus' prayer, "Give us this day our daily bread," until well after the coinage of the medieval English aphorism that bread is the "staff of life," wheat was the principal food in settled communities of Europe, temperate Asia, and the Mediterranean region. The bread demanded by the hungry mobs of Paris, and the cake supposedly offered them by Marie Antoinette, were both made from wheat. Even in modern societies with the varied diet associated with a high standard of living, wheat is the most important vegetable food.

There are about seven principal species of wheat, all annual or winter-annual grasses. Common wheat (*Triticum aestivum*, Fig. 28.1) is cultivated in many races, including truly annual (spring wheat) and winter-annual (winter or fall wheat) types, hard wheats with small grains rich in protein, soft wheats with larger grains rich in starch, and white-grained and red-grained forms. Durum wheat (*T. durum*, Fig. 28.1), with very hard, red grains rich in gluten (a sticky, proteinaceous substance), is used chiefly for macaroni and similar products. The other wheats, some of which more nearly resemble the putative wild ancestors of the group, are no longer of much direct importance, although they are sometimes used in breeding programs because of their hardiness and vigor.

Rice (*Oryza sativa*, Fig. 28.2) is the most important of all crops, from a worldwide standpoint, being the principal food of more than half the population of the earth. It is primarily a plant of hot, moist tropics, and is most extensively grown in the Orient, where it dominates the social and economic structure. Rice originated in southeastern Asia, being first extensively cultivated by the Chinese in prehistoric times. All the most important cultivars must be flooded for most of the growth period.

Fig. 28.1 Durum wheat (*left*) and common wheat (*right*). There are also bearded forms of common wheat. [New York Botanical Garden photo.]

Commercial polished rice, as sold in the United States, is very rich in carbohydrates but contains only minute amounts of proteins, minerals, and fats. Unpolished rice, with the embryo still included in the grain, is much more nutritious but is still relatively low in protein. In regions where rice is the staple food, soybeans are commonly eaten along with it as a principal source of protein.

Maize (*Zea mays*, Fig. 28.3) was the cereal of the American Indians, grown chiefly in the warmer regions, but also as far north as the St. Lawrence Valley and the Great Lakes region, and south to Chile and Argentina. It is now cultivated in all suitable parts of the world. The name

Fig. 28.2 Rice in cultivation on a flood plain near the Caspian Sea in Iran.

Indian corn, applied to it by the English colonists in America, has been shortened in the United States to corn; but in England corn still means any cereal grain, especially wheat. Although the protein in corn does not have an adequate supply of all the amino acids necessary for human nutrition, corn and beans collectively do, and these two foods have for centuries provided the basic diet for much of the population of Mexico.

Noncereal Crops

Potatoes (*Solanum tuberosum*) are the world's most important noncereal crop. Frequently called the Irish potato, because of its importance as a food crop in Ireland, *Solanum tuberosum* is actually native to the Andean region of tropical South America. Potatoes do best in rather cool, moist regions (at high altitudes, in the Andes). Although rich in carbohydrates, potatoes contain only a little protein and almost no fat, so that alone they do not provide anything approaching a balanced diet. In addition to their value as food, potatoes are used as a source of carbohydrate for the production of alcohol and other industrial products, especially in Europe.

Sweet potatoes (*Ipomoea batatas*), yams (*Dioscorea* spp.), and cassava (*Manihot esculenta*) are among the most impor-

Fig. 28.3 Maize. The male flowers form the tassel, at the top. The silk showing at the top of the ears consists of the styles of the individual pistillate flowers that collectively form the ear. Maize is wind pollinated. [Photo by Jesse Lunger, from National Audubon Society Collection/Photo Researchers, Inc.]

tant tropical crops. They all store food (mainly carbohydrate) in fleshy-thickened roots.

In temperate climates the fleshy fruits are important chiefly as a dietary filip and as a source of vitamins, but several tropical fruits are highly nourishing and form dietary staples in various regions. Among these are banana (*Musa paradisiaca*), date (*Phoenix dactylifera*), coconut (*Cocos nucifera*), and breadfruit (*Artocarpus utilis*). Captain Bligh's mission on the famous voyage of H. M. S. *Bounty* was to introduce the breadfruit (Fig. 28.4), a native of Malaya, into the British West Indies. (The people there anticlimactically found the fruit not to their liking.) The date palm, in desert regions from North Africa to India, and the coconut palm (see chapter-opening illustration), in the South Sea

Fig. 28.4   Breadfruit. [Photo by W. H. Hodge.]

Islands and along tropical coasts in general, are also noteworthy for their numerous nondietary uses, including basketry, mats, fibers, and construction timbers.

Saccharum officinarum, the *sugar cane,* is the traditional source of sucrose, and indeed sucrose has come to be called *cane sugar.* Sugar cane is a coarse grass, growing 2 to 5 m tall, with solid stems often 1 dm thick, containing up to about 20 percent sugar by weight. Sugar cane apparently originated in the East Indies, but it is now cultivated in tropical and subtropical regions throughout the world. In cooler regions special strains of the garden beet, *Beta vulgaris,* have been selected for sugar content and are grown as *sugar beets.* Commercial beet sugar is sucrose, chemically identical to cane sugar.

Spices

The most important of all spices is *pepper* (Fig. 28.5) derived from the small stone-fruits of *Piper nigrum,* a vine native to southeastern Asia. Black pepper is made by grinding the whole dried fruit, white pepper from the stone and seed. Pepper was one of the principal items imported into Europe along the overland trade route from the Orient in Roman and medieval times, and its high price was a great stimulus in the search for a sea route to the Orient. Red pepper, even more pungent than ordinary pepper, is produced from the fruit of *Capsicum,* a genus of the potato family native to tropical America. The whole fruits are also eaten by people with a taste for piquant food.

Nutmeg (Fig. 28.6), derived from the seed of the evergreen tree *Myristica fragrans,* was the principal spice of the Spice Islands (the Moluccas) of romantic fame. *Cinnamon* is one of the most widely used

Fig. 28.5 Pepper, leaves and fruit. [Courtesy of the American Spice Trade Association.]

Fig. 28.6 Nutmeg. The aril, shown partly covering the seed and separately at the lower right, is the source of mace. [Photo by W. H. Hodge.]

of the many other spices. It is made from the bark of *Cinnamomum zeylanicum*, an evergreen shrub or small tree native to Ceylon.

Drinks

Coffee and *tea* are the most important hot beverages, with *chocolate* third. Tea is used by fully half the people of the world, but coffee, used by fewer people, is consumed in greater quantities. Both owe their popularity to the stimulating effect of the caffeine they contain. Coffee is brewed from the ground-up seeds of several species of *Coffea*, chiefly *C. arabica* (see Fig. 2.11), a small tree native to Ethiopia and now cultivated in the tropics of both hemispheres. Tea is made from the dried leaves of *Camellia sinensis*, a small tree native to China or northeastern India, cultivated as a shrub in tropical and warm-temperate parts of Asia. Chocolate and cocoa are made from the seeds of *Theobroma cacao* (Fig. 28.7), a small tree native to the lowlands of tropical America, and now extensively cultivated in western tropical Africa as well. Unlike coffee, tea, and most other beverages, cocoa and chocolate are highly nutritious. The seeds contain 30–50 percent fatty oil, and about 15 percent each of protein and starch. They also contain theobromine (a mild stimulant) and traces of caffeine. Chocolate is a common ingredient of candy.

Fig. 28.7   Trunk, fruit, and flowers of cacao. [Photo courtesy of José Cuatrecasas.]

NARCOTICS AND HALLUCINOGENS

Tobacco is prepared from the dried leaves (and sometimes other parts) of *Nicotiana tabacum*, a coarse annual herb of the potato family, native to tropical America. Tobacco is now smoked over most of the world because of the effect of nicotine, its most important active principle, in easing nervous tension and promoting a sense of well-being. Although offensive to most of those not accustomed to its use, it is mildly habit-forming. There is a clear statistical association between the use of tobacco—especially cigarettes—and the occurrence of lung cancer, heart ailments, and other diseases, but the marked decline in its use among biologists and physicians in the last two decades is in contrast to its continued extensive use among the populace as a whole.

The opium poppy, *Papaver somniferum*, is the source of the drug *opium*. Crude opium, the dried exudate from maturing capsules that have been bruised or slit, contains a number of narcotic alkaloids, including *morphine* and *codeine*. *Heroin* is a chemical derivative of morphine. These drugs are habit-forming and have euphoric, analgesic, sedative, and toxic properties, heroin being the most dangerous and codeine the least. Continued use leads to addiction, physiological deterioration, and premature death; and withdrawal of the drug causes such severe pain that an addict will do literally anything in his power to obtain it.

The opium poppy is cultivated principally in China, India, Burma, Thailand, and the Middle East; addiction has long been common in those countries. China in particular has for centuries had a large number of addicts, but verifiable recent statistics are not available. The attempt of the Chinese government to prevent importation of opium from India, as part of a campaign against the use of the drug, led to the Opium War of 1840–1842, and the cession of Hong Kong to Great Britain. Heroin addiction appears to have increased markedly in the United States in the past two decades, spreading from the lower socioeconomic strata into the so-called counterculture group, especially among the young, despite vigorous governmental efforts to the contrary.

Fig. 28.8 *Cannabis sativa,* portion of female inflorescence, in Berlin Botanic Garden.

Hemp (*Cannabis sativa*, Fig. 28.8) is a coarse annual up to 3 or 4 m tall, with compound leaves and inconspicuous flowers. It is the source of fiber for cordage, and of a hallucinogenic drug. It is clearly of Eurasian origin, but no original wild population can be securely distinguished from the vigorous weedy forms escaped from cultivation. The species is divisible, with some difficulty, into a northern subspecies *sativa*, which has been cultivated and selected chiefly for fiber, and a more southern (south of about 30° north latitude) subspecies *indica*, which has been cultivated and selected chiefly for its drug content. The dried inflorescences are smoked as *marijuana*; the more potent, fat-soluble extracted product is chewed or eaten as *hashish*. It produces euphoria, accompanied by changes in time perception and a sense of the overriding importance of the immediate, without any obvious effect on physical coordination or motor ability. Our word *assassin* is a corruption of the Arabic *hashashin*, eaters of hashish. A significant number of people are hypersensitive to the drug, and its effects can recur unexpectedly some days after use. Chronic users tend to become withdrawn and lethargic, and many of them progress to addictive drugs such as heroin. *Cannabis sativa*, the true hemp, is sometimes called Indian hemp to distinguish it from other plants used for cordage, such as Manila hemp (*Musa textilis*, a relative of the banana).

MEDICINE

Angiosperms play a large but decreasing role in medicine. Once the source of practically all effective remedies for disease, they are gradually being replaced by synthetic drugs.

Digitalis is perhaps the most important therapeutic drug still obtained directly from angiosperms, the dried leaves being a standard treatment for certain types of heart disease. The drug became generally known through a report in 1785 by William Withering, an English physician, on its usefulness in treating dropsy, now known to be a symptom of circulatory disorder. Withering's interest was attracted by the success of an old woman of Shropshire in treating dropsy with an herb tea. He somehow managed to learn the components of the tea, and then, by cautious experimentation on his patients, he discovered that the active ingredient was foxglove (*Digitalis purpurea*, Fig. 28.9) and that the leaves are the most effective part of the plant.

Quinine is obtained from the very bitter bark of several species of *Cinchona*, native to the Andes of northern South America and subsequently extensively cultivated in the East Indies and nearby Asia. It consists of a mixture of similar alkaloids with slightly varying effects. As the only effective cure for malaria, it was once one of the most important drugs in the world. As recently as World War II, when the capture of Malaya and the East Indies by the Japanese cut off the main commercial source of quinine, the United States government found it necessary to set botanists to combing the forests of northwestern South America for the wild trees. For

Fig. 28.9 Foxglove, *Digitalis purpurea*. [Photo courtesy of the Boyce Thompson Institute for Plant Research.]

some two decades thereafter, quinine was largely replaced in medical practice by more potent synthetics developed under the stimulus of war-time need. More recently it has become necessary to revert to the use of quinine to treat the increasing number of cases that do not respond to the synthetic drugs.

INDUSTRIAL PRODUCTS

Fiber

Cotton (*Gossypium* spp.) is the most important fiber plant and the world's greatest industrial crop (Fig. 28.10). The most important species, *G. hirsutum* (upland cotton) is of tropical American origin, but

Fig. 28.10 Cotton plant in Arkansas. [USDA photo.]

species of Old World origin are also extensively grown in the warmer parts of Asia. The cotton boll is the opened ripe fruit, and the cotton fibers are hairs attached to the seeds. In 1793 Eli Whitney devised a machine, the cotton gin, for separating the fibers from the seeds (previously a time-consuming and expensive process), and thereafter cotton soon became the dominant textile fiber of the world.

The separated seeds from the cotton boll are also utilized. Cottonseed oil is used in salad oils and in the manufacture of oleomargarine and substitutes for lard, and the residue after the oil is expressed is used as livestock food and as a raw material in many industrial processes.

Wood

The wood of many angiosperms is of high commercial value. Maple, oak, beech, birch, walnut, and other hardwoods are used for flooring, fine furniture, and interior finish. These same species, as well as conifers, are used in plywood. Mahogany (*Swietenia*), of the New World tropics, is famous for its use in furniture, as is teak (*Tectona grandis*) of southeastern Asia. Many other tropical angiosperms are potentially valuable, but their exploitation is (perhaps fortunately) hindered by the fact that so many different species, and so few of each, commonly grow intermingled. White oak is the outstanding wood for tight cooperage, that is, for barrels that will hold liquids without imparting taste or odor. Sawmill wastes are used in destructive distillation, yielding acetic acid, wood alcohol, and other industrial chemicals.

Rubber

Rubber, a coagulated latex, is obtained from the bark of several kinds of trees belonging to different families of angiosperms. *Hevea brasiliensis*, member of the family Euphorbiaceae native to tropical South America, is vastly more important than all other rubber-producing species combined, and it is hevea rubber (also called Para rubber, from the state of that name in Brazil) that is generally intended when the word rubber is used without qualification.

The name *rubber*, first used by Priestley in 1770, refers to the use of the substance in erasing pencil marks. A process for waterproofing cloth with rubber was invented by Mackintosh in 1823, and vulcanization was discovered by Goodyear in 1839. Thereafter rubber rapidly became one of the world's most important industrial raw materials. Its plasticity, elasticity, resistance to abrasion and electric currents, and its impermeability to liquids and gases all contribute to its value. Automobile tires and tubes, which once took the bulk of the rubber crop, are now made mostly from synthetic rubber, as a result of the enforced improvement of quality and expansion of capacity for the synthetic product in the United States during World War II. For many other uses, including waterproofing and insulation, natural rubber remains supreme; and rubber promises to be an important crop for decades to come.

Rubber is the newest major cultivated crop. Until nearly the end of the nineteenth century, practically all commercial rubber was obtained from wild trees in the Brazilian jungles. The establishment of rubber plantations in Malaya, Ceylon, and the East Indies followed the smuggling of 70,000 seeds out of Brazil in 1876 by the British adventurer Henry Wickham. By 1914 rubber production from the cultivated crop had outstripped that from wild trees, and wild rubber has been of only negligible importance for the last several decades. Breeding programs, selection of high-yielding types, and grafting have increased the production per tree severalfold, and wild rubber can no longer compete in the market.

ORNAMENTALS

Angiosperms provide lawn grasses, shade trees for streets and homes, and ornamental shrubs and garden flowers. Bluegrass (*Poa pratensis*) is the most important lawn grass in cool-temperate regions. The choice of shade trees varies with the community and the climate. Various species of oak (*Quercus*) and maple (*Acer*) are persistent favorites in the United States. Thousands of species of angiosperms, belonging to many different families, are cultivated for their flowers. The Liliaceae (daffodil, jonquil, narcissus, hyacinth, tulip, day lily, Easter lily, tiger lily, amaryllis, crocus, etc.) and Asteraceae (aster, chrysanthemum, cosmos, dahlia, daisy, marigold, zinnia, etc.) stand out particularly as sources of garden ornamentals.

POISONOUS AND ALLERGENIC PLANTS

Many angiosperms are poisonous when eaten, and a smaller number are poisonous to the touch. The families Liliaceae, Apiaceae (carrot family) and Solanaceae (potato family), all of which have well-known edible species, are also notorious for their poisonous members. The Anacardiaceae (poison ivy family) are likewise infamous for their allergenic species.

Highly poisonous plants are seldom very palatable, and livestock usually avoid them if there is enough other forage. After cattle have been eating hay in the barn all winter, however, they are less choosy when first turned out to pasture. Water hemlock (*Cicuta*), a somewhat parsniplike member of the Apiaceae, causes many losses in the United States in the spring. The plant, growing in wet places along streams, is easily pulled up by an avid cow, and the shallow, tuberous-thickened roots are especially poisonous, a piece the size of a walnut being fatal.

Poison hemlock (*Conium maculatum*), a weedy, carrotlike member of the Apiaceae, occasionally poisons children who make whistles from the purple-spotted

Fig. 28.11 Philodendron climbing on a tree in a tropical forest in southern Mexico.

stem. It was used for execution in ancient Greece, Socrates being its most famous victim. Children also still die from eating the bulb of death camas (*Zygadenus*), mistaking it for the nutritious sego lily (*Calochortus nuttallii*), which served the Mormon pioneers in Utah as a source of emergency food.

Philodendron and *Dieffenbachia* (dumb cane), in the monocotyledonous family Araceae, are popular as house plants and in business offices because they do not need as much light as many other plants. Both are highly poisonous. *Philodendron* (Fig. 28.11), which destroys kidney function, is a prime cause of death of house cats, and *Dieffenbachia* can cause the human throat and tongue to swell to the point of strangulation.

Poison ivy (*Toxicodendron radicans* and related species, Fig. 21.4) is the best-known plant in America that is poisonous to the touch, the usual symptom being itching, watery blisters. It is a climbing vine or low shrub with trifoliolate leaves, the central leaflet on an evident stalk. The leaves are brightly colored in the fall, and scarcely a year goes by that some high school class does not use it to decorate a gymnasium for a dance. Susceptibility to the poison, a nonvolatile oil, varies greatly, and many people are apparently immune. The resistance is likely to disappear, however, after repeated exposure, and those who make a show of their immunity are courting hospitalization. The poisonous effect is thus evidently due to an allergy.

The importance of ragweed (*Ambrosia*, Fig. 28.12), as a cause of hay fever has been noted in Chapter 25. Many other plants, in a wide range of families, are allergenic to some people. The list of species to which at least a few people are allergic could be extended almost indefinitely, and probably any plant can be allergenic to someone, somewhere, under the proper conditions.

Fig. 28.12 Giant ragweed, *Ambrosia trifida,* one of the most important causes of hay fever in the United States. The wind-borne pollen is produced in the numerous tiny staminate flowers toward the top of the plant. [Photo by Arthur W. Ambler, from National Audubon Society Collection/Photo Researchers, Inc.]

SUGGESTED READING

Baker, H. G., *Plants and civilization,* 3rd ed., Wadsworth, Belmont, Calif., 1978.

Beadle, G. W., The ancestry of corn, *Sci. Amer.* **242**(1):112–117, Jan. 1980.

Dodge, B. S., *Plants that changed the world,* Little, Brown, Boston, 1959. An accurate popular account of some of the plant products that made history and of the men who sought them out.

Estes, J. W., and P. D. White, William Withering and the purple foxglove, *Sci. Amer.* **212**(6):110–119, June 1965.

Harlan, J. R., The plants and animals that nourish man, *Sci. Amer.* **235**(3):89–105, September 1976.

Harris, D. R., New light on plant domestication and the origins of agriculture; A review, *The Geographic Rev.* **57**:90–107, 1967.

Harris, D. R., The origins of agriculture in the tropics, *Amer. Sci.* **60**:180–193, 1972.

Heiser, C. B., *Seed to civilization; The story of man's food.* Freeman, San Francisco, 1973. Interesting and informative.

Mangelsdorf, P. C., *Corn: Its origin, evolution, and improvement,* Harvard University Press (Belknap Press), Cambridge, Mass., 1974.

Simmonds, N. W. (ed.), *Evolution of crop plants,* Longman, New York, 1976. A crop-by-crop symposium.

Spurr, S. H., Silviculture, *Sci. Amer.* **240**(2): 76–91, February 1979. Modern management and forest productivity: a wide ranging view.

Swain, T. (ed.), *Plants in the development of modern medicine,* Harvard University Press, Cambridge, Mass., 1972. A symposium.

Vavilov, N. I., *The origin, variation, immunity, and breeding of cultivated plants,* transl. by K. Starr Chester, Chronica Botanica, Waltham, Mass., 1951. Selected writings by a great Russian botanist who died in prison in 1943.

PART IV
POPULATIONS AND
COMMUNITIES

Sugar maple, *Acer saccharum*, being tapped in early April in Massachusetts.

Chapter 29

Classical Genetics

Mutant snapdragon flower, partly white and partly red, obtained after four months of irradiation of whole plants with gamma rays. The mutation affecting this flower evidently occurred during the development of the flower bud; only the cells derived from the mutated cell show the mutation. [Photo courtesy of A. H. Sparrow, Brookhaven National Laboratory, Upton, New York.]

The similarities and differences between parent and offspring have interested and puzzled men throughout history. The English proverb, *"Like father, like son"*, is a translation from classical Latin, yet recognition of the difference between brothers is as old as the story of Cain and Abel. All sorts of circumstances relating to conception, gestation, and birth have been called on to explain sibling differences, and such beliefs die hard. Shakespeare could realize that "The fault, dear Brutus, lies not in our stars," but daily newspapers still print zodiacal charts and astrological columns, and some prospective mothers still spend long hours at the piano, hoping to exert a prenatal influence.

The upsurge of scientific inquiry ushered in by the Renaissance led to experiments directed at unraveling the mystery of heredity. For several centuries the results were contradictory and confusing, adding little to the common knowledge that both parents usually contribute about equally. The discovery of sexuality in flowering plants by the German physician Rudolf Camerarius (1665–1721) in 1694 made plants available for genetic experiments. It was soon demonstrated that hereditary variation in plants is much like that in animals, and equally puzzling.

MENDELISM

Modern genetics is founded on the work of Gregor Mendel (1822–1884, Fig. 29.1), a monk at Brünn, Austria (now Brno, Czechoslovakia). His classic studies on inheritance in peas, begun in 1857, were published in an obscure Austrian journal in 1866 and escaped scientific attention until 1900. In that year DeVries in Holland, Correns in Germany, and von Tschermak in Austria reported in rapid

Fig. 29.1 Gregor Mendel (1822–1884), Austrian monk, who laid the foundation of genetics. [Photo courtesy of Verne Grant.]

succession their own independent rediscovery of Mendel's principles.

Mendel succeeded, where his predecessors for centuries had failed, because he studied characters that showed sharp differences, without intermediates, because he studied each character separately as well as in combination with others, because he kept accurate pedigree records, and because he worked with enough individuals to get statistically valid results. Furthermore, Mendel carefully chose a species that is normally self-pollinated but can easily be cross-pollinated. We shall see that self-pollination leads to genetic purity, and each of the original 22 cultivars that he used in his garden experiments was genetically pure for all the characters studied. Mendel was also lucky in his choice of characters to be studied. Another choice might have shown contradictory results, which we can now explain but which would have destroyed the beautifully consistent pattern he reported. Mendelian inheritance is the foundation of genetics, but not the whole structure; an understanding of Mendelism is essential to any further progress, but there are many genetic results that cannot be explained in simple Mendelian terms.

Mendel studied nine sets of characters, but three of these were so perfectly correlated that he correctly considered them as expressions of the same hereditary factor, thus reducing the number of separate characters to seven pairs. He found that the inheritance of each of the seven pairs followed the same consistent pattern, and that each was wholly independent of the others. The characters were:

1. Yellow versus green cotyledons
2. Round versus wrinkled cotyledons
3. a. Purple versus white flowers
   b. Purple versus green axils in the seedling
   c. Gray to brown or purplish versus white seed coat
4. Tall versus dwarf habit
5. Axial (distributed along the stem) versus terminal flowers
6. Green versus yellow pods (before maturity)
7. Inflated versus moniliform* pods

As a result of his experiments, Mendel concluded that each character studied was governed by a single pair of hereditary factors. In modern terms, we would

*Moniliform: constricted at intervals, like a string of beads.

say that a single pair of genes swings the balance in each case. Mendel concluded that each seed receives one factor of each pair from the ovule parent, and the other from the pollen parent. We agree, adding that the two factors of a pair are borne on homologous chromosomes. For example, if the factor in the sperm for pod color is on chromosome A, then the factor in the egg for pod color is also on chromosome A; and if the factor in the sperm for cotyledon color is on chromosome B, then the factor in the egg for cotyledon color is also on chromosome B. Each gene, in fact, has its own particular **locus** (site) on the chromosome.

Genetic Segregation

Mendel's **principle of segregation,** stated in modern terms, is that *the two genes of a pair do not contaminate each other but are segregated in reduction division and pass into different gametes.* It is a fundamental principle of heredity.

Mendel found that one character of each opposing pair is **dominant,** showing up whenever a factor (gene) for it is present, and that the other is **recessive,** showing up only in the absence of the dominant factor. In the above list the dominant character of each pair is given first. Thus the tall habit is dominant, and dwarf recessive: First generation hybrids between tall and short cultivars were uniformly tall, although these hybrids carried a recessive factor for dwarfness as well as a dominant factor for tallness. If the dominant factor, for tallness, is represented by T, and the recessive factor, for dwarfness, by t, then the original tall cultivar, being genetically pure for tallness, had the constitution TT, and the dwarf parent had the constitution tt. All gametes produced by the tall parent would necessarily carry the factor T, and all gametes produced by the dwarf parent would carry the factor t. The first generation hybrids, therefore, would all have the constitution Tt; and since T is dominant over t, the plants would be tall.

It should be noted that in genetic studies any cross between individuals differing in one or more genes under consideration is called a **hybrid.** This is in contrast to common usage, in which the term *hybrid* is applied only to crosses between different species. The first-generation offspring produced by hybridization (in a genetic sense) are commonly called the F_1 generation, the F standing for filial. Succeeding generations derived by self-pollination or sibling matings are called the F_2, F_3, etc. A backcross is a mating between a hybrid (usually an F_1) and one of the original parental types.

The F_1 hybrids referred to above are tall plants with the genetic constitution Tt. When these plants are allowed to set seed naturally (by self-pollination), 3/4 of the plants in the next generation (the F_2) are tall, and 1/4 are dwarf. This result can be explained by considering the processes of reduction division and subsequent fertilization. The two homologous chromosomes carrying the factors T and t, respectively, are pulled to opposite ends of the cell in the first division of meiosis, and each of the gametes ultimately produced therefore carries only one gene of the pair, either T or t; the two types are, of course, produced in equal numbers. The proportion of each genetic type to be expected in the F_2 generation can easily be seen by using a checkerboard device in which all the kinds of gametes produced by one parent are listed across the top, all the kinds produced by the other are listed along one side, and the possible matings

are shown in the squares. When the plants are self-pollinated, as in the present case, the list of gametes at the side of the chart is of course the same as the list at the top, and the chart appears as follows:

	T	t
T	TT	Tt
t	Tt	tt

Thus on the average, 1/4 of the F_2 plants have the constitution TT and are tall; 1/2 have the constitution Tt and are also tall, making a total of 3/4 tall; the remaining 1/4 have the constitution tt and are dwarf. Successive generations derived from these dwarf plants by self-pollination would all be dwarf, since they are genetically pure for dwarfness. Successive generations derived by selfing tall plants with the constitution TT would all be tall, being genetically pure for tallness, but the next generation derived from tall plants with the genetic constitution Tt (these being 2/3 of all the tall plants in the F_2) would again segregate in a ratio of 3:1.

It will be seen that the 3:1 ratio of tall to dwarf plants is based on a 1:2:1 ratio of genetic constitutions. This same sort of 1:2:1 ratio can be built up by shaking 2 pennies in a jar and rolling them onto a table. About 1/4 of the throws will show 2 heads, another 1/4 will show 2 tails, and the remaining 1/2 will show 1 head and 1 tail. This ratio is, of course, based on a large number of throws. One could not expect to get the correct ratio with only four throws (although it will sometimes happen), and the same is true of hybrid ratios.

The fact that plants of similar appearance may be genetically different necessitates some additional terms. The **phenotype** is the actual characteristic of the individual, as expressed in its form, structure, or physiology. The **genotype** is the

Fig. 29.2 Genetic segregation in corn. (*A, B*) Ears from pure lines, one of which has grains with pigmented endosperm. (*C*) The F_1 hybrid between them, heterozygous for pigment, but phenotypically like the pigmented parent. (*D*) The F_2 from this cross, showing a 3:1 segregation of pigmented and unpigmented endosperm. (*E*) A backcross from the F_1 to the pigmented parent; half the grains in this backcross are homozygous for pigmentation and half heterozygous, but phenotypically they are all alike. (*F*) A backcross from the F_1 to the unpigmented parent, showing a 1:1 ratio of pigmented (heterozygous) and unpigmented (homozygous recessive). The slight differences in intensity of color in the pigmented grains in this photo are due to environmental differences and to exposure to light after the ears were harvested. [Photo courtesy of G. W. Blaydes, from Transeau, Sampson, and Tiffany, *Textbook of botany*, rev. ed., Harper & Row, New York, 1953.]

genetic makeup. Peas with the genotype *TT* and those with *Tt* have the same phenotype, tall; those with the genotype *tt* all have the same phenotype, dwarf. Individuals with both factors of a pair alike are **homozygous** (Greek *homo-*, one and the same); those with the two factors of a pair opposed are **heterozygous** (Greek *hetero-*, other, unlike). Thus pea plants with the genotype *TT* are homozygous dominant; those with the genotype *Tt* are heterozygous; and those with the genotype *tt* are homozygous recessive.

A mating in which only a single pair of genes is considered is called a **monohybrid.** The possible monohybrid genetic ratios are 1:2:1, 1:1, and 1:0. The 1:2:1 ratio is obtained by crossing two heterozygous individuals, or by selfing a heterozygous individual, as shown in the previous checkerboard. Half the offspring from such a cross are heterozygous, and 1/4 belong to each of the two homozygous types.

The 1:1 ratio is obtained by crossing a heterozygous individual with a homozygous one. Half the offspring are heterozygous, like one parent, and half homozygous, like the other. Thus *Tt* × *tt* yields 1/2 *Tt* and 1/2 *tt*; and *Tt* × *TT* yields 1/2 *Tt* and 1/2 *TT*. It should be noted that phenotypically the parents and offspring in this last example are all tall; only the genotypes show a 1:1 ratio. The mating of heterozygous to homozygous may also be shown in the following checkerboards, with the gametes of the heterozygous parent shown at the top and the single kind of gamete produced by the homozygous parent shown at the side.

	T	t
t	Tt	tt

	T	t
T	TT	Tt

The 1:0 ratio is obtained by mating homozygous with homozygous. Thus *TT* × *TT* yields only *TT*, *tt* × *tt* yields only *tt*, and *TT* × *tt* yields only *Tt*. The next generation obtained by selfing the *Tt* individuals will, of course, segregate in a 1:2:1 genotypic ratio, like any heterozygous × heterozygous mating.

Independent Assortment

We noted in Chapter 10 that during reduction division one chromosome of each pair is pulled to one end of the cell, and the other to the other end. Each of the cells produced in meiosis thus has a complete, balanced set of chromosomes. Except for this limitation, the chromosomes are randomly assorted in the process; the maternal or paternal origin of the chromosomes is immaterial. The genes are likewise randomly assorted, as long as they are borne on separate chromosomes.

Each of the seven pairs of factors noted by Mendel in peas is borne on a different pair of chromosomes, and therefore he found a completely independent assortment of characters in the hybrid progeny. A modern restatement of Mendel's principle of independent assortment is that *insofar as they are borne on different chromosomes, each pair of genes is inherited independently of the others, and all combinations of genotypes occur in the offspring in the proportions governed by the mathematical laws of chance.* The first eight words of the principle as here stated introduce a limitation of which Mendel was unaware.

An example of independent assortment can be seen in dihybrids, in which two pairs of genes are considered. A typical dihybrid mating, such as those recorded by Mendel, shows a 9:3:3:1 phenotypic ratio in the F_2, with 9/16 of the individuals showing both dominant characters, 3/16 showing each of the two combinations of dominant and recessive, and 1/16 showing both recessive characters. This ratio can be understood by determining the possible gametes and using the checkerboard to show all possible genotypes in the offspring. For example, if a pea plant with yellow, round seeds (genotype YYRR) is crossed with a plant with green, wrinkled seeds (genotype yyrr), the F_1 has yellow, round seeds, with the genotype YyRr. These F_1 plants are heterozygous for both factors and can produce four kinds of gametes, YR, Yr, yR, and yr. The checkerboard showing the F_2 is as follows:

	YR	Yr	yR	yr
YR	YYRR	YYRr	YyRR	YyRr
Yr	YYRr	YYrr	YyRr	Yyrr
yR	YyRR	YyRr	yyRR	yyRr
yr	YyRr	Yyrr	yyRr	yyrr

It will be seen that nine different genotypes are represented on the checkerboard, and that these occur in definite proportions. The genotypes YYRR (1/16 of the total), YYRr (2/16 of the total), YyRR (2/16 of the total), and YyRr (4/16 of the total) produce the round, yellow phenotype. The genotypes yyRR (1/16 of the total) and yyRr (2/16 of the total) produce the green, round phenotype. The genotypes YYrr (1/16 of the total) and Yyrr (2/16 of the total) produce the yellow, wrinkled phenotype. And the genotype yyrr (1/16 of the total) produces the green, wrinkled phenotype.

These ratios can also be calculated by simple arithmetic. Thus 3/4 of the plants in the F_2 have yellow seeds, and 3/4 of that 3/4, or a total of 9/16, have round seeds as well. Of the 3/4 that have yellow seeds, only 1/4, or a total of 3/16, have wrinkled seeds. One fourth of the plants have green seeds. Three fourths of that 1/4 (or a total of 3/16) have round seeds, and only 1/4 of

that 1/4 (or a total of 1/16) have wrinkled seeds.

The genotypic ratios can also be calculated. One fourth of the individuals in the F_2 have YY; 1/2 have Yy; and 1/4 have yy. Of the 1/4 that have YY, only 1/4 (thus a total of 1/16) also have RR; etc.

It should be noted that in all these crosses it does not matter which is the pollen parent and which the ovule parent; nor does it matter how the characters are associated in the parents. A YYRR × yyrr cross gives the same results as a yyRR × YYrr cross. In the former case, the original gametes are YR and yr, and the genotype of the F_1 is YyRr; in the latter case, the original gametes are yR and Yr, and the genotype of the F_1 is still YyRr.

The F_2 derived from a **trihybrid** (three pairs of factors considered) normally shows phenotypic ratio of 27:9:9:9:3:3:3:1 if the parents were homozygous and differed in all three characters. This ratio can be demonstrated with a checkerboard or calculated mathematically, using the same principles that were set for dihybrids.

In working problems by the checkerboard method, it is of course necessary to make sure that none of the possible types of gametes has been overlooked. The number of kinds of gametes that can be produced by an individual is 2^n, if n is the number of heterozygous pairs. Thus an individual with the genotype TTYYRr has only one heterozygous pair, and can produce only two kinds of gametes, these being TYR and TYr; an individual with the genotype TTYyRr has two heterozygous pairs, and can produce four (2^2) kinds of gametes, these being TYR, TyR, TYr, and Tyr; an individual with the genotype TtYyRr has three heterozygous pairs, and can produce eight (2^3) kinds of gametes; etc. An individual with ten heterozygous pairs can produce 2^{10}, or 1064 kinds of gametes. The checkerboard method obviously has its limitations in working genetic problems.

Occurrence of Mendelian Characters

Most kinds of plants have a few characters that appear to be inherited in simple Mendelian fashion, and many others that are more complex. Summer squash has been much used to illustrate Mendelism: white fruit is dominant to yellow, and disk-shaped fruit is dominant to spherical.

There are no hard and fast rules as to which characters will be Mendelian and which will not, nor as to which character of a Mendelian pair will be dominant; each case must be investigated individually. Presence of a structure is usually dominant to its absence, but even here there are exceptions.

It is perhaps paradoxical that although Mendel's principles are essential to an understanding of heredity, it is only the exceptional character that shows typical Mendelian phenotypic ratios. The relative infrequency of simple Mendelian characters is a natural result of the mode of gene action described in the following paragraphs.

MECHANISM OF GENE ACTION

The mechanism of gene action was explored at some length in Chapter 9, but it is useful to consider some aspects of it here as well. We have seen that genes act by governing the kinds and amounts of molecules that are supplied to the remainder of the protoplasm. Some of the evidence supporting this concept came from experiments with *Neurospora* (especially

N. crassa), a mold belonging to the Ascomycetes.

The usefulness of *Neurospora* in genetic studies was demonstrated in the 1930s by the American botanist B. O. Dodge (Fig. 29.3). The vegetative stage is haploid, so that the complications of heterozygosity are avoided; the plant is easily grown in pure culture and has a short life cycle; vegetative reproduction occurs freely by conidia, and parts of the mycelium can also be broken off and established as new cultures; the species *N. crassa* has two mating types, and sexual reproduction occurs only when the two opposite types come into contact. The ascus produces eight ascospores in a row, and the spores are large enough so that they do not slip by each other in the ascus. By careful manipulation, the ascospores can be re-

Fig. 29.3   B. O. Dodge (1872–1960), American mycologist, the father of *Neurospora* genetics.

moved one at a time and separately cultured. It is possible, therefore, to study all the cells produced by reduction division from a single meiocyte (as it is not with higher plants, higher animals, or insects), and the principle of genetic segregation in meiosis can be demonstrated directly instead of merely statistically.

About 1940 the American botanists George W. Beadle (1903–) and E. L. Tatum (1909–1975) and their associates gave the use of *Neurospora* a new and important turn. By exposing the fungus to X-rays, they obtained mutant forms which were cultured and bred back to the wild type. Many of these mutants differed from the wild type in specific physiological traits that segregated in Mendelian fashion when bred back to the wild type, showing that only a single gene was involved. The wild type needs only inorganic salts, a carbohydrate such as sucrose, and the vitamin biotin; from these and water it can synthesize all the more complex substances it needs. More than 20 different mutants, each lacking the ability to produce some particular substance necessary for growth, have been found. Without the necessary substance, they die; when it is provided, they grow normally. The conclusion is that a mutation in a particular gene deprives the plant of the ability to produce a particular chemical. The substances required by these mutants include various vitamins, amino acids, and other compounds. One interesting mutant is unable to synthesize the vitamin riboflavin at temperatures above 28°C, whereas the wild type can make it at temperatures up to 40°C. This mutant would remain undetected at temperatures below 28°C.

As a result of their studies, Beadle and Tatum formulated the *one gene–one enzyme* concept. As we have seen in Chapter 9, this concept is now modified to *one gene–one polypeptide strand.*

The fact that genes act by governing the kinds of molecules supplied to the remainder of the protoplasm is vital to an understanding of heredity. Any change in a particular constituent of the protoplast has a chain of side effects, influencing the reactions that occur and the products that are formed, thus ultimately influencing the phenotype of the organism. Every characteristic of the organism is governed by its total genotype, in conjunction with the environment; and any change in any gene, as well as any environmental change, potentially has some effect on every character. Any particular gene affects some characters more than others. It may swing the balance as a Mendelian factor for one character, slightly modify the expression of another character, and have no detectable effect on another. Its effects on a particular character may be strikingly different against one genetic background than against another, or in one environment than in another.

The relative influence of heredity and environment ranges from one extreme, in which environmental conditions have so little effect that they may ordinarily be disregarded, to the other extreme, in which hereditary differences are overwhelmed by environmental differences. A character that is Mendelian in one population may be more complexly inherited in another population that differs in other genes, and a change in the environmental conditions may also modify or obscure the simple Mendelian ratios. In some cultivars of snapdragons, the environmental conditions sharply modify the effect of a particular pair of genes on flower color, and the effect of this pair of genes also

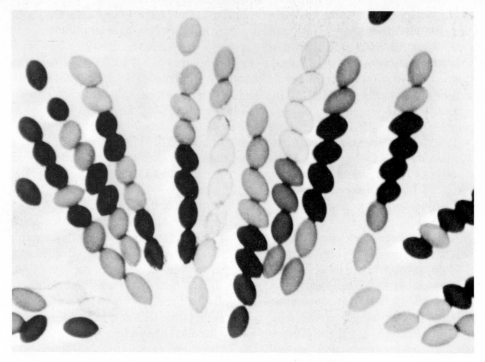

Fig. 29.4 Hybrid segregation and crossing over in *Sordaria*, an ascomycete, showing a part of a cluster of asci. (× 350.) These fruiting bodies resulted from mating of gray-spored and black-spored types, the difference in color being governed by a single gene. It should be recalled that each of the four nuclei formed by reduction division within the ascus divides again, mitotically, so that ordinarily eight spores are borne in an ascus. Some asci in this photograph show results of reduction division without crossing over, so that all black spores (derived from one of the two nuclei formed in the first division of meiosis) are at one end, and all gray spores (derived from the other of the two nuclei formed in the first division of meiosis) are at the other end. Other asci show the results of reduction division with crossing over. Reference to Fig. 10.3 will help to explain the arrangement of spores in these asci. [Photo courtesy of Lindsay S. Olive.]

differs strongly according to the other genes present.

The Mendelian characters are those which, in the particular population studied and under the conditions of the experiment, are governed by differences in a single pair of genes. When we say that the color of cotyledons in peas is governed by a single pair of factors, what we really mean is that among the genotypes studied and under the conditions of the experiment, differences in the color of the cotyledons are caused by differences in a single pair of factors (genes).

INHERITANCE NOT SHOWING TYPICAL MENDELIAN RATIOS

Mendel's principles of segregation and independent assortment are basic to an understanding of heredity, but we have seen that various things may modify or suppress the typical Mendelian phenotypic ratios. Four of these which remain to be discussed are incomplete dominance, linkage, cytoplasmic inheritance, and multiple factors.

Incomplete Dominance

Dominance exists in varying degrees. Heterozygous individuals may be phenotypically so similar to the homozygous dominant as to be indistinguishable, or they may resemble the homozygous dominant but deviate slightly toward the recessive type, or they may be obviously intermediate between the two homozygous types. When the two genes of a pair govern the formation of molecules with opposite effects, the effects may counterbalance each other, or the effect of one may be more or less submerged by the other.

Often the recessive factor really represents the absence of a gene, which has been destroyed by some genetic accident in the past; or the molecules governed by one of the genes may be essentially neutral in their influence on the character under study, so that the control is exerted by only one gene of the pair. In such cases, the single active gene of the pair may or may not be effective enough to make the plant look like one that is homozygous for the active gene.

Albinism in flowers furnishes good examples of both complete and incomplete dominance. Mendel found that the factor for purple flowers in peas was dominant to the factor for white flowers. The heterozygotes were phenotypically like the homozygous dominants. In some cultivars of snapdragons, on the other hand, the factor for red flowers is only

incompletely dominant over the factor for ivory flowers, and the heterozygotes have pink flowers. The F_1 hybrid between a homozygous red and a homozygous ivory has pink flowers, and the F_2 generation segregates in a 1:2:1 ratio (1/4 red, 1/2 pink, 1/4 ivory).

Linkage

Not long after the rediscovery of Mendel's work, it was found that not all Mendelian characters show independent assortment. The first satisfactory explanation of these exceptions was provided in 1910 by the American geneticist Thomas Hunt Morgan (1866–1945, Fig. 29.5). He suggested that genes which are linked in inheritance are borne on the same chromosome—the

Fig. 29.5   T. H. Morgan (1866–1945), American geneticist, outstanding contributor to the development of genetic theory. [Photo courtesy of Verne Grant.]

closer together the genes, the less chance of their being separated by the exchange of parts between homologous chromosomes at meiosis. This exchange of parts, called **crossing over,** is discussed in Chapter 10. Each individual has as many linkage groups as pairs of chromosomes. In the garden pea, for example, there are seven pairs of chromosomes and thus seven linkage groups, and in any list of genes, the total number that can be independently assorted in inheritance is not more than seven. If Mendel had studied just one more character, whatever it might have been, its governing gene would necessarily have been on the same chromosome as that governing one of the other characters, and thus it could not have been independently assorted.

Cytoplasmic Inheritance

A few characters are known to be inherited through the cytoplasm rather than through the nucleus. Because the egg carries so much more cytoplasm than the sperm, cytoplasmic characters are usually maternally dominated, and it is through this maternal effect that they are usually recognized. Some of the best-known cases of cytoplasmic inheritance involve plastids. In *Primula sinensis*, for example, there is a yellow-leaved form that contains less than the normal amount of chlorophyll. Pollen from yellow-leaved plants does not transmit the defect, but seeds from yellow-leaved plants produce yellow-leaved offspring, regardless of the color of the pollen parent.

Multiple Genes

It is customary to say that each hereditary character is governed by the whole genotype of the organism, in conjunction with the environment. The influence of a particular gene on a particular character may be so slight as to be negligible, however, or may become evident only under a particular set of environmental conditions. In simple Mendelian characters, a particular pair of genes regularly turns the balance governing the phenotypic expression. In many other characters the phenotypic expression is governed by several pairs of genes, each with its own effect, and the effect may be cumulative or opposed. Mendel himself discovered that the color of flowers in beans (*Phaseolus*) is governed by more than one pair of factors.

One of the simplest instances of multiple genes is provided by the color of the grain in wheat. Some crosses of red- to white-grained wheat show a simple 3:1 ratio of red to white grains in the F_2, but others show a 15:1 ratio, and still others a 63:1 ratio. Furthermore, the red grains vary in intensity of color. In the hybrid with a 3:1 ratio in the F_2, the F_1 grains are rather pale red, paler than the red parent, and the 3:1 ratio of red to white in the F_2 is really a 1:2:1 ratio, 1/4 of the grains being as dark as the original red parent, 1/2 being paler like the F_1, and 1/4 being white. In crosses showing 15:1 and 63:1 ratios, there are several degrees of redness, but it is not always possible to draw sharp lines between the classes. The 15:1 ratio is produced by the cumulative effect of two pairs of genes, and the 63:1 ratio by three pairs.

The 15:1 ratio of red to white grains in wheat may be used as a simple example of multiple factors in general. The original red parent has the genotype $R_1R_1R_2R_2$, and the white parent has $r_1r_1r_2r_2$. The red parent produces gametes of the genotype

R_1R_2, and the white parent produces gametes of the genotype r_1r_2. The F_1 has the genotype $R_1r_1R_2r_2$, and has red grains that are distinctly paler than those of the red parent. The F_1 produces four kinds of gametes, R_1R_2, R_1r_2 r_1R_2, and r_1r_2. The genotypes of the F_2 are shown on the checkerboard at the bottom of this page.

If the two factors of a pair have equal weight, then four classes of redness may be expected in the F_2. Only 1/16 of the individuals have all four factors for red, like the original red parent; 4/16 have three genes for red; 6/16 have two genes for red, and 4/16 have only one gene for red. The remaining 1/16 have no genes for red, and the grains are white.

One of the most important things about the operation of multiple factors is that some of the offspring may be more extreme than either parent. In the present example of wheat, only 6/16 of the F_2s have the same shade of red as the F_1, 5/16 have the grains more deeply colored than those of the F_1, and 5/16 have them less deeply colored. Most of the important characters of plants and animals are governed by multiple genes, with each individual being heterozygous for some genes and homozygous for others. In characters governed by multiple genes, the offspring are usually more or less intermediate between the two parents, but they differ among themselves. In any such character, an individual may be more nearly like one parent than the other, and it may even be more extreme than either parent.

	R_1R_2	R_1r_2	r_1R_2	r_1r_2
R_1R_2	$R_1R_1R_2R_2$	$R_1R_1R_2r_2$	$R_1r_1R_2R_2$	$R_1r_1R_2r_2$
R_1r_2	$R_1R_1R_2r_2$	$R_1R_1r_2r_2$	$R_1r_1R_2r_2$	$R_1r_1r_2r_2$
r_1R_2	$R_1r_1R_2R_2$	$R_1r_1R_2r_2$	$r_1r_1R_2R_2$	$r_1r_1R_2r_2$
r_1r_2	$R_1r_1R_2r_2$	$R_1r_1r_2r_2$	$r_1r_1R_2r_2$	$r_1r_1r_2r_2$

RESULTS OF INBREEDING AND OUTBREEDING

Pure Lines

Inbreeding tends to eliminate heterozygosity. If an individual heterozygous for a single pair of genes is selfed, the first generation hybrids segregate in a 1:2:1 genotypic ratio, 1/4 being of each of the homozygous types and 1/2 being heterozygous. If all these plants are again selfed, the homozygous ones of course produce only homozygous offspring, but the heterozygous ones again segregate in a 1:2:1 ratio. The proportion of heterozygous individuals is thus reduced by about 1/2 in each generation of inbreeding. In 10 generations, the theoretical proportion of heterozygotes would be $1/2^{10}$, or 1/1024, and in 100 generations it would be $1/2^{100}$, or 1/1,048,576. Theoretically there might always remain a continously diminishing fraction of heterozygous individuals, but for practical purposes 10 generations of inbreeding are usually considered enough to establish homozygosity.

The same situation holds true no matter how many pairs of genes are considered: After a few generations of inbreeding, the individuals are homozygous for all characters, with an insignificant and diminishing proportion of exceptions. Obviously, if the original individual was heterozygous for many genes, then many genetically different individuals could be derived from it by selfing. Any one such line that has become essentially homozygous by selfing is called a **pure line**. Establishment of pure lines in higher animals is more difficult, because an individual cannot be bred to itself, but by close inbreeding (father-daughter, mother-son, and brother-sister matings, etc.) combined with selection, lines can be established that are homozygous for many genes. The 22 cultivars of peas with which Mendel began his experiments were pure lines, since each line was self-pollinated. Pure lines are useful or even essential in many genetic studies.

Disclosure of Recessive Defects by Inbreeding

In any characteristically outbred population, many harmful or defective recessive genes occur in low frequencies. Only when two individuals with the same recessive defect mate does the character have a chance to be expressed. Given random mating, the chance that both individuals will carry the same recessive defect is the square of the frequency of the gene. Thus if a particular defective gene is carried by 1 percent of the population, only 1 mating in 10,000 (100 × 100) will be between parents that both carry the defective gene, and only 1/4 of the offspring of these matings (or 1/40,000 of all offspring) will be homozygous recessive. The elimination of recessive defects in an outbreeding population is thus a very slow process, and the defective genes accumulate in the population, through continued mutation, until the frequency of their elimination by natural selection equals the frequency of their origin.

If a normally outbred group is inbred, the number of homozygous recessives among the offspring is of course greatly increased, and many previously hidden defects are likely to appear. This is as true in plants as in animals, but its importance in domestic animals and man is greater than in plants, because of the difference in the number of offspring and the care given each. Plants generally produce large numbers of seeds, and the death of a few homozygous recessive defective pheno-

types causes no great difficulty. Comparable defective human offspring fill institutions and cause much mental agony as well as expense, even without inbreeding, and there would be many more of these unfortunates if incestuous marriage were common.

All higher animals are normally outbred, but many higher plants are normally inbred through self-pollination, or they combine inbreeding with outbreeding by mixtures of self- and cross-pollination. Seriously defective mutations in inbred populations are eliminated by natural selection soon after they are produced, and the frequency of such defects in naturally inbred groups is therefore very low.

Hybrid Vigor

It has long been noted that hybrids between different races of a species, or even between species, are often more vigorous than either parent (Fig. 29.6). The mule, although sterile, has more endurance than either a donkey or a horse. The popular ornamental shrub *Abelia grandiflora*, a hybrid between *A. chinensis* and *A. uniflora*, is hardier and flowers more freely than either of its parents. *Spiraea vanhouttei* (bridal wreath), another hybrid, is larger and more often cultivated than either of its parents. Many other such examples might be given. This **hybrid vigor,** or **heterosis,** is often even more conspicuous in crosses between different races within an outbreeding species, such as corn. Intercultivar crosses of characteristically inbreeding species, such as wheat, seldom show marked hybrid vigor, although there are a few examples. Mendel found that the F_1 hybrids between tall and dwarf cultivars of peas averaged a little taller than the homozygous tall par-

Fig. 29.6 Hybrid vigor in marigolds. The Climax marigold (*center*) is a hybrid between the cultivars shown at left and right. All of these plants belong to the same species, *Tagetes erecta*, and have the same chromosome number, 2n = 24. [Photo courtesy of W. Atlee Burpee Company.]

ent, but the difference was too slight to permit accurate identification of the heterozygotes in the garden.

Two theoretical explanations have been advanced for hybrid vigor, each backed by some experimental evidence. One explanation is that the hybrid simply gets the advantage of the better genes from each parent, with minor homozygous defects in each parent being covered by dominant normal genes from the other. The second explanation is that heterozygosity itself confers an advantage, having a higher survival value than either of the opposing homozygous types. One gene, for example, might help the plant to grow well under one set of environmental conditions, whereas its opposite number on the matching chromosome might help

under some other conditions, so that the range of conditions under which good growth might occur is greater for the heterozygote than for either of the homozygotes.

The relative importance of these two suggested mechanisms for heterosis remains to be determined, but present evidence emphasizes the first one. If heterozygosity were regularly useful for its own sake, hybrid vigor might be expected to be common in normally inbred as well as normally outbred groups.

Hybrid Corn

Corn (maize) is more sensitive to inbreeding than most crops, and it shows spectacular hybrid vigor. A few generations of inbreeding lead to stunted, unthrifty pure lines with a very low yield of grain. Hybrids between two scrawny inbred types that differ in a number of characters are often much larger and bear more heavily than the original strains from which the selections were made.

This discovery was made by the American botanist George H. Shull (1874–1954) during some experiments directed toward understanding the inheritance of the number of rows of kernels in the ear. Shull (Fig. 29.7) immediately recognized the potential importance of his discovery, and in 1908 and 1909 he suggested that such high-yielding F_1 hybrids could be used on farms in place of the usual cultivars. The difficulty with Shull's method is that the hybrid seed is necessarily produced on one of the scrawny parents, and the low yield makes the seed expensive. The seed set in the next generation by selfing the vigorous F_1 hybrid is not satisfactory, since it reverts toward homozygosity and loss of vigor. The plants produced from such seeds also differ greatly among themselves, having a wide variety of genotypes.

Several years later another botanist opened the door to successful commercial use of hybrid corn by replacing Shull's "single-cross" method with a "double-cross" method (Fig. 29.8). Two different F_1 hybrids produced by crossing inbred lines are themselves crossed. These plants, being vigorous and of full size, produce abundant seed. The offspring from this second cross are less uniform than single-cross hybrids, but still not so variable as ordinary open-pollinated cultivars; and if the proper crosses are made, there is no significant loss of vigor. Practically all the field corn grown in the United States today is from double-cross hybrid seed, and there are hundreds of cultivars, adapted to various climates and purposes. Sweet corn for canning or table use, in which uniformity is important, is more often grown as single-cross hybrids.

The use of hybrid corn is not without problems. The cultivar *Texas male—sterile*, widely used as a seed parent in hybrid corn breeding, turned out to be unusually susceptible to a fungus disease that was favored by the weather conditions in 1970, and there were severe losses in the susceptible cultivars derived from it.

One of the first men to develop cultivars of hybrid corn for regular farm use was Henry A. Wallace, who later became Secretary of Agriculture and then Vice-President of the United States.

Fig. 29.7   George H. Shull (1874–1954), father of hybrid corn. [Photo courtesy of Mrs. George H. Shull.]

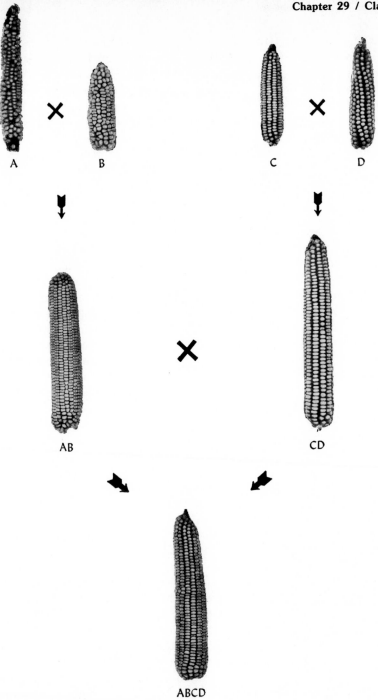

Fig. 29.8 The double-cross method of producing hybrid corn.
[Courtesy of William L. Brown, Pioneer Hi-Bred Corn Company, and
the Michigan State University Press.]

The fantastic success of hybrid corn led to the use of similar methods in breeding other crops and domestic animals. Watermelons, squashes, cucumbers, onions, and sugar beets are among the many crops in which hybrid vigor is being put to use; and hybrid chickens, pigs, and cattle are now also commonplace.

MUTATION

Inheritance in regular patterns, as discussed in the foregoing pages, depends on the precise and regular duplication of genes in connection with mitosis, as well as on a regular pattern of meiosis and fertilization. Anything that disturbs these processes may also disturb the pattern of inheritance.

Any inheritable change in the genetic material is called a **mutation.** In the broad sense the term includes such gross changes as addition or deletion of chromosomes, chromosome segments, or whole sets of chromosomes, as well as changes in individual genes; but when a geneticist speaks of mutation, without further qualification, he normally refers to gene mutation.

The chemical nature of gene mutation has been discussed in Chapter 9. Here we are concerned with the significance of mutations to the individual and to populations.

Many gene mutations with obvious phenotypic effects occur at the rate of about 1 or 2 per million gametes. A few are much more common, and in one set of experiments on maize, 273 mutations affecting a particular gene locus were detected in 554,786 gametes, a rate of nearly 1 in 2,000. A gene at this locus affects the occurrence, distribution, and abundance of red or purple pigment in the grain and herbage. No less than 22 different **alleles** (alternative genes at one site) are known for this locus, although not all of them turned up in this experiment.

Probably many gene mutations have such slight effects that they escape notice, especially when they influence characters governed by multiple factors. The frequency of such mutations is necessarily speculative, as is also their importance. They may well be very important in evolution.

The rate of mutation can be increased by exposure to extremes of heat and cold, certain chemicals, and high-frequency radiations, such as X-rays, and the rays given off by radioactive elements (see chapter-opening photograph). Ultraviolet light is in a special category as a mutagenic agent. The absorption spectrum of DNA has its peak near the middle of the ultraviolet range. The energy absorbed from ultraviolet light by DNA results in chemical and physical changes in the intimate details of gene structure. Such mutations are not mere losses, as are so many mutations induced by X-rays and other high-frequency radiation. Ultraviolet and cosmic rays are probably responsible for the increased rate of mutations observed at higher altitudes.

Mutation is usually said to be at random, but this is true only in the sense that it does not seem to be directed toward specific goals. Mutations induced by extreme environmental conditions do not in general help the organism to withstand such conditions, It is conceivable that a mutation induced by low temperature might make a plant more resistant to cold, but a mutation making it less resistant to cold would be just as likely; the cause does not govern the kind of effect. On the other hand, some mutations at a particular gene locus occur more often than others, and the rate in opposite directions is not the same. Furthermore, the frequency and kind of mutations occurring in some genes are influenced by other genes that are inherited in Mendelian fashion. One prominent evolutionary geneticist summed up this situation by saying that mutability, like other characteristics of the organism, is under partial genetic control.

Many mutations are obviously harmful; many others have little or no obvious importance in survival; and a few are distinctly beneficial. Some mutations have been shown to modify the optimum or limiting environmental conditions for the organism and are thus beneficial or harmful, according to the environment. In general, the greater the obvious effect of the mutation, the more likely it is to be harmful.

Inasmuch as existing organisms necessarily have delicately balanced genic systems, which are more or less closely adapted to the environment and have had the chance to incorporate beneficial mutations during ages past, it is only to be expected that most mutations from the existing type will be more or less harmful, and only a few beneficial. The situation might be compared to the results of turning a group of undisciplined children loose on an automobile. The crayon marks on the finish do little real harm. Rolling all the windows down makes the car more comfortable on a hot day, but less so on a cold one. The broken headlight is important only at night, and the broken windshield wipers will be missed only during rain or snow. The chewing gum, jam, and snags on the upholstery are deplorable, but minor. The extra water in the radiator is useful, but the sugar in the gas tank is disastrous.

POLYPLOIDY

We have noted repeatedly in earlier chapters that the gametophyte generation characteristically has one full set of chromosomes, and the sporophyte has two sets. The gametophytic (or gametic) number is called the n number, and the sporophytic (or zygotic) number, therefore, is $2n$. As early as 1907, it was discovered that some plants have more than two full sets of chromosomes in the sporophyte, and in 1917 the Danish botanist Oejvind Winge (1886–1964) showed that in many groups of plants the chromosome numbers of related taxa fall into a regular pattern of multiples of a basic (haploid) number. Plants with three or more sets of chromosomes in the sporophyte are now said to be **polyploid.** A polyploid may be triploid (3 sets), tetraploid (4 sets), pentaploid (5 sets), hexaploid (6 sets), etc. The most common polyploids are tetraploids, and even-numbered polyploids, such as tetraploids and hexaploids, are more common than odd-numbered ones, such as triploids and pentaploids. Numbers higher than octoploid are uncommon, but they do occur.

The dual concept of the symbol n as representing (1) one full set of chromosomes, and (2) the gametic number of chromosomes, thus fails in polyploids. After some initial confusion, it has become customary to retain the symbol n for the gametic number, regardless of how many sets are involved. The symbol x has been introduced to designate one full set of chromosomes. In most plants and nearly all animals, n and x represent the same number, but in a considerable proportion of plants n is a multiple of x.

Polyploidy influences both the characteristics of the individual and the genetic and evolutionary potentialities of the population. The evolutionary effects are discussed in Chapter 30.

A very common effect of polyploidy is an increase in size of the cells and often an increase in the size and vigor of the whole plant. This is particularly true in newly induced polyploids (Fig. 29.9). Natural polyploids that have existed for an indefinite time often closely resemble the diploids, and it has been thought that natural selection may operate to restore the polyploids to the size of diploids.

Other direct effects of polyploidy are not always predictable. Annuals may change into perennials. There may be subtle changes in physiological characters, altering the ecological niche to which the plant is best suited. Polyploids are sometimes less fertile than diploids, because the pairing and segregation of chromosomes at meiosis do not always proceed normally, and some pollen grains or megaspores thus do not receive the proper complement of chromosomes.

Polyploidy originates through the suppression of meiosis, so that unreduced gametes are formed, or through the failure of anaphase migration of chromosomes during mitosis or meiosis, the duplicated chromosomes all being incorporated into a single nucleus. Under natural conditions such an occurrence is a mere accident, favored by environmental extremes such as low temperatures. Colchicine, an alkaloid produced by the autumn crocus (*Colchicum officinale*), inhibits spindle formation in most other organisms and can be applied in low concentrations to induce polyploidy.

PLANT BREEDING

Since prehistoric times, farmers have selected seed from the strongest, most desir-

Fig. 29.9 Effect of autopolyploidy on blackberries. Polyploidy was induced by treatment with colchicine. Tetraploid (*above*) and diploid (*below*). [Photo courtesy of J. W. Hull and the Maryland Agricultural Experiment Station.]

able plants in the field, or they have selected the best seeds after harvesting. Over thousands of years, this selection, along with random cross-breeding in the fields, has had profound effects, so that, as noted in Chapter 28, it is often difficult to identify the wild ancestor of a crop plant. Selection for different purposes and in different regions led to the development of numerous cultivars of each of the more common crops, but except in self-pollinated groups, each of these cultivars was itself usually highly heterozygous, encompassing a wide range of variation.

The development of the science of genetics during the twentieth century gave a tremendous impetus to practical plant breeding. Hybridization and selection collectively provide the means to combine desirable genes from different parents into the offspring. Every important crop in the Western Hemisphere, as well as many of the minor ones, has been greatly improved by deliberate effort during this century. The northern limit of large-scale cultivation of wheat in Canada has been advanced hundreds of miles by the creation of new cultivars that mature a few days earlier. Crop improvement is now a major activity in the United States and many other countries. Agricultural experiment stations in each of the states, in cooperation with the United States Department of Agriculture, are continuously producing new cultivars of old crops, with higher yield, greater resistance to disease, and other desirable qualities.

The effects of hybrid vigor and polyploidy are also being exploited. The production of hybrid corn has already been discussed. Colchicine-induced tetraploids of some garden flowers, such as snapdragons (Fig. 29.10) and zinnias, have larger flowers than the natural diploids.

Similar efforts at crop improvement in tropical countries are as a rule not so far advanced, but recent work in the Philippine Islands has resulted in spectacular

Fig. 29.10 Diploid (*left*) and tetraploid (*right*) snapdragons, the tetraploidy induced by colchicine. [Photo courtesy of the W. Atlee Burpee Company.]

improvement in the yield of rice, and the American plant breeder Norman Ernest Borlaug (1914–   , Fig. 29.11) won the Nobel Peace Prize in 1970 for his work in developing improved kinds of wheat in Mexico.

These and other improvements in tropical crops during the past two decades, which may have important consequences in the race with Malthus' nightmare, have led agriculturists to speak of a "green revolution" in food production. As in other revolutions, the changes are not all net gain. The new, high-yielding cultivars require fertilization of the soil if they are

Fig. 29.11 Norman Ernest Borlaug (1914–   ), American plant breeder, an outstanding green-revolutionist. [Photo courtesy of Rockefeller Foundation.]

to produce well, and the need for fertilizer, with other changes in farming practices, may have far-reaching socio-economic effects. A drastic increase in the cost of the energy necessary to produce fertilizer now threatens an apocalyptic reversal.

SUGGESTED READING

Brown, L. R., *Seeds of change*, Praeger, New York, 1970. The nature and social consequences of the "green revolution" in tropical agriculture.

Feldman, M., and R. E. Sears, The wild gene resources of wheat, *Sci. Amer.* **244**(1):102–113, Jan. 1981.

Grant, V., *Genetics of flowering plants*, Columbia University Press, New York, 1975. A well-written textbook. The author assumes that the reader has some botanical background.

Harpstead, D. G., High-lysine corn, *Sci. Amer.* **225**(2):34–43, August 1971. Efforts to improve the lysine content, otherwise very low in corn.

Rothwell, N. V., *Understanding genetics*, Oxford University Press, New York, 1976.

Shine, I., and S. Wrabel, *Thomas Hunt Morgan: Pioneer of genetics*, University Press of Kentucky, Lexington, 1976.

Sigurbjörnsson, B., Induced mutations in plants, *Sci. Amer.* **224**(1):87–95, January 1971. Use of radiation and chemical mutagens in crop-improvement programs.

Stebbins, G. L., From gene to character in higher plants, *Amer. Sci.* **53**:104–126, 1965.

Wills, C., Genetic load, *Sci. Amer.* **222**(3):98–107, March 1970.

Zobel, B. J., The genetic improvement of southern pines, *Sci. Amer.* **225**(5):94–103, November 1971.

Chapter 30

Evolution

Burdock, *Arctium minus*, a member of the family Asteraceae that is adapted to seed dispersal by animals. Each flower head is subtended by a set of involucral bracts (specialized leaves) that are hooked at the tip. The hooks catch onto passing animals (including people) and the fruiting heads easily break off from the plant at maturity. [Photo by Lynwood M. Chace, from National Audubon Society Collection/Photo Researchers, Inc.]

HISTORICAL SUMMARY

The Pre-Darwinian Period

The human urge to define and explain things and events is one of the basic intellectual drives. Explanations must always, of course, be based on the information and attitudes of the time. As information increases and attitudes change, explanations that once seemed adequate cease to satisfy and must be modified or discarded in favor of newer ideas. Thunder and lightning are no longer generally regarded as the voice and weapon of the gods, nor crop failure as an expression of divine wrath.

The ancient Greek philosophers, with a relatively limited body of factual knowledge, and with no authoritarian doctrine to guide them, gave free reign to speculation. Empedocles (fifth century B.C.) suggested that the parts of animals had arisen separately and spontaneously from the earth and had assembled themselves at random into whole animals—thus explaining why some animals look so odd. Aristotle (384–322 B.C.), on the other hand, believed that the more complex forms of life had evolved from simpler ones, although he had no clear idea of an activating mechanism. Although Aristotle's compendia of knowledge became the recognized authority in Roman and medieval times, his idea of evolution was neglected, and for many hundreds of years men were content to believe that each kind of plant and animal had been separately created, remaining unchanged forever after. Apparently no conflict was felt between this belief and the also prevalent belief in spontaneous generation of mice, rats, fleas, flies, and other pests.

In the eighteenth century, a few biologists began to doubt the theory of special creation, and in 1802 the French biologist Jean Baptiste de Lamarck (1744–1829) put forward the first full-fledged theory of evolution. The activating principle of evolution, as seen by Lamarck, was the effect of use and disuse of body parts, through the inheritance of acquired characters: Giraffes have long necks because they keep reaching up into trees to browse, and a man will have brawnier children (according to Lamarckian theory), if he becomes a blacksmith than if he becomes a musician. Although Lamarck was one of the most respected biologists of his time, well known for his taxonomic contributions in both botany and zoology, his theory of evolution attracted relatively little interest and still less support during his own lifetime.

During the first half of the nineteenth century, the idea of organic evolution made little headway, but the increase in biological knowledge gradually prepared the way for its acceptance. DeCandolle's classification of angiosperms was implicitly, though not explicitly, evolutionary, as noted in Chapter 27. In 1830–1833 the British geologist Charles Lyell set forth, in his *Principles of Geology*, a correlation of the stratigraphic position and relative age of rocks with the fossils found in them. For 300 years before that the nature of fossils had been hotly contested: They had variously been considered as inventions of the devil, freaks of nature, relics of the Flood, or the remains of prehistoric plants and animals. The scientific side of this controversy was laid to rest by Lyell's work, and the religious opposition to the recognition of fossils as prehistoric remains eventually dwindled to insignificance.

Darwinism

In 1831 the young English dilettante Charles Darwin (1809–1882; Fig. 30.1) obtained the unsalaried post of naturalist

Fig. 30.1 Charles Darwin (1809–1882), English naturalist, who laid the foundation of modern evolutionary theory. [Photo from Hunt Institute for Botanical Documentation.]

on the cruiser *Beagle,* which was to circumnavigate the world principally for map-making purposes. Although he suffered from violent seasickness throughout the five-year journey, Darwin worked hard, sending home notes and collections of botanical, zoological, and geological specimens from various stopping points. The theory of special creation and immutability of species, which he had at first accepted in orthodox fashion, became progressively unsatisfactory to him during the voyage. The six-week stay on the Galapagos Islands, about 1000 km off the west coast of South America, was particularly disturbing to his earlier beliefs. Here he found a fauna of distinctly South American genera, but with many strictly endemic species. Often there was a different species on each island, obviously similar to, but also obviously different

from, the species on the other islands. He thought it irrational that so many very local species should have been created for such a small and desolate area, and puzzling that they should consistently belong to South American genera.

At the completion of his voyage, Darwin, now a thoroughly dedicated biologist, set to working up his collections and to reflecting on the possible explanation for his observations. Inclining toward an evolutionary explanation, he keenly felt the lack of a satisfactory mechanism but was wholly unable to accept the Lamarckian concepts of evolution through the effects of use and disuse. In search of information he studied the literature on the breeding of domestic animals and talked with many professional breeders. It seemed evident that, with human intervention, domestic varieties fully as distinct as natural species could be produced from an original form. Still Darwin lacked a mechanism to cause such changes in nature.

Then in 1838 Darwin read Malthus' book, *An Essay on the Principle of Population,* which had been published in 1798. Malthus' principal point, which is now (1981) receiving renewed and fearful attention, is that the human breeding potential far outstrips potential resources, and that the limited supply leads to competition for available goods and a struggle for existence. Here was the key to Darwin's problem, and he gladly adopted Malthus' terms *competition* and *struggle for existence.*

Adding Malthus' concept of population growth to his other information, Darwin developed a new theory of evolution, with these salient points:

(1) Organisms of the same species are not all alike but differ slightly among themselves; furthermore the offspring of a

particular set of parents tend to vary around the average of the parents.

(2) More individuals are produced than can possibly survive.

(3) These individuals must compete with each other in a struggle for existence.

(4) As a result of this competition, the individuals best adapted to the environment tend to survive (*survival of the fittest*).

(5) New species originate by this *natural selection* of the best-fitted individuals over a period of many generations.

Darwin now set himself to marshaling the evidence for his views, taking the greatest pains to substantiate each point with numerous examples as well as argument. He thoroughly explored the scientific literature of the time, and he laboriously accumulated new data of his own. He bred different races of domestic pigeons for many years, and from 1846 to 1854 he devoted much of his time to a detailed study of the taxonomy of barnacles. (There is a story, hopefully true, that his young son, after a tour of a companion's home, asked, "But where does your father do his barnacles?") Meanwhile, Darwin confided his views on evolution to only a few friends.

In 1858, as he was nearing the completion of his projected book, Darwin unexpectedly received from Alfred Russel Wallace (1823–1913) a paper expressing the very same ideas. Wallace had developed his theory from years of study in the East Indies, again with emphasis on differences in the flora and fauna between islands, and again with an assist from Malthus. Embarrassed, Darwin then sent Wallace's paper and a summary of his own evidence leading to the same conclusion to a meeting of the Linnean Society of London. The following year, in 1859, Darwin's magnum opus, *On the Origin of Species by Means of Natural Selection*, was published.

Darwin's theory of evolution took the scientific world by storm. Biologists such as Huxley and Hooker in England, Haeckel in Germany, and Gray in America, as well as the geologist Lyell, were almost immediate converts. Huxley in particular defended Darwinism with untiring vigor. Within a few years the theory of special creation was abandoned by biologists in favor of the theory of evolution.

The evolutionary concept, especially as applied to human beings, aroused violent religious opposition, just as Galileo's concepts about the movement of the earth with respect to the sun had done more than two centuries earlier. As great a Christian as St. Augustine had in the fourth century A.D. interpreted the first chapter of Genesis allegorically, to mean that God had bestowed on the first living thing the power to evolve into other forms, but many nineteenth-century and even twentieth-century theologians found this interpretation unacceptable. The teaching of evolution was even forbidden in some places both in the United States and Europe. As recently as 1925, John Scopes was tried, in Tennessee's notorious "monkey trial," for teaching contrary to the Bible. William Jennings Bryan acted as special prosecutor and was also called as a witness for the defense by Clarence Darrow, the most famous defense attorney of his time. Scopes was convicted, but Darrow won his case in the court of public opinion, and attempts to suppress the teaching of evolution waned rapidly in the United States after this episode.

The theory of evolution provided a unifying principle that explained many otherwise puzzling facts. The fossil record, with its appearance of progressively more complex types in later rocks; the useless and often vestigial structures of many plants and animals, from the appendix and ear muscles of humans to the micropyle of angiosperm ovules and the flagella of cycad sperms; the South American affinities of the Galapagos fauna; the succession of stages in the growth of embryos of humans and other animals; all of these and more fell into place as evidence for and results of evolution. The time was clearly ripe, and had it not been for Darwin, we would now be honoring Wallace for the discovery—although acceptance of the principle might have come more slowly. Darwin's exposition was so thorough and so carefully documented that to the modern reader it may seem to belabor the obvious, but it was exactly this approach that led to its early triumph.

Non-Darwinian Concepts of Evolution

The principle of evolution was accepted, but doubts were felt from the beginning as to the adequacy of natural selection alone as the mechanism. The principle of survival of the fittest explains the selection of the most useful characters from among those existing, but does it explain the origin of new characteristics? The weakest part of Darwin's theory was the assumption that change through selection has no limits, because in each successive generation the offspring vary around the average of the parents. The complicated and wholly speculative mechanism that he suggested to achieve this result turned out to be so wrong that it need not be discussed here. The *Origin of Species* was published before Mendel's work, and Mendel's results did not begin to influence biological thinking until 1900.

The triumph of Darwinism, strangely enough, led to a revival of interest in Lamarck's concept of evolution through the effects of use and disuse of organs. This concept captured the popular imagination, and it still survives in nonscientific circles in spite of the consistently negative results obtained in numerous careful experiments.

Paleontologists, dealing with fossil plants and animals, and taxonomists, working on modern groups, often discerned evolutionary trends that are not readily explainable in terms of survival value and selection. From these observations came the concept of **orthogenesis**—evolution predetermined to proceed along certain lines, regardless of natural selection. The possible mechanism of orthogenesis was most obscure, but many biologists became convinced of its existence.

The genetic studies of the early twentieth century threw even more doubt on natural selection as the driving force of evolution. It was shown that selection is ineffective, once a pure line has been established. Furthermore, attention was drawn to the major mutations that occur in a group of species of *Oenothera* (evening primrose). Spontaneous polyploidy and other chromosomal irregularities are unusually common in this group, causing the seeming formation of different species in single jumps.

The Modern Period

Meanwhile, the fruit fly *Drosophila melanogaster* was discovered as an ideal tool for genetic research. These hardy little creatures with $2n = 8$ chromosomes breed like flies, producing a new generation in the laboratory about every 2 weeks. They require no great care and dine happily on overripe fruit. They display a number of small, but definite, differences that are inherited in Mendelian fashion. When subjected to X-rays or other mutagenic agents, they undergo numerous gene mutations, some of which are evidently the same as in naturally occurring mutants. From the studies on *Drosophila* came not only the concepts of genes, linear sequence on the chromosomes, linkage, and gene mutations, but also the concept of population genetics.

Population genetics is the study of the changes in proportions of genotypes in an interbreeding population over a period of generations. In a fairly large population subjected to a uniform environment, the effects of selection are very evident. Many small characters of no obvious importance to the organism increase progressively (by differential reproduction or survival) under one set of conditions and decrease progressively under another set. Clearly these seemingly insignificant characters do have significance in survival, either directly or through other unrecognized effects of the genes that govern them. Such studies have strongly bolstered the selectionist interpretation of evolution.

The accidents of survival become progressively more important in small populations. The proportions of a particular gene fluctuate rather widely in successive generations when the population consists of only a few individuals, and a gene may reach 100 percent frequency—or conversely may be entirely lost—regardless of its survival value, so long as it is not too strongly deleterious.

By the mid-1930s a **neo-Darwinian** interpretation of evolution emerged, with gene mutation as the principal source of variation, and natural selection, tempered by genetic drift in small populations, as the directing force. Neo-Darwinism provides adequately for the degeneration of

Fig. 30.2   One-seeded fruit of beggar ticks, *Bidens*, adapted to distribution by animals. The barbed awns readily catch in fur and also in hair and clothing. [Photo by John H. Gerard, from National Audubon Society Collection/Photo Researchers, Inc.]

unused structures as well as for the growth and elaboration of useful ones, thus undercutting the Lamarckian ideas of inheritance of acquired characters. We have seen that most mutations are harmful rather than beneficial, and it is only the continuous pressure of natural selection that weeds out the defective mutants. Because each gene commonly has a number of phenotypic effects, a mutation adversely affecting a particular organ might even be beneficial in other respects.

Neo-Darwinism so clearly provides a satisfactory explanation for so much of organic evolution that one is tempted to consider it the sole explanation; and many modern students of evolution take exactly this view. Experiments with fruit flies and other organisms have shown that a character may have hidden significance in survival, either directly or through some other effect of the gene governing it. Therefore it is assumed that all evolutionary trends are based on survival value. Many ingenious and ingenuous explanations of hidden survival value have been advanced to explain evolutionary developments whose relationship to natural selection is obscure.

Not all contemporary biologists accept neo-Darwinism as the sole explanation of organic evolution. Many taxonomists, especially those dealing with angiosperms, observe apparent evolutionary trends that are not easily explained in terms of survival value, and they reject the strained explanations that have been proposed. Some molecular biologists are also recent recruits to the idea that selection may not be a *sine qua non* for all evolutionary trends. Some of the consistent differences in the amino acid composition of cytochrome *c* among different taxonomic groups have no functional significance of any sort, so far as present information shows.

It may well be that the nature of the supply of mutations is more important than had been realized, especially as regards characters of little or no selective significance. The terms **neutral evolution** and **autonomous evolution** have recently been proposed for what is in essence the old concept of orthogenesis, shorn of its aura of mystic vitalism. Proponents of the concept of neutral evolution emphasize the importance of genetic control over the supply of mutations and the potentiality for nonselective fixation of genes through the accidents of survival in small populations. Both of these factors are familiar to and accepted by the dominant, neo-Darwinian school of evolutionary theorists, but their possible significance in causing neutral evolution is still highly debatable. In any case, neutral evolution can be no more than an additional feature to be incorporated into evolutionary theory; it is not a replacement for neo-Darwinism.

SOME CURRENT CONCEPTS

Origin of New Structures

One of the aphorisms of evolutionary theory is that *organs do not arise functionless*. New structures generally arise by modification of pre-existing structures, rather than by starting as useless bumps that acquire a function thereafter. Some function must be retained through the evolutionary process, although the function, just as the structure, may be gradually modified. Thus petals may be modified stamens, or modified sepals, but they do not originate (phyletically) as useless bumps. Sepals are modified leaves, and stamens are modified microsporophylls, which are (at least in a broad sense) modified leaves. Leaves are modified stems, stems are modified parts of a thallus, and a thallus is a modified cell colony. Guard cells are modified epidermal cells, and epidermis is modified parenchyma.

Recapitulation and Neoteny

An important evolutionary concept that has emerged from embryological and other studies is that *ontogeny* (i.e., the developmental history of the individual) *is a recapitulation of phylogeny* (i.e., the evolutionary history of the group). The concept has more numerous and more obvious applications in mammals than in plants. The human fetus, for example, has a tail throughout much of its period of development; at a fairly early stage it has gill slits, which eventually become the eustachian tubes connecting the ears to the throat; later it briefly acquires a fairly dense coating of hair, which usually disappears before birth. The fact that in sexual reproduction each individual starts its life as a single cell recalls the unicellular ancestry of all multicellular organisms. Dodder (*Cuscuta*, Fig. 1.4), a parasitic plant, has no connection with the ground at maturity, but the seed germinates on the ground in the usual way, and the basal parts disappear only after the twining stems have developed haustoria attaching them to the host.

The concept that ontogeny recapitulates phylogeny should not be taken too literally, however. It is more nearly an aphorism expressing the fact that ancestral characters frequently persist in the developmental stages, so that the ontogeny often gives strong suggestions as to the phylogeny. Thus although the human child's liking for bananas is not to be taken as proof of his simian ancestry, yet it is probably significant that angiosperms with compound leaves generally produce a few simple leaves in the seedling stage before producing leaves of the mature type.

The reason that ancestral features are more likely to be expressed in the juvenile stages is fairly clear. Any organism necessarily has a complex genic balance more or less in harmony with the environment. The greater the phenotypic effect of a mutation, the more likely it is to disturb this balance and to be eliminated by natural selection. A gradual evolutionary change has much more chance of remaining in harmony with the environment

than an abrupt one. The later in ontogeny a mutation exerts its phenotypic effect, the less total effect it is likely to have, and the less likelihood there is that any possible beneficial effect will be outweighed by some harmful effect. Thus the early ontogenetic stages of an organism are the least likely to be affected by an evolutionary change, and these early stages therefore provide some evidence as to the ancestral condition.

Another way to look at recapitulation is to regard it as one aspect of **heterochrony**—evolutionary change in the relative time of appearance and rate of development of characters. Changes in developmental timing often produce parallels between ontogeny and phylogeny, as rough versions of adult ancestral features are rapidly passed through during the development of the individual. Heterochrony can also produce the opposite result, with juvenile ancestral features persisting into maturity, the adult features being postponed out of existence. Herbaceous angiosperms are now universally regarded as being derived from woody ones. They certainly do not recapitulate their phylogeny by passing through an early arborescent stage. Instead they have evolved into herbs by postponing and diminishing secondary growth, and advancing the time of reproductive maturity. This sort of evolutionary change, in which organisms attain sexual maturity while remaining juvenile in some other respects, is called **neoteny.** Neoteny has been regarded as an evolutionary means of escaping overspecialization. The phyletic change takes place by cutting off the mature, specialized stage, permitting development to proceed in another direction. Neoteny is now invoked to explain much of the major evolution among angiosperms. Current concepts of the regulation of gene action, as discussed in Chapter 9, are highly compatible with both neoteny and recapitulation.

SPECIFIC AND INFRASPECIFIC DIFFERENTIATION

The Role of Isolation

Most speciation probably takes place in small populations that are more or less isolated and thus have a chance to change without being swamped by a larger population with which they are interfertile. Climatic and geologic changes are constantly altering the size and outline of the area to which each species is adapted, isolating small populations in favorable locations separated from the remainder of their species. Many such local populations die out as the environmental conditions for them deteriorate; others are reunited with the main population of their species as conditions become more favorable again and the intervening area is reoccupied. A few such small populations, under pressure of selection and affected also by chance fixation of previously uncommon genes, change so much that they retain their distinctness even when brought back into contact with the parent species. These have become new species.

The amount and stability of the differences arising under isolation vary according to the circumstances. Morphological and physiological differentiation tend to go hand in hand with the development of barriers to interbreeding with the original population, but any one of these types of difference may fall behind or get ahead of the others. Thus an isolated small population may become morphologically and physiologically distinguishable from the parent population, without developing any internal barrier to cross-breeding with

it, so that the two breed together freely when brought together in the garden or when natural contact is reestablished. Similarly, a barrier to interbreeding may develop without much other change, or morphological differences and reproductive isolation may develop without much physiological change, or the population may remain morphologically almost unchanged after physiological changes and reproductive isolation have been established. There is necessarily some room for difference of opinion as to whether such partially differentiated types should, in a given instance, be regarded as separate species, subspecies, or varieties, or be denied any taxonomic recognition at all.

If a continuous large population is divided into two large populations (in similar habitats) by destruction of the geographically connecting part, the two do not usually diverge very rapidly. About the same kind and number of mutations will occur in each, and if the environmental conditions are similar, the effect of selection is about the same. Therefore although both populations may gradually change, the changes are rather similar and divergence is slow. Thus the deciduous forests of the southern Appalachian region of the United States contain a number of species so similar to species from eastern Asia that it is hard to decide whether we have two closely related species or a single species with two geographically isolated segments. Yet the two populations have been separated in each case for upward of half a million years, since the breakup of the Arcto-Tertiary forest at the beginning of the Pleistocene glaciation. The American tulip tree (*Liriodendron tulipifera*) and the Chinese tulip tree (*L. chinense*) are an example of such a species pair.

The southern Appalachian and Ozark

regions of the United States also show many closely related species pairs, or varieties of one species. These two areas, with somewhat similar climate and topography, have been continuously available for land plant habitation since the Cretaceous period, in contrast to the glaciated area to the north and the coastal plain (with its Mississippi embayment) to the south and between.

Ecological Influences

At the margins of its geographic range and ecological tolerance, a species is subjected to the most evolutionary pressure. If the proper mutations occur, it may be enabled to occupy new territory, showing a gradual change with the geography and habitat.

Species with a wide geographic range or ecological tolerance usually have many different genotypes and often show recognizable races correlated with the habitat. Thus *Solidago canadensis*, a common goldenrod, has one variety in the New England region and adjacent Canada, another in the deciduous forest region of the eastern United States, a third on the Great Plains, a fourth in the western cordillera, and a fifth along the coast of British Columbia and southern Alaska. Each of these varieties is adapted to the climate of its own region and differs from other varieties in minor morphological characters; but no sharp lines can be drawn among them. Such ecologically differentiated races within a species (Fig. 30.3) are often called **ecotypes.** A species may have different ecotypes on different soils or at different altitudes in the same region, as well as in different geographic regions.

An important factor in the development of evolutionary diversity is the existence of habitats available for exploitation. Some plant communities may be so perfectly adjusted to the environment that there is little or no room for an invader; each ecological niche and subniche is completely occupied by its own set of species, which exclude any less well-adapted immigrants. Climatic and geological changes do continue to disturb the balance between the vegetation and the environment, however, and if these changes occur rapidly enough, some habitats may be left temporarily unfilled, or they will be filled by species so poorly adapted to them as to allow invasion by outsiders.

Any such unfilled or imperfectly filled niche is a powerful stimulus to evolutionary change. When the vegetation is in harmony with the environment, almost any significant mutation will probably be harmful, but when some habitats are not properly filled, some of the newly produced mutants may be able to occupy them. An extreme example is provided by oceanic islands, such as the Galapagos and Hawaiian islands. When such islands first arise from the sea, all habitats are open. Sooner or later a few kinds of plants and animals accidentally reach them. Competition is at a minimum, and genotypes that would otherwise be eliminated have a chance to survive in one or another of the unfilled habitats. There is thus an evolutionary explosion to occupy the available habitats, and new species may perhaps arise even without initial geographic or reproductive isolation.

The competitive factors mentioned in the preceding paragraphs have been summarized into a **principle of competitive exclusion,** usually stated at the specific level ("in equilibrium communities no two species occupy the same niche"). Recent concepts about vertebrate taxonomy are permeated by the application of this same thought to higher taxa, and vertebrate taxonomists are accustomed to seeing a fairly good correlation between

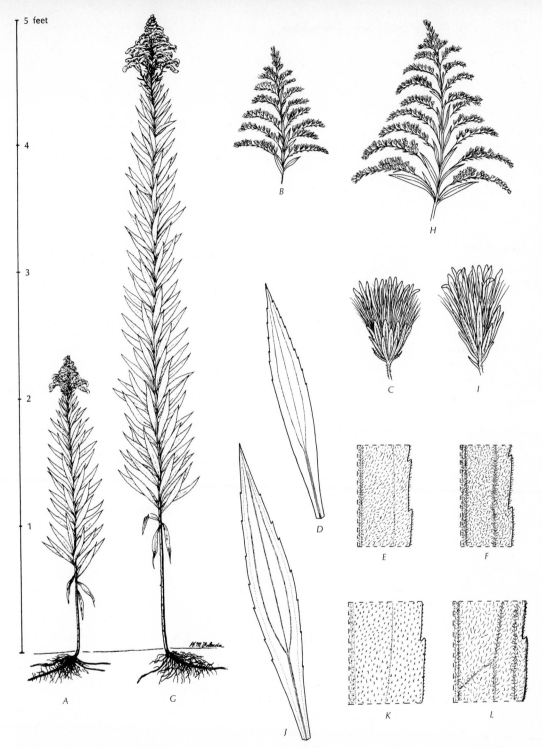

Fig. 30.3 Ecotypes of *Solidago canadensis*: A-F, var *gilvocanescens*, the characteristic phase of the Great Plains region; G-L, var. *scabra*, a characteristic phase of the eastern deciduous forest region. A, G, habit; B, H, inflorescence; C, I, flower head; D, J, leaf; E, K, upper side of leaf showing sparser, shorter, stiffer hairs on var. *scabra* than on var. *gilvocanescens*; F, L, lower side of leaf showing slightly longer hairs on var. *scabra*.

taxonomic groups and ecological niches or adaptive zones. The tightly integrated morphological-physiological system of an animal is closely bound to the way the animal makes its living—what it eats, how it gets what it eats, and how it keeps from being eaten. It is these correlations that permit vertebrate paleontologists to reconstruct whole animals (in concept) from a few miscellaneous bones. Any deviation from the standard pattern for the group is likely to decrease the competitive ability of the deviate within its own group, or to throw it into competition with a better adapted group in some other niche.

Similar correlations exist in plants, as we have seen, but they tend to be less stringent and less pervasive. The adoption of a particular way of life by a plant group does not necessarily use up and foreclose that way for something else. The mangrove habit does show competitive exclusion, but wind pollination by *Pinus* does not interfere with wind pollination by *Populus* (poplar), and animal transport of *Bidens* fruits does not interfere with animal transport of *Desmodium* fruits. Since multicellular plants do not move about from place to place, they are also more tolerant of casual variation in structure than are animals. The net result of this dilution of the competitive exclusion principle is that each of the obvious niches for land plants is occupied by species representing diverse families and orders, and there are many individual families and even genera that have species adapted to widely differing ecological niches.

Disturbance and Hybridization

Human activities also influence evolution. They disturb the habitat, upset ecological balances, and, in effect, create new ecolog-ical niches for exploitation. Many European weeds are known only around human habitations and can continue to exist only under the disturbed conditions created by humans. Some of these species have certainly originated by modification from species that continue to exist in less disturbed habitats.

One of the effects of disturbance is to increase the opportunity for hybridization between previously distinct species. The newly opened habitat may be invaded by species that formerly did not grow together, and the hybrids, which under natural conditions would have been quickly eliminated even if formed, may find suitable conditions for survival and reproduction. These hybrids may then breed with themselves or with either of the original parents, and genes of one species may thus be introduced into another species by this introgression, or **introgressive hybridization.**

The retreat of the glaciers after the height of the most recent glaciation (some 11,000 years ago) also provided ample opportunity for introgressive hybridization and rapid evolution on the areas laid bare. Species that are quite distinct south of the glacial boundary are sometimes difficult to distinguish north of it.

Origin of species by hybridization is most likely in disturbed conditions under which some niches are not adequately filled, or in dry areas where the soil is not completely covered by vegetation. Some of the hybrids or hybrid segregates may then turn out to be better adapted to the surroundings than either of the parents.

Polyploidy and Hybridization

Polyploidy can introduce or break down a barrier to interbreeding, depending on the circumstances, and it can make sterile hybrids fertile. The critical influence is commonly exerted at the time of meiosis. If every chromosome has one and only one homolog to pair with, meiosis proceeds normally, leading eventually to the formation of normal gametes. Otherwise there may be meiotic irregularities and subsequent abortion of the meiotic products.

A first-generation hybrid between a diploid and a tetraploid will be a triploid, with 3x chromosomes (Fig. 30.4). Because of meiotic problems, triploids produce only a small proportion of viable pollen and ovules. Triploids do not leave many offspring unless they can set seed apomictically or have effective means of vegetative reproduction. The commonly cultivated kinds of bananas are sterile triploids, and the small black dots along the middle are abortive seeds.

Too many homologous chromosomes can be almost as bad as not enough. Meiosis in simple autotetraploids (produced by doubling the number of sets of chromosomes within a species) is likely to be irregular, unless the cells have some internal mechanism favoring the formation of bivalents (simple pairs of chromosomes) instead of trivalents or tetravalents.

Sterile hybrids between species can become fertile through polyploidy. The famous Kew primrose (*Primula kewensis*) was the first plant observed to do so. The original *P. kewensis* (Fig. 30.5) was a sterile hybrid between *P. verticillata,* from Arabia, and *P. floribunda,* from the Himalayan region. It appeared as a chance hybrid in a greenhouse at the Kew Gardens in 1899 and was experimentally produced again in 1900. In 1905, 1923, and 1926, some flowers of the hybrid set seed. In each instance the seedlings were fertile and had twice as many chromo-

somes as the parent. Such fertile tetraploids are, in effect, potential new species of hybrid origin, interfertile among themselves, but not crossing readily with either of the parental species or the original hybrid. Many natural species are believed to have originated by hybridization and polyploidy, and a few have even been experimentally duplicated.

Whether a newly produced polyploid does actually evolve into a new species in nature depends on future events. It may do so, or it may fail to meet the test of competition, or it may fail for other reasons. Autopolyploids may take off on their own evolutionary path, or they may remain so much like the diploid ancestor that taxonomic segregation is impractical. Allopolyploids (produced by chromosome doubling in interspecific hybrids) may still be meiotically irregular, so that even though the plants are fertile they produce offspring that tend to revert toward one or another of the original parents. The allopolyploid of *Primula kewensis* breaks down in this way.

Species that contain both diploid and tetraploid races may be intersterile with related species at the diploid level, but interfertile at the tetraploid level. Here again meiosis is often the critical stage. The situation may be compared with that in *Primula kewensis,* except that the increase in number of chromosomes occurs before hybridization rather than after. Introgressive hybridization is relatively common in polyploid groups.

Although it is clear that new species can and do originate in nature through polyploidy, it is equally clear that polyploidy cannot be the usual method of speciation. There is a limit to the number of chromosomes that can effectively inhabit a single cell. Furthermore, the reduplication of genomes tends to decrease the

Fig. 30.4   Marigold hybrid (*below*) and parents. The "African" marigold *Tagetes erecta* (*top left*), a native of Mexico, is diploid, 2x = 24; the "French" marigold, *T. patula* (*top right*), also a native of Mexico, is tetraploid, 4x = 48. The hybrid is triploid, 3x = 36. In this instance there is no hybrid vigor, at least as regards the size of the flowers. [Photo courtesy of the W. Atlee Burpee Company.]

Fig. 30.5 *Primula kewensis,* the sterile hybrid from which the fertile tetraploid can arise. [New York Botanical Garden photo.]

effect of individual mutations. Polyploidy is often highly effective in producing new and successful combinations of existing characters, but in general it appears to be an evolutionary blind alley. It is the diploids that provide the continuing stream of evolution.

Evolutionary Pattern of Land Plants

The phylogeny of land plants, especially the vascular plants, is a story of increasing adaptation to a land habitat. The development of a vascular system is obviously necessary if the plant is to rise far above the ground. Stomates are also essential, permitting the exchange of gases necessary for photosynthesis in internal tissues, without allowing too much evaporation during periods of water shortage. The differentiation of an originally dichotomously branching axis into roots, stems, and leaves with particular functions in the economy of the plant is a further adaptation to growth on land.

The gametophyte, with none of the special adaptations of the sporophyte, is progressively reduced and is at first earth bound. The differentiation of the gametophytes into males and females, a specialization that ensures cross-fertilization, determines the direction of any further modifications of the gametophyte in relation to the land habitat. The female gametophyte, which must provide nourishment for the young sporophyte, cannot be reduced as much as the male gametophyte, which merely needs to produce sperms.

The next important step, after reduction in the male and female gametophytes seems to have reached the limits of efficiency, is the retention of the female gametophyte on the parent sporophyte, and the transfer of the whole male gametophyte (instead of merely the sperms) by wind or other means. The female gametophyte and its associated structures now constitute the ovule, and the male gametophyte is the pollen grain. Fertilization no longer depends on a film of water for the sperms to swim in, and the sperms, not having to swim, lose (phyletically) their flagella. The enclosure of the ovule within the ovary is a protective measure made possible by the fact that the male gametophyte now delivers the sperm to the egg through a pollen tube that absorbs nourishment from the surrounding tissue as it grows.

Another important adaptation to the land habitat, which has been made repeatedly but which has been most effectively exploited by angiosperms, is the xylem vessel. Vessels are much more effective conducting elements than tracheids. A vessel-bearing angiosperm tree growing in the same habitat as a vessel-*less* conifer provides a more reliable supply of water to the leaves, especially during periods of high evaporation, enabling the angiosperm to make fuller use of the environmental resources by lessening the need for xerophytic adaptations that restrict such exploitation.

SUGGESTED READING

Anderson, E., *Introgressive hybridization,* Wiley, New York, 1949. A clear and forthright presentation of hybridization as a factor in plant taxonomy and evolution.

Carlquist, S., *Island biology,* Columbia University Press, New York, 1974. Beautifully written and scientifically accurate.

Darwin, C., *Journal of researches into the geology and natural history of the various countries visited by H. M. S. Beagle,* 1839. Facsimile reprint, Hafner, New York, 1952. Good for browsing.

DeBeer, Sir G., *Charles Darwin,* Doubleday, Garden City, N.Y., 1964. Probably the best Darwin biography for the biologist.

de Camp, L. S., The end of the monkey war, *Sci. Amer.* **220**(2):15–21, February 1969. Darwinism, Darrow, Bryan, and the Scopes trial.

Ehrlich, P. R., and P. H. Raven, Butterflies and plants, *Sci. Amer.* **216**(6):104–113, June 1967.

Grant, V., *Organismic evolution,* Freeman, San Francisco, 1977.

Mayr, E., Darwin and natural selection, *Amer. Sci.* **65**:321–327, 1977.

Mayr, E. B., Evolution, *Sci. Amer.* **239**(3):47–55, September 1978. An introduction to an issue of *Scientific American* devoted to evolution.

Mulcahy, D. L., The rise of the angiosperms: a genecological factor, *Science* **206**:20–23, 1979.

Wallace, A. R., *The Malay Archipelago, the land of the orang-utan and the bird of paradise; a narrative of travel with studies of man and nature,* Dover, New York, 1962. An unabridged republication of the last revised edition of a work first published in 1869 by Macmillan, London. Good for browsing.

Chapter 31

Communities

Monarch butterfly (*Danaüs archippus*), adult and caterpillar, on milkweed (*Asclepias syriaca*). [Photos from National Audubon Society Collection/Photo Researchers, Inc., upper by Karl H. Maslowski, lower by Harry Brevoort.] Monarch butterfly caterpillars are especially adapted to feed on *Asclepias*, which is poisonous to most kinds of insects and other animals.

THE WEB OF LIFE

Everything depends on everything else. Just as "no man is an island," so no individual plant or animal, nor any kind of plant or animal, is independent of others. The cell, the organism, and the community represent different levels of integration, but in each the various parts are in constant interaction, influencing and being influenced by each other. The oft-repeated statement of the American ecologist Frederic E. Clements (1874–1945, Fig. 31.1) that "the community is an organism" is a perceptive aphorism and an essential guide to understanding, if not quite a literal truth.

Animals and saprobic organisms depend absolutely on green plants* as the immediate or ultimate source of food, but the plants in turn depend on the eaters and the decayers to replenish the pantry with raw materials. The reciprocal roles of photosynthesis and respiration in maintaining a balance of oxygen and carbon dioxide in the air have been noted in earlier chapters. Nutrient cycles may have all degrees of complexity, but cycles there must be if life is to go on.

The total volume of vegetable material in a biotic community is much greater than the volume of animal material, so that it is the plants which dominate the landscape. For some purposes it is useful to consider the plant communities alone, with little or no attention to the animals. For other purposes, especially for a more complete understanding, consideration of the animals is also essential.

Ecologists often find it useful to think of the biotic community as a pyramid (Fig. 31.2). Green plants form the broad base and the main bulk of the pyramid. Organisms that make their living directly from green plants form the next, necessarily smaller, segment of the pyramid. These first-level consumers include not only the grazing and browsing mammals, but also the seed-eating birds, the vast majority of insects, the bulk of the fungi and bacteria, and so on. Second-level consumers, which eat, parasitize, or decay first-level consumers, make up the third layer of the pyramid. The common carnivores and the

Fig. 31.1 Frederic E. Clements (1874–1945), sponsor of the climax concept, who dominated American ecological thought for several decades. [Photo from Hunt Institute for Botanical Documentation.]

*The term green *plants* is here taken to mean all plants that have the green pigment chlorophyll. Some of these, such as many of the marine algae, are not visually green, because the chlorophyll is masked by other pigments.

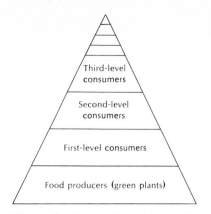

Fig. 31.2 The ecological pyramid, as discussed in the text.

insect-eating birds are familiar second-level consumers. Third-level consumers, which eat or parasitize (or decay) carnivorous or insectivorous animals, necessarily constitute a still smaller segment of the pyramid, and so on. Even fourth, fifth, and higher orders of consumers may be recognized, especially in the oceans, where the base of the pyramid consists mainly of plankton algae, and the orders of consumers run from small zooplankton, to larger zooplankton, to small fish, and to larger and larger predaceous fish.

The concept of the stratified pyramid is very useful, but should not be taken too rigidly. The same individual organism, or kind of organism, may belong to more than one layer. Man, for example, is a first-order consumer when he eats bread, a second-order consumer when he eats beef, and a consumer of some higher order when he eats salmon. Such details do not affect the principle that at each successive level most of the food is respired or otherwise degraded, rather than being converted into new protoplasm. A ratio of about 10 to 1 is not uncommon.

In some contexts it is useful to think simply of a food chain from plants through successive orders of consumers, without emphasizing the progressively smaller size of the successive links.

One of the important results of the food pyramid or chain is the progressive concentration, level by level, of certain substances that accumulate in the animal instead of being excreted or respired or effectively altered by metabolism. The decline and threatened extinction of large predaceous birds such as eagles, ospreys, and the peregrine falcon result from their being at the end of a long chain of accumulation of DDT, which upsets calcium metabolism. Only after the use of DDT was banned did some of these species begin to recover, and their existence is still precarious. The concentration of mercury in ordinary seawater is only about 0.0002 parts per million (ppm) and does not appear to have changed much during human time; but in swordfish, again at the end of a long food chain, the concentration often reaches 1.0 ppm, enough to raise questions about the safety of swordfish as human food.

Plants and animals affect each other's evolutionary development as well as their status in the biotic community. The ontogenetic pattern of grass leaves is clearly an evolutionary response to grazing. Monocot leaves characteristically mature from the tip downward, and in grasses this pattern has been intensified to produce an intercalary meristem at the base of the blade (see Chapter 18). The diverse mechanisms to make use of various kinds of animals (including insects) as agents of pollination and seed dispersal are of course evolutionary adaptations.

A less obvious adaptation is the production of poisons in the herbage, fruits, or seeds of the plant that ward off marauders. Almost inevitably, some small group of predators or parasites develops a concomitant tolerance to the poison, becoming progressively adapted and often eventually restricted to the particular group of hosts. The restriction of monarch butterfly caterpillars to milkweeds (*Asclepias* spp.) is a familiar example (see chapter-opening photograph).

Such reciprocal defenses and means to overcome them are particularly evident in the tropics. Tropical forests commonly consist of large numbers of kinds of trees, with relatively few and well-scattered individuals of each kind. It has been suggested that at least in some instances the minimum distance from one tree of a kind to another is the distance that gives a seed a reasonable chance to escape discovery by the specialized insects that eat it; and furthermore that the frequent tendency of individual trees to flower only in alternate or more distantly spaced years is a means to assure that the insects cannot just camp on the tree and await the next crop.

Another way for animals to meet the challenge of poisons in the food is to eat only a small amount of each of many kinds of foods, keeping the intake of each kind of poison below the tolerance level. This is sometimes the road to survival in tropical forests for squirrels, which can travel freely from tree to tree doing their diversified shopping in the supermarket.

Forests of temperate regions commonly have fewer species, with more individuals of each, and the tree seeds are not usually poisonous. Here the lean years and the years of heavy fruiting are commonly more synchronized within a region, probably partly in response to climatic signals. The lean years reduce the population of squirrels and other seed eaters, providing a chance for more seeds to survive in the years of heavy production.

Recognizing the complexity of cycles and interactions, we may proceed to consider why the flora and fauna differ in different parts of the world. Obviously the physical environment exerts the ultimate control under natural conditions. It is useful to think of the first dichotomy in the physical environment as that between aquatic and terrestrial habitats, and then to consider the effect of climate.

Aquatic and terrestrial habitats are fundamentally different. Large, multicellular plants can exist only on land, or in water shallow enough to permit them to be rooted to the bottom, or in relatively quiet marine waters such as the Sargasso Sea. Most of the open ocean is suitable only for plankton algae and the animals that depend directly or indirectly on them for food.

We have seen that the evolutionary history of the embryophytes is in large part a story of progressive adaptation to the land habitat. Submerged plants, whether attached to the bottom or not, do not need the special restrictions on transpiration that are necessary for terrestrial plants. On land, the water balance and the temperature are the dominant climatic factors, which other factors can modify but not supplant.

In a broad sense, it may be said that the climate controls the nature of the plants that grow in a region, and the fauna depends on the flora. Within this framework other factors, such as the nature of the soil or underlying rock (**edaphic factors**), the reciprocal influence of the animals and the plants on each other (**biotic factors**), irregular catastrophic disturbance, and the geologic and evolutionary history, govern the detailed composition of the biotic community. Climatic control of plant communities is further discussed later in this chapter.

THE NATURE OF PLANT COMMUNITIES

It is a common observation that certain kinds of plants are likely to grow together: oak and hickory; spruce and fir (Fig. 31.3); lawn grass and dandelions; beech, maple, and hemlock; buffalo grass (*Buchloe*) and little grama grass (*Bouteloua gracilis*). Each of these combinations occurs often enough to catch any observant eye. A little further study shows that certain kinds of shrubs and herbs are likely to be found where oak and hickory grow together; another group occurs with spruce and fir; another group with beech-maple-hemlock; and so on. Thus whole plant communities tend to repeat themselves in similar habitats. Any group of plants occupying a particular habitat at a particular place is called a **plant community,** and every plant in nature belongs to such a community. Communities may be narrowly or broadly defined, according to one's purposes of the moment. It may even be useful to think of communities within communities.

It does not require much closer observation to see that certain kinds of animals tend to go with certain plant communities. In a sagebrush desert, one may expect to find jackrabbits, ground squirrels, and

Fig. 31.3 Virgin spruce-fir forest in Quebec. [From Pierre Dansereau, *Biogeography*, 1957. Copyright by Ronald Press, New York.]

possibly rattlesnakes, but not snowshoe rabbits, flying squirrels, or cobras. In general, the animals are there because of the plants, rather than vice versa, but that concept is only a first-order approximation.

It will do no harm to repeat at this point that the organisms of a **biotic community** influence each other, as well as influencing and being influenced by the physical environment. Many of the shrubs and herbs of the forest floor require shade and could not survive if the trees around them were cut down. The bushy young second-growth forest that harbors a large population of deer will support only a much smaller number when the trees grow taller and there is less forage in reach. Many plants depend on animals (including insects) for the distribution of pollen or seeds, as well as being eaten by these same or other animals.

Just as it is true that certain kinds of plants repeatedly occur together, so it is also true that no two plant communities are ever exactly alike in the species represented and the proportions in which they occur. The more carefully two similar communities are analyzed, the more differences are likely to be found between them. Each species in the community has its own geographic limits and ecological amplitude, which do not exactly match those of any other species. Since the environment of two different plant communities is never the same in all details, each community differs to a greater or lesser degree from all others. Attempts to classify plant communities should therefore be taken with a certain reserve, to be used when they are useful, and ignored when they are not.

The transition from one plant community to the next may be gradual or abrupt, and the boundary zone between them (called the **ecotone**) is correspondingly broad or narrow. The ecotone is especially likely to be narrow when the two communities are of different growth habit, as a forest and a grassland (see Fig. 31.6), or when there is an abrupt change in the habitat, as at the shore of a lake. On the other hand, where the environmental change is gradual, and where the two communities are similar in habit (e.g., two grassland communities, or two coniferous forest communities) the ecotone may be very broad indeed.

The ecotone is a tension zone between two communities. Just as the last straw broke the camel's back, so may a slight environmental change in the ecotone have a great effect, causing one community to give way to another. Climatic changes, even slight ones, in an ecotone region may shift the position of the ecotone as one community expands into an area previously occupied by the other.

The tension between the forest community of the eastern United States and the grassland (prairie) community in the drier region of central and west central United States may serve as an example of shifting boundaries between communities. Fire was evidently a major factor in determining the boundary before the nineteenth century, and with the recent protection from fire, trees are now invading the prairie in parts of eastern Kansas (Fig. 31.4) and elsewhere. The boundary line was even farther eastward during a relatively warm, dry period (the postglacial xerothermic or hypsithermal period) some 5000 years ago, with a prairie peninsula reaching eastward into Ohio.

The Lawn as a Plant Community

A lawn may be taken as an example of a plant community with which the reader

Fig. 31.4 Trees coming up in native prairie near Manhattan, Kansas, following several decades of protection from fire. [Photo courtesy of Lloyd Hulbert.]

may be familiar. Many of the environmental factors governing the lawn are under human rather than natural control, but the essential principles of competition and response to environmental factors are the same as in natural communities.

Our lawn, being reasonably well kept, consists principally of bluegrass (*Poa pratensis*), with an admixture of white clover (*Trifolium repens*). If we were naming it in the same way as we often do natural communities, we would call it a bluegrass-clover community.

In the spring the lawn begins to grow vigorously and needs to be mowed frequently. One notices that this year the clover is especially abundant in a sloping area where the water runs off the driveway. Obviously there has been extensive leaching of nitrogen here, and the clover, which harbors nitrogen-fixing bacteria in its roots, has begun to crowd out the grass. In another year or so, at this rate, the clover will replace the grass in this spot completely and will begin to gain on the rest of the lawn too. So we apply fertilizer, being especially generous where the clover is most abundant. Since we are interested in vegetative growth, we use a fertilizer that is high in nitrogen, shying away from the high-phosphate fertilizers, such as bone meal, that encourage flowering and fruiting.

Early in the summer some seedlings of Norway maple come up in the lawn, the winged fruits having blown in from some trees across the street. There are even one or two silver maples among them. The next block has a row of silver maples on it, and we remember learning in general botany that *the number of seeds migrating to any given distance varies inversely with the square of the distance.* The maple seedlings soon get clipped off by the lawnmower and die. Without some such lethal factor as the lawnmower, these and other seedlings of woody plants would grow up in a few years to form a forest that would cut off so much light that the lawn would die out. Exactly this same sort of competition for light occurs in nature. If the climate is moist enough to support tree growth, grasses are shaded out and play only a minor role in the community, unless some special factor operates against the trees.

As summer comes on and the weather gets warmer, the bluegrass grows more slowly and does not need mowing so often. Then we begin to notice here and there some grass sprigs with broader, paler leaves among the deep green we have been so proud of. Crabgrass! This is a late-starting annual that grows vigorously all summer, especially if there is plenty of moisture. It will spread by stolons along the top of the ground, covering the bluegrass and starving it for light if allowed to continue.

Laziness tempts us to hope that if we just stop watering the lawn, all will be well. After all, bluegrass goes dormant in the summer anyway, given half a chance,

and then greens up again for fall. Our neighbors, whose lawn goes brown in summer from lack of water, doesn't seem to have crabgrass trouble. But then we remember the park lawn, a few blocks away, where this policy has been in effect for 20 years. Old-timers tell us that it used to be a good lawn, too. But each year it had a little more crabgrass, until now nothing else is left in the open places. It is green in the summer, and dead gray-brown all fall, winter, and spring. True, there is bluegrass under the scattered trees in the park, so in summer the lawn is green in the open and brown under the trees, and in spring and fall it is brown in the open and green under the trees. Obviously bluegrass stands shade better than crabgrass, but even bluegrass doesn't like it very well, and the turf under the trees is pretty skimpy. If the trees were clustered, instead of scattered, so that light did not get in from the side, the bluegrass would be shaded out, too.

But we are proud of our lawn, and we want it to be green all summer, as well as spring and fall, and we want it to be that way year after year. So we continue to water it, and we set the mower up a bit, to leave the grass longer so that the hot sun does not get through to the ground and parch it. We also get down on our knees and pull crabgrass. Naturally we work hardest on the front lawn, which everybody can see, and here our efforts are enough. Not only does it stay green, but when fall rolls around there is hardly a blade of crabgrass in it. Now the bluegrass starts to grow vigorously again, and we fertilize it and set the mower back down to cut more closely and stimulate growth and branching. The front lawn has survived another season and will be in good shape next spring after its long fall growth and winter dormancy.

The back lawn is another matter. Somehow, after the front lawn was weeded, there never seemed to be enough time to take proper care of the back lawn. We did water it and it stayed green, but by late summer there are big spreading patches of crabgrass. Under the crabgrass, the bluegrass looks pretty sick but is perhaps not yet dead. So, late in the summer, with the weather still warm and the crabgrass spreading wildly, we apply a chemical advertised as a crabgrass killer. We worry about this a bit, for we know that it is poisonous to bluegrass too, and part of its greater effect on the crabgrass is due merely to the fact that at this time the crabgrass is growing faster than the bluegrass. With a little bit of luck, the recommended dosage turns out to be all right, and the crabgrass turns yellow and dies. The bluegrass, wonder of wonders, is not yet dead, and, as the weather turns cooler, it comes back and fills in where the crabgrass has been. Only a few little bare patches remain to be reseeded, and we resolve to do better in the future. We remember that crabgrass is an annual, starting anew from seed each year. Perhaps next spring we shall try one of the chemicals that inhibits seed germination, and maybe the crabgrass problem will be licked for a while.

There were, of course, a few other weeds in the lawn besides crabgrass, mainly dandelions and plantain. These have broad leaves which shade out the grass but lie so close to the ground that the mower misses them. Fortunately, there were not very many, and the individual plants don't spread like crabgrass, so we were able to keep them down by digging them out. If we ever decide to do without clover in the lawn, we may try 2,4-D on these dicot weeds. This auxinlike chemical is much harder on most dicots than on

monocots such as grasses, and we shall have to be careful not to let any of it drift onto our flowers, or into our neighbors' gardens.

From this not wholly whimsical account of some problems involved in the maintenance of a lawn, one can see that the continued existence of a particular plant community requires a delicately balanced set of environmental factors. A change in one factor favors one or more species in the community at the expense of others, and a change in some other factor favors some other species. Similar environments call forth similar plant communities, but just as no two areas ever have exactly the same environment, so are no two communities exactly alike. Individual kinds of communities may be defined as broadly or as narrowly as one wishes. All lawns may be considered as one type of community, or all bluegrass lawns, or our front lawn may be classified as different from our back lawn. Communities thus vary in size from a few square meters, or even less, to hundreds of thousands of square kilometers; they may be of any shape, and there may be one or more examples of a particular kind, depending on the breadth of the definition and the geographic extent or recurrence of the necessary environmental conditions.

The foregoing discussion of the lawn has been from the standpoint of its existence as a *plant* community, and for some purposes this is sufficient. For a fuller understanding, we would have to consider it as a *biotic* community, with due attention to the earthworms that aerate the soil as they tunnel through it eating small bits of organic matter; to the larvae of June beetles, Japanese beetles, and other insects that may inhabit it; to the robins that eat the earthworms; to the moles and skunks that may be attracted by the beetle grubs; and so on. Yet here as elsewhere it remains true that it is the plants that largely determine the nature of the community, whereas the animals are there because of the plants.

PLANT SUCCESSION AND THE CLIMAX

Principles

All plant communities are transitory in terms of geologic time. Most of Canada and much of the northern United States, as well as northern Eurasia, were covered with ice only a few thousand years ago, just as the interior of Greenland is covered with ice today. There were, in fact, alternating glacial and interglacial periods during Pleistocene time, and we have no assurance that another glaciation is not in prospect. Even the areas too far south to be glaciated have undergone great climatic changes through the ages, with corresponding changes in vegetation. Organic evolution also affects the composition of plant communities. Many land habitats available during the Paleozoic era must have been physically similar to habitats existing today, but their vegetation was very different, because the angiosperms had not yet come into existence.

When time is considered in terms of decades or centuries, some plant communities are stable, perpetuating themselves without significant change, whereas others are transitory, giving way to a succession of other communities before a relative stability is achieved. Even the most stable communities are actually in a delicately balanced and constantly shifting equilibrium, as old organisms die and are replaced by new ones. The weather is never quite the same for two years in a row, and these year-to-year differences favor first some, then other elements of the community.

Replacement of one plant community by another on the same site with the passage of time has been observed by many people throughout past ages. The first man to observe and report an orderly, predictable succession leading to a stable community was apparently Anton Kerner (1831–1898, Fig. 31.5), an Austrian botanist whose paper in 1863 on the vegetation of the Danube basin is a neglected landmark of botany. Kerner's concepts of plant succession were adopted by some of his European contemporaries and successors, but it remained for two American botanists, Henry C. Cowles (1869–1939) and Frederic E. Clements (1874–1945), to bring them to full flower. Cowles' study

Fig. 31.5   Anton Kerner (1831–1898), Austrian botanist, father of the climax concept. [New York Botanical Garden photo.]

(1899) of the vegetation of the sand dunes along Lake Michigan greatly influenced botanical thought, especially in America, and helped to inspire Clements to develop the concept of the regional **climax,** the climatically determined, stable vegetation type toward which all successional types in a region lead.

Plant succession is most conspicuous and most easily observed when a bare area is newly made available for plant colonization. On dry land this happens after retreat of a glacier, after volcanism, after severe fire, and when a fallow field is abandoned. In water it happens when a lake or pond is formed behind a natural or man-made dam.

A series of plant communities starting on dry land (especially on bare rock) is known as a **xerosere,** or **xerarch succession,** and one starting in water is a **hydrosere,** or **hydrarch succession.** In each case the habitat becomes progressively more mesic (moist; neither very wet nor very dry) during the course of the succession.

In theory, the stable plant community achieved after a succession of transitory communities is determined by the climate alone, regardless of the nature of the area when first exposed. Thus all xeroseres and all hydroseres in a given region should lead to the same climatically determined climax vegetation, given sufficient time.

In actuality, various factors commonly intervene to prevent or substantially delay the realization of this theoretical ideal, and there are often two or more apparently stable vegetation types on different soils or on different exposures in the same region. The soil and moisture conditions, and therefore the vegetation, of a flood plain along a river are permanently different from those of the surrounding hills in terms of human, rather than geologic,

time. In regions where moisture is a critical factor, north or east slopes may be forested, whereas south or west slopes are covered with grass or sagebrush, with a sharp dividing line between the two types along the crest of a ridge (Fig. 31.6).

Xerosere

A typical xerosere begins with bare rock and leads to the climax community. The genera and species differ according to the climate and other conditions, but the broad picture is much the same in all climates. Crustose and foliose lichens (Fig. 31.7) are ordinarily the first inhabitants of the rock. These pioneers cause corrosion of the rock surface and catch a little dust from the wind. This first stage in the xerosere may last a very long time, especially in dry regions where the lichens must remain dormant for most of the year, growing only during the short periods when water is available.

Eventually disintegration of the rock and accumulation of small particles of debris progress to the point at which mosses can get a foothold among the pioneer lichens. Fruticose lichens may become established in the moss mat or among the crustose and foliose lichens. The accumulation of wind-blown dust and the decay of older parts of the plants in the moss mats lead to the formation of a thin layer of soil on the rock surface. Sooner or later small seed plants, especially annuals, become established in the moss mats, and the thin cover of soil gradually thickens. Then perennial herbs enter the community. In forested regions, the stage with herbs growing in the moss mat is generally followed by a shrub stage, and the shrubs in turn give way to trees. The first trees may be the climax domi-

Fig. 31.6   Forest-grass ecotone in northeastern Oregon. [U.S. Forest Service photo by Melvin H. Burke.]

nants, or they may be pioneers that later give way to other species before the climax is reached. In drier regions the herb or shrub stage may be final.

The moss mat, which played a vital role in the earlier stages of succession, is broken up and disappears during the later stages. In the climax community, mosses are largely restricted to special habitats such as the bark of trees, fallen logs, boulders that have not yet succumbed to the soil-building process, and so on.

Crevice plants also play a role in many xeroseres. Enough dust and debris may accumulate in a rock crevice, without the intervention of lichens and mosses, to permit the direct establishment of herbs, shrubs, and even trees. These species in turn give way to others as a mantle of soil is built up.

Completion of a xerosere may take from a few decades to many thousands of years, depending on the circumstances. On cliff faces and steep slopes, the early successional stages may be prolonged indefinitely as new bare surfaces continue to be formed by erosion. In general the process comes to a conclusion most quickly in

Fig. 31.7 Foliose lichens on a rock in Massachusetts.

moist regions, especially if these are also warm.

It will be seen that the critical factor in the xerosere is the formation of soil. The soil is necessary to provide anchorage for the plants and more especially to hold a supply of ground water that the plants can absorb throughout the growing season. As the depth and quality of the soil increase, larger and larger plants invade the habitat, overtopping and crowding out their predecessors. *Each stage in the series prepares the way for the next, bringing about its own downfall.*

Hydrosere

A hydrosere starts with open water and leads toward the climax community. All lakes and ponds are transitory; if other forces do not destroy them first, they are sooner or later filled by mineral sediments and organic remains. Given enough time,

the former lake surface becomes dry land that supports the climax vegetation.

Several stages in a hydrosere can often be seen as concentric zones around a pond or lake. The deep water in the middle, subject to disturbance by the wind, may support only plankton algae. In the shallower water near shore, there is a zone of submerged aquatics, usually with slender or dissected leaves. These plants may be rooted to the bottom, but often they are free floating, like the bladderwort (*Utricularia*). Shoreward from the submerged aquatics is a zone of aquatics with floating leaves, such as water lilies (Fig. 31.8). These are commonly rooted in the soil at the bottom, and their advance toward the middle of the pond is limited by the depth of the water. Their large leaves shade out the submerged aquatics. Nearest the shore there is a zone of emergent aquatics, such as cattails; these overtop and crowd out the floating-leaved aquatics. Moisture-loving shrubs or trees grow along the shore, and on the higher ground above the pond there is the climax vegetation.

As the sediments and organic remains accumulate in the pond, the water becomes more shallow, and each vegetation zone moves in toward the center. A trench dug along the shore will often disclose, at progressively deeper levels, remains from each of the earlier successional stages.

A special feature of the hydrosere in many lakes of glaciated regions is the sedge mat (Fig. 31.8), with shoots of sedges and other plants growing out of a densely tangled mat of rhizomes, peat, and debris. The inner margin of the sedge mat floats on the surface of the water, often for a distance of some yards, and encroaches progressively on the open water. A man can walk on the floating part

Fig. 31.8 A floating sedge mat in Michigan, with yellow water lily, *Nuphar*, a floating-leaved hydrophyte, in the open water. [Photo courtesy of Herman F. Becker.]

with caution, but he may also break through. Toward the shore, the sedge mat is invaded by bog shrubs, which in turn give way to bog trees. The climax forest follows behind these.

In tropical and subtropical regions the water hyacinth (*Eichhornia crassipes*, Fig. 31.9), an aquatic with the roots hanging free in the water and the leaves serving as floats, is a familiar feature of the hydrosere, often covering ponds and choking slow-moving streams.

It is often useful to think of a hydrosere in terms of progressive **eutrophication** of the lake or pond. Initially, the lake has a relatively low content of dissolved minerals and supports a relatively sparse growth of plankton algae to a depth of

several meters, as far as the light can penetrate strongly enough to permit photosynthesis. Algal photosynthesis keeps the oxygen content of the water at all it can hold. There is a normal food chain from plankton algae to zooplankton and eventually to fish, but none of these organisms is conspicuously abundant, and the water is clear. The lake is **oligotrophic** (Greek *oligos*, few, little, and *trophia*, nutrition).

If the lake or pond has little or no outflow, the supply of minerals carried in by drainage water gradually accumulates, and plant growth increases to the point at which respiration does not keep pace with production. Algae grow so abundantly at the surface that they shade out the

Fig. 31.9 Cows grazing on water hyacinth, *Eichhornia crassipes,* in a pond in Mexico.

plants of deeper water, and much of their own product of photosynthetic oxygen is lost to the air. Floating aquatic angiosperms mingle with the algae in covering the surface. The supply of dissolved oxygen in the deeper water is largely used up by decay bacteria, which still do not keep pace with the settling of dead material from above. The depletion of oxygen near the bottom, resulting from the decay of materials settling from above, reduces or eliminates the population of fish and other animals at that level. The water becomes murky with algae and decay products, and the pond fills in at an increasing rate. The pond has become **eutrophic.**

Eutrophication, like erosion, is a normal process. It is the vast acceleration of these processes by humans under some circumstances that upsets balances and makes life difficult. Accelerated eutrophication is discussed on a subsequent page.

Eutrophication may be diverted or short circuited under some circumstances, even in the absence of disturbance by man. In arid regions, especially, the dissolved minerals may accumulate to a point at which they inhibit algal growth, and the salt or alkali lake becomes nearly sterile.

A hydrosere may take from a few decades to many thousands of years, depending on the depth of the water, rate of sedimentation, and other factors. A small oxbow lake along the Mississippi may be filled in during the lifetime of a man, but we need not fear an imminent disappearance of Lake Superior.

Disturbance and Secondary Successions

The orderly progress of a xerarch or hydrarch succession may be interrupted or reversed by natural or artificial forces, with the vegetation reverting to an earlier successional stage or being wholly destroyed. Agriculture, fire, and overgrazing are three of the most important such disturbing factors. Successions that begin on soil after the natural vegetation has been destroyed are called **secondary successions,** in contrast to xeroseres and hydroseres, which are **primary successions.**

The Old-Field Succession

The old-field succession is a common secondary succession. The first plants to invade an abandoned field are usually annual weeds. Such plants depend on their ability to invade rapidly any soil not already vegetated. They generally produce large amounts of seed and often have especially effective means of distribution. In parts of the southeastern United States, horseweed (*Conyza canadensis*, Fig. 31.10) is a prominent inhabitant of newly abandoned fields. The single-seeded fruits of horseweed are tiny and have a tuft of hairs (a modified calyx) at one end that helps them to float in the breeze. In drier areas farther west, tumbleweeds, such as Russian thistle (*Salsola kali*), are common first invaders. A full-grown tumbleweed may blow along for miles in the fall, scattering seeds as it goes.

Although they "get there firstest with the mostest," the success of the annual weeds is only temporary. After one to several years, they begin to give way to perennial herbs. These typically have less effective means of seed distribution but are better competitors than the annuals. The perennial herb invaders in turn give way, usually within a few years, to other plants, either trees, shrubs, or other perennial herbs.

An old-field succession may reach its conclusion in as little as a decade, under the most favorable conditions, or may take more than a century. Establishment of the climax vegetation, whatever it may be, is of course hastened if the climax species still persist nearby. Climax species are more noted for their competitive ability than for rapid seed dispersal. The critical factor in the old-field succession, in contrast to the xerosere, is seed dispersal.

Fire

Fire is one of the most important catastrophic influences on vegetation. Natural fires, set by lightning, have occurred sporadically for millions of years. Furthermore, accidental and intentional fires were as much a product of aboriginal human societies as they are of our modern civilization. We have noted that in pre-Columbian America fire doubtless held the boundary between the forest and the prairie some distance east of where it would otherwise have been. Likewise in the Congo region the tropical forest would extend much farther south, onto present savanna land, had it not been for repeated fires set by the natives.

The effects of fire vary according to the severity of the fire, the nature of the plant community, and the habitat. A severe forest fire may destroy the whole biotic community, and if the soil is thereafter washed away, the area may never (in terms of human time) recover. In other circumstances some individuals (or seeds) survive, and the fire favors some elements of the community at the expense of others. In the Rocky Mountains, lodgepole pine (*Pinus contorta* var. *latifolia*) is known as a fire tree, which comes in rapidly after fire, but which eventually gives way to other trees in the absence of fire. The cones of lodgepole pine remain unopened on the tree for many years; they do not burn easily, but the heat of the fire

Fig. 31.10 Stages in the old-field succession in eastern United States. (*Above*) In Virginia, the annual weed stage, with horseweed, *Conyza canadensis* (the tall plants); rabbits-foot clover, *Trifolium arvense*; and crabgrass, *Digitaria* as dominant elements. (*Below*) A later stage, in Delaware, with loblolly pine, *Pinus taeda*, coming up in mixed herbaceous vegetation of goldenrod, *Solidago*; Queen-Anne's lace, *Daucus carota*; black-eyed Susan, *Rudbeckia hirta*; blackberry, *Rubus*; and other plants.

makes them open. The seeds released onto the bare ground after a fire find ideal conditions for their growth, and a new forest of lodgepole pine is established. Other pines are likewise fire trees in much of the southeastern United States.

The great Douglas fir (*Pseudotsuga menziesii*) forests of the Pacific Northwest (Fig. 31.11) owe their existence to fire. Douglas fir seeds, like those of lodgepole pine, germinate best in a mineral soil with little or no humus, and the seedlings do not tolerate shade. Western red cedar (*Thuja plicata*) and western hemlock (*Tsuga heterophylla*) tend to replace the Douglas fir in the absence of fire. Young trees of these species are much more shade tolerant than the Douglas fir, and the seeds germinate well in a soil rich in humus. Douglas fir, having a thicker, more fire-resistant bark than the cedar and hemlock, is more likely to survive a forest fire, and conditions after the fire are more favorable to the germination of its seeds and establishment of its seedlings.

Overgrazing

Overgrazing has had disastrous effects on much of the range land in the western United States and elsewhere in the world. An early effect of overgrazing is a decrease in abundance of the more palatable plants and an increase in the less palatable ones. More severe or long-continued abuse of the range forces livestock to eat the less palatable plants, including even poisonous ones. The plant cover diminishes, and weedy annuals become common (Fig. 31.12).

When the plant cover is diminished or destroyed, rain water tends to run off quickly, instead of sinking in. The periodic floods that are a natural feature of the land are exaggerated, and the streams may dry up in the summer. The water running off the ground carries some of the soil with it, and erosion is thus accelerated. In hilly or mountainous regions, an erosion pavement of pebbles and stones may cover the ground surface after the upper part of the soil is washed away. A torrential rain may then cause a mud flow with a semiliquid mixture of water, dirt, rocks, and large boulders rampaging down a canyon and spilling onto the plain below. Such mud-flows caused great damage in parts of Utah and California, especially during the 1930s, forcibly demonstrating the need for grazing control and stimulating crash programs to restore the vegetation in some regions.

Pollution

Pollution of the water and air have become increasingly important factors in upsetting natural vegetation balances in recent decades. One type of water pollution that is currently receiving much attention results in accelerated eutrophication of lakes and a variety of unpleasant associated consequences. The leaching and runoff of nitrates and other fertilizers from cultivated fields, combined with sewage disposal and the discharge of phosphates from detergents in household water, fertilize the streams and lakes so much that algal growth increases tremendously. The algae may even form an interwoven surface mat. Then the algae themselves are poisoned by their own wastes and they die. The water becomes thoroughly unpleasant, fit only to float boats on their way somewhere else. If the pollutants, especially sewage, are too concentrated, even the algae cannot grow, and all the oxygen in the water is used up by decay bacteria working on the sewage. Lake Erie, which is relatively shallow

Fig. 31.11 Mature Douglas fir, with an understory of hemlock, in western Oregon. [U.S. Forest Service photo by Leland J. Prater.]

Fig. 31.12   Results of overgrazing in the Uinta National Forest, Utah. The area to the left of the fence, heavily grazed, supports only small, scattered, weedy annuals; the fenced area at the right has a dense cover of perennial wheatgrass, *Agropyron trachycaulum*. [U.S. Forest Service photo by Howard K. Orr.]

(averaging less than 20 meters deep), has suffered particularly from eutrophication and other effects of pollution, and is now often hyperbolically referred to as a dead lake. Recent efforts to reduce the rate of eutrophication of waters in the United States have only begun to be useful. The ban on phosphate detergents in many areas has led to the use of other detergents, some with even less pleasant effects.

Thermal pollution is also playing an increasing role in American streams and lakes. Certain types of industrial facilities, most notably atomic power plants, find it convenient to use river water as a coolant. The hot water they turn back into the

stream is unsuitable for the growth of most kinds of organisms, and the water temperature remains unnaturally high for some miles downstream. Conversely, Lake Powell, behind Glenn Canyon Dam on the Colorado River, is so deep that the temperature of the subsurface water adjusts to that of the earth, and in summer the water emerges from the base of the dam at a much lower temperature than it entered. The consequences of this change have not yet been fully explored.

Many of the industrial pollutants of the air are captured and brought down by rain. Sulfur dioxide is one of the most ubiquitous industrial pollutants. Its direct effects on plant growth have been noted in

Chapter 26, but it also has indirect effects. Sulfur dioxide reacts with ozone and other atmospheric constituents, and is oxidized to form sulfates and eventually sulfuric acid, which is brought down in rain. This **acid rain** may fall hundreds of miles from the original source of the sulfur dioxide. The pH of water in streams and lakes is drastically lowered, often to as low as 5 (which indicates 100 times as many hydrogen ions as the pH 7 of pure water) or less, with a series of unpleasant consequences. One of the effects is to increase the solubility of aluminum ions. Although some kinds of plants do well in a high-aluminum environment, and even accumulate aluminum in their cells, most are poisoned by it. One of the physiological effects of aluminum ions is to harden the plasma membrane and reduce its permeability. The equally serious effects of acid rain on aquatic animal life are not here discussed.

Over a period of more than 200 million years until about two centuries ago, the earth may have had a relatively balanced atmospheric system, with about as much carbon dioxide being used in photosynthesis as was formed in respiration. The preservation of unrespired plant remains in the form of coal and oil was a continuing force toward decrease in carbon dioxide content of the air, but the release of carbon dioxide in volcanic activity was a force in the opposite direction. Formation of limestone (using carbon dioxide in the production of calcium carbonate) and its subsequent erosion and solution (releasing carbon dioxide) doubtless had some effect on balances, but in the long term these two reciprocal processes cancel out. The oceans, which in all hold about 60 times as much carbon dioxide as the atmosphere, form a permanent if slow-acting buffer against atmospheric changes.

Such, at least, is a commonly held current view. If there were any great fluctuations in the carbon dioxide content of the air during the time under consideration, we do not know about them.

The industrial revolution, beginning about two centuries ago, has progressively and now drastically shifted the balance, releasing much more carbon dioxide into the air (through the burning of coal and oil) than is used by green plants. The amount of carbon dioxide in the air has increased from about 290 parts per million (ppm) (0.029 percent) in 1850 to about 335 ppm in 1981, and the amount continues to increase by more than 1 ppm per year. It appears that the oceanic buffer is unable to keep up with the rate of change.

The effect of the increase in carbon dioxide content of the air up to the present is uncertain. There is no doubt that carbon dioxide in the air tends to act as a blanket, permitting sunlight to enter, but restricting the reradiation of heat. If we continue to use fossil fuels at present rates, the increase in carbon dioxide content of the air might cause a net increase of 2–3°C in mean annual temperature of the earth within 50 to 100 years. On the other hand, the particulate matter that is released into the air by industrial processes has the opposite effect. Furthermore, we are as yet unable to measure accurately the variation in amount of energy put out by the sun. The mean annual temperature of the earth increased by about 0.6°C between 1880 and 1940, but has since decreased by about 0.2°C. The relative significance of carbon dioxide content, particulate matter, and variation of solar radiation in these changes is still uncertain. We may wonder if the effects of the additional carbon dioxide and the particulate matter are now more or less canceling each other out, but in fact we do not know.

CLIMATIC CONTROL OF PLANT COMMUNITIES

The angiosperms, which dominate most of the land vegetation of the earth, evidently originated as a group of woody plants in moist tropical regions. The tropics still harbor by far the greatest number of genera and species of angiosperms. A great many species, with comparatively few individuals of each, commonly enter into tropical climax communities.

As one travels toward the poles, the number of species in the community tends to decrease, and (short of the most barren tundra) the number of individuals of each species increases. In temperate and boreal regions, it is often possible to pick out only a few species, or even a single species, as the dominants that give the community its essential character.

The most important variable factors in the physical environment of land plants are moisture and temperature. Direct resistance to low temperature is governed primarily by physiological features of the protoplasm which find little, if any, expression in the external morphology of the plant. The number of species of land plants that cannot withstand frost is much greater than the number that can, and the decreasing total number of species as one travels away from the tropics is largely a reflection of the difficulty of evolutionary adaptation to low temperature; but frost-resistant plants do not look obviously different as a group from frost-sensitive ones. Temperature does affect the aspect of the vegetation, but the effects may be indirect and are not always immediately obvious.

Adaptation to moisture conditions, in contrast to adaptation to low temperature, commonly does find a morphological expression, although the physiology of course is also involved. The size and longevity of the plant, and the size, shape, number, texture, duration, and position of the leaves are all more or less closely correlated with adaptation to moisture conditions. As we pointed out in Chapter 22, moisture conditions are also influenced by temperature; even winterkilling is often due to physiological drought, rather than directly to low temperature. The aspect of the natural vegetation provides a good key to moisture conditions, and to some extent also to temperature conditions, in any region.

Types of Vegetation

Most of the major formations of land vegetation can be classified as **forest, grassland, desert,** or **tundra.** The borders between these types, especially the first three, are not always sharp, and there are transitional types. **Savanna** (see Fig. 31.17), with widely scattered small trees, is intermediate between forest and grassland.

The tundra vegetation covers arctic and alpine areas that are too cold to permit forest growth. Of course, other factors such as soil and wind enter into the matter, but it is the long, cold winter and the short, cool growing season which put the plants at such a disadvantage that other factors can turn the balance.

The most characteristic features of the arctic tundra are the absence of trees and the presence of **permafrost** (permanently frozen soil) at a depth of a decimeter or so. to a meter or so. The depth to which roots penetrate is governed by the depth to which the ground thaws in the summer. Much of the ground surface is spongy or hummocky (Fig. 31.13) because of the winter freezing and summer thawing. In the moister areas the ground often cracks into rough polygons several meters across, the cracks being filled with deep ice wedges that thaw at the surface in the summer.

Perhaps the most characteristic single genus of the arctic tundra is *Cladonia,* the "reindeer moss," actually a fruticose lichen commonly about a decimeter tall. Among the true mosses, *Polytrichum* (hairy-cap moss) is common in dry places, and *Sphagnum* (peat moss) in wet ones. Sedges (family Cyperaceae) and grasses of various sorts are common, and there are many genera of herbaceous and shrubby dicots, in diverse families.

To the south, the arctic tundra gives way to the boreal forest in the lowlands and merges with the alpine tundra in the mountains. The alpine tundra (Fig. 31.14) resembles the arctic tundra at first glance, but there are significant differences. Alpine tundra is not generally underlain by permafrost, and the topography is generally rougher, so that much of the land is better drained. Mosses and lichens are much less prominent parts of the vegetation, and as one goes southward a progressively larger number of the species and even genera are derived from nearby taxa of lower elevations.

Outside of tundra regions, any area with adequate moisture usually supports some kind of forest community. The potential evaporation, as well as the amount of precipitation, is important in governing the moisture conditions. The most generally useful index to moisture conditions is the *p/e* ratio, in which *p* is the annual precipitation, and *e* is the potential annual evaporation from a free water surface. In general, a *p/e* ratio of about 1 will permit forest growth.

As the *p/e* ratio decreases, forest commonly gives way to grassland, or sometimes directly into desert. Grassland (Fig. 31.15) generally requires not only more moisture than a desert, but also a better seasonal distribution; there must be a fair supply of water through a reasonably long growing season. In temperate and warm-temperate regions with the rainfall mostly

Fig. 31.13 Hummocky tundra along the arctic coast of Norway.

Fig. 31.14   Alpine tundra and bits of timberline forest in the Beartooth Mountains on the Montana-Wyoming border. [U.S. Forest Service photo by Bluford W. Muir.]

concentrated in the winter months, one often sees an evergreen, sclerophyllous woody vegetation, passing into deciduous forest at the moister extreme, and through a chapparal (shrubby) community into desert at the drier extreme. Parts of California, the Mediterranean region, southern Africa, and southwestern Australia share a common climate (referred to as a Mediterrean climate) that fosters this sort of community.

Desert vegetation (see Chapter 22 opening photograph) consists chiefly of very deep-rooted plants (often shrubs), very shallow-rooted plants (often succulents, such as cacti), and ephemerals (see Fig. 22.7). The deep-rooted plants draw on a moisture supply not available to the others; the very shallow-rooted plants can benefit from even the lightest rain but require such potent means of transpiration control as to restrict also the rate of photosynthesis and growth; and the ephemerals complete their life cycle in a few weeks or months during the most favorable part of the year. Desert shrubs such as *Larrea* (see Fig. 22.8), which hold their small, firm leaves throughout the year, are very resistant to the effects of desiccation.

Most grasses require less moisture than most trees, but they are not notably deep

Fig. 31.15 Short-grass prairie in northwestern South Dakota.

rooted, and they generally lack the special adaptations of desert plants. If a fair amount of moisture is available during the growing season, the grasses blanket the ground and soak up the rain as it falls. They grow faster than the shallow-rooted desert plants, and they do not let much moisture sink to the level worked by the deeper-rooted ones. The desert ephemerals are also crowded out, finding no bare ground to get started in. Grasses also do well under moisture conditions suitable for forests, if the trees are kept down by fire or some other means. Forest and desert communities include some grasses, of course, but not as dominant elements.

We have noted that in regions of adequate moisture and temperature the climax land vegetation is generally a forest. In moist, tropical regions without a pronounced dry season, this forest consists of evergreen angiosperms. In tropical regions with alternating wet and dry seasons, the trees typically drop their leaves at the beginning of the dry season. To the north, winter is in effect the dry season, because the ground water is likely to be frozen and not readily available for absorption. This environmental challenge is likewise met by adoption of the deciduous habit, as long as there is an adequate growing season.

As one travels still farther north and the growing season gets progressively shorter, the deciduous habit becomes progressively less adequate to meet the challenge. To produce a new crop of leaves each year, only to lose them after a brief growing season, is a tremendous waste of energy. In these colder regions the coniferous forests come into their own. The trees are evergreen, and each season's crop of leaves lasts several years. Photosynthesis is not restricted to the brief summer but occurs on favorable days throughout the year. The firm, needlelike leaves withstand desiccation in the winter as ordinary leaves would not. (The relation of the sclerophyllous habit to drought resistance is discussed in Chapter 22.)

The conifers are not restricted to the far northern woods, but this is the largest region in which their features give them a competitive advantage over angiosperms. Elsewhere conifers are mostly found in areas a little too dry for broad-leaved trees, especially where there is a summer drought, or in areas subject to repeated fire, or in the mountains at altitudes where conditions approach those of the north woods. Each species, of course, has its own set of requirements and competitive abilities, and sometimes these permit conifers to grow intermingled with angiosperm trees, as in the beech-maple-hemlock forest of New England.

As we have noted in Chapter 30, the most important single feature that gives angiosperm trees a competitive advantage over coniferous trees in so much of the world is the presence of vessels in angiosperm wood. The vessel-bearing wood of angiosperms permits a more rapid delivery of water to the leaves than does the slow-acting tracheid-system of conifers. The angiosperms can thus expose a larger transpiring (and photosynthetic) surface to the sun without suffering fatal desiccation.

The effect of similar climate in producing vegetation of similar aspect in different parts of the world is evident to any observant traveler. The similarity of the vegetation of much of Armenian U.S.S.R. to that of much of the Great Basin region of the United States (Fig. 31.16) is obvious, and the savanna vegetation of parts of Tadjikistan, in south-central U.S.S.R. is very reminiscent of that in parts of south-

ern Arizona. Likewise the forests at middle altitude in the Caucasus mountains of Georgian U.S.S.R. look much like the forests of parts of the southern Appalachian Mountains in the United States, even to having a show of flowering herbs in the spring before the leaves of the trees have expanded.

What may not be so immediately evident is that plants of similar aspect may be of very different evolutionary affinity. Although the junipers (*Juniperus*) and sagebrushes (*Artemisia*) of Armenia belong to the same genera as the junipers and sagebrushes of the Great Basin, the *Pistacia* trees (Anacardiaceae, order Sapindales, subclass Rosidae) (Fig. 31.17) in the savannas of Tadjikistan have nothing to do with the oaks (*Quercus*, family Fagaceae, order Fagales, subclass Hamamelidae) of the savannas of southern Arizona.

The Cactaceae and Euphorbiaceae provide an even more striking example. The Cactaceae are an essentially American family, chiefly of desert regions, notable for their essentially leafless, succulent, photosynthetic stems that are protected from grazing animals by sharp spines. In African deserts the ecological role of the cacti is played by certain species of the large and ecologically diversified genus *Euphorbia*, in the family Euphorbiaceae, of quite different taxonomic affinity (Fig. 31.18).

These taxonomic differences among plants of similar appearance in different parts of the world reflect the evolutionary history of the vegetation and the taxonomic groups. Northern Eurasia and northern North America provide a nearly continuous pathway of migration for plants adapted to the climate, but farther south the climatically similar regions may be separated by oceans or broad stretches of

Fig. 31.16 Juniper-sagebrush community in Armenian U.S.S.R. (*above*) and northern Utah (*below*); *Juniperus excelsa* and *Artemisia fragrans* in Armenia, *J. osteosperma* and *A. tridentata* in Utah. [Lower photo courtesy of Arthur H. Holmgren.]

Fig. 31.17 Savanna community in western Tadjikistan, U.S.S.R. (*above*) and Arizona (*below*); *Pistacia vera* and *Acer regelii* in Tadjikistan, *Quercus oblongifolia* and *Q. emoryi* in Arizona.

Fig. 31.18 Convergent evolution to produce ecologically similar types. (*Left*) A species of *Cereus* (Cactaceae, subclass Caryophyllidae) in Mexico. (*Right*) A species of *Euphorbia* (Euphorbiaceae, subclass Rosidae) in Uganda. [*Euphorbia* photo courtesy of B. L. Turner, shown in foreground.]

land with very different climates. Therefore one finds marked floristic differences between ecologically similar communities in different regions.

When seeds are accidentally or deliberately carried from their native region to an ecologically similar one elsewhere in the world, the species often become established as an integral part of the flora. It is now difficult to be sure whether some species in southeastern Canada and adjacent United States are native there or represent very early post-Columbian introductions from Europe.

Weedy species, which quickly exploit disturbed habitats, are especially likely to do well in a new home ecologically comparable to their old one. We have noted that California has a Mediterranean climate, and it should be no surprise that many of the common weeds in California have been introduced from the Mediterranean region, where human disturbance over thousands of years has fostered the evolution of weeds. The traffic is not all one way. Europeans would gladly forego the cocklebur (*Xanthium strumarium*), which is of American origin.

Plant diseases may have dramatic and devastating effects when introduced from one part of the world to another. The American chestnut (*Castanea dentata*), once an important member of the deciduous forest community in much of eastern United States has been virtually eliminated by the chestnut blight, introduced from the Orient about 1904 (Fig. 31.19). The American elm (*Ulmus americana*) may also be on the way to destruction as a forest and street tree, because it is highly susceptible to the "Dutch" elm-disease, actually also of Oriental origin. In both cases the Oriental species are more resistant to the disease, although some individuals do succumb.

Plant geographers often find it conve-

Fig. 31.19   Mixed hardwood forest dominated by species of oak, *Quercus*, in former oak-chestnut region in the southern Appalachian mountains near Highlands, N.C. Dying chestnut snag in foreground.

nient to think in terms of floristic provinces, defined in terms of the taxonomic composition as well as the ecological aspect of the dominant regional communities.

A map showing the floristic provinces of the continental United States and Canada is presented in Fig. 31.20. The names of some of the provinces indicate the nature of the dominant vegetation. The Coastal Plain province has extensive pine forests (see Fig. 17.7), reflecting a history of repeated fires; in the absence of fires this area would probably have been angiospermous forest. The West Indian province is essentially tropical, with many

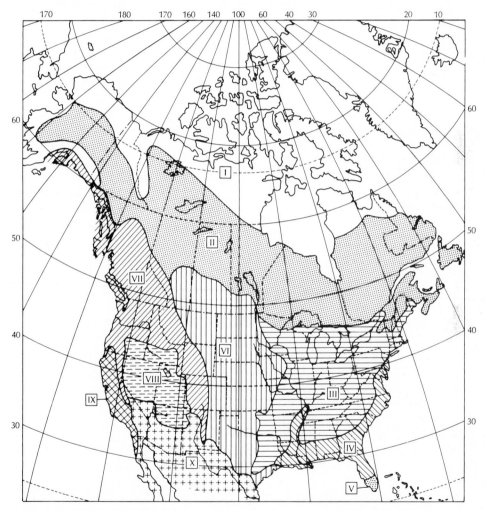

Fig. 31.20 The floristic provinces of the continental United States and Canada. I, Tundra province; II, Northern Conifer province; III, Eastern Deciduous Forest province; IV, Coastal Plain province; V, West Indian province; VI, Grassland province; VII Cordilleran Forest province; VIII, Great Basin province; IX, California province; X, Sonoran province. [From *The natural geography of plants*, by H. A. Gleason and Arthur Cronquist, copyright by Columbia University Press, 1964.]

frost-sensitive species. The Cordilleran Forest province is in some respects a southern extension of the Northern Conifer province, but with drier, grassy or semidesert intermontane valleys and plains toward the south. The Great Basin province is mainly cool desert, and the Sonoran province warm desert. The California province, with a Mediterranean climate, tends toward the broad sclerophyll vegetation, with an emphasis on the shrubby, chapparal type.

The Tundra and Northern Conifer provinces are essentially continuous with similar provinces in Eurasia, and the historical relationship of the Eastern Deciduous Forest province to a similar province in eastern Asia has been noted in Chapter 30. On the other hand, the Coastal Plain, West Indian, Grassland, Great Basin, Californian, and Sonoran provinces have relatively little in common, floristically, with ecologically similar provinces in the Old World.

EPILOGUE

As we have seen, the web of life is an intricate network of cause and effect, an endless intermingling of growth, predation, and decay, a myriad of cycles, each one potentially complete but none independent. But of all the organisms acting, reacting, and interacting, only man has the conscious means of breaking the web, of interrupting the cycle. With his power to manipulate his environment deliberately, man, alone among the organisms he depends on, bears a portentous responsibility. He must give careful thought to the far-reaching consequences of his behavior on the world ecosystem, and act accordingly if his civilization is to survive.

Man, like all other organisms, is a product of evolution. He is adapted to survive under the conditions in which he evolved. Until a few hundred or at most a few thousand years ago, man was not a very abundant animal. The environmental selective pressures that shaped his nature did not require him to conserve his environment; rather they called for him to bend the environment to his use. Our society, like its predecessors, reflects this basically exploitive attitude. There are differences in degree in different cultures, and our Western industrial society is more oriented toward exploitation than most, but it still does no more than reflect man's basically exploitive nature.*

Now the circumstances have changed. If we continue in our inherent ways, our civilization will certainly fall, although man as a species may well survive.

Our civilization itself may be compared to a species, just as a biotic community has been compared to an organism. We should recognize that we are making a comparison rather than asserting an identity, but the comparison is a useful aid to understanding.

As biologists we know the stringency of the evolutionary requirement of adaptation. No matter how well adapted a species may be to a particular set of conditions, it must change as the conditions change, if it is to survive. A few kinds of organisms, such as some of the photosynthetic bacteria, may find a restricted habitat that permits them to survive unchanged over billions of years, but for most species the story is entirely different. Sooner or later the habitat in which they arose changes enough so that they must change along with it if they are to survive.

*This and the three following paragraphs are taken almost unchanged from the paper "Adapt or die," by Arthur Cronquist, in *Bull. Jard. Bot. Nat. Belgique* **41**(1):135–144, 1971.

Ultimately, most of them face insurmountable evolutionary problems and become extinct. Our society, like a species, now faces a severe environmental stress, and it has the same alternative as a species: Adapt or die!

SUGGESTED READING

Bates, M., *The forest and the sea*, Random House, New York, 1960. Well written, provocative reflections of a mature biologist on broad-scale ecology.

Borman, F. H., and G. E. Likens, Catastrophic disturbance and the steady state in northern hardwood forests, *Amer. Sci.* **67**:660–669, 1979.

Daubenmire, R. F., *Plant geography, with special reference to North America*, Academic Press, New York, San Francisco, London, 1978.

Ehrenfeld, D., *Conserving life on earth*, Oxford University Press, New York, 1972. Literate, sound exposition emphasizing ecosystems and the importance of biological diversity to the maintenance of stability.

Gates, D. M., The flow of energy in the biosphere, *Sci. Amer.* **225**(3):88–103, September 1971.

Gleason, H. A., and A. Cronquist, *The natural geography of plants*, Columbia University Press, New York, 1964. Oriented toward North America, with many photographs.

Hardin, G., *The limits of altruism: An ecologist's view of survival*, Indiana University Press, Bloomington, 1977.

Hutchinson, G. E., Eutrophication, *Amer. Sci.* **61**:269–279, 1973.

Hutchinson, G. E., *The kindly fruits of the earth. Recollections of an embryo ecologist*, Yale University Press, New Haven, Conn., 1979.

Likens, G. E., R. F. Wright, J. N. Galloway, and T. J. Butler, Acid rain, *Sci. Amer.* **241**(4):43–51, October 1979.

Lorentz, W., The lessons of the dust bowl, *Amer. Sci.* **66**:560–569, 1978. Agriculture on the Great Plains is still not secure.

McCormick, J., *The life of the forest*, McGraw-Hill, New York, 1966. Nontechnical but accurate.

Martin, P. S., *The last 10,000 years*, University of Arizona Press, Tucson, 1963. Well written, focused on southwestern United States.

Singer, S. E., Human energy production as a process in the biosphere, *Sci. Amer.* **223**(3):174–190, September 1970.

Teal, J., and M. Teal, *Life and death of the salt marsh*, Little, Brown, Boston, 1969.

Chapter 32

Key to the Commoner Street and Yard Trees of the Contiguous United States

Ohio buckeye, *Aesculus glabra.* [U.S. Forest Service photo by W. D. Brush.]

1 Leaves slender, needlelike or scalelike, not more than about 3 mm wide, evergreen (for alternative, contrasting statement, see p. 601).

2 Leaves all needlelike, more than 1 cm long, and more or less spreading away from the twigs (for alternative, contrasting statement, see p. 600).

3 Needles (leaves), or most of them, in clusters.

4 Needles numerous in each cluster; cones unarmed, the cone-scales relatively thin and mostly falling away from the axis at maturity. **Cedrus. True cedar.**

There are 4 species of Cedrus, all sometimes cultivated: C. deodara (D. Don) G. Don, the Himalayan or Deodar cedar; C. libani A. Rich., the cedar of Lebanon; C. atlantica (Endl.) Manetti, the Atlas cedar; and C. brevifolia (Hook.f.) Dode, the Cyprus cedar. C. deodara, with drooping, densely short-hairy leader and twigs, and dark bluish-green needles mostly 2.5–5 cm long, is the most common and is illustrated.

4 Needles 2–5 in each cluster; cones woody, falling intact (or remaining on the tree), in many species each cone-scale with a prickle near the tip. **Pinus. Pine.**

Many species of this large genus are cultivated. The following 6 are perhaps the most common.

a Needles in sets of 5. **Pinus strobus L. Eastern white pine.**

a Needles in sets of 2 or 3.

b Needles all or mainly in sets of 3.

c Cones stout, only slightly longer than wide when open, tending to be borne near the branch tips; cultivated mainly in the western and plains states. **Pinus ponderosa Laws. Ponderosa pine.**

c Cones more slender, obviously longer than wide even when open, borne along the branches away from the tips; cultivated mainly in the southeastern states. **Pinus taeda L. Loblolly pine.**

b Needles all or mainly in sets of 2.

d Needles mostly 3–7 cm long; cones unarmed; bark of middle-sized trunks blistery-orange. **Pinus sylvestris L. Scotch pine.**

d Needles mostly 7–18 cm long; cone-scales each with a slender prickle on the back near the tip; bark various, but not blistery-orange.

e Young twigs pale and glaucous (covered with a powdery bloom that can be rubbed off); cones persistent on the branches for some years; bark of mature trees largely red-brown or yellow-brown; planted mainly in the southeastern states. **Pinus echinata Mill. Shortleaf pine.**

e Young twigs dark, not glaucous; cones falling soon after maturity; bark dark; planted mainly in the northeastern states. **Pinus nigra Arnold. Austrian pine.**

Fig. 32.1 *Cedrus deodara.* See also Fig. 17.8.

Fig. 32.2 *Pinus strobus.* See also Fig. 17.9.

Fig. 32.3 *Pinus sylvestris.*

3 Needles not clustered.

5 Seeds exposed, not borne in cones, fleshy, orange-red, berrylike; needles flat, sharp-pointed, with a short, twisted, slender base, spreading away from the twig in 2 ranks. **Taxus. Yew.**

Several closely related species of Taxus are cultivated. The most common is Taxus baccata L., the English yew.

5 Seeds borne in definite cones, dry, not berrylike; needles various, but not with characters combined as above.

6 Cones small, only 1–2 cm long; needles blunt, contracted to a short, twisted, slender base, tending to spread more or less in 2 ranks; leader nodding. **Tsuga canadensis (L.) Carr. Canadian hemlock.**

6 Cones larger, more than 3 cm long; needles various; leader erect.

7 Cones with an exserted, 3-parted bract behind each principal scale; needles with a slender, twisted base; winter buds shining red-brown, pointed. **Pseudotsuga menziesii (Mirb.) Franco. Douglas fir.**

7 Cone without exserted bracts behind the principal scales (at least in the commonly cultivated species); needles carrying their width to the base; winter buds dull brown, not pointed.

8 Needles quadrangular in cross section, sharp-pointed, tending to spread in all directions; twig with an elevated base supporting each needle; cones pendulous, falling intact. **Picea. Spruce.**

Several species of Picea are cultivated. The 2 most common are:

a Needles with a bluish cast; cones mostly 6–10 cm long. **Picea pungens Engelm. Blue spruce.**

a Needles dark green or somewhat yellowish-green; cones mostly 10–17 cm long. **Picea abies (L.) Karst. Norway spruce.**

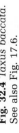

Fig. **32.4** *Taxus baccata.* See also Fig. 17.6.

Fig. **32.5** *Tsuga canadensis.*

Fig. **32.6** *Pseudotsuga menziesii.* See also Fig. 17.12.

Fig. **32.7** *Picea pungens (left)* and *P. abies (right).*

8 Needles flat, not sharp-pointed, tending to spread more or less in 2 ranks; twig not with elevated bases supporting the needles; cones erect, breaking up at maturity, the individual scales falling separately. **Abies. Fir.**

Several species of Abies are cultivated. The 2 most common are:

a Needles bright green, those of the lower branches 2–3 cm long; year-old twigs pubescent (provided with short hairs). **Abies balsamea (L.) Mill. Balsam fir.**

a Needles pale bluish-green, or with a whitish cast, those of the lower branches 3–6 cm long; year-old twigs glabrous (without hairs). **Abies concolor (Gord. & Glend.) Lindl. White fir.**

2 Leaves, or most of them, scalelike, more or less appressed to the twig, less than 1 cm long (leaves of young plants or of some branches may be more spreading.)

9 Leaves in well-separated whorls, the slender green twigs exposed for long distances between the whorls. **Casuarina. Australian pine.**

Casuarina is a highly specialized flowering plant, rather than a gymnosperm as the name might suggest. Several species are cultivated. The most common is C. equisetifolia J. R. Forst. & G. Forst., sometimes called horsetail tree.

9 Leaves opposite or ternate (whorled in sets of 3), clothing the twigs.

10 Twigs forming flattened sprays.

11 Cone-scales flattened, not stalked; leafy sprays large and conspicuous. **Thuja. Arbor vitae.**

The two most commonly planted species of Thuja are as follows:

a Twigs flattened in vertical planes; seeds wingless; cones mostly 2–2.5 cm long. **Thuja orientalis L. Oriental arbor vitae.**

a Twigs flattened in horizontal planes; seeds winged; cones mostly 1–1.5 cm long. **Thuja occidentalis L. American arbor vitae, or white cedar.**

11 Cone-scales peltate (umbrella-shaped or mushroom-shaped, with a stalk and a cap); leafy sprays small. **Chamaecyparis. False cypress.**

The three most commonly planted species are:

a Lateral leaves about the same size as the facial ones. **Chamaecyparis pisifera (Sieb. & Zucc.) Endl. Sawara cypress.**

a Lateral leaves much larger than the facial ones.

b Leaves acutish, with a conspicuous gland on the back. **Chamaecyparis lawsoniana (A. Murr.) Parl. Port Orford cedar.**

b Leaves obtuse, not glandular. **Chamaecyparis obtusa (Sieb. & Zucc.) Endl. Hinoki cypress.**

Fig. 32.8 Abies balsamea.

Fig. 32.9 Casuarina equisetifolia.

Fig. 32.10 Thuja occidentalis.

10 Twigs round or quadrangular, not forming flattened sprays.

  12 Cone woody, with peltate scales, opening at maturity, with many seeds under each scale.
  ***Cupressus* L. Cypress.**

  Several species are cultivated. One of the most common is *Cupressus macrocarpa* Hartweg, the Monterey cypress, with cones 2.5–4 cm thick.

  12 Cone dry but somewhat berrylike, not opening at maturity, with usually only 1–4 seeds in all, up to about 1 cm thick.
  ***Juniperus*. Juniper.**

  Several species of juniper are cultivated, most of them shrubby. Among those that reach tree size the two most common are:

  a Scale-leaves obtuse; juvenile leaves mostly ternate (whorled in sets of 3).
  ***Juniperus chinensis* L. Chinese juniper.**

  a Scale-leaves acute; juvenile leaves mostly opposite.
  ***Juniperus virginiana* L. Eastern red cedar.**

1 Leaves larger, not needlelike or scalelike, more than 3 mm wide, variously evergreen or deciduous.

  13 Plants with the characteristic growth habit of palms; trunk erect, unbranched, with a terminal crown of very large leaves; leaf blade, or its segments, with more or less parallel or gradually divergent main veins that are alternately elevated and depressed, as the ribs of a fan; plants all tropical or subtropical (for alternative, contrasting statement, see p. 603).
  Family Arecaceae. Palms.

  14 Fan-palms, i.e., the leaves nearly circular in outline, more or less palmately cleft.

    15 Leaves truly palmate, i.e., without a midrib; flowers, or most of them, unisexual; petiole toothed along the margins, but not spiny.
    ***Trachycarpus fortunei* (Hook.) H. Wendl. Windmill palm.**

    15 Leaves costapalmate, i.e., with a short but definite midrib; flowers perfect.

      16 Petiole smooth, unarmed.
      ***Sabal*. Palmetto.**

      Several species of *Sabal* are commonly cultivated in tropical America, but only two in the southern United States.

      a Leaves divided about two-thirds of the way to the main axis into deeply bifid segments; inflorescences usually surpassing the leaves.
      ***Sabal palmetto* (Walter) Lodd. Cabbage palmetto.**

      a Leaves divided about halfway to the main axis into simple segments; inflorescences shorter than the leaves.
      ***Sabal domingensis* Becc. Hispaniola palmetto.**

      16 Petiole spiny along the margins.

        17 Upper part of the trunk covered by a conspicuous shag of dead leaves.
        ***Washingtonia*. Washington palm.**

        There are only two species of *Washingtonia*, both cultivated in southwestern United States.

Fig. 32.11 *Juniperus virginiana.*

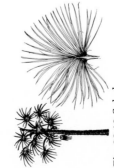

Fig. 32.12 *Sabal palmetto.*

a Leaf blades gray-green, without a tawny patch, provided with many long, slender, loose, filamentous fibers.

Washingtonia filifera **(H. Linden) H. Wendl. Desert fan-palm.**

a Leaf blades bright green, with a large tawny patch beneath, usually with only few or no loose fibers.

Washingtonia robusta **H. Wendl. Mexican Washington palm.**

17 Trunk without a conspicuous shag of dead leaves.

Livistona chinensis **(Jacq.) R. Br. Chinese fan-palm.**

14 Feather-palms, i.e., the segments, ribs, or veins pinnately arranged along a central axis or midrib.

18 Inflorescences borne beneath the leaves, at the base of the prominent crown-shaft, which is formed by the erect, closely appressed sheathing bases of the leaves.

19 Leaves mostly 1–2 m long; fruit mostly 3–5 cm long, orange to red.

Areca catechu **L. Betel palm.**

19 Leaves larger, mostly more than 2 m long; fruit smaller, up to about 2 cm long, dull dark red to purplish or black.

Roystonea. **Royal palm.**

Two species are frequently cultivated in southern Florida.

a Crown subglobose, the lower leaves drooping; primary segments of the leaf in 2 or more planes along most of its length.

Roystonea regia **(H.B.K.) O. F. Cook. Cuban royal palm.**

a Crown appearing nearly flat on the lower side; primary segments of the leaf all in about the same plane toward the base and tip of the leaf, but usually in 2 planes near the middle.

Roystonea oleracea **(Jacq.) O. F. Cook. South American royal palm.**

18 Inflorescences borne amongst the leaves; crownshaft mostly short or none.

20 Leaves unarmed, the petiole smooth, none of the segments transformed into spines; trunk bare.

21 Fruit very large, 20 cm or more thick, much larger than in other cultivated palms.

Cocos nucifera **L. Coconut.**

21 Fruit much smaller, less than 5 cm thick.

Arecastrum romanzoffianum **(Cham.) Becc. Queen palm.**

20 Leaves armed; trunk tending to be covered by old petiole bases.

22 Lower segments of the leaf transformed into long, stiff spines; petiole smooth.

Phoenix dactylifera **L. Date palm.**

22 Lower segments of the leaf, like the others, foliar, not transformed into spines; petiole with sharp, stout teeth along the margins.

Butia capitata **(Mart.) Becc. Jelly palm.**

Fig. 32.13 *Washingtonia filifera.*

Fig. 32.14 *Roystonea regia.*

Fig. 32.15 *Cocos nucifera.* See also Chapter 28 opening photograph.

13 Plants not palmlike in aspect, the trunk usually branched, the leaves mostly of more ordinary size.

23 Leaves compound (for alternative, contrasting statement, see p. 607).

24 Leaves opposite (for alternative, contrasting statement, see p. 604).

25 Leaves palmately compound with 5 or more leaflets. **Aesculus. Buckeye.**

Several species of Aesculus are cultivated. The following 3 are representative.

a Stamens long-exserted beyond the petals; fruit covered with small spines.

b Petals white, marked with red or yellow toward the base, the upper and lateral ones similar, with a broad blade that is rounded to a slender stalk, the fifth one (sometimes wanting) more tapering to the stalk.

Aesculus hippocastanum L. Horse-chestnut.

b Petals 4, greenish-yellow, gradually tapering to the base, the fifth (lowest) one wanting.

Aesculus glabra Willd. Ohio buckeye.

a Stamens equaling or shorter than the petals; fruit smooth.

Aesculus octandra Marsh. Sweet buckeye.

25 Leaves trifoliolate or pinnately compound.

26 Leaflets 3, seldom 5 on some leaves; fruit a double samara.

Acer negundo L. Boxelder.

26 Leaflets usually 7 or more, seldom only 5; fruit various, but not a double samara.

27 Flowers inconspicuous; fruit a single-winged samara, containing a single seed, not opening when ripe; leaflets 5–11; cultivated in temperate climates.

Fraxinus. Ash.

Several species of this rather large genus are cultivated. The 3 most common are:

a Fruit with a persistent small calyx.

b Lower surface of the leaflets glabrous (smooth, without hairs) and distinctly paler than the upper surface.

Fraxinus americana L. White ash.

b Lower surface of the leaflets pubescent (covered with short hairs) or glabrous, but not notably paler than the upper surface.

Fraxinus pennsylvanica Marsh.

The var. pennsylvanica, with the leaflets pubescent beneath, is called red ash. The var. subintegerrima (Vahl) Fern., with the leaflets smooth beneath, is called green ash. Both varieties are cultivated.

a Fruit (and the flowers as well) without a calyx.

Fraxinus excelsior L. European ash.

Fig. 32.16 Aesculus hippocastanum. See also Fig. 21.3.

Fig. 32.17 Acer negundo.

Fig. 32.18 Fraxinus americana.

27 Flowers large and showy; fruit a capsule (pod), opening when ripe to release the numerous small seeds; leaflets often more than 11; cultivated in tropical climates.

28 Leaves twice compound (the primary leaflets again compound); flowers blue. **Jacaranda mimosifolia D. Don. Jacaranda.**

28 Leaves once compound, with 9–19 leaflets; flowers bright orange-red. **Spathodea campanulata Beauvois. African tulip tree.**

24 Leaves alternate.

29 Leaves even-pinnate (without a terminal leaflet); fruit a pod of one or another sort.

30 Leaves once pinnate, with 10–18 pairs of leaflets up to about 2.5 cm long; flowers pale yellow, about 2.5 cm wide; pod woody, 5–15 cm long, constricted between the seeds and not opening; frost-sensitive tree, cultivated only in southernmost United States and in other tropical regions. **Tamarindus indica L. Tamarind.**

30 Leaves, or many of them, twice pinnate (the primary leaflets again pinnate).

31 Leaflets relatively large, mostly 5–8 cm long; pod woody, 8–15 cm long; hardy. **Gymnocladus dioica (L.) K. Koch. Kentucky coffee tree.**

31 Leaflets smaller, not more than about 4 cm long.

32 Trees usually armed with large, branched thorns on the trunk (but thornless cultivars are sometimes planted); some leaves once compound, with leaflets mostly 2–4 cm long, others twice compound, with leaflets 1–2 cm long; pods large, woody, mostly 10–30 cm long; hardy, widely cultivated. **Gleditsia triacanthos L. Honey locust.**

32 Trees unarmed; leaves all twice compound, with the leaflets not more than about 1 (1.5) cm long.

33 Midvein of the leaflets near the upper margin; flowers in nearly spherical heads, the petals not very conspicuous, the long filaments (of the stamens) making up the most showy part of the inflorescence; pods papery, mostly 8–15 cm long; not very hardy, cultivated mainly in the warmer parts of the United States. **Albizia julibrissin Durazz. Mimosa tree.**

33 Midvein of the leaflets central; flowers in elongate, slender inflorescences, the petals very large and showy; pods mostly 40–60 cm long; frost-sensitive tree, cultivated in southernmost United States. **Delonix regia (Bojer) Raf. Royal poinciana.**

Fig. 32.19 *Spathodea campanulata.*

Fig. 32.20 *Gleditsia triacanthos.*

Fig. 32.21 *Albizia julibrissin.*

29 Leaves odd-pinnate (with a terminal leaflet, but this sometimes poorly developed in *Ailanthus*); fruit of various sorts.

34 Leaves twice pinnate; fruit plumlike, less than 4 cm long; cultivated mainly in southern United States. **Melia azedarach L. Chinaberry tree.**

34 Leaves once pinnate, or in *Koelreuteria* some of them more or less distinctly twice pinnate; widely cultivated.

35 Leaflets entire (i.e., the margins smooth, not toothed); fruit a pod; flowers well developed, appearing in the summer.

36 Stipules modified into persistent, broad-based spines mostly 1–1.5 cm long; fruit a thin, flat pod, 6–10 cm long. **Robinia pseudoacacia L. Black locust.**

36 Stipules not modified into spines.

37 Leaflets small, mostly 3–5 cm long; pod cylindric, constricted between the seeds, up to about 10 cm long. **Sophora japonica L. Japanese pagoda-tree.**

37 Leaflets larger, at least the terminal one mostly 6–13 cm long; pod thin, flat, 7–10 cm long. **Cladrastis lutea (Michx. f.) C. Koch. Yellow-wood.**

35 Leaflets usually more or less strongly toothed (at least with 1 or 2 evident teeth toward the base); fruit of various types, but not a pod. (*Juglans regia*, with entire leaflets, differs from the foregoing group in the nature of the fruit and in its tiny, inconspicuous flowers, borne in catkins in early spring; it is keyed under the first leg of lead 42.)

38 Evergreen trees; leaflets relatively small, up to about 7 cm long and 1 cm wide; fruit a small, pea-sized, rose-colored stone-fruit; cultivated mainly in the southern United States. **Schinus molle L. Peruvian pepper-tree.**

38 Deciduous trees; leaflets often larger; fruit of diverse types, but not as above; widely cultivated.

39 Bark and twigs with the odor and taste of bitter almonds; fruits the size of a large pea, bright orange-red, with the structure of a miniature apple. **Sorbus. Mountain ash.**

Two species are commonly cultivated:

a Leaves persistently hairy beneath; winter buds densely hairy, not sticky. **Sorbus aucuparia L. European mountain-ash.**

a Leaves soon becoming glabrous (smooth) beneath; winter buds sticky, nearly glabrous. **Sorbus americana Marsh. American mountain-ash.**

Fig. 32.22 *Melia azedarach.*

Fig. 32.23 *Robinia pseudoacacia.* See also Fig. 21.3.

Fig. 32.24 *Schinus molle.*

Fig. 32.25 *Sorbus aucuparia.*

39 Bark without the odor and taste of bitter almonds; fruit quite different from that of Sorbus.

40 Leaflets entire or nearly so except for 1 or 2 large teeth near the base, these teeth (or 1 of them) with a prominent gland on the lower surface; fruit dry, remaining closed, winged, 3–5 cm long.

Ailanthus altissima (Mill.) **Swingle. Tree of heaven.**

40 Leaflets toothed for their whole length (except in *Juglans regia*), the teeth lacking prominent glands; fruits otherwise.

41 Leaflets relatively small, mostly 3–6 cm long; inflorescences appearing in early or midsummer, rather showy, yellow; fruit an inflated, 3-locular pod with 3 small black seeds.

Koelreuteria paniculata **Laxm. Golden-rain tree.**

41 Leaflets larger, mostly more than 6 cm long; inflorescences appearing in early spring, consisting of pendulous catkins without showy parts; fruit a nut enclosed by a fibrous or slightly fleshy hull.

42 Pith of twigs transversely partitioned; nut with a very rough, sculptured surface.

Juglans. **Walnut.**

Three species are commonly cultivated:

  a Leaflets nearly entire, mostly 7–9.

Juglans regia L. **English walnut.**

  a Leaflets evidently toothed, mostly 11–23.

  b Pith dark brown; bark with smooth ridges; fruit ovoid to short-cylindric; stellate hairs few or none.

Juglans cinerea L. **Butternut.**

  b Pith light brown; bark with very rough ridges; fruit subglobose; hairs, especially of the lower surface of the leaves, largely or wholly stellate (with several branches from a common base).

Juglans nigra L. **Black walnut.**

42 Pith not partitioned; nut smooth or nearly so.

Carya. **Hickory.**

Several species are cultivated, the following 4 perhaps the most frequently.

  a Leaflets mostly 11–17.

Carya illinoensis (Wang.) K. Koch. **Pecan.**

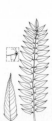

Fig. 32.26 *Ailanthus altissima.*

Fig. 32.27 *Koelreuteria paniculata.*

Fig. 32.28 *Juglans nigra.*

a Leaflets 5–9.

    b Leaflets persistently hairy beneath.

        Carya tomentosa (Poir.) Nutt. **Mockernut.**

    b Leaflets hairy beneath only when young, soon becoming glabrous or nearly so, sometimes with some hairs along the larger veins.

        c Bark soon separating into large, elongate, loose, relatively smooth plates.

            Carya ovata (Mill.) K. Koch. **Shagbark hickory.**

        c Bark rough and deeply furrowed, not separating into loose plates.

            Carya glabra (Mill.) Sweet. **Pignut hickory.**

23 Leaves simple, sometimes lobed or cleft, but without distinct leaflets.

43 Leaves fan-shaped, often cleft at the summit into two lobes, dichotomously veined (each vein repeatedly forking into two equal, very gradually divergent, nearly parallel veins, without cross-connections to other veins).

        Ginkgo biloba L. **Maidenhair tree.**

43 Leaves otherwise, neither distinctly fan-shaped nor dichotomously veined.

44 Leaves closely overlapping, evergreen, leathery, sharp-pointed; seeds borne in cones.

        Araucaria araucana (Mol.) C. Koch. **Monkey-puzzle.**

44 Leaves distinctly otherwise; seeds not borne in cones.

45 Leaves opposite, or whorled in 3s (for alternative, contrasting statement, see p. 609).

46 Leaves palmately lobed and veined; fruit a characteristic double samara (dry, closed, and with two broad wings). Many species of this large genus are cultivated. The following are among the most common.

    a Lower surface of the leaves gray or whitish, distinctly paler than the upper surface; wings of the fruit diverging at an angle of distinctly less than 120 degrees, often at about a right angle.

    b Leaves deeply 5-lobed to well below the middle, the central lobe strongly narrowed toward the base.

        Acer saccharinum L. **Silver maple.**

    b Leaves more shallowly 3- to 5-lobed, to about the middle or less, the central lobe not evidently narrowed toward the base.

Fig. 32.29 *Carya ovata.*

Fig. 32.30 *Ginkgo biloba.* See also Fig. 2.2.

Fig. 32.31 *Araucaria araucana.*

Fig. 32.32 *Acer saccharinum.*

c Axillary bud enclosed by the petiole; leaf lobes with many small and rather blunt teeth. **Acer pseudoplatanus L. Sycamore maple.**

c Axillary bud free from the petiole, readily visible; leaf lobes with sharp teeth.

d Lobes of the leaf with only a few rather large teeth, otherwise clean-edged as though cut by a razor; central lobe usually with nearly parallel sides from the base up to a pair of large teeth at about mid-length. **Acer saccharum Marsh. Sugar maple.**

d Lobes of the leaf with numerous, irregular, mostly small teeth, as if the margin had been hacked out with a dull knife; central lobe often more or less tapering from the base. **Acer rubrum L. Red maple.**

a Lower surface of the leaves green, about the same color as the upper surface (or red in red-leaved cultivars); wings of the fruit diverging at a wide angle, 120–180 degrees.

e Leaves 3- to 5-lobed to about the middle or less.

f Lobes and teeth of the leaves blunt or rounded, or the lobes virtually without teeth. **Acer campestre L. Hedge maple.**

f Lobes and teeth of the leaves very sharply acute. **Acer platanoides L. Norway maple.**

e Leaves 7- to 13-lobed, or if only 5-lobed then cleft to well beyond the middle; red-leaved cultivars common.

g Leaves lobed to about the middle or less. **Acer japonicum Thunb. Full-moon maple.**

g Leaves deeply lobed to well beyond the middle. **Acer palmatum Thunb. Japanese maple.**

46 Leaves not lobed; fruit various, but not a double samara.

47 Leaves large, mostly well over 15 cm long, heart-shaped.

48 Fruit an erect, short and broad pod, not more than about 5 cm long. **Paulownia tomentosa (Thunb.) Steud. Empress tree.**

48 Fruit elongate-cylindric, pendulous pod mostly 20–50 cm long and up to about 1.5 cm thick. **Catalpa.**

Two species of catalpa are commonly cultivated.

a Corolla about 5–6 cm wide across the summit; pod mostly 1–1.5 cm thick. **Catalpa speciosa Warder. Western catalpa.**

Fig. 32.34 Acer saccharum. See also Fig. 21.5 and Chapter 24 opening photograph.

Fig. 32.36 Acer platanoides. See also Fig. 2.1.

Fig. 32.38 Paulownia tomentosa.

Fig. 32.33 Acer pseudoplatanus.

Fig. 32.35 Acer rubrum. See also Fig. 2.1.

Fig. 32.37 Acer palmatum.

a Corolla about 3–4 cm wide across the summit; pod mostly 0.6–1 cm thick.   ***Catalpa bignonioides* Walter. Common catalpa.**

47 Leaves smaller (mostly less than 15 cm long), not heart-shaped.

49 Plants notably aromatic (crush the leaves or twigs and sniff), cultivated mainly in southern United States, as in Florida and California.
***Syzygium paniculatum* Gaertn. Australian brush-cherry.**

49 Plants not notably aromatic; widely cultivated in the United States, but seldom in the southernmost parts.   ***Cornus.* Dogwood.**
This genus includes many cultivated shrubs, and several cultivated trees. Trees cultivated in the United States mostly have 4 (or up to 7) large, white or pink, petal-like bracts beneath each compact cluster of small flowers (the whole simulating a single flower), and are known as flowering dogwoods.

a Petal-like bracts rounded at the end and with a distinct notch, as if a small piece had been bitten out of the tip; round-headed tree with very rough, finely and deeply checkered bark.
***Cornus florida* L. Common flowering dogwood.**

a Petal-like bracts tapering to a point, not notched; trees with a more elongate crown; bark relatively smooth.

b Petal-like bracts 4, broadest below the middle or near the base; fruits joined together into a globular, fleshy head.
***Cornus kousa* Hance. Kousa dogwood.**

b Petal-like bracts 4–7, broadest near or above the middle; fruits in dense clusters, but individually distinct.
***Cornus nuttallii* Audubon. Western flowering dogwood.**

45 Leaves alternate (often opposite on young or rapidly growing twigs of *Eucalyptus globulus*, and some of them often opposite in *Lagerstroemia*).

50 Leaves broadly and shallowly notched across the summit.
***Liriodendron tulipifera* L. Tulip tree.**

50 Leaves pointed, rounded, or minutely notched at the summit.

51 Plants with milky juice (break a petiole and squeeze the juice out).

52 Leaves pinnately veined, entire (not toothed).

53 Lower surface of the leaf covered with appressed, rusty-red hairs; plants not thorny, cultivated only in tropical climates.
***Chrysophyllum cainito* L. Star-apple.**

Fig. 32.39 *Catalpa bignonioides.*

Fig. 32.40 *Cornus florida.*

Fig. 32.41 *Liriodendron tulipifera.* See also Chapter 18 opening photograph and Fig. 21.5.

53 Lower surface of the leaf smooth, without hairs; plants usually somewhat thorny, cultivated in temperate climates.

Maclura pomifera (Raf.) Schneid. Osage-orange.

52 Leaves palmately or pinnipalmately veined; margins toothed or lobed.

54 Bark rather smooth, capable of being peeled off in layers; fruits orange-red; leaves relatively large, mostly 10–20 cm long.

Broussonetia papyrifera (L.) Vent. Paper mulberry.

54 Bark rough, not peeling; leaves smaller, mostly 7–12 cm long.

Morus. Mulberry.

Two species of mulberry are commonly cultivated.

a Leaves downy beneath; fruit dark purple to black.

Morus rubra L. Red mulberry.

a Leaves smooth beneath, or with a few hairs along the main veins; fruit variously white, pink, or pale purple to dark purple or black.

Morus alba L. White mulberry.

51 Plants with watery juice.

55 Branches thorny (see also *Elaeagnus*, lead 64).

Crataegus. Hawthorn.

Many species of this large genus are cultivated, but only 4 are fairly common.

a Leaves merely toothed, not at all lobed.

Crataegus crus-galli L. Cockspur thorn.

a Leaves, or many of them, shallowly to deeply lobed.

b Leaves deeply lobed, the deepest sinuses reaching distinctly more than halfway to the midrib; style 1; fruit with a single seed.

Crataegus monogyna Jacq. Single-seeded English hawthorn.

b Leaves more shallowly lobed, the deepest sinuses not reaching more than about halfway to the midrib; styles and seeds more than 1.

c Styles and seeds 2; leaf base somewhat cuneate (wedge-shaped).

Crataegus laevigata (Poir.) DC. Two-seeded English hawthorn.

c Styles and seeds 3–5; leaf base more or less truncate (cut straight across) or somewhat cordate (heart-shaped).

Crataegus phaenopyrum (L.f.) Medic. Washington thorn.

Fig. 32.42 *Maclura pomifera.*

Fig. 32.43 *Morus rubra.*

Fig. 32.44 *Crataegus monogyna.*

55 Branches unarmed.

56 Leaves of highly variable form, some simple and entire, others with a single large lateral lobe (the leaf thus asymmetrical) or with a large lateral lobe on each side, in any case smooth beneath.

Sassafras albidum (Nutt.) Nees. Sassafras.

56 Leaves all of basically the same form, except sometimes in *Populus alba*, this sometimes with some leaves palmately lobed and others merely toothed, but in this species the leaves are densely white-hairy beneath.

57 Leaves palmately lobed, more or less green beneath; fruits hanging in ball-like clusters.

58 Lobes of the leaf with numerous small teeth; bark rough, not patchy, not peeling.

Liquidambar styraciflua L. Sweet gum.

58 Lobes of the leaf with a few large, sharp teeth; bark of middle-sized trunks smooth, patchy, peeling in large layers.

Platanus. Sycamore, plane-tree.

The most commonly cultivated plane-tree in the United States is *Platanus* X *acerifolia* (Ait.) Willd., the London plane, a hybrid between *P. occidentalis* L., American plane-tree, and *P. orientalis* L., Oriental plane-tree. It is hardier than *P. orientalis* and more resistant to disease than *P. occidentalis*.

57 Leaves entire or toothed, or pinnately lobed; seldom some of them palmately lobed, then white beneath; fruits various, but not hanging in ball-like clusters.

59 Fruit an acorn (a nut cupped by a hull at the base); many species with pinnately lobed or cleft leaves, but others with the leaves entire or merely toothed (for alternative, contrasting statement, see p. 613). Many species of this large genus are cultivated. The following are among the most common. **Quercus. Oak.**

a Leaves entire or merely toothed, or in *Q. nigra* sometimes with a single pair of broad, blunt lobes near the tip.

b Leaves conspicuously round-toothed

Quercus prinus L. Chestnut oak.

Fig. 32.45 *Sassafras albidum.*

Fig. 32.46 *Liquidambar styraciflua.*

Fig. 32.47 *Platanus acerifolia.*

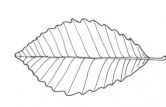

Fig. 32.48 *Quercus prinus.*

b Leaves entire or nearly so (or with a single pair of lobes).

  c Leaves densely and permanently short-hairy beneath.

    d Leaves evergreen, relatively small, mostly 4-8 cm long and 1-2 cm wide .... ***Quercus virginiana* Mill. Live oak.**

    d Leaves deciduous, larger, mostly 10–17 cm long and 3.5–7 cm wide .... ***Quercus imbricaria* Michx. Shingle oak.**

  c Leaves soon glabrous or nearly so beneath, except sometimes in the vein axils and along the midrib.

    e Leaves mostly linear-oblong to elliptic, broadest near the middle.

      f Leaves dull green beneath, mostly 4–6 times as long as wide .... ***Quercus phellos* L. Willow oak.**

      f Leaves glossy green beneath, mostly 2.5–3 times as long as wide .... ***Quercus laurifolia* Michx. Laurel oak.**

    e Leaves more or less obovate, broadest above the middle, 1.5–2.5 times as long as wide .... ***Quercus nigra* L. Water oak.**

  a Leaves pinnately lobed or cleft.

    g Leaf lobes not bristle-tipped.

      h Leaves auricled at the base (like an earlobe on each side) .... ***Quercus robur* L. English oak.**

      h Leaves more tapering at the base, not auricled.

        i Leaves evidently and persistently short-hairy beneath; larger leaf lobes evidently broadened above the base .... ***Quercus macrocarpa* Michx. Bur oak.**

        i Leaves soon becoming glabrous or nearly so beneath; leaf lobes not broadened upwards .... ***Quercus alba* L. White oak.**

Fig. 32.49 Quercus phellos.

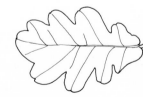

Fig. 32.50 Quercus robur.

Fig. 32.51 Quercus alba. See also Fig. 21.5.

g Leaf lobes shortly bristle-tipped.

j Terminal buds 4-angled, 7–10 mm long, densely short-hairy; leaves often more or less persistently short-hairy beneath.

Quercus velutina **Lam. Black oak.**

j Terminal buds ovoid, their scales glabrous or merely ciliate-margined; lower surface of the leaves (except in Q. *falcata*) soon becoming glabrous or nearly so, usually persistently hairy only in the vein axils.

k Leaves relatively shallowly lobed, the sinuses mostly not extending much if at all more than halfway to the midrib, the lobes mostly tapering from a broad base.

Quercus borealis **Michx. f. Northern red oak.**

k Leaves deeply lobed, much more than halfway to the midrib, the lobes relatively narrow and often not much tapering.

l Leaves persistently short-hairy beneath (use a hand lens); terminal leaf lobe tending to be notably elongate and often curved.

Quercus falcata **Michx. Southern red oak.**

l Leaves soon becoming smooth or nearly so beneath except in the vein axils; terminal leaf lobe not notably elongate.

m Acorn-cup shallow, covering up to a third of the nut; lower branches drooping.

Quercus palustris **Muenchh. Pin oak.**

m Acorn-cup deep, covering about half of the nut; lower branches not drooping.

Quercus coccinea **Muenchh. Scarlet oak.**

59 Fruit of various sorts, but not an acorn; leaves in most species entire or merely toothed, not lobed (some of the leaves of *Populus alba* often lobed).

60 Margins of the leaves entire, not toothed (for alternative, contrasting statement, see p. 618).

Fig. 32.52 *Quercus borealis.*

Fig. 32.53 *Quercus falcata.*

61 Leaves strongly aromatic (crush a part of a leaf and sniff cautiously); trees cultivated mainly in California and southwestern Oregon.

62 Fruit fleshy, plumlike, with a stone containing a single seed; stamens 9.

Umbellularia californica **(Hook. & Arn.) Nutt. California laurel.**

62 Fruit dry, firm, opening to release the more or less numerous small seeds; stamens numerous (more than 10). *Eucalyptus.*

Many species of this large, chiefly Australian genus are cultivated in California. The following are among the most common.

a Bark hard, dark, rough, persistent.

b Inflorescences large, branching, with many flowers. *Eucalyptus ficifolia* **F. J. Muell.**

b Inflorescences small, with 3–7 flowers.

Eucalyptus sideroxylon **A. Cunn. Red ironbark.**

a Bark pale and peeling off in large patches or strips, so that much of the trunk is patchy or ragged or seemingly devoid of bark.

c Flowers mostly borne singly, sessile or on a very short stalk, sometimes 2 together; leaves on young or rapidly growing shoots often opposite; most commonly planted Eucalyptus in the United States.

Eucalyptus globulus **Labill. Blue gum.**

c Flowers mostly 3–8 together, the short pedicels arising from the end of a short common stalk; leaves all alternate.

Eucalyptus viminalis **Labill. Manna gum.**

61 Leaves not strongly aromatic; trees widely cultivated.

63 Leaves heart-shaped, palmately veined; fruit a flat pod, 6–10 cm long and 1–1.5 cm wide.

Cercis canadensis **L. Redbud.**

Fig. 32.54 *Umbellularia californica.*

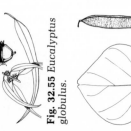

Fig. 32.55 *Eucalyptus globulus.*

Fig. 32.56 *Cercis canadensis.*

63 Leaves pinnately veined, not heart-shaped; fruit various, but not as above.

64 Leaves narrow, 3–8 times as long as wide, silvery on both sides with a dense covering of scales; fruit fleshy, 1 cm long, covered with scales like the leaves.   ***Elaeagnus angustifolia*** L. **Russian olive.**

64 Leaves and fruit various, but not as above, not covered with silvery scales.

65 Leaves small, up to about 6 cm long; petals with a distinct, slender stalk and an expanded, ruffled blade; cultivated in southern United States.   ***Lagerstroemia indica*** L. **Crape-myrtle.**

65 Leaves larger, many or all of them well over 6 cm long; petals not as above.

66 Trees with smooth, red-brown bark flaking off in thin plates or strips; leaves evergreen; cultivated mainly in the Pacific states.   ***Arbutus menziesii*** Pursh. **Madrone.**

66 Trees with various sorts of bark, but not as above; leaves variously evergreen or deciduous.

67 Frost-sensitive tropical trees, in the United States cultivated mainly in southern Florida.

68 Leaves relatively small, mostly 10–15 cm long; flowers borne singly, 2.5–3 cm long; fruit 6–10 cm long, wholly fleshy.   ***Annona squamosa*** L. **Custard-apple.**

Fig. 32.57 *Elaeagnus angustifolia.*

Fig. 32.58 *Lagerstroemia indica.*

68 Leaves larger, at least the larger ones 15–30 cm long; flowers small, less than 1 cm long, borne in slender spikes; fruit 4–7 cm long, with a large stone. *Terminalia catappa* L. Indian almond.

67 More or less frost-hardy trees, cultivated mainly farther north than southern Florida.

69 Leaves without stipules; receptacle not elongate.

70 Leaves relatively small, mostly 4–15 cm long; flowers tiny, well under 1 cm long and wide; fruit a dark blue or black stone-fruit, 1–1.5 cm long. *Nyssa sylvatica* Marsh. Sour gum.

70 Leaves larger, mostly 15–25 cm long; flowers 3–4 cm wide; fruit yellowish brown, wholly fleshy, 6–15 cm long. *Asimina triloba* (L.) Dunal. Pawpaw.

69 Leaves with stipules; flowers with an elongate receptacle to which the numerous distinct or coalescent ovaries are attached, each ripened ovary opening on the back to expose the two seeds. *Magnolia.*

The following are the most commonly cultivated species.

Fig. 32.59 *Terminalia catappa.*

Fig. 32.60 *Nyssa sylvatica.*

a Leaves evergreen, glossy-green above, densely rusty-hairy beneath.
Magnolia grandiflora L. **Southern magnolia.**

a Leaves deciduous (except often in *M. virginiana*), not rusty beneath.

b Leaves strongly glaucous beneath (i.e., coated with a powdery wax that can be rubbed off, like the surface of a prune).
Magnolia virginiana L. **Sweet bay.**

b Leaves green, not glaucous beneath.

c Flowers greenish-yellow, appearing with the leaves; large tree.
Magnolia acuminata L. **Cucumber tree.**

c Flowers white to pink or purple, appearing before the leaves; large shrub or small tree.

d Flowers white to pale pink, with widely spreading, slender tepals, about 6–10 cm across when wide open; leaves relatively small, mostly 5–12 cm long.
Magnolia stellata (Siebold. & Zucc.) **Maxim. Star magnolia.**

Fig. 32.61 *Magnolia grandiflora*. See also Chapter 27 opening photograph.

d Flowers white to often pink or purple, at first closed or cup-shaped, later opening and becoming 10–15 cm wide; leaves larger, mostly 10–15 cm long, ***Magnolia soulangiana*** **Soul.-Bod. Saucer magnolia.**

60 Margins of the leaves toothed (sometimes lobed in *Populus alba*).

71 Leaves evergreen, with a few pungently spiny teeth along the margins; fruit of bright red or orange-red, pea-sized berries. **Ilex. Holly.**

Two arborescent species are commonly cultivated.

a Flowers (and fruits) produced in clusters at the nodes on wood of the previous year.

***Ilex aquifolium* L. English holly.**

a Flowers borne in the axils of the leaves of the season, the pistillate ones (and fruits) solitary or 2–3 together.   ***Ilex opaca* Aiton. American holly.**

71 Leaves deciduous, not spiny-toothed; fruits not as in *Ilex.*

72 Leaves tending to be palmately or pinnipalmately veined, either with several main veins from the base, or with the first one or two pairs of lateral veins distinctly larger and longer and serving more of the leaf than the other lateral veins; leaves often but not always relatively broad (not much if at all longer than wide), or with a truncate or somewhat heart-shaped base (for alternative, contrasting statement, see p. 620).

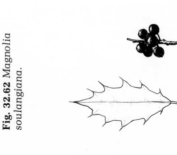

Fig. 32.62 *Magnolia soulangiana.*

Fig. 32.63 *Ilex aquifolium.*

73 Flowers and small, globose fruits borne in loose, open inflorescences, the stalk of the inflorescence seeming to be borne on a narrow, elongate bract of leafy texture that is unlike the obliquely heart-shaped foliage leaves.

Tilia. **Basswood, linden.**

Two species are commonly cultivated.

a Leaves small, mostly 5–10 cm long.

Tilia europaea L. **European linden.**

a Leaves larger, mostly 10–15 cm long.

Tilia americana L.

American linden, basswood.

73 Flowers and fruits distinctly otherwise.

74 Leaves distinctly oblique at the base, one side more broadly rounded than the other; fruit of sweet, dark red to black, pea-sized berries.

Celtis occidentalis L. **Hackberry.**

74 Leaves symmetrical at the base, both sides alike; fruit dry, either winged or with plumed seeds.

75 Bark peeling in thin layers, evidently marked with short, horizontal lines (lenticels). *Betula.* **Birch.**

Several species of birch are cultivated. The 2 following, both with conspicuously white bark, are the most common.

a Trees of weeping form, the smaller branches drooping; leaves smooth beneath.

Betula pendula L. **Weeping birch.**

a Trees of ordinary form, the branches ascending or spreading; leaves hairy in the vein axils beneath.

Betula papyrifera Marsh.

Paper birch.

Fig. 32.64 *Tilia americana.*

Fig. 32.65 *Celtis occidentalis.*

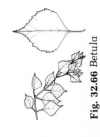

Fig. 32.66 *Betula pendula.*

75 Bark not peeling in thin layers, not marked with horizontal lines.
    Populus. Poplar, cottonwood.
The most commonly cultivated species are the following.

  a Leaves densely white-hairy beneath, the larger ones often palmately lobed; bark conspicuously whitish on young and medium-sized trunks.
    Populus alba L. White poplar.

  a Leaves smooth beneath, not hairy; bark sometimes pale, but not conspicuously whitish.

    b Trees of very slender, cylindrical form, the branches nearly erect.
      Populus nigra L. var. italica Duroi. Lombardy poplar.

    b Trees of more ordinary form, with ascending or spreading branches.

      c Leaves and twigs very resinous and aromatic; petioles more or less cylindric, not flattened.
        Populus balsamifera L. Balsam poplar.

      c Leaves and twigs not notably resinous nor aromatic; petioles flattened.   **Populus X canadensis Moench. Carolina poplar.**

72 Leaves distinctly pinnately veined, the lower lateral veins no larger (often smaller) than those above.

76 Leaves with 1 or 2 pairs of prominent, raised glands at the base of the blade, where it joins the petiole; bark of middle-sized trunks smooth, marked with lenticels; bark of young twigs with the odor and taste of bitter almonds.   **Prunus.**

Fig. 32.67 *Populus alba.*

Fig. 32.68 *Populus nigra* var. *italica.*

This large genus includes prunes, plums, cherries, peaches, apricots, and almonds. The only commonly cultivated street tree is P. *serrulata* Lindl., the Japanese flowering cherry.

76 Leaves without prominent, raised glands at the base of the blade; bark various, but not as in Prunus.

77 Fruit a miniature apple, persistent all summer and into the winter; flowers showy, white to pink or purple, with separate petals. ***Pyrus.***

The genus Pyrus includes apples, crabapples, and pears. Only some of the crabapples are common street trees. Many of the cultivated trees are of hybrid origin. The most commonly cultivated species are the following:

a Many of the leaves shallowly lobed as well as toothed.

Pyrus coronaria L.
American crab-apple.

a Leaves all merely toothed.

b Calyx lobes persistent on the yellow fruit.

Pyrus spectabilis Aiton.
Chinese crab-apple.

b Calyx soon deciduous; fruit red or yellow.

c Styles mostly 5; leaf folded face to face in bud.

Pyrus pulcherrima Aschers. &
Graebner. **Showy crab-apple.**

c Styles mostly 4; leaf laterally rolled in bud.

Pyrus baccata L.
Siberian crab-apple.

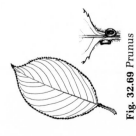

Fig. 32.69 *Prunus serrulata.*

77 Fruit entirely otherwise; flowers either without petals, or with the petals united into a lobed tube.

78 Leaves conspicuously toothed and very prominently and regularly veined, the main veins raised on the lower surface, running directly from the midrib to a marginal tooth.
Fagus sylvatica L. **European beech.**

79 Trunk with notably smooth, gray bark.

79 Trees with more or less roughened bark.
Ulmus. **Elm.**

The 3 following species are the most commonly cultivated elms.

a Leaves simply (not doubly) toothed, relatively small, seldom any of them more than about 8 cm long.
Ulmus pumila L. **Siberian elm.**

a Leaves doubly toothed (with small teeth on the large teeth).

b Graceful tree with a characteristic form, the main branches closely ascending below, gradually arching out upwards; leaves relatively large, mostly 8–14 cm long.
Ulmus americana L. **American elm.**

b Tree with an ordinary round crown; leaves smaller, mostly 5–10 cm long.
Ulmus procera Salisb. **English elm.**

78 Leaves finely toothed, less prominently and less regularly veined, the main veins mostly curved-ascending and not closely approaching the marginal teeth.

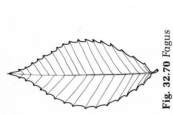

Fig. 32.70 *Fagus sylvatica.*

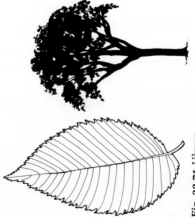

Fig. 32.71 *Ulmus americana.* See also Fig. 21.5.

80 Flowers with well-developed petals joined to form a sympetalous corolla, perfect, not borne in catkins; seeds not plumed; leaves mostly more than 2 cm wide.

81 Flowers small, the corolla less than 1 cm long; fruit less than 1 cm long, opening to release the numerous tiny seeds.

Oxydendrum arboreum **(L.) DC. Sourwood.**

81 Flowers larger, the corolla 2–2.5 cm long; fruit 3–5 cm long, 4-winged, remaining closed.

***Halesia monticola* (Rehder) Sargent. Silver-bell tree.**

80 Flowers without petals, borne in catkins, the male and female on separate trees; seeds tiny, plumed; leaves narrow, up to about 1.5 cm wide (to 3.5 cm in *S. fragilis*).

***Salix*. Willow.**

Most willows are shrubs. The following are the most commonly cultivated trees in the group.

a Trees of weeping form, the smaller branches conspicuously drooping.

***Salix babylonica* L. Weeping willow.**

a Trees of ordinary form, the branches ascending, spreading, or only slightly drooping.

b Leaves white-silky beneath.

***Salix alba* L. White willow.**

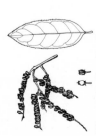

Fig. 32.72 *Oxydendrum arboreum.*

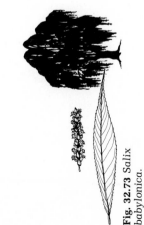

Fig. 32.73 *Salix babylonica.*

A form with yellow twigs (resembling the weeping willow in this regard) is called Golden willow.

b Leaves at maturity green and smooth beneath.

c Leaves mostly 1–1.5 (2) cm wide; petiole not glandular. ***Salix nigra* L. Black willow.**

c Leaves mostly 2–3.5 cm wide; petiole glandular above at the tip.   ***Salix fragilis* L. Crack-willow.**

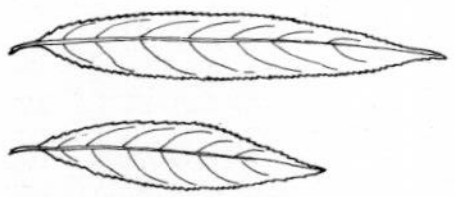

Fig. 32.74 *Salix nigra.*

Glossary

Note on derivations and definitions: Many botanical terms are derived from Greek, often through Latin. The Greek antecedent, being the older, is usually given instead of the Latin. When some latitude in transliteration of classical Greek to English is permissible, the form that more nearly coincides with the spelling of the modern English word is generally given. The definitions are in all cases for botanical usage. Many of the terms have a broader popular definition or a different specialized definition in some other field. For terms not included in the glossary, see the index.

Å. Ångstrom unit, one tenth of a nanometer.

absorption spectrum. A comparative measure of absorption of different wavelengths of radiant energy by a pigment or other substance.

ADP. Adenosine diphosphate, a compound that can add another phosphate group (forming ATP) in important metabolic transfers of energy.

adventitious (L. *adventitius,* from *ad,* to + *venire,* to come). Originating from mature nonmeristematic tissues, especially if such a development would not ordinarily be expected.

aecium (Gr. *aikia,* injury, or *oikidion,* a little house). Among rusts, a cup at the surface of the host, in which dikaryotic spores called aeciospores are borne.

akinete (Gr. *a,* not; *kinēsis,* movement). An algal spore produced by transformation of a whole vegetative cell, with the original cell wall forming part or all of the spore wall.

algae (L. seaweed). A broad artificial group of plants, characterized by the presence of chlorophyll *a* and the absence of the characteristic multicellular sex organs or specialized water-conducting tissues of higher plants.

allele (Gr. *allēlon,* of one another). One of the two or more genes which in different genomes can occupy a particular locus on a chromosome.

allelopathy (Gr. *allēlon,* of one another + *pathos,* suffering). The chemical inhibition of growth of one plant by another.

amino acid. An organic acid having one or more amine (NH_2^-) groups in place of hydrogen atoms.

amitosis (Gr. *a,* not + mitosis). Nuclear division or cell division (as in prokaryotes) that does not follow the pattern of mitosis.

amylopectin (Gr. *amylon,* starch + pectin). One of the constituents of starch, consisting of a branching polymer of α-glucose.

amyloplast (Gr. *amylon,* starch + plastid). A leucoplast in which starch is stored.

amylose (Gr. *amylon,* starch). One of the constituents of starch, consisting of an unbranched polymer of α-glucose.

anabolism (Gr. *ana,* up + [meta]bolism). Those phases of metabolism which make more complex substances from simpler ones, requiring an input of energy.

anaphase (Gr. *ana,* up + phase). The stage in mitosis in which the two chromatids of each chromosome are separated and move toward the respective poles of the spindle.

angiosperm (Gr. *angeion,* vessel + *sperma,*

seed). A member of the group of plants (Magnoliophyta) characterized by having the ovules enclosed in an ovary.

anion (Gr. *anios*, going up). A negatively charged ion.

annual (L. *annus*, year). A plant that completes its life cycle and dies in a year.

anther (Gr. *anthos*, flower). The part of a stamen, consisting of one or usually two pollen sacs (and a connective layer between them), that bears the pollen.

antheridium (Gr. *anthos*, flower + *eidos*, resemblance). A specialized cell, or a multicellular structure, within which one or more sperms are produced.

anthocyanin (Gr. *anthos*, flower + *kyanos*, dark blue). A chemical class of water-soluble pigments, ranging in color from blue or violet to purple or red, often found in the central vacuole of a cell, especially in flower petals.

apomixis (Gr. *apo*, away + *mixis*, mingling). The setting of seed without fertilization.

archegonium (Gr. *arche*, beginning + *gonos*, offspring). A specialized structure, composed of more than one cell, within which an egg is produced.

ascocarp (Gr. *askos*, bladder + *karpos*, fruit). The fruiting body of an ascomycete.

ascospore. A spore produced in an ascus.

ascus (Gr. *askos*, bladder). The characteristic sporangium of the ascomycetes, within which the sexual spores (ascospores) are borne.

ATP. Adenosine triphosphate, a compound that releases metabolic energy when it gives up its third phosphate group.

autoecious (Gr. *autos*, self + *oikos*, house). Among rusts, completing the life cycle on a single host.

autotroph (Gr. *autos*, self + *trophē*, nourishment). An organism that makes its own food from raw materials; opposite of *heterotroph*.

auxin (Gr. *auxein*, to grow). A group of growth regulators that promote cell elongation and also have other effects.

axil (L. *axilla*, armpit). The point of the angle formed by the leaf or petiole with the upward internode of the stem.

basidiocarp (basidium + Gr. *karpos*, fruit). The fruiting body of a basidiomycete.

basidiospore. A spore borne on a basidium.

basidium (L. diminutive from Gr. *basis*, base). A characteristic structure of certain fungi (the Basidiomycetes) which bears sexual spores externally.

biennial (L. *biennium*, a period of two years). A plant that completes its life cycle and dies in two years (or more than one year but less than two).

blepharoplast (Gr. *blepharon*, eyelid + *plastos*, formed, molded). The basal body, in the outer part of the cytoplasm, to which a flagellum is attached.

budding. Unequal cell division, as in yeasts; also, a type of grafting in which a bud is used as the scion.

calorie (L. *calor*, heat). The amount of energy required to raise the temperature of 1 gr. of water by 1°C, under certain specified conditions.

calyptra (Gr. *kalyptēr*, veil). In bryophytes, the structure derived partly or wholly from the archegonium, which caps or immediately surrounds the developing sporophyte.

calyx (Gr. *kalyx*, cup). All the sepals of a flower, collectively.

cambium (L. *cambiare*, to exchange). A lateral meristem; specifically, the vascular cambium, which produces xylem internally and phloem externally.

capsid (L. *capsa*, box, case). The proteinaceous sheath of a virion.

capsule. The sporangium of a bryophyte; in angiosperms, a common type of dry dehiscent fruit, composed of more than one carpel.

carbohydrate. A food whose molecules consist entirely, or almost entirely, of carbon, hydrogen, and oxygen, with the ratio of hydrogen to oxygen being 2:1.

carotene (*Daucus carota*, the carrot). A fat-soluble, yellow hydrocarbon, associated with chlorophyll in the photosynthetic process.

carpel (Gr. *karpos*, fruit). One of the female organs of a flower, i.e., the megasporophyll.

casparian strip. A waxy band encircling an endodermal cell along its radial walls.

catabolism (Gr. *kata*, down, against, + [meta]bolism). Those phases of metabolism which make simpler substances from more complex ones, resulting in a liberation of stored energy.

catalyst (Gr. *katalysis*, dissolution). A substance that facilitates a chemical reaction between other substances, without itself being used up in the process.

cation (Gr. *kation*, going down). A positively charged ion.

caulidium (diminutive of L. *caulis*, stem). The stemlike organ of a gametophyte of a bryophyte.

cell. An organized unit of protoplasm, bounded by a membrane or wall, and usually divisible into a nucleus and some cytoplasm; also applied to the cell wall, after the death of the protoplast.

cell plate. An initial partition between two sister cells, formed *de novo* in the cytoplasm in association with the spindle fibers, as part of the terminal stage in typical mitotic cell division in most plants.

cell sap. The watery contents of the central vacuole of a cell.

cellulose (L. *cellula*, cell). A complex polysaccharide that is the principal constituent of the cell walls of most plants.

central cylinder. The primary vascular structure of a stem or root, together with any tissue (such as the pith) that may be enclosed.

centriole (diminutive of L. *centrum*, center). A cytoplasmic organelle of some algal flagellates, motile plant cells (such as sperms or zoospores), and most animals, from which microtubules of the mitotic spindle appear to originate.

centromere (Gr. *meros*, part, portion). A specialized part of the chromosome to which the tractile fibers are attached during mitosis.

chlorenchyma (Gr. *chlōros*, green + *enchein*, to pour in). A tissue (generally parenchymatous) characterized by having chloroplasts.

chlorophyll (Gr. *chlōros*, green + *phyllon*, leaf). A group of magnesium-porphyrins, green in color, that are essential for photosynthesis.

chloroplast. A plastid that bears chlorophyll.

chromatid (Gr. *chrōma*, color). One of the two longitudinal halves that make up a chromosome during prophase and metaphase.

chromatin (Gr. *chrōma*, color). DNA plus its associated protein.

chromonema (Gr. *chrōma*, color + *nēma*, thread). One of the threadlike, DNA-bearing structures within the nucleus, which give rise to the chromosomes during mitosis; an interphase chromosome.

chromoplast (Gr. *chrōma*, color + plastid). A colored plastid, other than a chloroplast.

chromosome (Gr. *chrōma*, color + *soma*, body). A body within the nucleus characterized by the presence of DNA and usually also histone proteins; the term is conventionally applied to such bodies during the contracted state that they assume in mitosis; by extension, the major DNA-bearing body (or bodies) of prokaryotic cells. Cf. chromonema.

class. A group of related orders; a particular rank in the taxonomic hierarchy, between the order and the division.

coenocyte (Gr. *koinos*, in common + *kytos*, a hollow vessel). An organism which has the nuclei scattered in a continuous (usually filamentous) protoplast that is not divided into definite cells.

coenzyme. A biological catalyst, of smaller molecular size than an enzyme, which must cooperate with an enzyme to produce its effect.

collenchyma (Gr. *kolla*, glue + *enchein*, to pour in). A strengthening tissue composed of living cells with thick primary walls that consist largely of cellulose.

colloid (Gr. *kolla*, glue). A suspension (usually in water) in which the suspended particles are larger than ordinary molecules but not large enough to settle out.

companion cell. A modified parenchyma cell of the phloem tissue, associated with a sieve element and having a direct ontogenetic relationship with it.

conidium (Gr. *konos*, offspring, or *konis*, dust). An asexual spore; in fungi, a spore (asexual) produced by differentiation and abscission of a hyphal tip.

conjugated protein (L. *conjugatus*, united). A protein that yields one or more other compounds in addition to amino acids when hydrolyzed.

conjugation (L. *conjugare*, to unite, or *conjugatus*, united). Fusion of isogametes; in bacteria, a parasexual process in which genetic material is passed directly from one cell to another.

cork. A waterproofing protective tissue, dead when functional, characterized by suberization of the cell walls.

corolla (L. diminutive of *corona*, crown). All the petals of a flower, collectively.

cortex (L. bark or rind). The tissue between the central cylinder (stele) and the epidermis of a stem or root.

cotyledon (Gr. *kotylēdon*, a cup-shaped hollow). A primary leaf of the embryo of a seed.

cryptogam (Gr. *kryptos*, hidden + *gamos*, marriage). A plant belonging to the non–seed-bearing general group; opposite of *phanerogam*.

cultivar. A recently coined word for cultivated variety.

cutin (L. *cutis*, skin). A waxy material, characteristic of epidermal cells of embryophytes, that retards evaporation of water.

cyclosis (Gr. *kyklosis*, an encircling). The characteristic streaming motion of cytoplasm.

cytochrome (Gr. *kytos*, a hollow vessel + *chrōma*, color). A group of iron-porphyrins that function in energy transfer in both photosynthesis and respiration.

cytokinesis (Gr. *kytos*, a hollow vessel + *kinesis*, movement). The division of the cytoplasm (as contrasted to the nucleus) during cell division.

cytokinin (from *cytokinesis*). A group of growth regulators that promote cell division.

cytology (Gr. *kytos*, a hollow vessel + *logos*, word, discourse). The study of the protoplasmic structure of cells, including the changes that occur during cell division.

cytoplasm (Gr. *kytos*, a hollow vessel + *plasma*, a thing molded or formed). The non-nuclear part of protoplasm.

deciduous (L. *decidere*, to fall off). Falling off; dropping its leaves in the autumn.

dehiscent (L. *dehiscere*, to gape, to yawn). Opening at maturity, releasing or exposing the contents.

dicot (Gr. *dis*, two + cotyledon). Short for dicotyledon; the dicotyledons (class Magnoliopsida) are one of the two great groups of angiosperms.

dictyosome (Gr. *dictyon*, net + *soma*, body). Same as golgi body.

dictyostele (Gr. *dictyon*, net + stele). A siphonostele with numerous and elongate leaf gaps and branch gaps, so that the primary vascular tissue forms a ring of vascular bundles.

digestion. The partial breakdown of foods, making them more soluble or more readily diffusible without significantly changing the amount of stored energy.

dikaryotic (Gr. *dis*, two + *karyon*, nut). Having two nuclei (typically haploid) of different origin in each segment or cell of a filament.

dioecious (Gr. *dis*, two + *oikos*, house). Producing male and female structures (in seed plants the microsporophylls and megasporophylls, or stamens and pistils) on separate individuals.

diploid (Gr. *diploos*, double). Having two full chromosome complements per cell.

disaccharide (Gr. *dis*, two + *sakchar*, sugar). A sugar whose molecules consist of two chemically linked molecules of simple sugar; the two simple sugars may be alike or unlike.

division. A group of similar classes; the highest category regularly used in the taxonomic hierarchy.

DNA. Deoxyribonucleic acid, the bearer of hereditary information.

dorsiventral (L. *dorsum*, back + *venter*, belly). Flattened, with the two flattened sides unlike; having a back side and a belly side.

ecology (Gr. *oikos*, house + *logos*, word, discourse). The study of the relation between organisms and their environment, the influence of each on the other.

egg. A nonmotile gamete that can fuse with a sperm to form a zygote; the female gamete.

EM. Abbreviation for electron micrograph or electron microscope.

embryo (Gr. *embryon*, fetus). The young sporophyte, before it has begun to take on its mature form.

embryo sac. The female gametophyte of angiosperms, within which the embryo begins to develop.

embryophyte (Gr. *embryon*, fetus + *phyton*, plant). A member of the plant subkingdom that is characterized by the early development of the sporophyte as a parasite on the gametophyte.

EMP pathway. The Embden-Meyerhof-Parnas pathway for the respiratory breakdown of glucose to pyruvate.

enation (L. *e*, from +*natus*, born). A structure that takes its evolutionary origin as an outgrowth from another structure.

endodermis (Gr. *endon*, within + *derma*, skin). A layer of specialized cells in many roots and some stems, delimiting the inner margin of the cortex.

endoplasmic reticulum. A set of irregularly joined and perforated double membranes that permeate the cytoplasm and connect to the nuclear membrane.

endosperm (Gr. *endon*, within + *sperma*, seed). In the broadest sense, the food-storage tissue of a seed, other than the embryo or the seed coat; narrowly, the food-storage tissue developed from the triple-fusion nucleus of an angiosperm ovule.

endospore (Gr. *endon*, within + spore). A spore that is formed *within* a sporangium or cell wall, rather than externally.

enzyme (Gr. *en*, in + *zymē*, yeast). A catalyst that is a simple or conjugated protein.

epicotyl (Gr. *epi*, upon + cotyledon). The part of the embryo of a seed that gives rise to the shoot.

epidermis (Gr. *epi*, upon + *dermis*, skin). The characteristic outermost tissue of leaves and of young roots and stems.

epigynous (Gr. *epi*, upon + *gynē*, woman). Having the perianth and stamens attached at or near the top of the ovary, rather than beneath it.

eukaryotic (Gr. *eu*, *eus*, well, good + *karyon*, nut). Having a characteristic set of intracellular organelles (vesicular nucleus, mitochondria, golgi bodies, endoplasmic reticulum, often also plastids) indicating a relatively complex organization of the protoplast.

eutrophication (Gr. *eu*, *eus*, well, good + *trophia*, nutrition). The sequential process by which a pond or lake acquires an increased supply of nutrients, fills in, and becomes murky with algae and decay products.

exoenzyme (Gr. *exo*, outside + enzyme). An enzyme that characteristically functions outside the organism that produced it; applied especially to digestive enzymes produced by fungi and bacteria.

family. A group of related genera; a particular rank in the taxonomic hierarchy, between the genus and the order.

fat. A food whose molecules consist entirely, or almost entirely, of carbon, hydrogen, and oxygen, with the ratio of hydrogen to oxygen being greater than 2:1.

fertilization. Fusion of a sperm with an egg.

fiber. A long, slender, thick-walled strengthening cell, usually dead when functional.

flagellate. A unicellular or colonial organism, other than a bacterium, that moves by means of flagella.

flagellum (L. whip). A slender, motile, protoplasmic projection from a cell.

flavonoid pigments. A chemical group of pigments, consisting of anthocyanins and anthoxanthins; flavone is a kind of anthoxanthin.

flower. A specialized short shoot with modified leaves, some of which characteristically produce sexual reproductive structures, leading ultimately to the formation of seeds.

food. Any chemical compound that can be broken down by living organisms to release energy.

free energy. The capacity (in terms of energy) to do work.

fructose (L. *fructus*, fruit). A sugar, found especially in fruit, which, like glucose, has the formula $C_6H_{12}O_6$, but with the atoms somewhat differently arranged.

fruit (L. *fructus*). A ripened ovary, together with any other structures that ripen with it and form a unit with it.

fruiting body. In fungi, the compact mycelium, usually of definite form and structure, associated with the perfect stage in the life cycle.

fundamental tissue. The pith, the cortex, and the primary tissue separating the vascular bundles.

gametangium (Gr. *gamos*, marriage + *angeion*, vessel). Any structure in which gametes are borne.

gamete (Gr. *gamos*, marriage). Any cell that is capable of fusing with another cell to form a new individual.

gametophore (Gr. *gamos*, marriage + *phorein*, to bear). In mosses, the main body of the gametophyte, on which the archegonia and/or antheridia are borne.

gametophyte (Gr. *gamos*, marriage + *phyton*, plant). The generation that has n chromosomes and produces gametes as reproductive cells.

gene (Gr. *genos*, stock, race). A segment of a DNA double helix that acts as a template for the production of a particular kind of molecule of RNA; more generally, one of the individual bits of chromatin that governs the hereditary characteristics of an organism.

genera. Plural of genus.

genome. One complete set of chromosomes, i.e., a chromosome complement.

genotype. The genetic make-up of an individual. Cf. *phenotype*.

genus (L. a general kind). A group of related species; a particular rank in the taxonomic hierarchy, between the species and the family.

geotropism (Gr. *gē*, earth + *tropē*, a turning). A growth response to gravity.

germination (L. *germinare*, to sprout). The resumption of growth by any spore, seed, or other propagule, after a period of dormancy.

gibberellin (*Gibberella*, the fungus that causes the foolish seedling disease of rice). A group of growth regulators that promote shoot growth and have other effects as well.

glucose (Gr. *glucus*, sweet). A common simple sugar with the formula $C_6H_{12}O_6$.

golgi body (Camillo Golgi, Italian zoologist). A cytoplasmic organelle of eukaryotes, concerned with the packaging of diverse carbohydrates and other substances.

guard cell. One of a pair of specialized epidermal cells bounding a stomate.

gymnosperm (Gr. *gymnos*, naked + *sperma*, seed). A member of the group of plants characterized by having ovules that are not enclosed in an ovary.

haploid (Gr. *haploos*, single). Having only one full set of chromosomes (one chromosome complement) per cell.

haustorium (L. *haustor*, a drawer, from *haurire*, *haustum*, to draw, to drink). A specialized organ or organelle through which a parasite extracts nourishment from its host.

heme (Gr. *haema*, blood). A group of iron-porphyrins, including most notably cytochromes, which are important in respiratory processes of both plants and animals.

hemicellulose (Gr. *hemi*, half + cellulose). A chemical group of substances allied to cellulose but more readily soluble, including pentosans (cellulosans), pectin, and some similar compounds.

herbarium (L. *herba*, herb). A collection of plant specimens, so dried and preserved as to illustrate as far as possible their characteristics.

heteroecious (Gr. *heteros*, other + *oikos*, house). Among rusts, requiring two different kinds of hosts to complete the full sexual life cycle.

heterogamete (Gr. *heteros*, other + gamete). A gamete that can fuse only with a gamete of different form; ordinarily, an egg or a sperm.

heterosporous (Gr. *heteros*, other + spore). Producing two different kinds of spores, one of which gives rise to male gametophytes, the other to female.

heterotroph (Gr. *heteros*, other + *trophē*, nourishment). An organism that does not make its own food; opposite of *autotroph*.

heterozygous (Gr. *heteros*, other + zygote). Having opposing genes on homologous chromosomes; opposite of *homozygous*.

hexose (Gr. *hex*, six). A sugar, such as glucose, with six atoms of carbon in the molecule.

holophytic (Gr. *holos*, whole + *phyton*, plant). Autotrophic, making its own food.

holozoic (Gr. *holos*, whole + *zoon*, animal). Eating (ingesting) rather than making or absorbing food.

homosporous (Gr. *homos*, one and the same + spore). Having the spores all about alike, each giving rise to a gametophyte that produces both antheridia and archegonia; opposite of *heterosporous*.

homozygous (Gr. *homos*, one and the same + zygote). Having identical (rather than opposing) genes on homologous chromosomes; opposite of *heterozygous*.

host. An organism that, while still alive, provides food or lodging for another organism.

hyaloplasm (Gr. *hyalos*, glass + *plasma*, a thing molded or formed). The relatively clear, fluid part of the cytoplasm, in which the endoplasmic reticulum and the cytoplasmic organelles are dispersed.

hybrid. A cross between different species; in genetics, a cross between individuals differing in one or more genes.

hydrolysis (Gr. *hydōr*, water + *lysis*, a loosening). The chemical breakdown of larger molecules into smaller ones by insertion of the components of water at the point of breakage.

hydrophyte (Gr. *hydōr*, water + *phyton*, plant). A plant that grows in water.

hymenium (Gr. *hymēn*, membrane). In fungi, a layer of asci or basidia, often intermingled with sterile hyphae.

hypha (Gr. *hyphē*, web). Any single filament of a fungus.

hypocotyl (Gr. *hypo*, under + cotyledon). The part of the embryo of a seed that lies just beneath the cotyledons, connecting the epicotyl to the radicle.

hypogynous (Gr. *hypo*, under + *gynē*, woman). Having the perianth and stamens attached directly to the receptacle, beneath the ovary.

inflorescence. A cluster of flowers.

initial. One of the relatively few cells in an apical meristem that remains permanently meristematic; also, a cell that by cell division gives rise to a particular structure.

integument (L. *integumentum*, a covering). One of the one or two layers that partly enclose the nucellus of an ovule; the forerunner of the seed coat.

internode (L. *inter*, between + node). The part of a stem between two successive nodes.

interphase. The time or stage between mitotic divisions of a cell.

involucre (L. *involucrum*, a covering). Any structure that surrounds the base of another structure.

ion (Gr. *ion*, going). An atom or submolecular group of atoms that has lost or gained one or more electrons and therefore has an electric charge.

isogamete (Gr. *isos*, equal + gamete). A gamete that can fuse with another of similar appearance to form a zygote.

karyogamy (Gr. *karyon*, nut + *gamos*, marriage). Sexual fusion of nuclei (as opposed to the cytoplasm).

latex (L. fluid). A colorless to more often white, yellow, or reddish liquid, produced by some plants, characterized by the presence of colloidal particles of terpenes dispersed in water.

laticifer (L. *latex*, *laticis*, fluid + *ferre*, to bear). A latex-bearing tube.

leaf. One of the primary organs of most vascular plants, typically being the principal photosynthetic organ.

leaf gap. A parenchymatous opening into a stele, left by the departure of a leaf trace.

leaf trace. A vascular bundle, from the point where it leaves the stele to the point where it enters the leaf.

lenticel (L. *lens, lentis,* a lentil). A slightly raised area in the bark of a stem or root, consisting of loosely arranged, nearly or quite unsuberized cells.

leucoplast (Gr. *leukos,* white + plastid). A colorless plastid.

lignin (L. *lignum,* wood). A high polymer of several compounds derived from phenyl propane; an essential strengthening component of the cell walls of many cells, especially xylem.

lipid (Gr. *lipos,* fat, grease). Fats and some other chemical compounds that resemble fats in being insoluble in water but soluble in certain organic solvents.

lumen (L. an opening for light). The space enclosed by a cell wall: used especially in speaking of dead cells, from which the protoplast has disintegrated.

mastigoneme (Gr. *mastigion,* a diminutive of *mastix,* whip + *nema,* thread). One of the slender, threadlike lateral appendages present on some kinds of flagella.

medullary ray. A parenchymatous connection between the cortex and the pith of a stem.

megaphyll (Gr. *megas,* large + *phyllon,* leaf). A leaf derived from a branch system, often associated with a leaf gap in the stele and having a branching vein system; opposite of *microphyll.*

megasporangium (Gr. *megas,* large + sporangium). A sporangium bearing one or more megaspores.

megaspore (Gr. *megas,* large + spore). A spore that develops into a female gametophyte.

megasporophyll (megaspore + Gr. *phyllon,* leaf). A leaf (sporophyll) bearing or subtending one or more megasporangia.

meiosis (Gr. *meion,* smaller, less). Reduction division.

meristem (Gr. *meristos,* divided). A tissue characterized by cell division.

mesophyll (Gr. *mesos,* in the middle + *phyllon,* leaf). The tissue, other than vascular tissue, between the upper and the lower epidermis of a leaf.

metabolism (Gr. *metabolē,* change). The complex set of interrelated chemical processes characteristic of life.

metaphase (Gr. *meta,* beyond + phase). The stage in mitosis during which the chromosomes become (and for a time remain) arranged in an equatorial plate.

micelle (L. *micella,* a small crumb). A minute, slender rod; applied especially to the cellulose rods of which cell walls are commonly composed.

microbody. Any of several ellipsoid cytoplasmic organelles in the size range of 0.2–0.6 micron long, bounded by a unit (not double) membrane.

micron (Gr. *mikros,* small). One millionth of a meter; one thousandth of a millimeter.

microphyll (Gr. *mikros,* small + *phyllon,* leaf). A leaf, usually small, with an unbranched midvein whose departure from the stele does not leave a gap; variously thought to have originated as an enation or a single telome; opposite of megaphyll.

micropyle (Gr. *mikros,* small + *pylē,* gate). The opening through the integuments of an ovule to the nucellus.

microsporangium (Gr. *mikros,* small + sporangium). A sporangium containing microspores.

microspore (Gr. *mikros,* small + spore). A spore that develops into a male gametophyte.

microsporophyll (microspore + Gr. *phyllon,* leaf). A leaf (sporophyll) that bears or subtends one or more microsporangia.

microtubule. A cytoplasmic organelle in the form of an elongate, slender tube.

middle lamella. The thin layer, composed of pectic substances, that joins two adjacent cells.

mitochondrion (Gr. *mitos,* thread + *chondrus,* cartilage). A cytoplasmic organelle of eukaryotes, concerned especially with TCA-cycle respiration.

mitosis (Gr. *mitos,* thread). A complex, orderly process of nuclear division that ordinarily results in the formation of two daughter nuclei with identical hereditary potentialities; loosely, cell division associated with mitotic nuclear division.

monocot (Gr. *monos*, one + cotyledon). Short for monocotyledon; the monocotyledons (class Liliopsida) are one of the two great groups of angiosperms.

monoecious (Gr. *monos*, one + *oikos*, house). Bearing reproductive structures of both sexes on the same plant, but (in angiosperms) not in the same flower.

monokaryotic (Gr. *monos*, one + *karyon*, nut). Having uninucleate segments, or having multinucleate segments with the nuclei not associated in pairs. Cf. *dikaryotic*.

monosaccharide (Gr. *monos*, one + *sakchar*, sugar). Any simple sugar, i.e., a sugar that cannot be broken down into smaller molecules without ceasing to be a sugar.

morphology (Gr. *morphē*, shape, form + *logos*, word, discourse). Form and structure, or the study thereof, often taken to include the changes of form occurring during the life cycle.

mutation (L. *mutare*, to change). An inheritable change in a chromosome or gene; applied especially to gene mutations.

mycelium (Gr. *mykēs*, mushroom). A mass of branching hyphae.

mycorhiza (Gr. *mykēs*, mushroom + *rhiza*, root). A symbiotic, fungus-root combination; by extension, a similar association of a fungus and some other underground part of a plant.

NAD. Nicotinamide-adenine dinucleotide, an important coenzyme in respiratory transfer of energy.

NADP. Nicotinamide-adenine dinucleotide phosphate, an important coenzyme in both respiratory and photosynthetic transfer of energy.

nanometer (L. *nanus*, dwarf + meter). One millionth of a millimeter; one thousandth of a micron.

nm. Abbreviation for nanometer.

node (L. *nodus*, knot). A place on a stem where a leaf is (or has been) attached.

nucellus (L. a small kernel). The tissue sur-rounding the female gametophyte of a seed plant, i.e., the megasporangium wall of a seed plant.

nucleolus (L. diminutive of *nucleus*, kernel). A specialized, more or less spherical body within the nucleus.

nucleus (L. a kernel). A protoplasmic vesicle (or body) characterized by the presence of DNA, which governs hereditary characteristics of the cell; the noncytoplasmic component of a protoplast.

Occam's Razor (William of Occam, fourteenth-century philosopher). The philosophical principle that explanations should be no more complicated than necessary.

ontogeny (Gr. *onta*, existing + *genesis*, origin). The developmental history of an individual.

oögamous (Gr. *oon*, egg + *gamos*, marriage). Producing sperms and eggs (rather than isogametes or anisogametes).

oögonium (Gr. *oon*, egg + *gonos*, offspring). A specialized single cell within which one or more eggs are produced.

operculum (L.). A little lid.

order. A group of related families; a particular rank in the taxonomic hierarchy, between the family and the class.

organ. A tissue or group of tissues that make up a morphologically and functionally distinct part of an organism.

osmosis (Gr. *osmo*, push, thrust). The diffusion of water through a differentially permeable membrane.

ovary (L. *ovum*, egg). The structure that in angiosperms encloses the ovules.

ovule (L. diminutive of *ovum*, egg). A young seed; the megasporangium, plus the enclosing integuments, of a seed plant.

oxidation. A type of chemical reaction, typically involving the use of oxygen, in which the net positive valence of an element is increased.

palmate (L. *palma*, palm of the hand). Having several similar parts spreading from a common point.

paraphysis (Gr. *para*, beside + *physis*,

growth). One of the sterile filaments associated or intermingled with reproductive filaments in some fungi and algae; it differs from a pseudoparaphysis in being free at one end.

parasexual. Sexlike in that new character combinations are achieved, but not strictly sexual because nuclear fusion and reduction division are not involved.

parasite (Gr. *parasitos*, eating beside another). An organism that obtains its food from another living organism.

parenchyma (Gr. *para*, beside + *enchein*, to pour in). A tissue composed of relatively unspecialized cells.

parthenogenesis (Gr. *parthenos*, virgin + *genesis*, origin). Reproduction by means of unfertilized eggs.

pectic substances. Pentose or hexose polymers with incorporated -COOH groups which make them hydrophilic and permit them to combine with certain metals.

pectinate (L. *pecten*, comb, *pectinatus*, comblike). Having a single row of lateral appendages, like the teeth of a comb.

perennial (L. *perennis*). A plant that lives more than two years.

perfect. In flowers, having both stamens and pistils; in fungi, having or relating to the structures associated with nuclear fusion.

perianth (Gr. *peri*, around + *anthos*, flower). The calyx and corolla, collectively.

pericarp (Gr. *peri*, around + *karpos*, fruit). The ovary wall of a fruit.

pericycle (Gr. *peri*, around + *kyklos*, ring, circle). A tissue, usually parenchymatous, between the endodermis and the vascular cylinder.

perigynous (Gr. *peri*, around + *gynē*, woman). Having the perianth and stamens united into a basal saucer or cup (the hypanthium) distinct from the ovary.

periplast (Gr. *peri*, around + *plastos*, formed, molded). A differentiated bounding layer of a protoplast, firm but usually flexible.

peroxisome (peroxide + Gr. *soma*, body). A microbody which contains enzymes that break down hydrogen peroxide, and in which the glyoxylate cycle and photorespiration can occur.

petal (Gr. *petalon*, a leaf). A member of the second set of floral leaves (i.e., the set just internal to the sepals), usually colored or white and serving to attract pollinators.

petiole (L. *petiolus*, a little foot, a fruit stalk). A leaf stalk.

phanerogam (Gr. *phaneros*, visible, manifest + *gamos*, marriage). A seed plant; opposite of *cryptogam*.

phelloderm (Gr. *phellos*, cork + *derma*, skin). The tissue produced internally by the phellogen (as opposed to the cork or phellem, which is produced externally).

phellogen (Gr. *phellos*, cork + *genesis*, origin). A lateral meristem that produces cork; the cork cambium.

phenotype (Gr. *phaino*, shining, *phainein*, to show). The actual character of an individual, as expressed in its form, structure, or physiology. Cf. *genotype*.

phloem (Gr. *phloos*, bark). The characteristic food-conducting tissue of higher plants.

-phore (Gr. *phorein*, to bear). Suffix meaning a supporting stalk.

photon (Gr. *phōs*, *phōtos*, light). A quantum of light.

photosynthate. Carbohydrates and other complex substances produced directly by photosynthesis.

phototropism (Gr. *phōs*, *phōtos*, light + *tropē*, a turning). A directional growth response to light.

phycobilin (Gr. *phykos*, seaweed + L. *bilis*, bile). A class of accessory pigments, found in certain algae, consisting of open-chain tetrapyrroles tightly bound to a protein.

phycocyanin (Gr. *phykos*, seaweed + *kyanos*, dark blue). A blue phycobilin.

phycoerythrin (Gr. *phykos*, seaweed + *erythros*, red). A red phycobilin.

phyllidium (Gr. diminutive of *phyllos*, leaf). A leaflike organ of bryophyte gametophytes.

phylogeny (Gr. *phylon*, tribe + *genos*, lineage). The evolutionary history of a group.

phylum (Gr. *phylon*, tribe). A taxonomic group of high rank, used in zoology for a group equivalent to the *division* of botanical classification. The term is often loosely carried over into botanical usage.

physiology (Gr. *physis*, nature + *logos*, word, discourse). The study of the processes occurring in living organisms and of the functions of the different parts of the organism.

phytochrome (Gr. *phyton*, plant + *chroma*, color). A growth-regulating pigment, chemically allied to the phycobilins.

pigment. A substance which absorbs different proportions of different wavelengths of light, especially if that substance occurs in relatively small amounts in another substance or body whose color it affects.

pinnate (L. *pinna*, feather). Having two rows of lateral branches or appendages, like the barbs on a feather.

pistil (L. *pistillum*, pestle). The female structure of a flower, consisting of one carpel or of several carpels joined together in a single unit.

plankton (Gr. *planktos*, drifting). Microscopic or barely megascopic plants and animals that float freely in water, being carried about by currents and waves.

plasma membrane. The living membrane at the outer edge of the cytoplasm.

plasmodesma (Gr. *plasma*, anything formed or molded + *desmos*, band or chain). A cytoplasmic strand connecting two adjacent cells.

plasmodium (Gr. *plasma*, anything formed or molded). A naked, multinucleate mass of protoplasm that is not divided into separate cells.

plasmogamy (Gr. *plasma*, anything formed or molded + *gamos*, marriage). The fusion of protoplasts that precedes karyogamy.

plastid (Gr. *plastos*, formed, molded). A specialized cytoplasmic body, usually associated with the manufacture or storage of food, or obviously pigmented, or both.

plumule (L. *plumula*, a little feather). The epicotyl of an embryo.

pollen (L. fine flour, dust). The mass of young male gametophytes (pollen grains) of a seed

plant, at the stage when they are released from the anther.

pollination. In angiosperms, the transfer of pollen from the anther to the stigma; in gymnosperms, from the microsporangium to the micropyle.

polyglucan (Gr. *polys*, many + glucose). A glucose polymer.

polymer (Gr. *polys*, many + *meros*, part). A compound formed by the chemical linkage of a number of molecules of the same or closely related kinds.

polypeptide (Gr. *polys*, many + *peptos*, digested). An amino acid polymer.

polyploid (Gr. *polys*, many; *ploid* is of recent origin from the common endings of haploid and diploid). Having three or more sets of chromosomes.

polysaccharide (Gr. *polys*, many + *sakchar*, sugar). A substance formed by the chemical linkage of several or many monosaccharide molecules.

porphyrin (Gr. *porphyros*, purple). A tetrapyrrole that forms a closed ring.

primary tissue. A tissue derived directly by differentiation from an apical (or intercalary) meristem.

primary wall. In cells, the outer or first-formed wall layer, in contrast to the secondary wall.

primordium (L. the beginning). A group of cells which is destined to develop into a particular structure, e.g., a leaf primordium.

procambium (Gr. *pro*, before + cambium). The young tissue of a root or shoot that is destined to develop into vascular tissue; more narrowly, the potential cambium between the xylem and phloem of a vascular bundle.

prokaryotic (Gr. *prōtos*, first + *karyon*, nut). Lacking a vesicular nucleus and other double-membrane organelles, indicating a relatively primitive organization of the protoplast.

prophase (Gr. *pro*, before + phase). An early stage in mitosis, in which the chromonemata give rise to the chromosomes.

prosthetic group. The non-amino-acid part of a conjugated protein, either loosely or tightly bound to the remainder of the molecule.

protein (Gr. *proteios*, primary). A food containing nitrogen (and often other elements) in addition to carbon, hydrogen, and oxygen.

prothallial cell. A cell, or one of several cells, found in some male gametophytes, which may represent an evolutionary vestige of the vegetative body of the gametophyte.

prothallus (Gr. *pro*, before + *thallus*). The gametophyte of a pteridophyte.

protonema (Gr. *prōtos*, first + *nēma*, thread). The filament (or flat thallus) formed by germination of a moss spore.

protoplasm (Gr. *prōtos*, first + *plasma*, anything formed or molded). Living substance.

protoplast. An organized unit of protoplasm; typically, the living content of a cell.

protostele (Gr. *prōtos*, first + stele). A stele with a solid core of xylem, lacking a pith.

protoxylem (Gr. *prōtos*, first + xylem). Primary xylem that matures while the surrounding tissues are still elongating; the first xylem to mature at any particular level of the stem or root.

pycnium (Gr. *pyknos*, crowded, dense). Among rusts, a pocket at the surface of the host, in which monokaryotic spores, called pycniospores, are borne.

pyrenoid (Gr. *pyrēn*, stone of a fruit, kernel). A specialized, proteinaceous part of a chloroplast, found chiefly in algae, on or around which starch or other food-storage products accumulate.

pyrrole (Gr. *pyrrhos*, red). A chemical ring of four carbon atoms and one nitrogen atom, each with an attached hydrogen atom.

quantasome (L. *quantum*, how much + Gr. *soma*, body). The ultimate photosynthetic unit, containing some 200–300 molecules of chlorophyll plus all of the associated light-gathering apparatus.

quantum (L. how much). An ultimate, indivisible unit of radiant energy.

rachis (Gr. backbone). A main axis, such as that of a compound leaf.

radicle (L. diminutive of *radix*, *radicis*, root).

The part of the embryo of a seed that gives rise to the root.

receptacle (L. *receptaculum*, reservoir). A structure that bears or contains other parts; in flowers, the end of the stem, to which the other flower parts are attached.

reduction. A type of chemical reaction in which the net positive valence of an element is decreased (or the negative valence increased).

reduction division. The process by which the number of chromosomes in a cell is reduced from $2n$ to n.

respiration. The chemical breakdown of foods by living organisms, resulting in release of metabolically useful energy.

rhizoid (Gr. *rhiza*, root). A structure of rootlike form and function, but of simple anatomy, lacking xylem and phloem.

rhizome (Gr. *rhiza*, root). A creeping underground stem.

ribosome (ribose + Gr. *soma*, body). A cytoplasmic organelle that contains some RNA and is involved in protein synthesis.

RNA. Ribonucleic acid, an essential participant in the synthesis of proteins.

root. One of the primary organs of most vascular plants, typically serving to anchor the plant in the soil and to absorb water and minerals.

saprophyte (Gr. *sapros*, rotten + *phyton*, plant). A plant that absorbs its food from dead organic matter, rather than making it or eating it.

schizogenous (Gr. *schizein*, to split + *genesis*, origin). Originating by splitting or separation of tissue.

sclerenchyma (Gr. *sklēros*, hard + *enchein*, to pour in). A nonvascular strengthening tissue, usually dead when functional, in which the cells have a definite secondary wall that is often lignified.

secondary tissue. A tissue derived from the cambium or other lateral meristem.

secondary wall. In cells, an inner-wall layer, formed after the primary wall and often of different composition.

seed. A ripened ovule; the characteristic resting body in the reproductive cycle of many plants.

SEM. Abbreviation for scanning electron microscope.

sepal (Gr. *skepē*, a covering). A member of the outermost set of floral leaves, typically green or greenish and more or less leafy in texture.

septum (L. hedge, enclosure). A partition.

sessile (L. *sessilis*, sitting). Lacking a stalk.

sexual (L. *sexus*, sex). Relating in some way to nuclear fusion or reduction division.

shoot. The stem and leaves of a plant, collectively.

sieve element. The fundamental type of cell in phloem, being long, slender, and thin walled, and having cytoplasm but no nucleus at maturity.

sieve plate. A perforated end wall connecting two sieve elements.

sieve tube. A phloem tube formed from several sieve elements set end to end.

siphonostele (Gr. *siphōn*, tube + stele). A stele in which the vascular tissue surrounds a central pith.

solute. A component of a liquid solution, whose particles are dispersed separately from each other. Cf. *solvent*.

solution. A liquid or gaseous mixture in which the dispersed particles are of ordinary molecular or ionic size.

solvent. The continuous component of a liquid solution. Cf. *solute*.

sorus (Gr. *sōros*, a heap). A cluster of sporangia or externally produced spores.

species (L. a particular kind). A particular kind of plant or animal, which maintains its distinctness from other kinds in nature over a period of many successive generations.

sperm (Gr. *sperma*, seed). A motile gamete that can fuse with an egg to form a zygote; the male gamete.

spindle. A structure, formed during mitosis, which is associated with the movement of the chromosomes to the poles.

sporangium (spore + Gr. *angeion*, vessel). A case or container for spores.

spore (Gr. *spora*, seed). A one-celled (seldom two-celled) reproductive structure other than a gamete or zygote; zygotes that become thick walled and go into a resting stage, thus resembling many spores, are called zygospores or oöspores.

spore mother cell. A cell capable of undergoing reduction division to form spores.

sporophyll (spore + Gr. *phyllon*, leaf). A leaf that bears or subtends one or more sporangia.

sporophyte (spore + Gr. *phyton*, plant). The generation that has 2n chromosomes and produces spores (meiospores) as reproductive bodies.

stamen (Gr. *stēmōn*, thread, fiber). The male organ of a flower, i.e., the microsporophyll.

starch. A carbohydrate food reserve consisting of a mixture of amylose, amylopectin, and perhaps other substances.

stele (Gr. post, pillar). Same as *central cylinder*.

stem. One of the primary organs of vascular plants, typically serving to bear the reproductive structures and display the leaves.

stigma (Gr. the mark of a pointed instrument; also, L. a mark or brand). The part of a pistil which is receptive to pollen.

stipe (L. *stipes*, stalk, branch, trunk). Any stemlike structure on which some other structure is borne.

stipule (L. diminutive of *stipes*, stalk). One of a pair of basal appendages found on many leaves.

stomate (Gr. *stoma*, mouth). A special kind of intercellular space in epidermal tissue, bounded by a pair of guard cells which, under certain conditions, close off the space by changing shape.

strobilus (Gr. *strobilos*, cone). A cluster of sporophylls on an axis; a cone.

style (Gr. *stylos*, pillar, column). The part of a pistil connecting the stigma to the ovary.

suberin (L. *suber*, the cork oak). A waterproofing substance characteristic of the walls of cork cells.

succulent (L. *succulentus*, juicy). A plant that accumulates reserves of water in the fleshy stems or leaves, due largely to the high proportion of hydrophilic colloids in the protoplasm and cell sap.

sucrose (Fr. *sucre*, sugar). The most common disaccharide; it has the chemical formula $C_{12}H_{22}O_{11}$ and yields equal amounts of glucose and fructose when hydrolyzed.

suspension. A mixture of water and solid or liquid particles that are of larger than molecular size.

suspensor. A cell or organ, derived from the zygote but not part of the embryo proper, which in some vascular plants pushes (by its growth) the young embryo deeper into the tissue of the gametophyte.

symbiosis (Gr. *symbiōsis*, a living together). A close physical association between two different kinds of organisms, typically with benefit to both.

sympodial (Gr. *syn*, with + *pous*, *podos*, foot). With the apparent main axis actually consisting of a series of usually short branches.

syngamy (Gr. *syn*, with + *gamos*, marriage). Fusion of gametes.

taxon (Gr. *taxis*, order, arrangement). Any taxonomic unit of classification.

taxonomy (Gr. *taxis*, arrangement + *nomos*, law). Classification according to presumed natural (i.e., evolutionary) relationships.

TCA cycle. The tricarboxylic acid cycle (Krebs cycle) for the respiratory breakdown of pyruvate.

telome (Gr. *telos*, end). An ultimate branch of a dichotomously branching stem.

telophase (Gr. *telos*, end + phase). The stage in mitosis during which daughter nuclei are formed.

TEM. Abbreviation for transmission electron microscope.

tetrad (Gr. *tetras*, four). A group of four.

tetrapyrrole (Gr. *tetras*, four + *pyrrhos*, red). A group of four chemically linked pyrroles.

thallophyte (thallus + Gr. *phyton*, plant). A plant in which the body is a thallus; a plant group consisting of the algae, fungi, and bacteria.

thallus (Gr. *thallos*, a sprout). A complete plant body that lacks specialized conducting tissues, especially if it is multicellular but relatively simple in form, not being divided into parts which resemble roots, stems, and leaves.

thylakoid (Gr. *thylakos*, pouch). An individual photosynthetic lamella, these often stacked to form grana within the chloroplast.

tissue. A group of cells, making up part of a multicellular organism, with similar or related function(s).

tracheid (L. *trachea*, the windpipe). The most characteristic type of cell in xylem, being long, slender, tapered at the ends, with a lignified secondary wall and without living contents at maturity.

transpiration (L. *trans*, across + *spirare*, to breathe). Evaporation and consequent loss of water from a living plant.

triple-fusion nucleus. The nucleus formed by fusion of a sperm with the two polar nuclei of the embryo sac in angiosperms; it becomes the primary nucleus of the developing endosperm.

tropism (Gr. *tropē*, a turning). A directional growth response to an environmental stimulus.

turgor (L. *turgere*, to swell out). A condition of rigidity, normal in plant cells, resulting from the osmotic intake or retention of so much water that the cytoplasm is firmly pressed against the cell wall.

vacuole (L. diminutive of *vacuus*, empty). A watery vesicle within a protoplast, chemically relatively inactive and usually regarded as nonliving.

vascular (L. *vasculum*, a small vessel). Having or pertaining to a conducting system; having xylem and phloem.

venter (L. belly). The swollen base of an archegonium, containing an egg.

vessel. A xylem tube formed from several vessel elements (modified tracheids with imperfect or no end walls) set end to end.

virion (L. *virus*, poison + Gr. *-on*, fundamental particle, unit). The basic structural unit of a virus, the form in which it is transferred from one host (or cell) to another.

xanthophyll (Gr. *xanthos*, yellow + *phyllon*,

leaf). A carotenelike yellow pigment that contains some oxygen in addition to carbon and hydrogen.

xerophyte (Gr. *xeros*, dry + *phyton*, plant). A plant that grows in very dry places.

xylem (Gr. *xylon*, wood). The characteristic water-conducting tissue of higher plants.

zoospore (Gr. *zoon*, animal + spore). A swimming (motile) spore.

zygospore. A zygote, formed by fusion of isogametes, which secretes a thick wall and goes into a resting stage.

zygote (Gr. *zygōtos*, yoked). A cell formed by fusion of gametes.

Index

Numbers in boldface indicate pages containing illustrations.

Å, defined, 29
Abies, 325, 334
  A. balsamea, 600
  A. concolor, 600
Abnormal growth, 485–486, **487**
Abscisic acid, 476
  structural formula for, **477**
Abscission, defined, 386
  role of auxin in, 475
Absorption, 370, 400–403
  by aerial organs, 327, 403, 421
  through mycorhizae, 380–381
  of solutes, 91, 418
Absorption of food
  in bacteria, 176
  in fungi, 242
Absorption of minerals, 417
  mechanics of, 418–420
Absorption spectrum, 41, 480
  of β-carotene, **42**
  of chlorophylls, **96**
  of phycobilin, **99**
  of phytochrome, **481**
Accessory chlorophylls, 97
Accessory pigments, 97–100
Acer, 18, 607. *See also* Red ma-
      ple, Norway maple, etc.
  A. campestre, 608
  A. japonicum, 608
  A. negundo, 603
  A. palmatum, 608
  A. platanoides, 608; leaf, **15**
  A. pseudoplatanus, 608
  A. regelii, **590**
  A. rubrum, 608; leaf, **15**
  A. saccharinum, 607
  A. saccharum, **358**, **427**, 608.
      See also Sugar maple
  fossil fruit, **502**
  fruit, **458**
  seed germination in, 464
  twig, **354**
Acetyl-CoA, 119, 124
Acetylene, effect on flowering,
      445

Acid
  acetic, 33
  abscisic, 476; structure, **477**
  amino, 38; RNA code for, **143**
  aspartic, 433
  fatty, 37
  formation of, 33
  formic, 33
  gibberellic, structure of, 476
  lactic, 135
  nucleic, 66
  organic, 33
  oxalacetic, 432
  phosphoglyceric, 111
  proprionic, 33
  pyruvic, 135
Acid rain, 583
Actinomycetales, 170
Actinomycetes, antibiotic, 267
Action spectrum of light, 430,
      480
Adaptation, 549–550, 556, 594
  for drought, 412–413
  to land habitat, 561
  for seed dispersal, 458–460
Adapt or die, 594–595
Adenine, 139
  structural formula, **140**
Adenosine diphosphate. *See*
      ADP
Adenosine triphosphate. *See*
      ATP
Adiantum pedatum, **299**
ADP, 40, 126. *See also* ATP
Advanced vs primitive char-
      .acters in angiosperms, 500
Adventitious buds, 354
Adventitious root system, 370,
      371
Adventitious tissues, defined,
      280, 354
Aecia, in rusts, 257, **259**
Aeciospores, 257
Aerial organs
  mineral absorption by, 421

water absorption by, 327, 403
Aerial roots, 381
Aerobic respiration, 119
Aesculus, 603
  A. glabra, 603
    leaves and flowers, **597**
  A. hippocastanum, 603
    leaf, **387**
  A. octandra, 603
African tulip tree, 604
African violet, 18
  leaf cutting, **396**
Afterripening in seeds, 463
Agar, 236, 237
Agaricaceae, 261
Agarics, 261–262
Agaricus brunnescens, 262
Agaricus campestris, 262, 379
Agathis, 326, 330, 334
Agave, **3**
  A. horrida, **412**
Age, of trees, 325, 359
Age of Cycads, 337
Age of Ferns, 336
Agropyron trachycaulum, **582**
Ailanthus altissima, 606
Akinete
  in blue-green algae, 195
  in green algae, 207
Albinism, 535
Albizia julibrissin, 604
Alcohol, 134, 250
Algae, 20, 21–23, 192
  accessory pigments in, 98
  blue-green, 192–199
  brown, **23**, 231–233
  chlorophylls in, 97
  chloroplast in, 103
  evolutionary development in,
      204
  green, 201–216
  number of species, 27
  plankton, 215, 218, 220
  in polluted water, 580
  red, 218, 233–238

81 82 83 84 9 8 7 6 5 4 3 2 1